T0398048

GREEN'S FUNCTIONS AND LINEAR DIFFERENTIAL EQUATIONS

Theory, Applications, and Computation

Published Titles

Advanced Differential Quadrature Methods, Zhi Zong and Yingyan Zhang
*Computing with hp-ADAPTIVE FINITE ELEMENTS, Volume 1, One and Two Dimensional
Elliptic and Maxwell Problems*, Leszek Demkowicz
*Computing with hp-ADAPTIVE FINITE ELEMENTS, Volume 2, Frontiers: Three
Dimensional Elliptic and Maxwell Problems with Applications*, Leszek Demkowicz,
Jason Kurtz, David Pardo, Maciej Paszyński, Waldemar Rachowicz, and Adam Zdunek
CRC Standard Curves and Surfaces with Mathematica®: *Second Edition*,
David H. von Seggern
Discovering Evolution Equations with Applications: Volume 1-Deterministic Equations,
Mark A. McKibben
*Exact Solutions and Invariant Subspaces of Nonlinear Partial Differential Equations in
Mechanics and Physics*, Victor A. Galaktionov and Sergey R. Svirshchevskii
Geometric Sturmian Theory of Nonlinear Parabolic Equations and Applications,
Victor A. Galaktionov
*Green's Functions and Linear Differential Equations: Theory, Applications,
and Computation*, Prem K. Kythe
Introduction to Fuzzy Systems, Guanrong Chen and Trung Tat Pham
Introduction to non-Kerr Law Optical Solitons, Anjan Biswas and Swapan Konar
Introduction to Partial Differential Equations with MATLAB®, Matthew P. Coleman
Introduction to Quantum Control and Dynamics, Domenico D'Alessandro
Mathematical Methods in Physics and Engineering with Mathematica, Ferdinand F. Cap
Mathematical Theory of Quantum Computation, Goong Chen and Zijian Diao
Mathematics of Quantum Computation and Quantum Technology, Goong Chen,
Louis Kauffman, and Samuel J. Lomonaco
Mixed Boundary Value Problems, Dean G. Duffy
Multi-Resolution Methods for Modeling and Control of Dynamical Systems,
Puneet Singla and John L. Junkins
Optimal Estimation of Dynamic Systems, John L. Crassidis and John L. Junkins
Quantum Computing Devices: Principles, Designs, and Analysis, Goong Chen,
David A. Church, Berthold-Georg Englert, Carsten Henkel, Bernd Rohwedder,
Marlan O. Scully, and M. Suhail Zubairy
A Shock-Fitting Primer, Manuel D. Salas
Stochastic Partial Differential Equations, Pao-Liu Chow

CHAPMAN & HALL/CRC APPLIED MATHEMATICS
AND NONLINEAR SCIENCE SERIES

GREEN'S FUNCTIONS AND LINEAR DIFFERENTIAL EQUATIONS

Theory, Applications, and Computation

Prem K. Kythe

University of New Orleans
Louisiana, USA

CRC Press
Taylor & Francis Group
Boca Raton London New York

CRC Press is an imprint of the
Taylor & Francis Group an **informa** business

A CHAPMAN & HALL BOOK

Chapman & Hall/CRC
Taylor & Francis Group
6000 Broken Sound Parkway NW, Suite 300
Boca Raton, FL 33487-2742

© 2011 by Taylor and Francis Group, LLC
Chapman & Hall/CRC is an imprint of Taylor & Francis Group, an Informa business

No claim to original U.S. Government works

Printed in the United States of America on acid-free paper
10 9 8 7 6 5 4 3 2 1

International Standard Book Number: 978-1-4398-4008-5 (Hardback)

**Visit the Taylor & Francis Web site at
http://www.taylorandfrancis.com**

**and the CRC Press Web site at
http://www.crcpress.com**

Contents

Preface

Boundary value problems associated with ordinary and partial differential equations have been an integral part of mathematics, mathematical physics, and applied sciences, and Green's functions for these problems have become an important subject, particularly appealing to mathematicians, physicists and applied scientists. Although Green's functions were first introduced by George Green in 1828, with the physical interpretation as an 'influence function' and 'potential' in certain mechanical problems, specially the string problem, these functions have been developed and widely used during the past 60 years, firstly as part of an exciting research area and later as part of a curriculum in courses on partial differential equations offered by mathematics, engineering and physics departments to their senior and graduate students. In most textbooks on ordinary and partial differential equations and boundary value problems there is generally a single chapter on Green's functions that provides a selected portion of this topic with a few basic results and examples, while in others there exists some sort of limited material scattered throughout the book. Although the significance of these functions in solving boundary value problems is justifiably indicated in many books, the material in such a single chapter becomes either very difficult or just a collection of a few well known boundary value problems that are solved to justify the presence of this topic.

There are very few books written on the subject of Green's functions, although some of the notables are those by Roach [1970], Stakgold [1979], and Sagan [1989]. These books were written to provided theory and examples to justify the development of the subject, but they are, however, not suitable as textbooks. There are other highly specialized and research-oriented books on Green's functions that confine to certain restricted fields, such as potentials, diffusion and waves, solid state physics, quantum mechanics, lattice Schrödinger operators, and boundary element method. Such books certainly do not count as standard textbooks.

Overview

Our aim is to provide a kind of textbook on Green's Functions that satisfies the following criteria:

1. It is simple enough for the average senior or graduate student to understand and grasp the beauty and power of the subject.

2. It has enough mathematical rigor to withstand criticism from the experts.

3. It has a large number of examples and exercises from different areas of mathematics, applied science and engineering.

4. It provides motivation to appeal to students and teachers alike, by providing sufficient theoretical basis which leads to the development of Green's function method which is applied to solve initial and boundary value problems involving different linear ordinary and partial differential equations.

5. It possesses a robust self-contained text, full of explanation on different problems, graphical representations where necessary, and a few appendices for certain background material.

6. It contains about one hundred solved examples and and twice as many exercises with hints and answers and difficult ones with adequate hints and sometimes complete solution.

7. It describes and uses the following methods for solving initial and boundary value problems, which are explained with clarity and detail: (i) classical method of variation of parameters; (ii) a generalization of method of variation of parameters to construct one-sided Green's functions for initial value problems involving linear ordinary differential equations; (iii) a variation of the above method, called the Wronskian method, to find one-sided Green's functions for initial value problems involving linear ordinary differential equations; (iv) a variation of the above method, called the Green's function method, to construct Green's functions for boundary value problems involving linear ordinary differential equations, with a step-by-step procedure and some useful shortcuts; (v) Bernoulli's separation method for linear partial differential equations; (vi) integral transform method to determine Green's functions for linear parabolic, hyperbolic, and elliptic equations; (vii) method of images for linear elliptic equations and related Green's functions; and (viii) conformal mapping method for determining Green's functions for linear elliptic equations; and (ix) an interpolation method for numerical construction of Green's functions for convex and starlike regions.

8. The subject material of this textbook is arranged in a manner that tries to eliminate the two detrimental effects on students, namely, to impart the impression to mathematically inclined students that they are not left out, that they are not dwelling in a textbook that consists of some rather dull manipulations with integrals or infinite series; and to technically oriented students that the subject of linear ordinary and partial differential equations is not to be treated simply by the method of variation of parameters, or separation of variables, or integral transforms.

Outline

In most books on the subject, the approach to introduce the concept of Green's functions has been firstly to provide a definition of the Dirac delta function, secondly to solve a boundary value problem involving a nonhomogeneous ordinary differential equation, like the one (Example 2.9) solved in this book, and finally to define the kernel of the solution, obtained in the form of an integral equation as Green's function for the boundary value problem in question. This inverse approach is also seen in the case of linear partial differential equations.

The approach adopted in this book is more direct. The first task has been to explain and define the Dirac delta function precisely and to unravel the mystery surrounding this generalized function. Then the concept of Green's function and its relationship with the Dirac delta function has been carefully established. After these two hurdles are overcome, the mystique of the development of the entire subject of Green's functions and its application in solving linear ordinary and partial differential equations has been unfolded in a manner that makes the subject simply appealing and useful.

The book starts in Chapter 1 with an introduction to some basic results and definitions from topics such as Euclidean space, specially the metric space and the concept of inner product; classes of continuous functions and those that are infinitely differentiable with compact support; convergence of sequences, their weak and strong convergence, convergence in the mean, and convergence of infinite series; linear functionals, and linear transformations; Cramer's rule; divergence theorem and Green's identities; Leibniz's rule for differentiation of integrals and for the nth derivative of product of two functions; formulas for integration by parts; Bessel's inequality for Fourier series and for square-integrable functions; Schwarz's inequality for infinite sequences; and Parseval's equality or completeness relation.

The concept of a Green's function is presented in Chapter 2 by first introducing the 'generalized function' known as the Dirac delta function through a limiting process for certain admissible 'test functions' which are infinitely differentiable with compact support. This approach to define the Dirac delta function seems to be the simplest as compared to more advanced definitions, available prior to 1945, that arose out of Schwarz's theory of distributions, namely, (i) $\delta(x) = \dfrac{d}{dx} H(x)$, due to Heaviside; (ii) $\delta(x) = \lim_{n\to\infty} f_n(x)$ or $\delta = \sum_{n=0}^{\infty} f_n$ for suitable functions f_n, due to Fourier, Kirchhoff, Heaviside, Jordan and Pauli; (iii) $\delta(x) = 0$ for $x \neq 0$, and $\int_{-\infty}^{\infty} \delta(x)\, dx = 1$, due to Dirac and Heaviside; and (iv) $\int_{-\infty}^{\infty} \delta(x-a) f(x)\, dx = f(a)$, or $\int_{-\infty}^{\infty} \delta(x) f(x)\, dx = f(0)$, due to Fourier, Heaviside, and Dirac. These and other variations of such definitions did not provide any simple mathematical basis for operating with this generalized function. In fact, the theory of distributions evolved without making the δ-function as its starting point, and then played a significant role in future research on partial differential equations, and especially on the theory of Green's functions. Basic theorem for the construction of Green's functions and their important properties are established in this chapter, which ends with a very simple introduction to fundamental solutions for some differential operators. These solutions, also known as Green's functions in the large, or 'free-space' Green's functions, or singular solutions, are very useful in the development and application of the boundary element method.

A thorough review and detailed description of the construction of one-sided Green's functions for initial value problems, and that of Green's functions for boundary value problems for linear ordinary differential equations are presented in Chapter 3. The method of variation of parameters is generalized to produce an effective method and its variation, called the Wronskian method, to generate one-sided Green's functions for initial value problems. A step-by-step procedure for the Green's function method to construct Green's functions for boundary value problems in general and Sturm-Liouville systems in particular is included in this chapter, together with periodic and singular Sturm-Liouville systems. This chapter ends with a detailed presentation of eigenvalue problems for Sturm-Liouville systems.

Linear partial differential equations begin in Chapter 4, which is devoted to a complete description of Bernoulli's separation method. This method has been very prominent since 1735 when Daniel Bernoulli formulated the principle of coexistence of small oscillations, which, after using Taylor's and John Bernoulli's theory of vibrating string, led him to believe that the general solution of this problem could be found in the form of a trigonometric series. But no method for determining the coefficients of this series solution was available until about 25 years later when in 1812 Fourier wrote and later published his extensive memoir *Théorie du mouvement de la chaleur dans les corps solides* in the French journal 'Mémoires de l'Académie', 1819-1822, wherein he considered the following five problems: (i) one-dimensional flow of heat; (ii) two-dimensional flow of heat in a rectangle; (iii) three-dimensional flow of heat in a rectangular parallelopiped; (iv) flow of heat in a sphere when the temperature depends only on the distance from its center; and (v) flow of heat in a right circular cylinder when the temperature depends only on the distance from the axis. These problems also included the question of radiation. In the first three problems the series solution, when one or more dimensions become infinite, was shown to degenerate into what we now call 'Fourier's integrals'. On the question of determination of the coefficients in the theory of trigonometric series, it was for the first time that the real importance of this series was shown and Fourier's name was associated with its development. Six problems and fourteen examples are solved in this chapter by using Bernoull's separation method.

To make this book self contained, a detailed description of integral transforms, especially the Laplace, Fourier, Fourier sine and cosine, finite Fourier, multiple Laplace and Fourier, and Hankel transform, is presented in Chapter 5 with properties of these transforms, along with many examples.

Parabolic equations are dealt in Chapter 6, which describes methods to construct Green's functions and solve Dirichlet and other problems for the diffusion equation, Schrödinger diffusion equation, axisymmetric diffusion equation in one-, two-, and three-dimensions, and one-dimensional fractional diffusion and Schrödinger diffusion equations. Hyperbolic equations are studied in Chapter 7 where detailed construction of Green's functions for wave equation, axisymmetric wave equation, Schödinger wave equation in one-, two- and three-dimensions, and wave equation in a cube are presented; one-dimensional fractional wave equation is studied; and application to the

Cauchy problem and its d'Alembert solution is studied. Other applications, like the free vibrations of a large circular membrane, are included. The problem of diffraction in quasi-optics with both Fraunhoffer and Fresnel approximations for diffraction of monochromatic waves is presented. The investigation into the mathematical aspect of the hydrogen atom, in conjunction with Schödinger's wave equation, makes a special topic presented in this chapter.

Chapter 8 deals with elliptic equations, and Green's functions for Laplace's equation in two and three dimensions, including Laplace's equation in a rectangular parallelopiped, are constructed. Harmonic functions are introduced, and Green's functions for two- and three-dimensional Helmholtz's and Poisson's equations are determined. Some important applications of elliptic equations, namely, the Dirichlet, Neumann and Robin problems, and the vibration equation for the unit sphere, are presented in detail.

Spherical harmonics, which are the angular portion of a set of Laplace's equation in spherical coordinates, are discussed in Chapter 9. They are useful in many theoretical and physical applications in physics, seismology. geodesy, spectral analysis, magnetic fields, and quantum mechanics. Both solid and surface spherical harmonics are analyzed in detail, including the Condon-Shortley phase factor, and discrete energy spectrum for the hydrogen atom are studied in some detail.

The conformal mapping method to determine Green's and Neumann's functions for the circle, ellipse, infinite strip, and annular region is presented in Chapter 10. The discussion starts with some useful definitions and results from conformal mapping, and the relationship between an analytic function and Green's function is established, which is used to present an interpolation method for numerically constructing Green's functions for different types of convex and starlike regions. The computational part has been simplified to the extent that a good quality digital calculator is sufficient to produce the required numerical values, although a Mathematica® notebook greens.nb and some projects are available in Kythe et al. [2003].

Six appendices, one each on adjoint operators, fundamental solutions, list of harmonics, tables of integral transforms used in the book, fractional derivatives, and one-sided Green's functions for systems of linear first-order ordinary differential equations, are presented in Appendix A through F, which are followed by the Bibliography and the Index.

Layout of the Material

The material of this book ranges from average to challenged sections, which can be differently suitable for students and readers depending on their background. The basic prerequisite, however, is a thorough knowledge of differential and integral calculus, advanced calculus, and an elementary course on linear ordinary differential equations. This background assumes that the readers are at least at the senior undergraduate level. The readers with these skills can easily go through the first three chapters. Readers with advanced knowledge of elementary linear partial differential equations of the parabolic, hyperbolic and elliptic types together with some understanding of the method of separation of variables (called Bernoulli's separation method in this book) and integral transforms will be able to go through the first nine chapters, in part or whole, depending on their interest and academic requirements. It is assumed that such readers are at the level of graduate students in mathematics, applied mathematics, or physics. Some portions of the book, if found too demanding, can be omitted, depending on the discretion of the reader or instructor. Books dealing with the kind of broad subject as this one cater to the needs of students and readers from different branches of science, engineering and technology. The simplest rule to get full use of such a book is to decide on the requirements and choose the relevant topics.

The author takes this opportunity to thank Mr. Robert B. Stern, Executive Editor, CRC Press/Taylor & Francis Group, for encouragement to complete this book; his editorial staff for a very efficient job and cooperation; the two reviewers who took time to communicate some valuable suggestions; and finally my friend Michael R. Schäferkotter for help and advice freely given whenever needed.

Shreveport, LA
Northfield, MN

Notations and Definitions

A list of the notations, definitions, and abbreviations used in this book is given below.

a, thermal conductivity

$a_0 = \dfrac{\hbar^2}{mc^2} = 0.529$ Å, first Bohr radius

a.e., almost everywhere

$[a_{ij}]$, elements of an $n \times n$ matrix A

a_0, a_n, b_n, Fourier coefficients, $n = 1, 2, \ldots$

$\arg\{z\}$, argument of a complex number z

$A \backslash B$, complement of a set B with respect to a set A

$A \times B$, product of the sets A and B

$\overline{A}$, closure of a set A

$A = [a_{ij}]$, $1 \le i \le m$, $1 \le j \le n$, or $A = [c_1 \,|\, c_2 \,|\, \cdots \,|\, c_j \,|\, \cdots \,|\, c_n]$, a matrix with elements a_{ij}, or a matrix A where c_j, $i \le j \le n$, denotes the jth column of A

A^T, transpose of a matrix A

$\overset{\circ}{\text{A}}$, Ångström, angstrom unit is a unit of length equal to 0.1 nanometre or $a \times 10^{-10}$ metres; named after Anders Jonas Ångström (1814-1874), a Swedish physicist; $1\,\overset{\circ}{\text{A}} = 100.00 \times 10^{-12}$ m or 0.1 nm

$\qquad = 328.10^{-12}$ ft $= 3.9370 \times 10^{-9}$ in (US/Imperial units)

$B(r, \mathbf{x}_0)$, open ball of radius r centered at a point $\mathbf{x}_0$

$B(\varepsilon, x_0)$, neighborhood of a point $x_0 \in \mathbb{R}$, or $\varepsilon < x - x_0 < \varepsilon$, for an arbitrary small $\varepsilon > 0$

$B(r, a)$, open disk of radius r and center at a in $\mathbb{C}$

$\overline{B}(r, a)$, closed disk of radius r and center at a in $\mathbb{C}$

$B(1, 0) \equiv U$, open unit disk in $\mathbb{C}$

B_1, B_2, linear differential operators of order ≤ 1 defining boundary conditions

c, wave speed, or wave velocity

const, constant

cof (A_{ji}), cofactors of an $n \times n$ matrix A

$[c_1 \,|\, c_2 \,|\, \cdots \, c_j \,|\, \cdots \, c_n]$, representation of an $n \times n$ matrix A, where c_j, $1 \le j \le n$ denote the columns of the matrix A

C, capacity of a conductor; cross-section of a pipe or channel

$C^k(D)$, class of real-valued functions continuous together with all the derivatives up to order k inclusive, $0 \le k < \infty$, in a domain $D \in \mathbb{R}^n$.

$C^\infty(D)$, class of functions infinitely differentiable in D, i.e., their continuous partial derivatives of all orders exist

$C_0^\infty(D)$, class of functions which are infinitely differentiable on D and vanish outside some bounded region (index 0 indicates compact support)

$C_0^\infty(\mathbb{R})$, class of functions with compact support on $\mathbb{R}$ and infinitely differentiable on $\mathbb{R}$

$\mathbb{C}$, complex plane

$\det[A]$, determinant of a square matrix A

$d(x, y)$, metric or distance on $\mathbb{R}$

ds, line element

dS, surface element

$d\mathbf{x} = dx\, dy\, dz$, volume element in $\mathbb{R}^3$

D, domain; also, differential operator d/dx

$\bar{D} = D \cup \partial D$, closure of a domain D

$D^n \equiv \dfrac{d^n}{dx^n}$ or $\dfrac{d^n}{dt^n}$, nth derivative with respect to x or t

$D_t^p \equiv \dfrac{\partial^p}{\partial t^p}$, $0 < p \le 1$, Caputo time derivative

$D^\beta[f(x)]$, Riemann-Liouville derivative of order β, $0 \le \beta \le 2$

$1/D$, inverse operator of D

∂D, boundary of a domain D

$\mathcal{D}$, class of admissible functions in $C_0^\infty(\mathbb{R})$ which have compact support over the intervals of the form $[-a, a]$ and which approach the Dirac δ-function (or δ-distribution) as $a \to 0$; that is, the space of all distributions on $C_0^\infty(\mathbb{R}^n)$

e.g., for example (Latin, *exempli garatia*)

$\mathrm{erf}(x) = \dfrac{2}{\sqrt{\pi}} \displaystyle\int_0^x e^{-t^2}\, dt$, error function

$\mathrm{erfc}(x) = 1 - \mathrm{erf}(x) = \dfrac{2}{\sqrt{\pi}} \displaystyle\int_x^\infty e^{-t^2}\, dt$, complementary error function

E, modulus of elasticity; kinetic energy; total energy

$\mathrm{Ext}(\Gamma)$, interior of a simple closed contour Γ

E_{jn}, energy levels of the Hydrogen atom

EI, flexural rigidity of a beam

$E(x, y, z) = u(x, y, z)\, e^{ikz}$, harmonic wave function

$E(r) = \Re\{\Psi(r)\}$, radial wave, where $\Psi(r)$ is a complex wave

$E_N(r)$, radial wave for N slits

$E_{p,q}(z) = \displaystyle\sum_{m=0}^\infty \dfrac{z^m}{\Gamma(pm+q)}$, $p, q > 0$, Mittag-Leffler function

Eq(s), equation(s) (when followed by an equation number)

$\mathbf{E}$, electric field

Fig. (Figs.), abbreviation for Figure (Figures), followed by numerical tag(s)

$F(s) = \mathcal{L}\{f(t)\}$, Laplace transform of $f(t)$

$\|f\|$, norm of f, defined by $\langle f, f \rangle^{1/2}$

$\{f_n(x)\}$, a sequence of real-valued functions

$\widetilde{f}_s(n)$, $\widetilde{f}_s(n, y)$, finite Fourier sine transform

$\dfrac{2}{a} \displaystyle\sum_1^\infty \widetilde{f}_s(n) \sin\left(\dfrac{n\pi x}{a}\right)$, inverse finite Fourier sine transform

$\widetilde{f}_c(n)$, $\widetilde{f}_c(n, y)$, finite Fourier cosine transform

$\dfrac{\widetilde{f}_c(0)}{a} + \dfrac{2}{a} \displaystyle\sum_1^\infty \widetilde{f}_c(n) \cos\left(\dfrac{n\pi x}{a}\right)$, inverse finite Fourier sine transform

$f \circ g$, composition of functions f and g such that $(f \circ g)(x) = f(g(x))$

F, field

$\mathcal{F}_c\{f(x)\} \equiv \widetilde{f}_c(\alpha)$, Fourier cosine transform

$\mathcal{F}^{-1}\{\widetilde{f}_c(\alpha)\} \equiv f(x)$, inverse Fourier cosine transform

$\mathcal{F}_s\{f(x)\} \equiv \widetilde{f}_s(\alpha)$, Fourier sine transform

$\mathcal{F}^{-1}\{\widetilde{f}_s(\alpha)\} \equiv f(x)$, inverse Fourier sine transform

$\mathcal{F}\{f(x)\} \equiv \widetilde{f}(\alpha)$, Fourier complex trannsform

$\mathcal{F}^{-1}\{\widetilde{f}(\alpha)\} \equiv f(x)$, inverse Fourier complex transform

g, electric charge; total distribution charge density

$g_\varepsilon(x) = \dfrac{1}{\sqrt{2\pi\varepsilon}}\, e^{-x^2/(2\varepsilon)}$, Gaussian functions, or Gaussian distribution

$F \star G = \int_0^t f(t-u)g(u)\,du = \int_0^t f(u)g(t-u)\,du$, convolution of F and G

$g(x, s)$, one-sided Green's function for initial value problems

$g_{ij}(x, s)$, one-sided Green's function for a system of n first-order ordinary differential equations

$G(t, t')$, or $G(x, x')$, also $G(t, s)$ or $G(x, s)$, Green's function

$G(\mathbf{x}, \mathbf{x}')$, Green's function, also written as $G(\mathbf{x} - \mathbf{x}')$

$G(\mathbf{x}, \mathbf{x}'; t, t') \equiv G(\mathbf{x} - \mathbf{x}'; t - t')$, Green's function for a space-time operator

h, Planck's constant, $h = 2\pi\,\hbar$

$\hbar = 1.054 \times 10^{-34}$ joules-sec, or $= 6.625 \times 10^{-27}$ erg-sec, Planck's constant

$h_n^{(1)}(x) = \sqrt{\dfrac{\pi}{2x}}\, H_{n+1/2}^{(1)}(x)$, spherical Bessel function of first kind and order n

$H(t)$ or $H(x)$, Heaviside unit step function

$H(z)$, or H_{pq}, Fox H-function $H(z)$, introduced by Fox [1961], is defined as

$$H(z) = H_{p,q}^{m,n}\left(z \,\middle|\, \begin{matrix} (a_1, \alpha_1), \ldots, (a_p, \alpha_p) \\ (b_1, \beta_1), \ldots, (b_p, \beta_p) \end{matrix}\right)$$

$$= \dfrac{1}{2\,i\,\pi} \int_C \dfrac{\prod_{j=1}^m \Gamma(b_j - \beta_j\, s)\, \prod_{j=1}^n \Gamma(1 - a_j + \alpha_j\, s)}{\prod_{j=n+1}^p \Gamma(a_j + \alpha_j\, s)\, \prod_{j=m+1}^q \Gamma(1 - b_j + \beta_j\, s)},$$

where $0 \leq n \leq p$, $0 \leq m \leq q$; $\alpha_j, \beta_j > 0$, s complex, and a_j, b_j are complex numbers such that no pole of $\Gamma(b_j - \beta_j s)$ for $j = 1, \dots, m$ coincides with any pole of $\Gamma(1 - a_j + \alpha_j s)$ for $j = 1, \dots, n$; and C is a contour in the complex s-plane from $\gamma - i\infty$ to $\gamma + i\infty$ such that $\dfrac{b_j + k}{\beta_j}$ and $\dfrac{a_j - 1 - k}{\alpha_j}$ lie to the left and right of C, respectively (Prudnikov et al. [1990:626]).

$H_n(x) = (-1)^n e^{x^2} \dfrac{d^n}{dx^n} e^{-x^2}$, Hermite polynomials of degree n

$H_0^{(1)}(kr), H_0^{(2)}(kr)$, Hankel functions of the first and second kind, respectively, and of order n; $H_0^{(1,2)}(kr) = J_0(kr) \pm i Y_0(kr)$

H_{12}^{20} , see $H(z)$ Fox H-function

$\mathcal{H}_n\{f(x)\} \equiv \hat{f}_n(\sigma)$, Hankel transform of order n

$\mathcal{H}_0\{f(x)\} \equiv \hat{f}_0(\sigma)$, zero-order Hankel transform

$\mathcal{H}_n^{-1}\{\hat{f}_n(\sigma)\} \equiv f(x)$, inverse Hankel transform of order n

$\mathbf{H} = -\dfrac{\hbar^2}{2m}\dfrac{d^2}{dt^2} + 2\pi^2 \nu^2 m t^2 = -\dfrac{\hbar^2}{2m}\dfrac{d^2}{dt^2} + \frac{1}{2}m\omega^2 t^2$, Hamiltonian operator

i.e., that is (Latin *id est*)

iff, if and only if

$\mathbf{i}, \mathbf{j}, \mathbf{k}$, unit vectors along the rectangular coordinates axes x, y, and z, respectively

I, moment of inertia; current intensity

Int(Γ) , interior of a simple closed contour Γ

$I(x)$, intensity of a wave

$I_0(z)$, modified Bessel function of the first kind and of order zero

$I_n(x)$, modified Bessel function of the first kind and order n, defined by $I_n(x) = e^{-in\pi/2} J_n\left(e^{i\pi/2}x\right)$

$I(z, z_0)$, index or winding number of a simple closed contour with respect to a point z_0 in the complex plane

$\Im$, imaginary part of a complex quantity

j, total angular momentum eigenvalue

J, Jacobian, defined by $J\left(\dfrac{(x,y,z)}{(u,v,w)}\right) = \dfrac{\partial(x,y,z)}{\partial(u,v,w)} = \begin{vmatrix} \dfrac{\partial x}{\partial u} & \dfrac{\partial x}{\partial v} & \dfrac{\partial x}{\partial w} \\[6pt] \dfrac{\partial y}{\partial u} & \dfrac{\partial y}{\partial v} & \dfrac{\partial y}{\partial w} \\[6pt] \dfrac{\partial z}{\partial u} & \dfrac{\partial z}{\partial v} & \dfrac{\partial z}{\partial w} \end{vmatrix}$

$J_n(x)$, Bessel function of first kind and order $n = 0, 1, 2, \dots$

k, thermal diffusivity; spring constant

$K_0(z)$, modified Bessel function of the third kind and of order zero

$K_n(x)$, modified Bessel function of the third kind and order n, defined by

$K_n(x) = \dfrac{\pi \left[I_{-n}(x) - I_n(x)\right]}{2 \sin n\pi}$

$l_2 = \{\mathbf{x} \in X : \sum_{i=1}^{\infty} |x_i|^2 < \infty\}$, vector space

$\log z = \ln|z| + i \arg\{z\}$, (multiple-valued) logarithm function in $\mathbb{C}$

$L_1([a,b])$, vector space of first-order integrable functions on $\mathbb{R}$

$L_2([a,b])$, vector (Hilbert) space of square integrable functions on $\mathbb{R}$ or $\mathbb{C}$

$L_p([a,b])$, $p \geq 1$, vector space of p-order integrable functions on $\mathbb{R}$

L, linear differential operator; induction coefficient of a conductor

L^*, adjoint operator to a differential operator L

$$L[u] \equiv \frac{d}{dx}\left[p(x)\frac{du}{dx}\right] + q(x)u, \ a < x < b, \ \text{Sturm-Liouville operator}$$

$L\left[D,\dfrac{\partial}{\partial t}\right]$, transient operator

L^{-1}, inverse (integral) operator

$L_n(x)$, Laguerre polynomials of order n

$L_n^m(x)$, associated Laguerre polynomials of order n and degree m

$L_n(t,\phi,t_1,\phi_1)$, Laplace's coefficients, $t = \cos\theta$

$\mathcal{L}\{f(t)\} \equiv F(s)$, Laplace transform

$\mathcal{L}^{-1}\{F(s)\} \equiv f(t)$, inverse Laplace transform

n, outward normal perpendicular to the boundary of a curve or surface; azimuthal (orbital angular) quantum number

n_x, n_y, n_z, components of the outward normal n along x, y, and z axis, respectively

$N(z,z')$, Neumann's function

$\mathbb{N} = \mathbb{Z}^+$, set of natural numbers

m, mass; mass of a quantum particle; magnetic quantum number

p.v., Cauchy's principal value of an integral

p, partial derivative u_x, or $\dfrac{\partial u}{\partial x}$; also, pressure

$P(D)$, ordinary differential operator, defined by a polynomial of degree n of the form $P(D) = a_0(x)\,D^n + a_1(x)\,D^{n-1} + \cdots + a_{n-1}(x)\,D + a_n(x)$, $a_0(x) \neq 0$

$P(r)$, probability function

$P(\mathbf{x})$, gravitational potential

$P_n(x)$, Legendre polynomials of degree $n = 0,1,2,\ldots$; Legendre's coefficients, or surface zonal harmonics

$P_n(\cos\theta)$, Laplace's coefficient of degree n

$P_n^m(\cos\theta)$ or $P_n^m(t)$, associated Legendre polynomials of order $m = 0,\pm1,\pm2,\ldots$ and degree $n = 0,1,2,\ldots$

q, partial derivative u_y, or $\dfrac{\partial u}{\partial y}$

$Q_n(x)$, Legendre function of the second kind of order n

r, partial derivative u_{xx}, or $\dfrac{\partial^2 u}{\partial x^2}$; also, radial axis

(r,θ), polar coordinates: $x = r\cos\theta$, $y = r\sin\theta$, $r = \sqrt{x^2+y^2}$, $\theta = \arctan\dfrac{y}{x}$

(r,θ,z), polar cylindrical coordinates: $x = r\cos\theta$, $y = r\sin\theta$, $r = \sqrt{x^2+y^2}$, $\theta = \arctan\dfrac{y}{x}$, $z = z$

(r, θ, ϕ), spherical coordinates: $x = r \sin \theta \cos \phi$, $y = r \sin \theta \sin \phi$, $z = r \cos \theta$, $r = \sqrt{x^2 + y^2 + z^2}$, $\theta = \arccos \dfrac{z}{r}$, $\phi = \arctan \dfrac{y}{x}$

$r^n P_n(\cos \theta)$ or $r^{-n-1} P_n(\cos \theta)$, solid zonal harmonics

$r^n Y_n^m(t, \phi)$ or $r^{-n-1} Y_n^m(t, \phi)$, solid spherical harmonics of degree n

$|\mathbf{r}|$, average probability distance for the hydrogen atom

$R = \{0 < x < a, \ 0 < y < b\}$, rectangle

$R_n^m(r)$, radial wave function

$\mathbb{R}$, real line

$\mathbb{R}^n$, Euclidean n-space; $\mathbb{R}^1 \equiv \mathbb{R}$

$\mathbb{R}^+$, set of positive real numbers

$\Re$, real part of a complex quantity

s, variable of the Laplace transform; spin quantum number

$s(\mathbf{x}, \mathbf{x}')$, singularity function

$\mathrm{sgn}(x) = \begin{cases} 1, & x > 0, \\ 0, & x = 0, \\ -1, & x < 0, \end{cases}$ signum function

$\mathrm{sinc}(x) = \dfrac{\sin x}{x}$, sinc function

$\mathrm{supp}\, f$, support of a continuous function f

$S = \sum\limits_{n=1}^{\infty} s_n$, sum of an infinite sequence $\{s_n\}$

$S_n = s_1 + s_2 + \cdots + s_n$, partial sum of an infinite sequence $\{s_n\}$

$S(r, \mathbf{x}_0)$ boundary (surface) of the open ball $B(r, \mathbf{x}_0)$, or $\{\mathbf{x}, |\mathbf{x} - \mathbf{x}_0| = r\}$

$S_n(1) = \dfrac{2\pi^{n/2}}{\Gamma(n/2)}$, surface area of the unit ball $B(1, 0)$ in $\mathbb{R}^n$

t, partial derivative u_{yy}, or $\dfrac{\partial^2 u}{\partial y^2}$; also, time

t', source point ; singularity

T, linear transformation; also, kinetic energy

u, dependent variable; displacement; temperature

$u_c(x)$ or $u_c(t)$, complementary function for an ordinary differential equation

$u_p(x)$ or $u_c(t)$, particular integral for an ordinary differential equation

$u_{1,s}, u_{2,s}, u_{3,s}$, first, second, and third state of radial energy of the hydrogen atom, respectively

$u^*(x, x')$, fundamental solution, or 'free-space' Green's function, or Green's function in the large

$U \equiv B(1, 0)$, open unit disk

V, volume; potential energy

$W(t)$, Lambert W-function, or Omega function, or Product Log function. It is the inverse function of $f(w) = w\, e^w$, where w is a complex number, i.e., $z = W(z)\, e^{W(z)}$; this function is multiple-valued except at $z = 0$. If we restrict it to real numbers and keep w real, then the function is defined only for $x \geq -1/e$; it is double-valued on the interval $(-1/e, 0)$. If we impose an additional restriction of $w \geq -1$, then the function is single-valued, and is denoted by $W_0(x)$, where $W_0(0) = 0$ and $W_0(-1/e) = -1$. The other branch on $[-1/e, 0]$ with $w \leq -1$ is denoted by $W_{-1}(x)$ and decreases from $W_{-1}(-1/e) = -1$ to $W_{-1}(0^-) = -\infty$. This function satisfies the differential equation $\dfrac{dW}{dz} = \dfrac{W(z)}{z(1 + W(z))}$ for $z \neq -1/e$.

$$W(x) = W(u_1, \ldots, u_n) = \begin{vmatrix} u_1(x) & \cdots & u_n(x) \\ u_1'(x) & \cdots & u_n'(x) \\ \cdots & \cdots & \cdots \\ u_1^{(n-1)}(x) & \cdots & u_n^{(n-1)}(x) \end{vmatrix}, \text{ Wronskian}$$

$W_j(x)$ or $W_j(t)$, determinant obtained from the Wronskian $W(x)$ or $W(t)$ by replacing the jth column by $[0\ 0\ \cdots\ 1]^T$

$W(z, p, q) = \sum\limits_{n=0}^{\infty} \dfrac{z^n}{n!\,\Gamma(pm + q)}$, $p, q > 0$, Wright function

$\{x_n\}$, a sequence of real numbers

$\{x_{n_k}\}$, a subsequence of real numbers

(x, y, z), cartesian coordinates

X, metric space

$\mathbf{x}$, a point $(x_1, x_2, \ldots, x_n)$ in $\mathbb{R}^n$; a field point

$\mathbf{x}'$, source point; singularity

$|\mathbf{x} - \mathbf{y}|$, Euclidean distance between the points $\mathbf{x}, \mathbf{y} \in \mathbb{R}^n$

$y_n(x) = \sqrt{\dfrac{\pi}{2x}}\, Y_{n+1/2}(x) = (-1)^n \sqrt{\dfrac{\pi}{2x}}\, J_{n-1/2}(x)$, spherical Bessel function of order n

$Y_0(x)$, Bessel function of second kind of order zero

$Y_n(x)$, Bessel function of the second kind of order n, defined by

$$Y_n(x) = \dfrac{\cos n\pi\, J_n(x) - J_{-n}(x)}{\sin n\pi}; \text{ sometimes denoted by } N_n(x)$$

$Y_n^m(\theta, \phi)$, spherical harmonics

z, a complex number $z = x + iy$

z^*, inverse (symmetrical) point of z with respect to the unit circle in $\mathbb{C}$

$|z| = \sqrt{x^2 + y^2}$, $z = x + iy$, modulus of a complex number z

$\mathbb{Z}$, set of integers

$\mathbb{Z}^+$, set of positive integers

α, variable of Fourier transform, of Fourier sine and cosine transforms

$\gamma = 0.577215665$, Euler gamma

Γ, simple contour, or path; boundary of a domain

$\Gamma(z) = \int_0^\infty t^{\alpha-1}(1-t)^\alpha\, dt$, $\Re\{z\} > 0$, gamma function

$\delta(x)$, $\delta(x, x')$, $\delta(x, s)$, Dirac delta function; also denoted by $\delta(x - x')$

δ_{mn}, Kronecker delta, equal to 1 if $m = n$ and 0 if $m \neq n$

$\boldsymbol{\zeta} = (\alpha, \beta, \gamma)$, variable of 3-D Fourier transform

α, wavelength

λ_n, eigenvalues

$\lambda_\varepsilon(x) = \dfrac{1}{\pi}\dfrac{\varepsilon}{x^2 + \varepsilon^2}$, Cauchy densities, or Lorentz curves

ν, vibration frequency

ϕ, phase; latitude (azimuth)

ϕ_n, eigenfunctions

$\phi_\varepsilon(t)$, 'cap' function

(λ_n, ϕ_n), eigenpairs

ν, vibration frequency

$\rho(x)$, weight function

$\rho(\mathbf{r})$, charge density function

σ, variable of zero-order Hankel transform

$\varsigma = (\alpha, \beta)$, variable of 2-D Fourier transform

θ, colatitude (polar angle)

ϑ-function, Riemann theta function is defined as

$$\vartheta(z; \tau) = \sum_m \exp\left\{ 2\pi i \left(\tfrac{1}{2} m^T \tau m + m^T z \right) \right\},$$

where $z \in \mathbb{C}^n$ is a complex vector, $\tau \in H_n$ is the Siegel upper half-plane, and T denotes the transpose. For $n = 1$, H is the upper half-plane in $\mathbb{C}$. It is related to the Jacobi theta functions, where for our interest ϑ_{00} is defined by

$$\vartheta_{00}(z; \tau) = \vartheta(z; \tau) = \vartheta_3(z; q), \text{ where } \vartheta_{00}(w; q) = \sum_{n=-\infty}^{\infty} \left(w^2\right)^n q^{n^2}, \text{ where }$$

$w = e^{i\pi z}$ is called the argument and $q = e^{i\pi\tau}$ the nome. A useful defini-

tion is: $\vartheta_{00}(z; \tau) = -i \displaystyle\int_{i-\infty}^{i+\infty} \dfrac{e^{i\pi\tau u^2} \cos(2uz + \pi u)}{\sin(\pi u)}\, du$. For more details, see

Abramowitz and Stegun [1968: §16.27ff]; Akhizer [1990], Pierpont [1959], and Dubrovin [1981].

$\chi_D(x)$, characteristic function of a domain D

$\Psi(r)$, monochromatic complex wave

ω, frequency; angular frequency of vibrations

$\partial B(r, a) = \{z \in \mathbb{C} : |z - a| = r\}$, circle of radius r and center a

∂D, boundary of the domain D

1-D, 2-D, 3-D, and 4-D, one-dimensional, two-dimensional, three-dimensional, and four-dimensional (three space dimensions and one time), respectively

$\mathbf{0} = (0, 0, \dots, 0)$, zero vector, null vector, or origin in $\mathbb{R}^n$

$\mathbf{1} = (1, 1, \dots, 1)$, unit vector in $\mathbb{R}^n$

$\dbinom{n}{k} = \dfrac{n!}{k!\,(n - k)!}$, $\dbinom{n}{0} = 1$, binomial coefficients

$n! = 1 \cdot 2 \cdot \dots (n - 1)n$, $0! = 1! == (-1)! = 1$, factorial n

$\langle x, y \rangle$, inner product of $x, y \in \mathbb{R}$

$\| \cdot \|$, norm

$\langle f, g \rangle = \int_a^b f(x) \, g(x) \, dx$, inner product of functions f and g

$\oint$, line or surface integral

$\iint$, double (surface) integral

$\iiint$, triple (volume) integral

∇, grad, defined by $\nabla = \mathbf{i} \, \dfrac{\partial}{\partial x} + \mathbf{j} \dfrac{\partial}{\partial y} + \mathbf{k} \dfrac{\partial}{\partial z}$

∇^2, Laplacian $\dfrac{\partial^2}{\partial x^2} + \dfrac{\partial^2}{\partial y^2} + \dfrac{\partial^2}{\partial z^2}$

$\nabla^2 + k^2$, Helmholtz operator

∇^4, biharmonic operator, defined by $\nabla^4 u = \nabla^2(\nabla^2 u)$

$\dfrac{\partial}{\partial t} - k\nabla^2$, diffusion operator; heat conduction operator

$\dfrac{\partial}{\partial n}$, partial derivative with respect to n

$\dfrac{\partial}{\partial t} - \dfrac{\partial}{\partial x}\left(\dfrac{\partial}{\partial x} + x\right)$, Fokker-Plank operator

$\Box_c \equiv \dfrac{\partial^2}{\partial t^2} - c^2 \nabla^2$, d'Alembertian, or wave operator; also, $\Box \equiv \Box_1$

$\blacksquare$, end of an example or of a proof

$\boxed{!!}$, attention sign

1

Some Basic Results

In this chapter we discuss some basic definitions and present results which are needed to study Green's functions and linear ordinary and partial differential equations. Proofs of these results can be found in standard textbooks on advanced calculus and real analysis. The notation used in this book, although standard, is presented prior to this chapter. Readers familiar with the topics covered in this chapter may still like to read it; others are advised to study them thoroughly.

1.1. Euclidean Space

A real finite dimensional vector space on which an inner product is defined is called the *Euclidean space*, which is denoted by $\mathbb{R}^n$, $n = 1, 2, 3, \ldots$. Let F be a field, let "+" denote a mapping of $\mathbb{R}^n \times \mathbb{R}^n$ into $\mathbb{R}^n$, and let "·" denote a mapping of $F \times \mathbb{R}^n$ into $\mathbb{R}^n$. The elements $\mathbf{x} \in \mathbb{R}^n$ are called *vectors*, such that $\mathbf{x} = (x_1, x_2, \ldots, x_n)$ represents a (position) vector of a point with cartesian coordinates $(x_1, x_2, \ldots, x_n)$. The elements of the field F are called *scalars*, while the operation "+" defined on $\mathbb{R}^n$ is called *vector addition* and the mapping "·" the *scalar multiplication* or *multiplication of vectors by scalars*. Then for each $\mathbf{x}, \mathbf{y} \in \mathbb{R}^n$ there is a unique element $\mathbf{x} + \mathbf{y} \in \mathbb{R}^n$, called the *sum* of $\mathbf{x}$ and $\mathbf{y}$, and for each $\mathbf{x} \in \mathbb{R}^n$ and $\alpha \in F$ there is a unique element $\alpha \cdot x = \alpha \mathbf{x} \in \mathbb{R}^n$, called the *multiplication* of $\mathbf{x}$ by α. The non-empty set $\mathbb{R}^n$ and the field F along with the above two mappings of vector addition and scalar multiplication constitute a *vector space* or a *linear space* if the following axioms are satisfied: (i) $\mathbf{x} + \mathbf{y} = \mathbf{y} + \mathbf{x}$ for every $\mathbf{x}, \mathbf{y} \in \mathbb{R}^n$; (ii) $\mathbf{x} + (\mathbf{y} + \mathbf{z}) = (\mathbf{x} + \mathbf{y}) + \mathbf{z}$ for every $\mathbf{x}, \mathbf{y}, \mathbf{z} \in \mathbb{R}^n$; (iii) There is a unique vector in $\mathbb{R}^n$, called the *zero vector* or the *null vector* or the *origin*, denoted by $\mathbf{0}$ such that $\mathbf{0} + \mathbf{x} = \mathbf{x} + \mathbf{0} = \mathbf{x}$ for all $\mathbf{x}, \mathbf{y} \in \mathbb{R}^n$; (iv) $\alpha(\mathbf{x} + \mathbf{y}) = \alpha\mathbf{x} + \alpha\mathbf{y}$ for every $\alpha \in F$ and for every $\mathbf{x}, \mathbf{y} \in \mathbb{R}^n$; (v) $(\alpha + \beta)\mathbf{x} = \alpha\mathbf{x} + \beta\mathbf{x}$ for all $\alpha, \beta \in F$ and for every $\mathbf{x} \in \mathbb{R}^n$; (vi) $(\alpha\beta)\mathbf{x} = \alpha(\beta\mathbf{x})$ for all $\alpha, \beta \in F$ and for every $\mathbf{x} \in \mathbb{R}^n$; and (vii) $\mathbf{0}\mathbf{x} = \mathbf{x} + \mathbf{0} = \mathbf{0}$ for all $\mathbf{x} \in \mathbb{R}^n$; (viii) $\mathbf{1}\mathbf{x} = \mathbf{x}$ for all every $\mathbf{x} \in \mathbb{R}^n$, where $\mathbf{1} = (1, 1, \ldots, 1)$ denotes the unit vector directed from the origin $\mathbf{0}$ along the coordinate axes.

Since n denotes the dimension of the space, and since we will be mostly dealing with $n = 1, 2, 3$, we will use the following notation: We denote $\mathbb{R}^1$ simply by $\mathbb{R}$ which is the set of all points on the real axis. The 2-D space $\mathbb{R}^2$ is the set of all points (x, y) in the real plane, and the 3-D space $\mathbb{R}^3$ represents the set of all points (x, y, z). The point $\mathbf{0}$ represents the origin of coordinates, and the vector $\mathbf{1}$ is represented by $\mathbf{i}, \mathbf{j}, \mathbf{k}$ in $\mathbb{R}^3$, which are the unit vectors along the rectangular coordinates axes x, y, and z, respectively.

We will denote by $\mathbb{R}^+$ the set of nonnegative real numbers, by $\mathbb{Z}$ the set of integers, and by $\mathbb{N}$ the set of natural numbers. Notice that $\mathbb{N} = \mathbb{Z}^+$. Since $|\mathbf{x} - \mathbf{y}|$ defines the Euclidean distance between points $\mathbf{x}$ and $\mathbf{y}$ in $\mathbb{R}^n$, where $\mathbf{x} = (x_1, \ldots, x_n)$ and $\mathbf{y} = (y_1, \ldots, y_n)$, we define an open ball of radius r centered at a point $\mathbf{x}_0 \in \mathbb{R}^n$ by $\{\mathbf{x} : |\mathbf{x} - \mathbf{x}_0| < r\}$, and denote it by $B(r, \mathbf{x}_0)$. The boundary (surface) of the open ball $B(r, \mathbf{x}_0)$ is denoted by $S(r, \mathbf{x}_0) = \{\mathbf{x} : |\mathbf{x} - \mathbf{x}_0| = r\}$, where $S_n(1) = \dfrac{2\pi^{n/2}}{\Gamma(n/2)}$ is the surface area of the unit ball $B(1, 0)$ in $\mathbb{R}^n$. The ε-neighborhood of a point $x_0 \in \mathbb{R}$ is defined by $B(\varepsilon, x_0)$, or $-\varepsilon < x - x_0 < \varepsilon$ for an arbitrarily small $\varepsilon > 0$. The complement of a set B with respect to a set A is denoted by $A \backslash B$, the product of the sets A and B by $A \times B$, and the closure of a set A by $\overline{A}$.

1.1.1. Metric Space. Since $\mathbb{R}$ is a metric space, let d be a real-valued function such that $d : \mathbb{R} \mapsto \mathbb{R}$, where d has the following properties: (i) $d(x, y) \geq 0$ for all $x, y \in \mathbb{R}$, and $d(x, y) = 0$ iff $x = y$; (ii) $d(x, y) = d(y, x)$ for all $x, y \in \mathbb{R}$ (symmetry property); and (iii) $d(x, y) \leq d(x, z) + d(z, y)$ for all $x, y, z \in \mathbb{R}$ (triangular inequality). The function d is called a *metric* (or distance) on $\mathbb{R}$, and $d(x, y) = |x - y|$ for all $x, y \in \mathbb{R}$; obviously, $d(x, y) = |x - y| = 0$ iff $x = y$, and $d(x, y) = |x - y| = |(x - z) + (z - y)| \leq |x - z| + |z - y| = d(x, y) + d(z, y)$ for all $x, y, z \in \mathbb{R}$. These properties also hold on the space $\mathbb{R}^n$

1.1.2. Inner Product. For every $x, y \in \mathbb{R}$ the inner product $\langle x, y \rangle$ possesses the following three properties: (i) $\langle x, y \rangle > 0$ for all $x \neq 0$ and $\langle x, y \rangle = 0$ if $x = 0$; (ii) $\langle x, y \rangle = \langle y, x \rangle$ for all $x, y \in \mathbb{R}$; (iii) $\langle \alpha x + \beta y, z \rangle = \alpha \langle x, z \rangle + \beta \langle x, y \rangle$ for all $x, y, z \in \mathbb{R}$ and for all $\alpha, \beta \in F$. The inner product for vectors in $\mathbb{R}^n$ is defined by $\langle \mathbf{x}, \mathbf{y} \rangle = \sum_{i=1}^{n} x_i y_i$ for any vectors $\mathbf{x} = (x_1, \ldots, x_n)$ and $\mathbf{y} = (y_1, \ldots, y_n)$ in $\mathbb{R}^n$. For real-valued functions $f(x)$ and $g(x)$ defined on an interval $(a, b) \in \mathbb{R}$, where a can be $-\infty$ and b can be ∞, the inner product of f and g is denoted by $\langle f, g \rangle$ and defined by

$$\langle f, g \rangle = \int_a^b f(x)\, g(x)\, dx. \tag{1.1}$$

The weighted inner product of f and g with weight function $\rho > 0$ on the interval (a, b) is defined by

$$\langle f, g \rangle_\rho = \int_a^b f(x) g(x) \rho(x)\, dx. \tag{1.2}$$

1.2. Classes of Continuous Functions

A real- (or complex-) valued function f is said to belong to the class $C^k(D)$ if it is continuous together with all the derivatives up to order k inclusive, $0 \leq k < \infty$, in a domain $D \in \mathbb{R}^n$. The functions f in the class $C^k(D)$ which admit continuous continuations in the closure $\bar{D} = D \cup \partial D$, where ∂D denotes the boundary of the domain D, form the class of functions $C^k(\bar{D})$. The class $C^\infty(D)$ consists of functions f which are infinitely differentiable in D, i.e., their continuous partial derivatives of all orders exist. These classes are linear sets; thus, every linear combination $\lambda f + \mu g$, where λ and μ are arbitrary real or complex numbers, also belongs to this class. Further, a function defined on $D \subset \mathbb{R}$ is said to belong to the class $C_0^\infty(D)$ if it is infinitely differentiable on D and vanishes outside some bounded region. The *support* of a continuous function f (written supp f) is the closure of the set $\{x \in D : f(x) \neq 0\}$. Then the class $C_0^k(D)$ denotes the set of functions in $C^k(D)$ that have compact support, where the index 0 indicates compact support. A function is said to be of compact support if it is equal to zero outside a given bounded set in its domain. A function f with compact support on $\mathbb{R}$ is said to belong to the class $C_0^\infty(\mathbb{R})$ if it is infinitely differentiable on $\mathbb{R}$.

1.3. Convergence

Various results on convergence of sequences and infinite series in $\mathbb{R}$ are discussed.

1.3.1. Convergence of Sequences.
A sequence $\{x_n\}$ in a set $X \subseteq \mathbb{R}$ is a function $f : \mathbb{Z}^+ \mapsto X$. Thus, if $\{x_n\}$ is a sequence in X, then $f(n) = x_n$ for each $n \in \mathbb{Z}^+$. Let $d(x, x_n)$ denote the distance (or metric) between x and x_n. If $\{x_n\}$ is a sequence of points in X, and if x is a point of X, then the sequence $\{x_n\}$ is said to converge to x if for every $\varepsilon > 0$ there is an integer N such that for all $n \geq N$, $d(x, x_n) < \varepsilon$, i.e., x_n belong to the ε-neighborhood of the point x for all $n \geq N$. In general, N depends on ε, i.e., $N = N(\varepsilon)$, and we write $\lim_n x_n = x$, or alternatively, $x_n \to x$ as $n \to \infty$. If there is no $x \in X$ to which the sequence converges, then we say that the sequence $\{x_n\}$ diverges. Further, if the range of $f(n)$ is bounded, then the sequence $\{x_n\}$ is said to be bounded. In this definition the range of $f(n)$ may consists of a finite number or an infinite number of points. If the range of f consists of one point, then we say that the sequence is a constant sequence, Obviously, all constant sequences are convergent. Let $\{x_n\}$ be a sequence in X, and let $n_1, n_2, \ldots, n_k, \ldots$ be a sequence of positive integers which is strictly increasing, i.e., $n_j > n_k$ for all $j > k$. Then the sequence $\{x_{n_k}\}$ is called a subsequence of $\{x_n\}$. If the subsequence $\{x_{n_k}\}$ converges, then its limit is called a subsequential limit of $\{x_n\}$. Let $\{x_n\}$ be a sequence in X. Then:

(i) There is at most one point $x \in X$ such that $\lim_n x_n = x$;

(ii) If $\{x_n\}$ is convergent, then it is bounded;

(iii) $\{x_n\}$ converges to a point $x \in X$ iff every ball (neighborhood) about x contains all but a finite number of terms in $\{x_n\}$;

(iv) $\{x_n\}$ converges to a point $x \in X$ iff every subsequence of $\{x_n\}$ converges to x;

(v) If $\{x_n\}$ converges to $x \in X$ and if $y \in X$, then $\lim_n d(x_n, y) = d(x, y)$;

(vi) If $\{x_n\}$ converges to $x \in X$ and if the sequence $\{y_n\}$ of X converges to $y \in X$, the $\lim_n d(x_n, y_n) = d(x, y)$; and

(vii) If $\{x_n\}$ converges to $x \in X$, and if there is a $y \in X$ and a $c > 0$ such that $d(x_n, y) \le c$ for all $n \in \mathbb{Z}^+$, then $d(x, y) \le c$.

A sequence $\{x_n\} \in X$ is said to be a *Cauchy sequence* if for every $\varepsilon > 0$ there is an integer N such that $d(x_n, x_m) < \varepsilon$ whenever $n, m \ge N$. Then (i) every convergent sequence in a metric space is a Cauchy sequence; (ii) if $\{x_n\}$ is a Cauchy sequence, then $\{x_n\}$ is a bounded sequence; (iii) if a Cauchy sequence $\{x_n\}$ contains a convergent subsequence $\{x_m\}$, then the sequence $\{x_n\}$ is convergent.

A function $f : X \mapsto Y$, where $X, Y \subseteq \mathbb{R}$, is continuous at a point $x_0 \in X$ iff for every sequence $\{x_n\}$ of points in X which converges to x_0 the corresponding sequence $\{f(x_n)\}$ converges to the point $f(x_0)$ in Y; i.e., $\lim_{n \to \infty} f(x_0) = f\left(\lim_{n \to \infty} x_n\right) = f(x_0)$ whenever $\lim_{n \to \infty} x_n = x_0$. Let f be a mapping from X into Y. Then (i) f is continuous on X iff the inverse image of each open subset of $\{Y, d_y\}$ is open in $\{X, d_x\}$; and (ii) f is continuous on X iff the inverse image of each closed subset of $\{Y, d_y\}$ is closed in $\{X, d_x\}$. Moreover, let f be a mapping from X into Y, and let g be a mapping from Y into Z. If f is continuous on X and g is continuous on Y, then the composite mapping $f \circ g$ of X into Z is continuous on X.

Let $X, Y \subseteq \mathbb{R}$, and let $\{f_n\}$ be a sequence of functions from X into Y. If $\{f_n(x)\}$ converges at each $x \in X$, then we say that $\{f_n\}$ is *pointwise convergent*, and write: $\lim_n f_n = f$, where f is defined for every $x \in X$. In other words, we say that the sequence $\{f_n\}$ is pointwise convergent to a function f if for every $\varepsilon > 0$ and every $x \in X$ there is an integer $N = N(\varepsilon, x)$ such that $d_y(f_n(x), f(x)) < \varepsilon$ whenever $n \ge N(\varepsilon, x)$. In general, $N(\varepsilon, x)$ is not necessarily bounded. However, if $N(\varepsilon, x)$ is bounded for all $x \in X$, then we say that the sequence $\{f_n\}$ *converges to f uniformly* on X. Equivalently, let $M(\varepsilon) = \sup_{x \in X} N(\varepsilon, x) < \infty$. Then we say that the sequence $\{f_n\}$ converges uniformly to f on X if for every $\varepsilon > 0$ there is an $M(\varepsilon)$ such that $d_y(f_n(x), f(x)) < \varepsilon$ whenever $n \ge M(\varepsilon)$ for all $x \in X$. Further, if the sequence $\{f_n\}$ converges uniformly to f on X, then f is continuous on X. Also, if f is continuous on X and if Z is a compact subset of X, then (i) f is uniformly continuous on Z; (ii) f is bounded on Z; and (iii) If $Z \ne \emptyset$, f attains its infimum and supremum on Z; i.e., there exists $x_0, x_1 \in Z$ such that $f(x_0) = \inf\{f(x : x \in Z\}$ and $f(x_1) = \sup\{f(x) : x \in Z\}$.

1.3.2. Weak Convergence. A sequence $\{x_n\}$ of elements in X is said to *converge weakly* to an element $x \in X$ if for every $x' \in X$ the inner product $\langle x_n, x \rangle \to \langle x, x' \rangle$. If the sequence $\{x_n\}$ converges to $x \in X$, i.e., if $\|x_n - x\| \to 0$

as $n \to \infty$, then we call this convergence *strong convergence*, or the sequence $\{x_n\}$ *convergences in the norm*, to distinguish it from weak convergence. Let $\{x_n\}$ be a sequence in X which converges in the norm to $x \in X$. Then $\{x_n\}$ converges weakly to x. This result states that convergence in the norm (or strong convergence) implies weak convergence. Note that every weakly convergent sequence in $\mathbb{R}^n$ is convergent.

Example 1.1. For each $n \in \mathbb{N}$, define f_n on $[0, 2\pi]$ by $f_n(x) = \cos nx$. Then the sequence $\{f_n\} \subset L_p([0, 2\pi])$ for each $p \geq 1$. The Riemann-Lebesgue lemma[1] shows that $f_n \to 0$ weakly for all $p \geq 1$. ∎

Example 1.2. The vector space $L_p([a, b])$ of p-order integrable functions, $p \geq 1$, consists of all continuous functions f and g on the interval $[a, b]$ with metric defined by $d(f, g) = \left(\int_a^b |f(x) - g(x)|^p \, dx \right)^{1/p}$. For $p = 1$ it represents the space of first-order integrable function and for $p = 2$ of the square-integrable functions on the interval $[a, b]$. Let D be a subset of $\mathbb{R}$, and $\chi_D(x) = \begin{cases} 1 \text{ if } x \in D, \\ 0 \text{ if } x \in \mathbb{R} \backslash D, \end{cases}$ be the characteristic function of D. Take $f_n = n \chi_{[0,1/n]}$. Then $\{f_n\} \subset L_p([0, 1])$ for all $p \geq 1$, and $f_n \to 0$ a.e., but $f_n \not\to 0$ weakly. ∎

1.3.3. Metric. Let $X = \mathbb{R}^n$, and define the vector space $l_2 = \{x \in X : \sum_{i=1}^\infty |x_i|^2 < \infty\}$. If $y \in l_2$ is defined by $y = (d_1, d_2, \ldots, d_n, \ldots)$, then the metric for this space is defined by $d(x, y) = \left[\sum_{i=1}^\infty |c_i - d_i|^2 \right]^{1/2}$ for every $x, y \in l_2$.

Example 1.3. Consider the sequence $\{x_n\}$ defined in the vector space l_2 by the vectors $x_1 = (1, 0, \ldots, 0, \ldots), x_2 = (0, 1, 0, \ldots, 0, \ldots), x_3 = (0, 0, 1, \ldots, 0, \ldots),$ $\ldots$. To show that this sequence converges weakly, note that if $x = (c_1, c_2, \ldots, c_n \ldots),$ then every $x' \in l_2$ can be represented as the inner product with some fixed vector $y = (d_1, d_2, \ldots, d_n, \ldots)$, i.e., $\langle x, x' \rangle = \sum_{i=1}^\infty c_i \, d_i$. For our sequence $\{x_n\}$ we have $\langle x_n, x' \rangle = d_n$, and since $d_n \to 0$ as $n \to \infty$ for every $y \in l_2$, we find that $\langle x_n, x' \rangle \to 0$ as $n \to \infty$ for every $x' \in l_2$. Thus, $\{x_n\}$ converges to 0 weakly. However, $x_n \not\to 0$ strongly, because $\|x_n\| = 1$. ∎

1.3.4. Convergence of Infinite Series. The sum of an infinite sequence $\{s_n\}$, denoted by

$$S = \sum_{n=1}^\infty s_n = s_1 + s_2 + \ldots + s_n + \ldots, \tag{1.3}$$

is called an *infinite series*. With each series there is associated a sequence of partial sums: $S_n = s_1 + \ldots + s_n$. If $\lim_{n \to +\infty} S_n = S$ is a finite number, then the series (1.3) is said to *converge* and S is called its *sum*. If $\lim_{n \to +\infty} S_n$ does not exist, the series

---

[1] This lemma states that if $f_n \in L_1([0, 2\pi])$, then $\lim_{|n| \to \infty} f_n(x) = 0$.

(1.3) is said to *diverge*. There are two cases in which a series diverges:
(a) $\lim\limits_{n\to+\infty} S_n = \infty$, or (b) as n increases, S_n increases and decreases (i.e., oscillates) without reaching a limit. For convergence we have the following useful results:

(i) A convergent (divergent) series remains convergent (convergent) even after the removal or alteration of any or all of its first n terms.

(ii) The sum of a convergent series is unique.

(iii) If $\sum s_n$ converges to S, then $\sum k s_n$, where k is any constant, converges to kS; if $\sum s_n$ diverges, so does $\sum k s_n$.

(iv) If $\sum s_n$ converges, then $\lim\limits_{n\to\infty} s_n = 0$. The converse is not true; for example, for the harmonic series $1 + \frac{1}{2} + \frac{1}{3} + \frac{1}{4} + \cdots + \frac{1}{n} + \cdots$ we have $\lim\limits_{n\to\infty} s_n = 0$, but the series diverges.

(v) If $\lim\limits_{n\to\infty} s_n \neq 0$, then $\sum s_n$ diverges. The converse is not true, as is obvious from the harmonic series.

1.3.5. Tests for Convergence of Positive Series. A series $\sum s_n$, all of whose terms are positive, is called a *positive series*. A positive series is convergent if the sequence $\{S_n\}$ of its partial sums is bounded. The following tests for convergence of positive series are based on the fact that the sequences of their partial sums is always nondecreasing.

(a) INTEGRAL TEST. Let $f(n)$ denote the general terms s_n of a positive series $\sum s_n$. If $f(x) > 0$ and never increases on the interval $x > x_0$, where x_0 is some positive integer, then the series $\sum s_n$ converges or diverges according as $\int_{x_0}^{\infty} f(x)\,dx$ exists or does not exist.

(b) COMPARISON TEST FOR CONVERGENCE. A positive series $\sum s_n$ is convergent if each term (perhaps, after a finite number of terms) is less than or equal to the corresponding term of a known convergent positive series $\sum c_n$.

(c) COMPARISON TEST FOR DIVERGENCE. A positive series $\sum s_n$ is divergent if each term (perhaps, after a finite number of terms) is equal or greater than the corresponding term of a known divergent positive series $\sum d_n$.

(d) RATIO TEST. A positive series $\sum s_n$ converges if $\lim\limits_{n\to\infty} \frac{s_{n+1}}{s_n} < 1$, and diverges if $\lim\limits_{n\to\infty} \frac{s_{n+1}}{s_n} > 1$. If $\lim\limits_{n\to\infty} \frac{s_{n+1}}{s_n} = 1$, the test fails to indicate either convergence or divergence.

1.4. Functionals

Let $\langle f, \phi \rangle = \int_{\mathbb{R}^n} f(\mathbf{x})\phi(\mathbf{x})\,d\mathbf{x}$ denote a real number associated with each $\mathbf{x} \in \mathbb{R}^n$ for every test function $\phi \in C_0^\infty(\mathbb{R}^n)$. Then f is said to be a *functional* on $\mathbb{R}^n$. For example, the Fourier series of $f \in C^1[0, \pi]$, defined by

$$f(x) = \sum_{n=1}^{\infty} b_n \sin nx, \quad \text{where} \quad b_n = \frac{2}{\pi} \int_0^\pi f(x) \sin nx \, dx, \qquad (1.4)$$

is a functional on $\mathbb{R}$, with test functions in the set $\{\sin x, \sin 2x, \cdots\}$. Some useful properties of test functions are:

(i) If $\phi_1(x)$ and $\phi_2(x)$ are test functions on $\mathbb{R}^n$, so is their linear combination $c_1\phi_1(x) + c_2\phi_2(x)$, where c_1 and c_2 are real numbers;

(ii) If $\phi(x) \in C_0^\infty(\mathbb{R}^n)$, so do all partial derivatives of $\phi(x)$ belong to the class $C_0^\infty(\mathbb{R}^n)$;

(iii) If $\phi(x) \in C_0^\infty(\mathbb{R}^n)$ and $\alpha(x)$ is infinitely differentiable, then the product $\alpha(x)\phi(x)$ belongs to the class $C^\infty(\mathbb{R}^n)$; and

(iv) If $\phi(x_1, \cdots, x_m) \in C_0^\infty(\mathbb{R}^m)$ and $\psi(x_{m+1}, \cdots, x_n) \in C_0^\infty(\mathbb{R}^{n-m})$, then $\phi(x_1, \cdots, x_m)\psi(x_{m+1}, \cdots, x_n) \in C_0^\infty(\mathbb{R}^n)$.

A functional f on $\mathbb{R}^n$ is said to be *linear* if $\langle f, \lambda\phi_1 + \mu\phi_2 \rangle = \lambda\langle f, \phi_1 \rangle + \mu\langle f, \phi_2 \rangle$ for all real numbers λ, μ and all $\phi_{1,2} \in C_0^\infty(\mathbb{R}^n)$. Note that $\langle f, 0 \rangle = 0$, and

$$\langle f, \sum_{n=1}^m \alpha_n\phi_n \rangle = \sum_{n=1}^m \alpha_n\langle f, \phi_n \rangle.$$

A linear functional f on $C_0^\infty(\mathbb{R}^n)$ is said to be continuous if the numerical sequence $\langle f, \phi_m \rangle \to 0$ as $m \to \infty$, where $\{\phi_m(x)\}$ is a null sequence in $C^\infty(\mathbb{R}^n)$, i.e., $\text{supp}\,\phi_m$, $m = 1, 2, \cdots$, is contained in a sufficiently large ball, and

$$\lim_{m \to \infty} \max_{x \in \mathbb{R}^n} |D^k\phi_m(x)| = 0 \quad \text{for every multi-index } k, \ |k| \leq n.$$

A continuous linear functional f on $C_0^\infty(\mathbb{R}^n)$ is said to be a *distribution*. The number $\langle f, \phi \rangle$ is called the value of f at ϕ, or the action of f on ϕ. The space $\mathcal{D}$ of all distributions on $C_0^\infty(\mathbb{R}^n)$ is a linear space. A locally integrable function $f(x) \in \mathbb{R}^n$ generates an n-dimensional distribution f such that for all $\phi \in C_0^\infty(\mathbb{R}^n)$

$$\langle f, \phi \rangle = \int_{\mathbb{R}^n} f(x)\phi(x) \, dx$$
$$= \int_{-\infty}^{\infty} \cdots \int_{-\infty}^{\infty} f(x_1, \cdots, x_n) \, \phi(x_1, \cdots, x_n) \, dx_1 \cdots dx_n. \qquad (1.5)$$

Hence, every locally integrable function f can be regarded as a distribution.

Let $f_1(x)$ and $f_2(x)$ be two different continuous functions. Then each generates a different distribution such that there exists a ϕ in $C_0^\infty(\mathbb{R}^n)$ for which $\langle f_1, \phi \rangle \neq \langle f_2, \phi \rangle$, i.e., $\langle f_1 - f_2, \phi \rangle \neq 0$. Two functions f_1 and f_2 are said to be equal almost everywhere (a.e.) on a bounded domain Ω if $\int_\Omega |f_1 - f_2| \, dx = 0$. Hence, two locally integrable functions which are equal a.e. generate the same distribution. A distribution of the form (1.5), where $f(x)$ is locally integrable, is said to be *regular*. All other distributions are called *singular*, although formula (1.5) can be used formally for such distributions. Note that if $f \in \mathcal{D}$, then the distribution f vanishes in any region lying outside supp f, i.e.,

$$\langle f, \phi \rangle = 0, \quad \text{supp } f \cap \text{supp } \phi = \emptyset. \tag{1.6}$$

1.4.1. Examples of Linear Functionals. Some examples of linear functionals are as follows.

Example 1.4. Consider the space of functions $C[a, b]$. Then the mapping $f_1(x) = \int_a^b x(s)\, ds$, $x \in C[a, b]$ is a linear functional on $C[a, b]$. Also, the function defined by $f_2(x) = x(s_0)$, $x \in C[a, b]$, $s_0 \in [a, b]$, is a linear functional on $C[a, b]$, and so is the function $f_3(x) = \int_a^b x(s)\, x_0(s)\, ds$, where x_0 is a fixed element of $C[a, b]$ and x any element in $C[a, b]$. ∎

Example 1.5. Let X be a vector space over a field F. Let $X = F^n$, and denote $\mathbf{x} \in X$ by $\mathbf{x} = \{x_1, \ldots, x_n\}$. The mapping $f_4(\mathbf{x}) = \mathbf{x}$ is a linear functional on X. Also, let $\boldsymbol{\alpha} = \{\alpha_1, \ldots, \alpha_n\} \in X$ be fixed and let $\mathbf{x} = \{x_1, \ldots, x_n\}$ be an arbitrary element of X. Then the function $f_5(\mathbf{x}) = \sum_{i=1}^{n} \alpha_i\, x_i$ is a linear functional on X. ∎

1.5. Linear Transformations

A mapping T of a linear space X into a linear space Y, where X and Y are vector spaces over some field F, is called a *linear transformation* or *linear operator* if

(i) $T(x + y) = T(x) + t(y)$ for all $x, y \in X$; and

(ii) $T(\alpha x) = \alpha\, T(x)$ for all $x \in X$ and for all $\alpha \in F$.

Often we find it convenient to write $T \in L(X, Y)$ to specify that T is a linear transformation from a linear space X into a linear space Y. This means that $L(X, Y)$ denotes the set of all linear transformations from a linear space X into a linear space Y. This definition implies that T is a linear transformation from a linear space X into a linear space Y iff $T\left(\sum_{i=1}^{n} \alpha_i x_i \right) = \sum_{i=1}^{n} \alpha_i\, T(x_i)$ for all $x_i \in X$ and for all $\alpha_i \in F$, $i = 1, \ldots, n$. This result is also known as the *principle of superposition*. We cite some examples of linear transformations.

Example 1.6. Let $X = Y$ denote the space of real-valued function of the class $C[a, b]$. Let $T : X \mapsto Y$ be defined by $T[x](t) = \int_a^t x(s)\, ds$, $a \leq t \leq b$, where integration is in the Riemann sense. Then T is a linear transformation. ∎

Example 1.7. Let X denote the set of function $x(t)$ in the class $C[a, b]$, and let vector addition and scalar multiplication be defined respectively by $(x + y)(t) = x(t) + y(t)$ and $(\alpha x)(t) = \alpha x(t)$ for all $t \in [a, b]$. Then $[T]x(t) = \dfrac{dx(t)}{dt}$ is a linear transformation. ∎

Example 1.8. Let X denote the space of all complex-valued function $f(t)$ defined on the interval $[0, \infty)$ such that $f(t)$ is Riemann-integrable and such that $\lim_{t \to \infty} |f(t)| < k\,e^{\alpha t}$, where k is some positive constant and α any real number. Let vector addition and scalar multiplication be defined as in Example 1.7. Let Y denote a linear space of complex functions of a complex variable $s = \sigma + i\omega$. Then the mapping $T : X \mapsto Y$ defined by $[T]f(s) = \int_0^\infty f(t)\,e^{-st}\,dt$ is a linear transformation, and is called the *Laplace transform* of $f(t)$ (see Chapter 5). ∎

Example 1.9. Let X be the space of real-valued functions $f(t) \in C[a, b]$. Let $k(s, t)$ be a real-valued function defined for $a \le s \le b, a \le t \le b$, such that for each $x \in X$ the Riemann integral $\int_a^b k(s, t)\, f(t)\, dt$ exists and defines a continuous function of s on $[a, b]$. Let $T_1 : X \mapsto X$ be defined by $[T_1]f(s) = y(s) = \int_a^b k(s, t)\, f(t)\, dt$. Then T_1 is a linear transformation, and this equation is called the *Fredholm integral equation of the first kind*. ∎

1.6. Cramer's Rule

An $n \times n$ matrix A is said to be *nonsingular* (or *invertible*) iff there exists an $n \times n$ matrix B such that $AB = I$, where I is the $n \times n$ identity matrix. Consider a system of m linear algebraic equations in n unknowns:

$$
\begin{aligned}
a_{11}x_1 + \cdots + a_{1n}x_n &= b_1, \\
a_{21}x_1 + \cdots + a_{2n}x_n &= b_2, \\
&\vdots \\
a_{m1}x_1 + \cdots + a_{mn}x_n &= b_m,
\end{aligned}
\tag{1.7}
$$

where the coefficients $a_{ij}, 1 \le i \le m, 1 \le j \le n$, and right-hand side $b_i, 1 \le i \le m$, are all given real (or complex) numbers. Here the index i in the coefficients a_{ij} refers to the row and the index j the column in which the coefficient occurs. In matrix notation the system (1.7) is represented by $Ax = b$, where the left side is

$$
Ax = \begin{bmatrix} a_{11} & \cdots & a_{1n} \\ \vdots & \vdots & \vdots \\ a_{n1} & \cdots & a_{mn} \end{bmatrix} \begin{Bmatrix} x_1 \\ \vdots \\ x_m \end{Bmatrix} = \begin{Bmatrix} \sum_{j=1}^n a_{1j}x_j \\ \vdots \\ \sum_{j=1}^n a_{mj}x_j \end{Bmatrix}.
\tag{1.8}
$$

The matrix A is also represented by $A = [a_{ij}]$, or by $A = [c_1 \,|\, c_2 \,|\, \cdots \,c_j \,|\, \cdots \,|\, c_n]$, where $c_j, 1 \le j \le n$, denotes the jth column of A.

Let $A = [a_{ij}]$ be an $n \times n$ matrix. If the ith row and jth column of A are deleted, the remaining $(n-1)$ rows and $(n-1)$ columns form another matrix M_{ij}, called the *minor* of a_{ij}. The cofactor of $[a_{ij}]$ is defined as $c_{ij} = (-1)^{i+j} \det[M]$, such that $\sum_{i=1}^n a_{ij}c_{ik} = 0$ for $j \ne k$, and $\sum_{i=1}^n a_{ij}c_{kj} = 0$ for $i \ne k$. These relations can also be written as $\sum_{i=1}^{n} a_{ij} c_{ik} = \det[A]\,\delta_{jk}$ and $\sum_{i=1}^{n} a_{ij} c_{ik} = \det[A]\,\delta_{ik}$, where δ_{ik} is the Kronecker delta.

If an $n \times n$ matrix A is invertible, the associated matrix operator A, represented by (1.8), has its precise range, and the matrix equation $Ax = b$ has the solution $x = Bb$, where the $n \times n$ matrix B whose ith column is b_i has the property $AB = I$. The matrix B is called the *inverse* of the matrix A and is denoted by A^{-1}. Note that (i) $A^{-1}A = AA^{-1} = I$; (ii) A^{-1} is nonsingular (invertible), and (iii) $(AB)^{-1} = B^{-1}A^{-1}$.

The solution of an $n \times n$ linear system $Ax = b$ exists and is unique when A is nonsingular; it is given by $x = A^{-1}b$. If $b = 0$, then the equation $Ax = 0$ has only the trivial solution $x = 0$. When A is singular, the equation $Ax = b$ cannot be solved for all b, and if a solution exists, it is not unique; but in this case the equation $Ax = 0$ has a nontrivial solution. For a nonsingular matrix A the inverse A^{-1} is given by

$$A^{-1} = \frac{1}{\det[A]}\, A^T, \quad \det[A] \neq 0, \tag{1.9}$$

where A^T denotes the transpose of A.

Creamer's Rule. Let A be an $n \times n$ matrix with $\det[A] \neq 0$, and denote the columns of A by $c_1, c_2, \ldots, c_n$. Then the unique solution of the equation $Ax = b$ is

$$x_i = \frac{\det\left[c_1 \mid \cdots \mid b \mid \cdots \mid c_n\right]}{\det[A]}, \quad i = 1, 2, \ldots, n. \tag{1.10}$$

PROOF. Since $x = A^{-1}b$, let the entries of A^{-1} be denoted by d_{ij} so that $A^{-1} = [d_{ij}]$. Then by (1.9) we get $d_{ij} = \dfrac{\text{cof}\,A_{ji}}{\det[A]}$, where $\text{cof}\,A_{ji}$ denotes the cofactor of A_{ji}. Hence, for each $j = 1, 2, \ldots, n$, we have

$$x_i = \sum_{j=1}^{n} \frac{\text{cof}\,A_{ji}}{\det[A]}\, b_j = \frac{\det\left[c_1 \mid \cdots \mid b \mid \cdots \mid c_n\right]}{\det[A]}. \quad \blacksquare$$

1.7. Green's Identities

Let D be a finite domain in $\mathbb{R}^n$, $n = 1, 2, 3$, bounded by a piecewise smooth, orientable surface (or boundary) ∂D, and let F be a scalar function and $\mathbf{G}$ a vector function in the class $C^3(D)$. Then

$$\text{Gradient theorem:} \quad \iint_D \nabla F\, dD = \oint_{\partial D} \mathbf{n}\, F\, dS,$$

$$\text{Divergence theorem:} \quad \iint_D \nabla \cdot \mathbf{G}\, dD = \oint_{\partial D} \mathbf{n} \cdot \mathbf{G}\, dS,$$

$$\text{Stokes's theorem:} \quad \iint_D \nabla \times \mathbf{G}\, dD = \oint_{\partial D} \mathbf{G} \cdot \mathbf{t}\, dS,$$

where $\mathbf{n}$ is the outward normal to the surface ∂D, $\mathbf{t}$ is the tangent vector at a point on ∂D, $\oint$ denotes the surface or line integral, and dS (or ds) denotes the surface (or line) element depending on the dimension of D. The divergence theorem in the above form is also known as the *Gauss theorem*. This theorem implies that $\nabla \cdot [G\,\nabla h - h\nabla g] = G\nabla^2 h - h\nabla^2 G = G\nabla^2 h$, since $\nabla^2 G = 0$. Thus,

$$\iint_D G\nabla^2 h \, d\xi \, d\eta = \oint_{\partial D} \left[G\frac{\partial h}{\partial n} - h\frac{\partial G}{\partial n} \right] ds. \tag{1.11}$$

The Stokes's theorem in $\mathbb{R}^2$ is a generalization of *Green's theorem* which states that if $\mathbf{G} = (G_1, G_2)$ is a continuously differentiable vector field defined on a region containing $D \cup \partial D \subset \mathbb{R}^2$ such that ∂D is a Jordan contour, then

$$\iint_D \left(\frac{\partial G_2}{\partial x} - \frac{\partial G_1}{\partial y} \right) dx \, dy = \oint_{\partial D} G_1 \, dx + G_2 \, dy. \tag{1.12}$$

The gradient and the divergence theorems are valid in R^3 if the surface integral on the left side is replaced by a volume integral and the line integral on the right is replaced by a surface integral. Let the functions $M(x,y,z), N(x,y,z)$, and $P(x,y,z)$, where $(x,y,z) \in D$, be the components of the vector $\mathbf{G}$ in $\mathbb{R}^3$. Then, by the divergence theorem

$$\iiint_D \left(\frac{\partial M}{\partial x} + \frac{\partial N}{\partial y} + \frac{\partial P}{\partial z} \right) d\mathbf{x}$$
$$= \iint_{\partial D} [M \cos(\mathbf{n}, x) + N \cos(\mathbf{n}, y) + P \cos(\mathbf{n}, z)] \, dS, \tag{1.13}$$

where $d\mathbf{x} = dx \, dy \, dz$, dS denotes the surface element, ∂D denotes the boundary of D, and $\cos(\mathbf{n}, x), \cos(\mathbf{n}, y)$, and $\cos(\mathbf{n}, z)$ the direction cosines of $\mathbf{n}$. If we take $M = u\frac{\partial v}{\partial x}, N = u\frac{\partial v}{\partial y}$, and $P = u\frac{\partial v}{\partial z}$, then (1.13) yields

$$\iiint_D \left(\frac{\partial u}{\partial x}\frac{\partial v}{\partial x} + \frac{\partial u}{\partial y}\frac{\partial v}{\partial y} + \frac{\partial u}{\partial z}\frac{\partial v}{\partial z} \right) d\mathbf{x} = \iint_{\partial D} u\frac{\partial v}{\partial n} \, dS - \iiint_D u\nabla^2 v \, d\mathbf{x}, \tag{1.14}$$

which is known as *Green's first identity*. Moreover, if we interchange u and v in (1.13), we get

$$\iiint_D \left(\frac{\partial u}{\partial x}\frac{\partial v}{\partial x} + \frac{\partial u}{\partial y}\frac{\partial v}{\partial y} + \frac{\partial u}{\partial z}\frac{\partial v}{\partial z} \right) d\mathbf{x} = \iint_{\partial D} v\frac{\partial u}{\partial n} \, dS - \iiint_D v\nabla^2 u \, d\mathbf{x}. \tag{1.15}$$

If we subtract (1.14) from (1.15), we obtain *Green's second identity*:

$$\iiint_D \left(u\nabla^2 v - v\nabla^2 u \right) d\mathbf{x} = \iint_{\partial D} \left(u\frac{\partial v}{\partial n} - v\frac{\partial u}{\partial n} \right) dS, \tag{1.16}$$

which is also known as *Green's reciprocity theorem*. This result also holds in $\mathbb{R}^2$. Note that Green's identities are valid even if the domain D is bounded by finitely many closed surfaces; however, in that case the surface integrals must be evaluated over all surfaces that make the boundary of D, and in $\mathbb{R}^2$ the line integrals must be evaluated over all paths that make the boundary of D. If f and g are real and harmonic in $D \subset \mathbb{R}^2$ and $\Gamma = \partial D$, then from (1.16)

$$\int_\Gamma \left(f \frac{\partial g}{\partial n} - g \frac{\partial f}{\partial n} \right) ds = 0. \tag{1.17}$$

Let D be a simply connected region in the complex plane with boundary Γ. Let z_0 be any point inside D, and let D be the region obtained by indenting from D a disk of radius ε and center at z_0, where $\varepsilon > 0$ is small (Fig. A.1 (a)). Then ∂D consists of the contour Γ together with the contour Γ_ε.

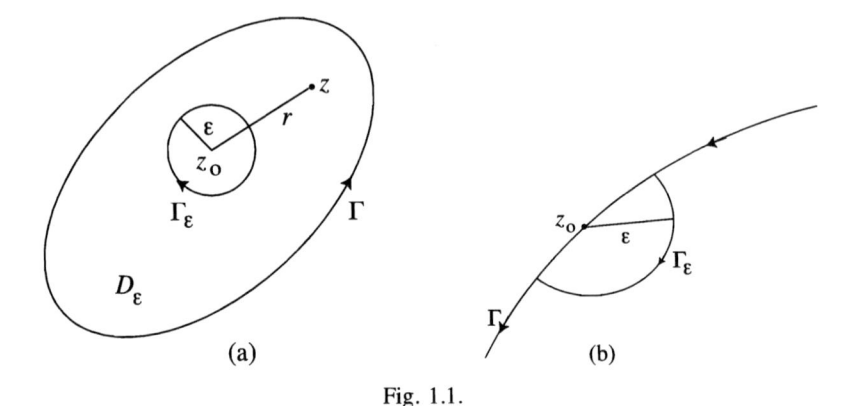

(a)   (b)

Fig. 1.1.

If we set $f = u$ and $g = \log r$ in (1.17), where $z \in D$ and $r = |z - z_0|$, then, since $\frac{\partial}{\partial n} = -\frac{\partial}{\partial r}$ on Γ_ε, we get

$$\int_\Gamma \left[u \frac{\partial (\log r)}{\partial n} - \log r \frac{\partial u}{\partial n} \right] ds - \int_{\Gamma_\varepsilon} \left(\frac{u}{r} - \log r \frac{\partial u}{\partial r} \right) ds = 0. \tag{1.18}$$

Since

$$\lim_{\varepsilon \to 0} \int_{\Gamma_\varepsilon} \frac{u}{r} \, ds = \lim_{\varepsilon \to 0} \int_0^{2\pi} u(z_0 + \varepsilon \, e^{i\theta}) \frac{1}{\varepsilon} \varepsilon \, d\theta = 2\pi \, u\,(z_0),$$

$$\lim_{\varepsilon \to 0} \int_{\Gamma_\varepsilon} \log r \frac{\partial u}{\partial r} \, ds = \lim_{\varepsilon \to 0} \int_0^{2\pi} \log \varepsilon \frac{\partial u}{\partial \varepsilon} \varepsilon \, d\theta = 0,$$

we let $\varepsilon \to 0$ in (1.18) and obtain

$$2\pi \, u(z_0) = \int_\Gamma \left[u \frac{\partial (\log r)}{\partial n} - \log r \frac{\partial u}{\partial n} \right] ds, \tag{1.19}$$

which is known as *Green's third identity*. Note that Eq (1.19) gives the value of a harmonic function u at an interior point in terms of the boundary values of u and $\dfrac{\partial u}{\partial n}$.
If the contour Γ is continuously differentiable (has no corners) and if the point z_0 is on Γ, then, instead of indenting the entire disk of radius ε, we indent from D a half disk with center at the point z_0 (Fig. 1.1(b)), and Green's third identity becomes

$$\pi\, u(z_0) = \text{p.v.} \int_\Gamma \left[u\, \frac{\partial (\log r)}{\partial n} - \log r\, \frac{\partial u}{\partial n} \right] ds \qquad (1.20)$$

where p.v. denotes the principal value of the integral, i.e., it is the limit as $r \to 0$ of the integral over the contour Γ obtained by deleting that part of Γ which lies within the circle of radius ε and center z_0.

1.8. Differentiation and Integration

Some useful formulas on Leibniz's rule and integration by pats in $\mathbb{R}^n$, $n = 1, 2$, are presented.

1.8.1. Leibniz's Rules.

(a) For differentiation of integrals:

$$\frac{d}{dx} \int_{\phi_1(x)}^{\phi_2(x)} F(x, t)\, dt = \int_{\phi_1(x)}^{\phi_2(x)} \frac{\partial F}{\partial x}\, dt + F(\phi_2, x)\frac{d\phi_2}{dx} - F(\phi_1, x)\frac{d\phi_1}{dx}. \qquad (1.21)$$

(b) For the nth derivative of a product of two functions u and v:

$$D^n(uv) = u(D^n v) + \binom{n}{1}(Du)(D^{n-1}v) + \binom{n}{2}(D^2 u)(D^{n-2}v) + \cdots + (D^n u), \qquad (1.22)$$

where $\dbinom{n}{k}$ are the binomial coefficients.

1.8.2. Integration by Parts.
Let $f(x)$ and $g(x)$ be piecewise continuous functions in an interval (a, b) with their first derivatives continuous in (a, b). Then using integration by parts we get

$$\int_a^b g\, \frac{df}{dx}\, dx = -\int_a^b f\frac{dg}{dx}\, dx + fg \Big|_a^b. \qquad (1.23)$$

Further, if $f(x)$ and $g(x)$ be piecewise continuous functions in an interval (a, b) with their first and second derivatives piecewise continuous in (a, b), and $h(x)$ is a continuous function in (a, b), then using integration by parts we get

$$\int_a^b g \frac{d}{dx}\left[\frac{df}{dx}\right] dx = \int_a^b g \frac{d^2 f}{dx^2} dx = -\int_a^b \frac{df}{dx}\frac{dg}{dx} dx + g \frac{df}{dx}\Big|_a^b, \tag{1.24}$$

$$\int_a^b g \frac{d^2}{dx^2}\left[h(x)\frac{d^2 f}{dx^2}\right] dx = \int_a^b h(x)\frac{d^2 f}{dx^2}\frac{d^2 g}{dx^2} dx$$
$$+ g \left[\frac{d}{dx}\left(h(x)\frac{d^2 f}{dx^2}\right) - h(x)\frac{d^2 f}{dx^2}\frac{dg}{dx}\right]_a^b. \tag{1.25}$$

Let a finite domain $D \subset \mathbb{R}^2$ be bounded by a smooth closed curve ∂D, and let w and F be scalar functions continuous on D. Then the gradient and divergence theorems lead to two useful identities in $\mathbb{R}^2$, which in component form are:

$$\iint_D w \frac{\partial F}{\partial x} dx\,dy = -\iint_D \frac{\partial w}{\partial x} F\,dx\,dy + \oint_{\partial D} n_x\,w\,F\,ds,$$
$$\iint_D w \frac{\partial F}{\partial y} dx\,dy = -\iint_D \frac{\partial w}{\partial y} F\,dx\,dy + \oint_{\partial D} n_y\,w\,F\,ds. \tag{1.26}$$

1.9. Inequalities

Some useful inequalities are presented for ready reference.

1.9.1. Bessel's Inequality for Fourier Series. A piecewise continuous function $f(x)$ can be represented as the Fouries series $f(x) = \frac{a_0}{2} + \sum_{n=1}^{\infty} (a_n \cos nx + b_n \sin nx)$, where the series on the right converges pointwise, uniformly and in the mean to $f(x)$ on the interval $[-\pi, \pi]$. The following inequality, known as *Bessel's inequality*, holds:

$$\frac{a_0^2}{2} + \sum_{n=1}^{\infty} (a_n^2 + b_n^2) \leq \frac{1}{\pi}\int_{-\pi}^{\pi} f^2(x)\,dx. \tag{1.27}$$

1.9.2. Bessel's Inequality for Square-Integrable Functions. Let a given function $f(x)$, defined on an interval (a, b), be represented by a uniformly continuous series of the form $f(x) = \sum_{n=1}^{\infty} c_n \phi_n(x)$, where the coefficients c_n are constants and ϕ_n are square-integrable functions orthogonal with respect to a positive weight function $\rho(x)$. Then $c_n = \dfrac{\displaystyle\int_a^b f \phi_n \rho\,dx}{\displaystyle\int_I \phi_n^2 \rho\,dx}$, which yields *Bessel's inequality*:

$$\sum_{n=1}^{\infty} c_n^2 \int_a^b \phi_n^2 \rho \, dx \le \int_a^b f^2 \rho \, dx. \tag{1.28}$$

If this series converges in the mean to $f(x)$, that is, if

$$\lim_{n \to \infty} \int_a^b \left[f(x) - \sum_{k=1}^{n} c_k \, \phi_k(x) \right]^2 \rho(x) \, dx = 0, \tag{1.29}$$

then

$$\sum_{n=1}^{\infty} c_n^2 \int_a^b \phi_n^2 \rho \, dx = \int_a^b f^2 \rho \, dx, \tag{1.30}$$

which is known as *Parseval's equality*, or *Parseval's completeness relation* in the sense that if the limit (1.29) holds for every function f for which $\int_a^b f^2 \rho \, dx$ is finite, then the set of functions $\{\phi_1, \phi_2, \dots\}$ is said to be complete.

1.9.3. Schwarz's Inequality for Infinite Sequences. Let c_n and d_n be two sequences of numbers . Then

$$\left| \sum_{n=1}^{\infty} c_n \, d_n \right| \le \left\{ \sum_{n=1}^{\infty} c_n^2 \right\}^{1/2} \left\{ \sum_{n=1}^{\infty} d_n^2 \right\}^{1/2}, \tag{1.31}$$

which is known as *Schwarz's inequality* for the two infinite sequences.

1.10. Exercises

1.1. Compute $S_n(1)$ for $n = 1, 2$, and 3, and interpret your results in $\mathbb{R}$, $\mathbb{R}^2$, and $\mathbb{R}^3$.

ANS. $2, 2\pi$, and $4\pi/3$, respectively.

1.2. Prove that the limit of a convergent sequence is unique.

HINT. Suppose that $\lim_{n \to \infty} s_n = s$, and $\lim_{n \to \infty} s_n = t$, where $|s - t| > 2\varepsilon > 0$. Then show that there are contradictory properties in the ε-neighborhoods of s and t, namely, these neighborhoods have no points in common, and each contains all but a finite number of terms of the sequence. Hence $s = t$.

1.3. Prove that the harmonic series diverges.

HINT. $S_1 > 2$, $S_8 > 2.5$, $S_{16} > 3$, $S_{32} > 3.5$, $S_{64} > 4$, $\dots$.

1.4. If $\{s_n\}$ is a sequence of non-zero terms and if $\lim_{n \to +\infty} s_n = \infty$, then show that $\lim_{n \to +\infty} 1/s_n = 0$.

HINT. Let $\varepsilon > 0$. Then, $\lim_{n \to +\infty} s_n = \infty$ implies that for any $M > 1/\varepsilon$, there exists an integer $m > 0$ such that whenever $n > m$ then $|s_n| > M > 1/\varepsilon$. For this m, $|1/s_n| < 1/M < \varepsilon$ for $n > m$.

1.5. Determine whether or not each of the following sequences is bounded, whether convergent or divergent, and whether oscillating: (a) $\left\{n + \dfrac{3}{n}\right\}$; (b) $\left\{\sin \dfrac{n\pi}{4}\right\}$; (c) $\left\{\sqrt[3]{n^2}\right\}$; (d) $\left\{\dfrac{n!}{10^n}\right\}$; (e) $\left\{\dfrac{\ln n}{n}\right\}$.

1.6. For $a > 1$ show that $\lim\limits_{n \to +\infty} a^n = +\infty$.

HINT. Let $M > 0$ and take $a = 1 + b$, $b > 0$. Then use binomial expansion of $a^n = (1 + b)^n$ to show that $a^n > M$ when $n > M/b$.

1.7. Show that $\lim\limits_{n \to +\infty} \sqrt[n]{1/n^p} = 1$, $p > 0$.

HINT. Use $n^p = e^{(p \ln n)/n}$.

1.8. Show that the infinite arithmetic series $a + (a + d) + (a + 2d) + \cdots + [a + (n - 1)d] + \cdots$ diverges for $a^2 + d^2 > 0$.

HINT. $S_n = \frac{1}{2} n [2a + (n - 1)d]$, and $\lim\limits_{n \to +\infty} S_n = \infty$ unless $a = d = 0$.

1.9. Show that the infinite geometric series $a + ar + ar^2 + \cdots + ar^{n-1} + \cdots$, $a \neq 0$, converges to $\dfrac{a}{1 - r}$ if $|r| < 1$ and diverges if $|r| \geq 1$.

HINT. $S_n = \dfrac{a - ar^n}{1 - r}$.

1.10. Use integral test to check convergence of the following series:

(a) $\dfrac{1}{\sqrt{3}} + \dfrac{1}{\sqrt{5}} + \dfrac{1}{\sqrt{7}} + \dfrac{1}{\sqrt{9}} + \cdots$;

(b) $1 + \dfrac{1}{2^p} + \dfrac{1}{3^p} + \dfrac{1}{4^p} + \cdots$;

(c) $\sin \pi + \frac{1}{4} \sin \frac{1}{2}\pi + \frac{1}{9} \sin \frac{1}{3}\pi + \frac{1}{16} \sin \frac{1}{4}\pi + \cdots$.

HINT. (a) $f(n) = s_n = \dfrac{1}{\sqrt{2n + 1}}$; take $f(x) = \dfrac{1}{\sqrt{2x + 1}}$. ANS. Diverges.

(b) $f(n) = s_n = \dfrac{1}{n^p}$; take $f(x) = \dfrac{1}{x^p}$. Discuss the cases: (i) $p > 1$, (ii) $p = 1$, and (iii) $p < 1$. ANS. Case (i) the series converges; cases (ii) and (iii) it diverges.

(c) $f(n) = s_n = \dfrac{1}{n^2} \sin \dfrac{\pi}{n}$. Consider $\int_2^\infty f(x)\, dx$ and show that it is equal to $\dfrac{1}{\pi}$, and the series converges.

1.11. Use the comparison test to check convergence of the following series:

(a) $\dfrac{1}{2} + \dfrac{1}{5} + \dfrac{1}{10} + \dfrac{1}{17} + \cdots + \dfrac{1}{n^2 + 1} + \cdots$;

(b) $1 + \dfrac{2^2 + 1}{2^3 + 1} + \dfrac{3^2 + 1}{3^3 + 1} + \dfrac{4^2 + 1}{4^3 + 1} + \cdots$;

(c) $1 + \dfrac{1}{2!} + \dfrac{1}{3!} + \dfrac{1}{4!} + \cdots$.

HINT. (a) $s_n = \dfrac{1}{n+1} < \dfrac{1}{n^2}$; ANS. Converges; (b) $s_n = \dfrac{n^2+1}{n^3+1} \geq \dfrac{1}{n}$; ANS. diverges. (c) $s_n = \dfrac{1}{n!}$, and use $n! \geq 2^{n-1}$, so $\dfrac{1}{n!} \leq \dfrac{1}{2^{n-1}}$, and compare with the convergent geometric series. ANS. Converges.

1.12. Use the ratio test to check convergence of the following series:

(a) $\dfrac{1}{3} + \dfrac{2}{3^2} + \dfrac{3}{3^3} + \dfrac{4}{3^4} + \cdots$; (b) $\dfrac{1}{3} + \dfrac{2!}{3^2} + \dfrac{3!}{3^3} + \dfrac{4!}{3^4} \cdots$.

(c) $1 + \dfrac{2^2+1}{2^3+1} + \dfrac{3^2+1}{3^3+1} + \dfrac{4^2+1}{4^3+1} + \cdots$.

HINT. (a) $s_n = \dfrac{n}{3^n}$, $s_{n+1} = \dfrac{n+1}{3^{n+1}}$; ANS. Converges; (b) $s_n = \dfrac{n!}{3^n}$, $s_{n+1} = \dfrac{(n+1)!}{3^{n+1}}$. ANS. Diverges; (c) $s_n = \dfrac{n^2+1}{n^3+1}$, $s_{n+1} = \dfrac{(n+1)^2+1}{(n+1)^3+1}$; ANS. Test fails.

2

The Concept of Green's Functions

Green's function for a differential equation is its solution when the forcing term is the Dirac delta function due to a unit point source (or sink) in a given domain. This solution provides us with a method to generate solutions of linear ordinary and partial differential equations subject to different kinds of boundary conditions and internal sources. This method is important in a variety of physical problems and applications, and has a special role in the study of certain problems of mathematical physics.

2.1. Generalized Functions

The construction of Green's functions for different types of boundary value problems is an important subject in itself since these functions are used in a wide range of applications. Before developing the subject of Green's functions we must have a clear understanding of the mathematical behavior of generalized functions. To this end we will first study certain elementary aspects of generalized functions, followed by the concept of Green's function.

A point source is represented by the Dirac delta function, which belongs to a class of functions known as *generalized functions* (or *distributions*, as they are often called, although care must be taken to distinguish them from ordinary distributions). The existence of distributions in the generalized sense is based on relations which are not functions in the traditional sense. The physical examples we want to consider are related to the determination of the intensity of a concentrated force F acting at time $t = 0$ or at the origin $\mathbf{x} = \mathbf{0}$ of a rectangular cartesian coordinate system.

Example 2.1. Consider the vibrating system of a spring, and suppose that an external force $F(t)$ is applied to the interval $-t_0 \leq t \leq t_0$, such that

$$F(t) = \begin{cases} 0, & t_0 < |t|, \\ f(t), & t_0 \geq |t|, \end{cases} \tag{2.1}$$

where the measure $m(t)$ of the strength of this force at the point $t = 0$ can be represented as $m(t) = \int_{-\infty}^{\infty} F(t)\,dt = \int_{-t_0}^{t_0} f(t)\,dt$. Now, suppose that $f(t)$ is increased while t_0 is decreased in such a way that $m(t)$ remains constant. Then we have what is known as an *impulsive force*, and for $m(t) = 1$ (unit force) we have $\lim_{t_0 \to 0} \int_{-t_0}^{t_0} f(t)\,dt = 1$. There are many functions $f(t)$ which satisfy the requirement (2.1), and one such function is $f(t) = \dfrac{1}{2t_0}$. Using this function, let

$$F(t) = \begin{cases} 0 & \text{if } t_0 < |t|, \\ \dfrac{1}{2t_0} & \text{if } t_0 \geq |t|. \end{cases} \tag{2.2}$$

This function $F(t)$ is an ordinary function which belongs to the class $C_0^{\infty}(I)$, where $I = [-t_0 \leq t \leq t_0]$. However, $F(t) \to \delta(t)$ as $t_0 \to 0$; that is, in the limit as $t_0 \to 0$ the function $F(t)$ converges to the Dirac delta-function (henceforth written δ-function)[1] which is not a function in the ordinary sense; it is called a *generalized function*. This limit is taken in the weal sense, which means that for any continuous function $\phi(t)$ the weak limit of $F(t)$ is

$$\lim_{t_0 \to 0} \int_{-t_0}^{t_0} F(t)\phi(t)\,dt = \int_{-t_0}^{t_0} \delta(t)\phi(t)\,dt = \phi(0).$$

Physically, the δ-function signifies a force of strength unity, a sudden impulse, at time $t = 0$ which is called the *source point* or *singularity* of the δ-function. If the source point is translated to t', then the function (2.2) becomes

$$F(t) = \begin{cases} 0 & \text{if } t_0 < |t - t'|, \\ \dfrac{1}{2t_0} & \text{if } t_0 \geq |t - t'|, \end{cases}$$

such that the weak limit of $F(t)$ as $t_0 \to 0$ is $\delta(t - t')$. Note that if the time variable t in Eq (2.2) is replaced by the space variable x, then $\lim_{x_0 \to 0} F(x) = \delta(x - x')$, which will signify a sudden impulse of unit strength at the source point $x = x'$. The δ-function in either case is also written as $\delta(t - t')$ or $\delta(x - x')$.

Example 2.2. Consider the 2-D case of a concentrated force F of unit strength applied at the origin $\mathbf{x} = \mathbf{0}$, $\mathbf{x} = (x, y)$, positioned at the boundary of a region D of the (x, y)-plane and directed along the z-axis. To measure F let this force be

---

[1] This function was introduced by P. A. M. Dirac in 1926-27. The development of the analysis in this and next two sections has been reiterated by various authors, but it can be traced back to examples and ideas found in Dirac [1926-27; 1947].

uniformly distributed over a circular area of radius ε and center at the origin. Then the mean value of the force F over the region D is given by the function

$$
F_\varepsilon(x, y) = \begin{cases} \dfrac{1}{\pi\,\varepsilon^2} & \text{if } x^2 + y^2 < \varepsilon, \\ 0 & \text{if } x^2 + y^2 > \varepsilon. \end{cases} \tag{2.3}
$$

Note that $\iint_D F_\varepsilon\, dx\, dy = 1$. Let the pointwise limit value of the function $F_\varepsilon(x, y)$ as $\varepsilon \to 0$ be denoted by $\delta(x, y)$. Then

$$
\delta(x, y) = \begin{cases} +\infty & \text{if } x^2 + y^2 = 0, \\ 0 & \text{if } x^2 + y^2 \neq 0. \end{cases} \tag{2.4}
$$

Also, since the measure of intensity of the force F is equal to 1, we must have $\iint_D \delta(x, y)\, dx\, dy = 1$. But the function $\delta(x, y)$ defined by (2.4) does not reproduce the value of the force. However, this situation is avoided if instead of the pointwise limit of F_ε we take the weak limit such that for any continuous function $\phi(x, y)$ the weak limit of F_ε is

$$
\lim_{\varepsilon \to 0} \iint_D F_\varepsilon(x, y)\phi(x, y)\, dx\, dy \equiv \iint_D \delta(x, y)\phi(x, y)\, dx\, dy = \phi(0, 0).
$$

In fact, the weak limit of $F_\varepsilon(x, y)$ is a functional which assigns the value $\phi(0, 0)$ to a function $\phi(x, y)$. Thus, we say that the Dirac δ-function is a weak limit of the sequence $\{F_\varepsilon\}$, i.e., $\lim_{\varepsilon \to 0}\{F_\varepsilon(x, y)\} = \delta(x, y)$, and we have established $\iint_D \delta(x, y)\, dx\, dy = 1$. ∎

This example can be extended to higher dimensions. If the origin is translated (shifted) to a point (x', y'), then the δ-function is represented by $\delta(x, y; x', y')$, and $\iint_D \delta(x, y; x', y')\, dx\, dy = 1$. The origin or the point $\mathbf{x}' = (x', y')$ is called the source point or singularity of the Dirac δ-function.

There are other descriptions of this function, but they all lead to the same properties in $\mathbb{R}$, which are

$$
\delta(t - t') = \begin{cases} 0 & \text{if } t \neq t', \\ +\infty & \text{if } t \to t', \end{cases} \quad \text{and} \quad \int_{t'-\varepsilon}^{t'+\varepsilon} \delta(t - t')\, dt = 1, \tag{2.5}
$$

where $\varepsilon > 0$ is an arbitrarily small real number. This property implies that

$$
\int_{t'-\varepsilon}^{t'+\varepsilon} f(t)\, \delta(t - t')\, dt = f(t'), \tag{2.6}
$$

which is known as the *translation* or *shifting property* of the δ-function. Similar results hold in higher dimensions.

We will denote by $\mathcal{D}$ the class of those functions in $C_0^\infty(\mathbb{R})$ which have compact support over the intervals of the form $[-a, a]$ and which approach the Dirac δ-function (or δ-distribution) as $a \to 0$. The members of the class $\mathcal{D}$ are also called *test functions*. To legitimize the δ-function, we will consider some of these test functions defined in the ordinary sense on an interval $I = [-a, a]$, where $a > 0$. Note that the choice of a test function is made from the class $\mathcal{D}$ of functions which satisfy the following two conditions: (i) It must belong to the class $C_0^\infty(\mathbb{R})$, and (ii) it must approach the δ-function as $a \to 0$.[2]

Example 2.3. Let us modify the definition of the function (2.2) by writing

$$f_\varepsilon(t) = \begin{cases} 0 & \text{for } |t| > \varepsilon, \\ \dfrac{1}{2\varepsilon} & \text{for } |t| < \varepsilon. \end{cases} \tag{2.7}$$

As $\varepsilon \to 0$, the rectangles become narrower and taller (see Fig. 2.1), but their areas remain constant, since $\int f_\varepsilon(t)\, dt = 1$ for any $\varepsilon > 0$. This means that the δ-function can be defined as the limit $\delta(t) = \lim\limits_{\varepsilon \to 0} f_\varepsilon(t)$. Obviously, this limit is zero everywhere except at the point $t = 0$. This has led to the commonly applied mathematical definition of the Dirac δ-function as

$$\delta(t) = \begin{cases} \infty & \text{for } t = 0, \\ 0 & \text{for } |t| > 0, \end{cases} \tag{2.8}$$

subject to the additional normalization condition that the area under this function is equal to 1. ∎

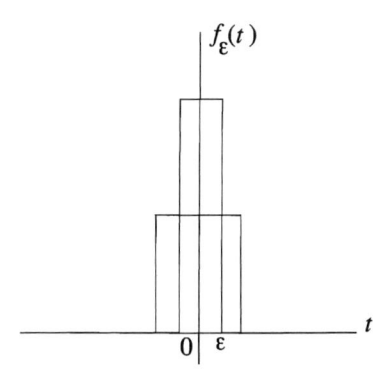

Fig. 2.1. Delta Function as a Limit of Rectangular Functions.

Properties of the δ-function are:

(i) $\delta(x) = 0$ for $x \neq 0$,

(ii) $\displaystyle\int_{-\infty}^{\infty} \delta(x)\, dx = 1$ $\qquad$ (2.9)

(iii) $\displaystyle\int_{-\infty}^{\infty} f(x)\delta(x)\, dx = f(0)$.

---

[2] The class $\mathcal{D}$ is defined in § 1.2.

These properties of the Dirac delta-function $\delta(x)$ show that it is not a function: It is zero for every x with the exception of $x = 0$ where it is infinite and its integral over the entire real axis is equal to 1. This function is not analytic, but can be obtained as a limiting case of either analytic continuous or piecewise continuous functions with finite support, i.e., belonging to the class $\mathcal{D}$ of functions in $C_0^\infty(\mathbb{R})$. The δ-function is not a function in the strict sense; it has been legitimized with the above defining properties. However, this function is very useful only as part of an integrand and never as an end result. By shifting the singularity (source point) from $x = 0$ to $x = x' \neq 0$, this function is written as $\delta(x - x')$. Then the third result in (2.9) becomes

$$\int_{-\infty}^{\infty} f(x)\delta(x - x')\, dx = f(x'). \tag{2.10}$$

which is known as the translation or shifting property of the δ-function. In the 3-D case and using spherical coordinates we have for the singularity (or source) at the origin

$$\int_0^\infty \int_0^\pi \int_0^{2\pi} \delta(r)r^2 \sin\theta\, dr\, d\theta\, d\phi = \int_{-\infty}^\infty \int_{-\infty}^\infty \int_{-\infty}^\infty \delta(x)\delta(y)\delta(z)\, dx\, dy\, dz = 1,$$
$$\tag{2.11}$$

whereas for the singularity at $r = r_1$, Eq (2.11) becomes

$$\int_0^\infty \int_0^\pi \int_0^{2\pi} \delta(r_2 - r_1)r_2^2 \sin\theta_2\, dr_2\, d\theta_2\, d\phi_2 = 1. \tag{2.12}$$

Note that $\delta(r_1 - r_2) = \delta(r_2 - r_1)$.

The definition (2.8) is consistent with the classical definition of a function and automatically satisfies the second property in (2.9). An important consequence of this property is that if $f(t)$ is any continuous function, then

$$\int_{-\infty}^\infty \delta(t)f(t)\, dt = \lim_{\varepsilon \to 0} \int_{-\varepsilon}^\varepsilon \delta(t)f(t)\, dt = f(0).$$

To prove, note that by definition (2.7)

$$\int_{-\infty}^\infty \delta(t)f(t)\, dt = \lim_{\varepsilon \to 0} \int_{-\varepsilon}^\varepsilon \frac{1}{2\varepsilon}f(t)\, dt = \lim_{\varepsilon \to 0} \frac{1}{2\varepsilon}2\varepsilon f(t') = f(0),$$

where t' is a point at which $f(t)$ takes its average value in $(-\varepsilon, \varepsilon)$, such that $t' \in (-\varepsilon, \varepsilon)$, and therefore, $t' \to 0$ as $\varepsilon \to 0$.

The dimension of the δ-function, which is important to physicists and engineers but mostly ignored by mathematicians, can be determined from the fact that this function is one of those few self-similar functions whose argument can be a space variable or a time variable, and therefore, depending on the dimension of its argument, the δ-function has a nonzero dimension. For example, the dimension of the δ-function of time $\delta(t)$ is equal to the inverse time, i.e., the dimension of frequency because, by definition, $\int_{-\infty}^{\infty} \delta(t)\, dt = 1$, which is dimensionless. In other words, the dimension of $\delta(t)$ is equal to the dimension of the inverse function $1/t$.

The Dirac δ-function is used in electronics to represent the unit impulse and sampled signals. It also represents a unit mass ideally concentrated at the origin; if concentrated elsewhere, e.g., at the integer n, it is denoted by δ_n, and the δ-function translated at n is defined by $\delta_n(t) = \delta(t - n)$. Although the above definition of the δ-function is heuristically relevant, it is not mathematically correct. In the Riemann sense the integral of the δ-function defined by (2.10) is not well-defined, or in the Lebesgue sense it equals zero. However, for each $\varepsilon > 0$ the integral

$$T_\varepsilon[\phi] = \int_{-\infty}^{\infty} f_\varepsilon(t)\phi(t)\, dt \tag{2.13}$$

exists for any fixed continuous function ϕ, and it converges to the value $\phi(0)$ as $\varepsilon \to 0_+$; that is, $\lim_{\varepsilon \to 0_+} T_\varepsilon[\phi] = T[\phi] = \phi(0)$. This definition can provide us with a valid definition of δ-function as a limit of the integrals of type (2.13). In fact, the integral (2.13) defines a linear functional on the test function $\phi(t)$, generated by the function $f_\varepsilon(t)$ which is known as the kernel of the functional T_ε. The notion of a functional is more general than that of a function because a functional depends on a variable which is a function itself but its values are real numbers. The linear functional $T_\varepsilon[\phi]$ on the test function ϕ is generated by the integral (2.13) with the kernel $f_\varepsilon(t)$, where the test function ϕ belongs to the class $\mathcal{D}$ of smooth functions in $C_0^\infty(\mathbb{R})$ with compact support. The functional that assigns to each test function its value at 0 will correspond to the Dirac delta distribution. The choice of test functions can be any function in the class $\mathcal{D}$.

To show that the value of a continuous function f at any point t is determined by the value of the functional (2.13) on all test functions in the class $\mathcal{D}$, we will discuss by an example.

Example 2.4. Consider the function

$$\omega(t) = \begin{cases} C \exp\left\{ -\dfrac{1}{1 - t^2} \right\}, & \text{for } |t| < 1, \\ 0, & \text{for } |t| \geq 1, \end{cases} \tag{2.14}$$

where the constant C is determined from the normalization condition $\int_{-\infty}^{\infty} \omega(t)\, dt = 1$, which gives $C \approx 2.25$. The function $\omega(t) \in C^\infty$; it vanishes outside the interval $[-1, 1]$ which is bounded, and it is differentiable everywhere including the endpoints -1 and 1 such that at -1 (or at 1) all the right (or left) derivatives are zero and the left (or right) derivatives are identically zero. The function $\omega(t)$, after rescaling for each $\varepsilon > 0$, gives a new function

$$\omega_\varepsilon(t) = \frac{1}{\varepsilon}\, \omega\left(\frac{t}{\varepsilon}\right), \tag{2.15}$$

which has compact support since it vanishes outside the interval $[-\varepsilon, \varepsilon]$, and it is infinitely differentiable which can be checked by applying the chain rule. Also, $\int_{-\infty}^{\infty} w_\varepsilon(t)\, dt = 1$. Applying the generalized mean value theorem for integrals and assuming that the function $f(t)$ is continuous, we find the value of the functional as

$$T_\varepsilon[f] = \int_{-\infty}^{\infty} f(t)\, w_\varepsilon(t)\, dt \to f(0) \quad \text{as } \varepsilon \to 0.$$

Thus, the value of f can be recovered by evaluating the functionals T_ε at f. ∎

We have shown that any linear functional $T[\phi]$ which is continuous on the set $\mathcal{D}$ of functions in C^∞ is a distribution. A functional T is said to be linear on $\mathcal{D}$ if $T[\alpha\phi + \beta\psi] = \alpha T[\phi] + \beta T[\psi]$ for any test functions ϕ and ψ in $\mathcal{D}$, where α and β are any real (complex) numbers. A functional T on $\mathcal{D}$ is said to be continuous if for any sequence of functions $\phi_k(t) \in \mathcal{D}$ which converge to a test function $\phi(t)$, the numbers $T[\phi_k]$ which are the values of the functional T on ϕ_k converge to the number $T[\phi]$. The convergence in this definition is understood in the following sense: The supports of all ϕ_k are all contained in a fixed bounded set in their domain, and as $k \to \infty$, the functions ϕ_k, including all their derivatives ϕ_k^m, $m = 1, 2, \ldots$, converge uniformly to the corresponding limit test function ϕ and its derivatives, i.e., $\phi_k^{(m)} \to \phi^{(m)}$ for each $m = 0, 1, 2, \ldots$.

It is interesting to note that the support of the function $\delta(t)$ is the single point $t = 0$. For this reason the δ-function has the physical significance of the source density for a unit source concentrated at a single point known as the source point.

Besides the functions defined in (2.2), (2.7) and (2.15), there are several other test functions that belong to the class $\mathcal{D}$. They are given below, and their graphs are presented in Fig. 2.2:

(a) $$\phi_\varepsilon(t) = \begin{cases} e^{\varepsilon^2/(|t|^2 - \varepsilon^2)}, & |t| < \varepsilon, \\ 0, & |t| > \varepsilon, \end{cases}$$

such that $\phi_\varepsilon(0) = e^{-1}$ and $|\phi_\varepsilon(t)| \le e^{-1}$ (known as the 'cap' function);

(b) $$\phi_j(t) = \frac{1}{\varepsilon^j} w_\varepsilon(t), \quad j = 1, 2, \cdots, \ |t| < \varepsilon,$$

where $w_\varepsilon(t)$ is defined in (2.15);

(c) $$\phi_j(t) = \frac{j}{\pi} \frac{1}{1 + j^2 t^2}, \quad j = 1, 2, \cdots, \ |t| < \varepsilon;$$

(d) $$\phi_j(t) = \frac{j}{\sqrt{\pi}} e^{-j^2 t^2}, \quad j = 1, 2, \cdots, \ |t| < \varepsilon;$$

(e) $$\phi_j(t) = \frac{1}{j\pi} \frac{\sin^2 jt}{t^2}, \quad j = 1, 2, \cdots, \ |t| < \varepsilon.$$

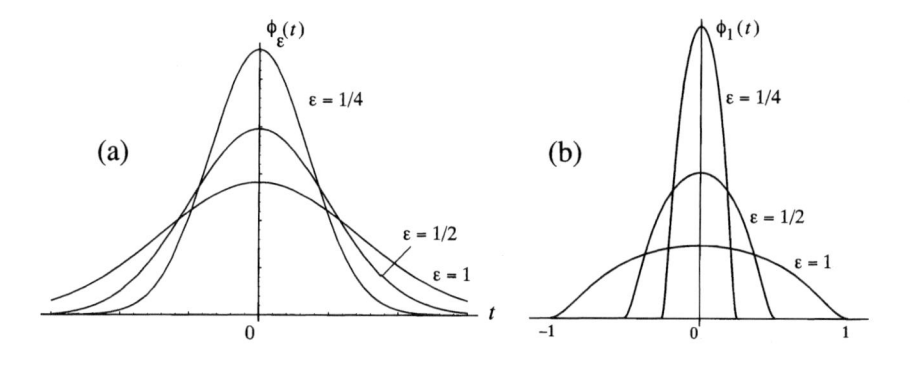

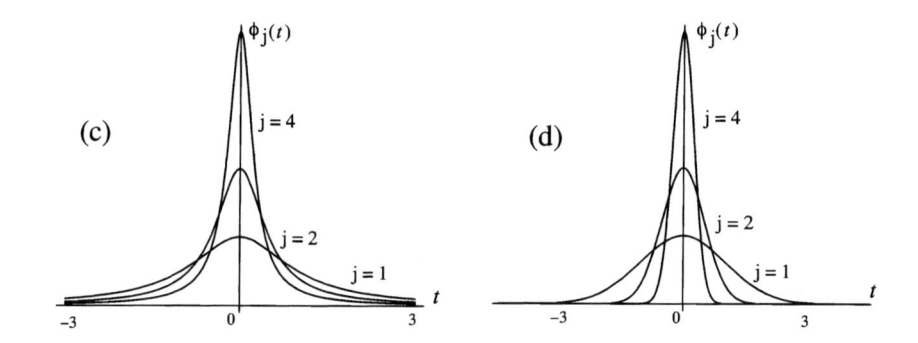

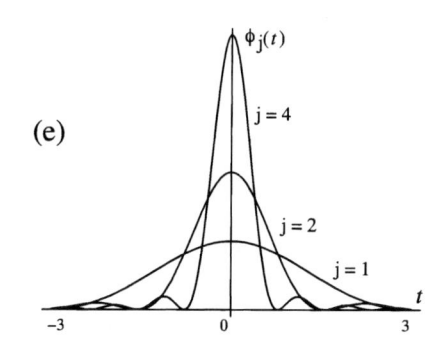

Fig. 2.2. Graphs of Test Functions (a)–(e).

Moreover, the functions

$$\frac{1}{2\sqrt{\pi\varepsilon}}\,e^{-x^2/4\varepsilon}, \qquad \frac{1}{\pi}\sin\frac{x}{\varepsilon}, \qquad \frac{1}{\pi}\frac{\varepsilon}{x^2+\varepsilon^2}, \qquad \frac{\varepsilon}{\pi x^2}\sin^2\frac{x}{\varepsilon},$$

with their domains of definition restricted to a compact set in $\mathbb{R}$, say $|x| < \varepsilon$, are other examples of test functions belonging to the class $\mathcal{D}$. All test functions in the class $\mathcal{D}$ approach $\delta(t)$ in the limit as $\varepsilon \to 0$.

The 2-D examples the test functions are:

(i) The 'cap' function

$$w_\varepsilon(x,y) = \begin{cases} C_\varepsilon\, e^{-\varepsilon^2/(\varepsilon^2-|\mathbf{x}|^2)}, & \text{if } |\mathbf{x}|^2 = x^2 + y^2 \le \varepsilon, \\ 0, & \text{if } x^2 + y^2 > \varepsilon, \end{cases}$$

where the constant C_ε is determined from the condition that $\iint_D w_\varepsilon(x,y)\, dx\, dy = 1$.

(ii) The function $w_\varepsilon(x,y) = \dfrac{1}{\varepsilon\sqrt{2\pi}}\, e^{-|\mathbf{x}|^2/(2\varepsilon^2)}$, $|\mathbf{x}|^2 = x^2 + y^2$, where $\iint_D w_\varepsilon(x,y)\, dx\, dy = 1$.

To summarize, some important properties of the δ-function are as follows:

1. $\displaystyle\int_{-\infty}^{\infty} f(t)\,\delta(t-t')\, dt = f(t')$ (Fundamental property);

2. $\displaystyle\int_{-\infty}^{\infty} f(t')\,\delta(t-t')\, dt = f(t') \int_{-\infty}^{\infty} \delta(t-t')\, dt = f(t')$;

3. $f(t)\,\delta(t-t') = f(t')\,\delta(t-t')$;

4. $t\,\delta(t) = 0$;

5. $\delta(t-t') = \delta(t'-t)$;

6. $\displaystyle\int_{-\infty}^{\infty} \delta(s)\, ds = \begin{cases} 1 & \text{for } t > 0, \\ 0 & \text{for } t < 0 \end{cases} = H(t)$;

7. $\dfrac{d}{dt} H(t) = \delta(t)$,

where $H(t)$ is the Heaviside unit step function (Fig. 2.3). Note that the last property is not true in the classical sense since $H(t)$ is not differentiable at $t = 0$. In the above results the point $t = t'$ is the source point (or singularity) for the δ-function.

The δ-function can be similarly defined for the space variable $\mathbf{x}$, and all the results given above hold for $\delta(\mathbf{x})$, where $\mathbf{x} \in \mathbb{R}^n$, or for $\delta(\mathbf{x} - \mathbf{x}')$ where $\mathbf{x}'$ is the source point for the δ-function. As in the 1-D case, the δ-function in $\mathbb{R}^2$ is formally written as $\delta(\mathbf{x}) = \delta(x)\,\delta(y)$, and $\delta(\mathbf{x};t) = \delta(x)\,\delta(y)\,\delta(t)$; similarly, in $\mathbb{R}^3$ it is formally written as $\delta(\mathbf{x}) = \delta(x)\,\delta(y)\,\delta(z)$, and $\delta(\mathbf{x};t) = \delta(x)\,\delta(y)\,\delta(z)\,\delta(t)$.

2.1.1. Heaviside Function. A very useful application of the property (iii) in (2.9) provides us with an alternative definition of the function $\delta(t)$: If $\langle f,g\rangle = \displaystyle\int f(t)g(t)\, dt = f(0)$ for continuous or piecewise continuous functions on $\mathbb{R}$, then the function $g(t)$ must be the δ-function. We will use this result in the following example to discuss the Heaviside function $H(t)$.

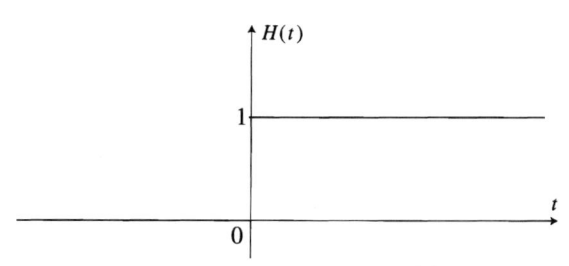

Fig. 2.3. Heaviside Function $H(t)$.

Example 2.5. An alternative definition of the δ-function is based on the relationship between the Heaviside function $H(t)$, defined in the above Property 6, and the function $\delta(t)$. This can be established as follows: The Heaviside function $H(t)$ defines the distribution $\langle f, H \rangle = \int_0^\infty f(t)\, dt$, and so, in the generalized sense, its derivative is defined by $\langle f, H' \rangle = \langle H, -f' \rangle = -\int_0^\infty f'(t)\, dt = f(0)$; thus,

$$H'(t) = \delta(t). \tag{2.16}$$

Alternately, since $H' = \lim\limits_{\varepsilon \to 0} \dfrac{H(t+\varepsilon) - H(t-\varepsilon)}{\varepsilon}$, by property (iii) in (2.9) we have

$$\langle f, H' \rangle = \lim_{\varepsilon \to 0} \frac{1}{\varepsilon} \left[\int_{-\varepsilon}^\infty f(t)\, dt - \int_0^\infty f(t)\, dt \right]$$

$$= \lim_{\varepsilon \to 0} \frac{1}{\varepsilon} \int_{-\varepsilon}^0 f(t)\, dt = f(0), \quad \text{since } f \text{ is continuous at } t = 0,$$

which implies (2.16). Since $H(t)$ is not differentiable at $t = 0$ in the classical sense, the distributional definition (2.16) for $H'(t)$ when formally integrated gives (2.16). The function $H(t - t')$ is similarly defined by

$$H(t - t') = \begin{cases} 1, & t > t', \\ 0, & t < t'. \end{cases} \tag{2.17}$$

The Heaviside function is generally used to restrict a time-dependent solution for initial value problem to the interval $t > t'$. ∎

2.1.2. Delta Function in Curvilinear Coordinates. The δ-function in the polar coordinates (r, θ) in $\mathbb{R}^2$ is given by

$$\delta(x)\,\delta(y) = \frac{\delta(r)}{2\pi r}, \quad r^2 = x^2 + y^2, \tag{2.18}$$

and in the spherical coordinates (r, θ, ϕ) in $\mathbb{R}^3$ by

$$\delta(x)\,\delta(y)\,\delta(z) = \frac{\delta(r)}{4\pi r^2}, \quad r^2 = x^2 + y^2 + z^2. \tag{2.19}$$

For definition of these coordinate systems, see §4.1. In general, for the two-dimensional case, if the cartesian coordinates $\mathbf{x} = (x, y)$ are transformed into general coordinates $\mathbf{u} = (u, v)$, then

$$\delta\left(\mathbf{x} - \mathbf{x}'\right) = \delta\left(x - x'\right)\delta\left(y - y'\right) = \delta\left(\mathbf{u} - \mathbf{u}'\right)\left[J\left(\frac{(x', y')}{(u', v')}\right)\right]^{-1}, \qquad (2.20)$$

provided the Jacobian $J\left(\dfrac{(x', y')}{(u', v')}\right) \neq 0$, where $\mathbf{x}' = (x', y')$ and $\mathbf{u}' = (u', v')$. Similarly, for the three-dimensional case, where $\mathbf{x} = (x, y, z)$, $\mathbf{x}' = (x', y', z')$, $\mathbf{u} = (u, v, w)$, and $\mathbf{u}' = (u', v', w')$, the transformation is given by

$$\begin{aligned}
\delta\left(\mathbf{x} - \mathbf{x}'\right) &= \delta\left(x - x'\right)\delta\left(y - y'\right)\delta\left(z - z'\right) \\
&= \delta\left(\mathbf{u} - \mathbf{u}'\right)\left[J\left(\frac{(x', y', z')}{(u', v', w')}\right)\right]^{-1},
\end{aligned} \qquad (2.21)$$

provided $J\left(\dfrac{(x', y', z')}{(u', v', w')}\right) \neq 0$. We will ignore the proofs and also the singular case when the Jacobian J is zero; they can be found in Stakgold [1979]. Some useful results in the polar and the spherical coordinates are:

$$\delta\left(x - x'\right)\delta\left(y - y'\right) = \frac{\delta\left(r - r'\right)\delta\left(\theta - \theta'\right)}{r} \quad \text{in } \mathbb{R}^2,$$

$$\delta\left(x - x'\right)\delta\left(y - y'\right)\delta\left(z - z'\right) = \frac{\delta\left(r - r'\right)\delta\left(\theta - \theta'\right)\delta\left(\phi - \phi'\right)}{r^2 \sin\phi} \quad \text{in } \mathbb{R}^3.$$

The δ-function in all dimensions has the following properties:

$$\int_D f\left(\mathbf{x}\right)\delta\left(\mathbf{x} - \mathbf{x}'\right) d\mathbf{x} = \begin{cases} f\left(\mathbf{x}'\right) & \text{if } \mathbf{x} \in D, \\ 0 & \text{if } \mathbf{x} \notin D, \end{cases} \qquad (2.22)$$

and

$$\int_D \delta\left(k\left(\mathbf{x} - \mathbf{x}'\right)\right) d\mathbf{x} = \int_D \frac{1}{k}\delta\left(\mathbf{x} - \mathbf{x}'\right) d\mathbf{x}, \qquad (2.23)$$

where the integral is single, double, or triple, depending on the domain D.

Note that $\delta\left(x - x'\right)$ is also written as $\delta\left(x, x'\right)$, and has the units $[L^{-1}]$. Also, we often write $\delta\left(\mathbf{x}, \mathbf{x}'\right)$ for $\delta\left(\mathbf{x} - \mathbf{x}'\right)$, $\delta\left(\mathbf{x}, t\right)$ for $\delta\left(\mathbf{x}\right)\delta(t)$, and $\delta\left(\mathbf{x}\right)$ for $\delta\left(\mathbf{x}, 0\right)$.

$\boxed{!\,!}$ For the sake of simplicity we will henceforth use s in stead of t' or x' when we discuss Green's functions in $\mathbb{R}$ in this and the following chapter.

2.2. Singular Distributions

We will define singular distributions as limits of regular functions. As we have seen, the Dirac delta distribution can be represented as a limit of a sequence of integral functionals $T_k[\phi] = \int_{-\infty}^{\infty} f_k(x)\,\phi(x)\,dx$ with respect to the kernels that are regular functions. This leads to the representation of $\delta(x)$ in the limit such that $T_k[\phi] \to \delta(\phi)$ as $k \to \infty$ for each $\phi \in \mathcal{D}$. The convergence in this situation is always the weak convergence (see §1.3.2 and §2.1). We will consider the cases of three regular functions, namely the Gaussian functions, Cauchy densities, and complex-valued oscillating functions, in the following examples.

Example 2.6. Consider the family of Gaussian functions

$$g_\varepsilon(x) = \frac{1}{\sqrt{2\pi\varepsilon}}\, e^{-x^2/(2\varepsilon)}, \quad \varepsilon > 0, \tag{2.24}$$

which are normalized by the condition that $\int_{-\infty}^{\infty} g_\varepsilon(x)\,dx = 1$. Let $f_k(x) = g_{1/k}(x)$, $k = 1, 2, \ldots$ be a weakly approximating sequence. Then $\varepsilon \to 0$ as $k \to \infty$, and the graphs of the approximating Gaussian functions $g_k(x)$ show higher and higher peaks and move more and more close toward $x = 0$, but the area under each one of them remains constant (see Fig. 2.4.). ∎

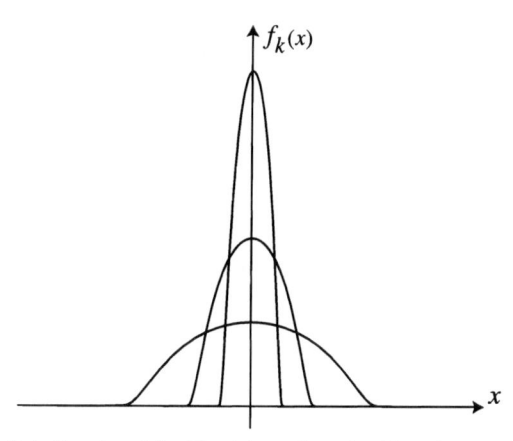

Fig. 2.4. Graphs of the First Three Gaussian Functions $g_k(x)$.

Example 2.7. Consider the Cauchy densities (or the Lorentz curves, as the physicists call them)

$$\lambda_\varepsilon(x) = \frac{1}{\pi}\,\frac{\varepsilon}{x^2 + \varepsilon^2}, \tag{2.25}$$

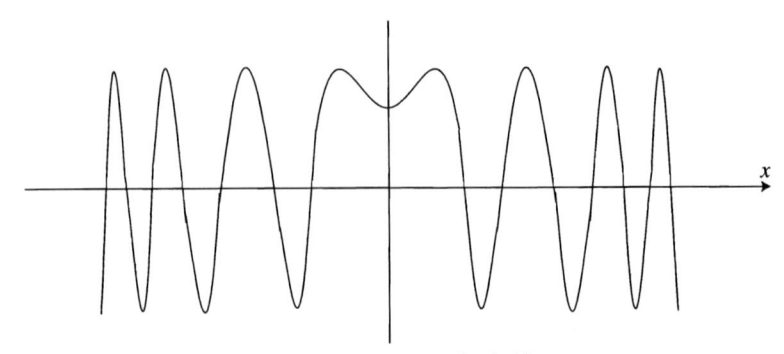

Fig. 2.5. Graphs of $\Re\{f_\varepsilon(x)\}$.

and define the kernels $f_k(x) = \lambda_{1/k}(x)$. These kernels, which look almost like the Gaussian functions (Fig. 2.4) but differ from them significantly, weakly converge to the δ-function. In fact, both the Gaussian functions and the Cauchy densities are infinitely differentiable and integrable functions for all values of $x \in \mathbb{R}$, since the integral $\displaystyle\int \frac{dx}{x^2 + 1} = \arctan x$ has finite limit at $x = -\infty$ and $x = \infty$, but at $x = 0$

$$g_\varepsilon(0) = \frac{1}{\sqrt{2\pi\varepsilon}}, \quad \text{and} \quad \lambda_\varepsilon(0) = \frac{1}{\pi\varepsilon},$$

and thus both go to $+\infty$ as $\varepsilon \to 0$. But these two functions differ from each other as follows: The Gaussian function decays exponentially to 0 as $x \to \pm\infty$, whereas the Cauchy densities, being asymptotic to $\dfrac{\varepsilon}{\pi x^2}$, decay more slowly to 0 as $x \to \pm\infty$ than the Gaussian, and the area under them are less concentrated around $x = 0$ than in the case of the Gaussian functions. In particular, for $f(x) = x^2$

$$T_f[g_\varepsilon] = \int_{-\infty}^{\infty} \frac{x^2}{\sqrt{2\pi\varepsilon}}\, e^{-x^2/(2\varepsilon)}\, dx = \varepsilon, \quad T_f[\lambda_\varepsilon] = \frac{1}{\pi} \int_{-\infty}^{\infty} x^2\, \frac{\varepsilon}{x^2 + \varepsilon^2}\, dx = \infty,$$

which shows that the functional with the Gaussian function are well defined while those with the Cauchy densities are not. ∎

Example 2.8. Consider the complex-valued functions

$$f_\varepsilon(x) = \sqrt{\frac{i}{2\pi\varepsilon}}\, e^{-ix^2/(2\varepsilon)}, \quad \varepsilon > 0. \tag{2.26}$$

These functions are found in quantum mechanics and quasi-optics.[3] The graph of $\Re\{f_\varepsilon(x)\}$ is given in Fig. 2.5. Note that $|f_\varepsilon(x)| = \dfrac{1}{\sqrt{2\pi\varepsilon}} = \text{const}$, and diverges to

---

[3] In quasi-optics $\Re\{f_\varepsilon(x)\}$ turn out to be Green's functions of a monochromatic wave in the Fresnel approximation (see §7.12.1(b)).

∞ as $\varepsilon \to 0$. However, these functions converge weakly to the δ-function as $\varepsilon \to 0$. It follows from the fact that these functions oscillate at higher and higher speed as ε becomes smaller and smaller. ∎

2.3. The Concept of Green's Functions

Green's function $G(\mathbf{x}, \mathbf{x}')$ for a linear ordinary or partial differential operator L satisfies the following system: $L[G(\mathbf{x}, \mathbf{x}')] = \delta(\mathbf{x} - \mathbf{x}')$ in a domain D, and subject to the boundary conditions $BG(\mathbf{x}, \mathbf{x}') = 0$ on ∂D, where B is a linear ordinary or partial differential operator whose order is less than that of L. If the domain D is the entire space, the solution to this system is known as the *fundamental* or *(singularity) solution* for the operator L (see §2.5).

To understand the concept of Green's functions[4] we will take some examples and firstly discuss Green's functions for ordinary differential equations.

Example 2.9. Consider the boundary value problem of the forced, transverse vibrations of a taut string of length l, defined by

$$L[u] = \frac{d^2 u}{dt^2} + k^2 u = -f(t), \quad 0 < t < l, \tag{2.27}$$

where L denotes the linear differential operator, and subject to the boundary conditions $u(0) = 0 = u(l)$, which imply that the endpoints of the string are kept fixed. The general solution of Eq (2.27) is

$$u(t) = A(t) \cos kt + B(t) \sin kt,$$

where the functions $A(t)$ and $B(t)$ are to be determined by the method of variation of parameters. Thus, we solve the equations

$$A'(t) \cos kt + B'(t) \sin kt = 0,$$
$$-kA'(t) \sin kt + kB'(t) \cos kt = -f(t)$$

to find $A'(t)$ and $B'(t)$, which are given by

$$A'(t) = \frac{f(t) \sin kt}{k}, \quad \text{and} \quad B'(t) = -\frac{f(t) \cos kt}{k}.$$

Integrating these functions with respect to t to obtain $A(t)$ and $B(t)$, we find the formal solution of Eq (2.27) as

$$u(t) = \frac{\cos kt}{k} \int_a^t f(s) \sin ks \, ds - \frac{\sin kt}{k} \int_b^t f(s) \cos ks \, ds, \tag{2.28}$$

---

[4] George Green (1793–1841) was an English miller and self-taught mathematician who also made contributions to the theory of electricity and magnetism, including the term 'potential'.

where the constants a and b are determined from the two boundary conditions. Applying the boundary condition $u(0) = 0$ to Eq (2.28) we find that a must be chosen such that $\int_a^0 f(s) \sin ks\, ds = 0$. Since $f(s)$ is arbitrary, we must choose $a = 0$. Similarly, on applying the boundary condition $u(l) = 0$ to Eq (2.28), we must have

$$u(l) = \frac{\cos kl}{k} \int_0^l f(s) \sin ks\, ds - \frac{\sin kl}{k} \int_b^l f(s) \cos ks\, ds = 0,$$

or

$$\frac{\sin kl}{k} \int_b^0 f(s) \cos ks\, ds = \frac{1}{k} \int_0^l f(s) \sin k(s - l)\, ds. \tag{2.29}$$

Combining (2.28) and (2.29), the solution of this boundary value problem is given by

$$\begin{aligned}
u(t) &= \frac{1}{k} \int_0^t f(s) \sin k(s - t)\, ds - \frac{\sin kt}{k \sin kl} \int_0^l f(s) \sin k(s - l)\, ds \\
&= \int_0^t f(s) \frac{\sin ks \sin k(l - t)}{k \sin kl}\, ds + \int_t^l f(s) \frac{\sin kt \sin k(l - s)}{k \sin kl}\, ds \tag{2.30} \\
&= \int_0^l f(s)\, G(t, s)\, ds,
\end{aligned}$$

where

$$G(t, s) = \begin{cases}
\dfrac{\sin ks \sin k(l - t)}{k \sin kl}, & \text{for } 0 \le s \le t, \\[2mm]
\dfrac{\sin kt \sin k(l - s)}{k \sin kl}, & \text{for } t \le s \le l.
\end{cases} \tag{2.31}$$

This function $G(t, s)$ is a two-point function of position known as Green's function for the boundary value problem considered here.[5] This function exists provided $\sin kl \ne 0$. Notice that Green's function defined by (2.31) is independent of the forcing term $f(t)$, but it depends only on the associated homogeneous differential equation and the boundary conditions. This implies that in all such problems with different function $f(t)$ the knowledge of Green's function allows us to construct the solution $u(t)$ of the boundary value problem (2.27). ∎

Example 2.10. Consider the following boundary value problem:

$$L[u] = u'' - k^2 u = f(x), \quad a < x < b; \quad u(a) = 0 = u(b). \tag{2.32}$$

The solution for this problem, using the method of variation of parameters, is

$$\begin{aligned}
u(x) &= \frac{1}{2k} \int_a^x \frac{\cosh k(b - x - s + a) - \cosh k(b - x + s - a)}{\sinh k(b - a)} f(s)\, ds \\
&\quad + \frac{1}{2k} \int_x^b \frac{\cosh k(b - x - s + a) - \cosh k(b + x - s - a)}{\sinh k(b - a)} f(s)\, ds.
\end{aligned}$$

---

[5] This function was called the 'influence function' by George Green in 1828 when he solved the string problem.

This can be simplified as

$$u(x) = -\frac{1}{k}\left[\int_a^x \frac{\sinh k(b-x)\sinh k(s-a)}{\sinh k(b-a)}f(s)\,ds \right.$$
$$\left. + \int_x^b \frac{\sinh k(x-a)\sinh k(b-s)}{\sinh k(b-a)}f(s)\,ds\right].$$

Now, if we define

$$G(x,s) = \begin{cases} -\dfrac{\sinh k(x-a)\sinh k(b-s)}{k\sinh k(b-a)} & \text{if } a \leq x < s, \\[4mm] -\dfrac{\sinh k(b-x)\sinh k(s-a)}{k\sinh k(b-a)} & \text{if } s < x \leq b, \end{cases} \qquad (2.33)$$

we can express the solution (2.32) as

$$u(x) = \int_a^b G(x,s)f(s)\,ds. \qquad (2.34)$$

The function $G(x,s)$ is known as Green's function for the problem (2.32). ∎

In these examples we notice that if the function $f(x)$ is taken with a minus sign, as in (2.27), Green's function defined by (2.31) does not have a minus sign; this situation has, however, reversed in Example 2.10. Mathematically it does not matter whether $f(x)$ is taken with a minus or a plus sign. Now, from these two examples we can derive some conclusions which should help us solve the problem of constructing Green's functions for more general initial and boundary value problems. We notice the following properties of Green's function $G(x,s)$:

1. This function satisfies the given differential equation, i.e., $G'' \pm k^2 G = 0$ in each intervals $0 \leq s < t,\ t < s \leq l$ (in Example 2.9), or $a \leq x < s,\ s < x \leq b$ (in Example 2.10);

2. In Examples 2.9 and 2.10, the function G is continuous at the point $s = x$ since

$$\lim_{s\to x^-} G(x,s) = \frac{\sin kx \sin k(l-x)}{k\sin kl} = \lim_{s\to x^+} G(x,s),$$

$$\lim_{s\to x^-} G(x,s) = \frac{\sinh k(x-a)\sinh k(b-x)}{k\sinh k(b-a)} = \lim_{s\to x^+} G(x,s).$$

3. The derivative dG/ds is discontinuous at $s = x$, since in Example 2.9

$$G'(x,x^-) = \lim_{s\to x^-} G'(x,s) = \frac{\cos kx \sin k(l-x)}{\sin kl},$$

$$G'(x,x^+) = \lim_{s\to x^+} G'(x,s) = -\frac{\sin kx \cos k(l-x)}{\sin kl}.$$

Thus, $G'(x,x^+) - G'(x,x^-) = -1$. Similar results hold for Example 2.10;

4. Since $G(x,0) = G(x,l) = 0$, the function G satisfies the boundary conditions;

5. The function G is symmetric in its arguments, i.e., $G(x,s) = G(s,x)$.

2.4. Linear Operators and Inverse Operators

Definition of linear operators, their inverse and adjoint operators, and their role in determining Green's functions for linear differential equations with prescribed boundary conditions are discussed.

2.4.1. Linear Operators and Inverse Operators.
The linear differential operator L was introduced in Examples 2.9 and 2.10. We will now introduce the inverse operator L^{-1} and express the solution of boundary value problems as

$$u(x) = L^{-1}[f]\,(x) = \int_a^b G(x,s)f(s)\,ds, \qquad (2.35)$$

or symbolically as $L^{-1}[\ ](\cdot) = \int_a^b G(x,s)\,(\cdot)\,ds$. We will now show that $G(x,s)$ is the solution of the boundary value problem

$$\begin{aligned} u'' - k^2 u &= \delta(x-s), \quad a < x < b, \\ u(a) &= u(b) = 0. \end{aligned} \qquad (2.36)$$

Note that in some textbooks, Green's function for boundary value problems is defined as the solution of the equation $u'' - k^2 u = -\delta(x-s), \quad a < x < b; \quad u(a) = u(b) = 0$, which assumes an impulsive force of strength -1 at the source point (or singularity) $x = s$. In Eq (2.36) we are assuming this force to be of strength $+1$. Before we solve this differential equation, we will prove the following result:

Theorem 2.1. *The solution of the differential equation*

$$P(x)\,u'' + Q(x)\,u' + R(x)\,u = \delta(x-s), \quad a < x < b, \qquad (2.37)$$

subject to the boundary conditions $u(a) = u(b) = 0$, *satisfies the condition* $u'(x_+) - u'(x_-) = \dfrac{1}{P(x)}$, *provided* $P(x) \neq 0$.

PROOF. Define $F(x) = e^{\int Q/P\,dx}$. Then Eq (2.37) can be written as

$$F\,u'' + \frac{Q}{P}\,F\,u' + \frac{R}{P}\,F\,u = \frac{F}{P}\,\delta(x-s),$$

where we have dropped the arguments in the functions F, P, Q and R, or as

$$\frac{d}{dx}\,(F\,u') + \frac{R}{P}\,F\,u = \frac{F}{P}\,\delta(x-s),$$

Integrating both sides of this equation from $x - \varepsilon$ to $x + \varepsilon$, we get

$$\int_{x-\varepsilon}^{x+\varepsilon} \frac{d}{dx}(F\,u')\,dx + \int_{x-\varepsilon}^{x+\varepsilon} \frac{R}{P}\,F\,u\,dx = \int_{x-\varepsilon}^{x+\varepsilon} \frac{F(x)}{P(x)}\,\delta(x-s)\,dx = \frac{F(x)}{P(x)}.$$

Now, if we take the limit as $\varepsilon \to 0$, the second term in the above equation vanishes and we have

$$\lim_{\varepsilon \to 0}\left[F(x+\varepsilon)\,u'(x+\varepsilon) - F(x-\varepsilon)\,u'(x-\varepsilon)\right] = \frac{F(x)}{P(x)},$$

or

$$F(x)\left[u'(x_+) - u'(x_-)\right] = \frac{F(x)}{P(x)},$$

which yields the required result. ∎

Example 2.11. Consider the problem (2.36) whose solution is given by

$$u(x) = \begin{cases} A\sinh k(x-a) & \text{if } x < s, \\ B\sinh k(b-x) & \text{if } x > s, \end{cases} \tag{2.38}$$

where the solution (2.38) satisfies the boundary conditions and the constants A and B satisfy the condition of continuity for $u(x)$ and the jump condition for $u'(x)$ (see Theorem 2.1). Thus,

$$A\sinh k(s-a) = B\sinh k(b-s),$$
$$-Bk\cosh k(b-s) - Ak\cosh k(s-a) = 1.$$

Solving this set of equations for A and B and substituting them into Eq (2.38) and simplifying, we get

$$G(x,s) = \begin{cases} -\dfrac{\sinh k(x-a)\sinh k(b-s)}{k\sinh k(b-a)} & \text{if } a \le x < s, \\[2ex] -\dfrac{\sinh k(s-a)\sinh k(b-x)}{k\sinh k(b-a)} & \text{if } s < x \le b. \end{cases}$$

Notice that this Green's function is the same as (2.33) and it satisfies the same five conditions given at the end of the previous section. ∎

2.4.2. Adjoint Operators. If L is a linear differential operator with dependent variable u and independent variable x, then an operator L^* which satisfies the relation

$$\int_a^b v\,L[u]\,dx = \int_a^b u\,L^*[v]\,dx + \left[M(u,v,u',v',x)\right]_a^b, \tag{2.39}$$

where $M(u, v, u', v', x)$ represents the boundary terms obtained after integration by parts, is called the *adjoint operator* of L. An operator L is said to be *self-adjoint* if $L = L^*$. For more on adjoint operators, see Appendix A.

The basic result in the theory of Green's functions is as follows:

Theorem 2.2. *If $G(\mathbf{x}, \mathbf{x}')$ is Green's function for the linear operator L and $G^*(\mathbf{x}, \mathbf{x}')$ is Green's function for its adjoint operator L^*, then $G(\mathbf{x}, \mathbf{x}') = G^*(\mathbf{x}', \mathbf{x})$.*

Before we prove this theorem, we will discuss two boundary value problems involving second-order differential operators. This will not only clarify the concept of Green's functions but also show the significance of this theorem.

Example 2.12. (Problem I): Let L be a second order ordinary differential operator. Consider the boundary value problem:

$$L[u] = f(x) \quad \text{in } (a, b), \text{ such that } B_1 u\Big|_{x=a} = 0, B_2 u\Big|_{x=b} = 0,$$

where $L[u] = A(x)u'' + B(x)u' + C(x)u$, and B_1 and B_2 are linear differential operators of order ≤ 1 defining the boundary conditions. Then integration by parts yields

$$\int_a^b v\, L[u]\, dx = \Big[A(x)vu' - \big(A(x)v\big)'u + B(x)vu\Big]_a^b + \int_a^b u L^*[v]\, dx,$$

where $L^*[v] = A(x)v'' + [2A'(x) - B(x)]\, v' + [A''(x) - B'(x) + C(x)]\, v$. If we require v to satisfy the boundary conditions $B_1^* v\Big|_{x=a} = 0$ and $B_2^* v\Big|_{x=b} = 0$, where the operators B_1^* and B_2^* are chosen such that $\Big[A(x)vu' - \big(A(x)v\big)' + B(x)vu\Big]_a^b = 0$, then the pair of the boundary conditions $B_1^* v\Big|_{x=a} = 0$ and $B_2^* v\Big|_{x=b} = 0$ become the adjoint boundary conditions for the given pair of the boundary conditions $B_1 u\Big|_{x=a} = 0$ and $B_2 u\Big|_{x=b} = 0$. Next, consider the following boundary value problem:

$L[u] = f(x)$ in (a, b) with the homogeneous boundary conditions $B_1 u\Big|_{x=a} = 0$, $B_2 u\Big|_{x=b} = 0$. Let $G^*(x, s)$ be the solution of the problem:

$$L^*[v] = \delta(x - s) \quad \text{in } (a, b), \text{ such that } B_1^* v\Big|_{x=a} = 0, B_2^* v\Big|_{x=b} = 0.$$

Then if we replace v by $G^*(x, s)$ in Eq (2.39), it reduces to

$$\int_a^b G^*(x, s)\, L[u]\, dx = \int_a^b u\, L^*[G^*]\,(x, s)\, dx = \int_a^b u\, \delta(x - s)\, dx = u(s).$$

But since $\int_a^b G^* (x, s) \, L[u] \, dx = \int_a^b G^* (x, s) \, f(x) \, dx$, the above equation reduces to

$$u(s) = \int_a^b G^* (x, s) \, f(x) \, dx.$$

Moreover, since s is an arbitrary point, we interchange x and s in the above result and get

$$u(x) = \int_a^b G^* (s, x) \, f(s) \, ds. \tag{2.40}$$

In this solution Green's function for the adjoint operator L^* turns out to be such that $G^*(s, x) = G(x, s)$, where s is used for x'. ∎

Example 2.13. (Problem II): Consider a second-order partial differential operator L, and let $u(\mathbf{x})$ and $v(\mathbf{x})$ be two C^2-functions. Then, integrating by parts, we get

$$\iiint_D v(\mathbf{x}) \, L[u](\mathbf{x}) \, d\mathbf{x} = \iint_{\partial D} M(u, v) \, dS + \iiint_D u(\mathbf{x}) L^*[v](\mathbf{x}) \, d\mathbf{x}, \tag{2.41}$$

where $M(u, v)$ is a differential operator of the first order, ∂D is the boundary of the region D, L^* is the adjoint operator of L, and $d\mathbf{x} = dx \, dy \, dz$ is the volume element. Consider the following boundary value problem:

$$L[u] = f(\mathbf{x}) \quad \text{in } D, \text{ such that } Bu = 0 \text{ on } \partial D,$$

where B is a linear partial differential operator of the first order. To find the solution of this problem, we use Eq (2.41) and the solution of the following problem:

$$L^*[v] = \delta(\mathbf{x}, \mathbf{y}) \quad \text{in } D, \text{ such that } B^* v = 0 \text{ on } \partial D, \tag{2.42}$$

where B^* is a linear partial differential operator of the first order such that $M(u, v) = 0$ on ∂D. The boundary condition $B^* v = 0$ on ∂D is known as the adjoint boundary condition for $Bu = 0$ on ∂D. The solution $v(\mathbf{x}, \mathbf{x}')$ of the problem (2.42) is denoted by $G^*(\mathbf{x}, \mathbf{x}')$ and is known as Green's function for the problem (2.42). Using (2.41), we get

$$\iiint_D G^* (\mathbf{x}, \mathbf{x}') \, L[u](\mathbf{x}) \, d\mathbf{x} = \iint_{\partial D} M(u, G^*) \, dS + \iiint_D u(\mathbf{x}) \, L^* [G^*] (\mathbf{x}, \mathbf{x}') \, d\mathbf{x}, \tag{2.43}$$

which gives

$$\iiint_D G^* (\mathbf{x}, \mathbf{x}') \, f(\mathbf{x}) \, d\mathbf{x} = \iiint_D u(\mathbf{x}) \, \delta (\mathbf{x} - \mathbf{x}') \, d\mathbf{x} = u(\mathbf{x}').$$

Since $\mathbf{x}'$ is arbitrary, we interchange $\mathbf{x}$ and $\mathbf{x}'$ and obtain

$$u(\mathbf{x}) = \iiint_D G^*(\mathbf{x}', \mathbf{x}) \, f(\mathbf{x}') \, d\mathbf{x}'.$$

Then using Theorem 2.2, we obtain

$$u(\mathbf{x}) = \iiint_{D'} G(\mathbf{x}, \mathbf{x}') \, f(\mathbf{x}') \, d\mathbf{x}'.$$

It turns out that even when the boundary condition on u is nonhomogeneous, we can find the solution from Eq (2.43). In the above solution we needed Green's function $G^*(\mathbf{x}, \mathbf{x}')$ for the adjoint operator L^* defined in the problem (2.42). ■

   Proof of Theorem 2.2. In this proof we will show that $G^*(\mathbf{x}, \mathbf{x}') = G(\mathbf{x}', \mathbf{x})$, where $G^*(\mathbf{x}, \mathbf{x}')$ is solution of the equation $L^*[G^*](\mathbf{x}, \mathbf{x}') = \delta(\mathbf{x} - \mathbf{x}')$ subject to homogeneous boundary conditions. Clearly, $G(\mathbf{x}, \mathbf{z}_1)$ satisfies the differential equation $L[G](\mathbf{x}, \mathbf{z}_1) = \delta(\mathbf{x} - \mathbf{z}_1)$ in D, and the boundary condition $G(\mathbf{x}_s, \mathbf{z}_1) = 0$ on the boundary ∂D, where $\mathbf{x}_s$ is any point on ∂D, while $G^*(\mathbf{x}, \mathbf{z}_2)$ satisfies the differential equation $L^*[G^*](\mathbf{x}, \mathbf{z}_2) = \delta(\mathbf{x} - \mathbf{z}_2)$ in D and the boundary condition $G^*(\mathbf{x}_s, \mathbf{z}_2) = 0$ on ∂D. Multiplying the first differential equation by $G^*(\mathbf{x}, \mathbf{z}_2)$ and integrating over the whole region D, we get

$$\iiint_D G^*(\mathbf{x}, \mathbf{z}_2) \, L[G](\mathbf{x}, \mathbf{z}_1) \, d\mathbf{x} = \iiint_D G^*(\mathbf{x}, \mathbf{z}_2) \, \delta(\mathbf{x} - \mathbf{z}_1) \, d\mathbf{x} = G^*(\mathbf{z}_1, \mathbf{z}_2).$$

If we integrate $\displaystyle\iiint_D G^*(\mathbf{x}, \mathbf{z}_2) \, L[G](\mathbf{x}, \mathbf{z}_1) \, d\mathbf{x}$ by parts, we get

$$\iiint_D G^*(\mathbf{x}, \mathbf{z}_2) \, L[G](\mathbf{x}, \mathbf{z}_1) \, d\mathbf{x} = \iiint_D G(\mathbf{x}, \mathbf{z}_1) \, L^*[G^*](\mathbf{x}, \mathbf{z}_2) \, d\mathbf{x},$$

$$= \iiint_D G(\mathbf{x}, \mathbf{z}_1) \, \delta(\mathbf{x} - \mathbf{z}_2) \, d\mathbf{x} = G(\mathbf{z}_2, \mathbf{z}_1).$$

The boundary terms vanish due to the boundary conditions. Hence,

$$G(\mathbf{z}_2, \mathbf{z}_1) = G^*(\mathbf{z}_1, \mathbf{z}_2). \tag{2.44}$$

Replacing $\mathbf{z}_2$ by $\mathbf{x}$ and $\mathbf{z}_1$ by $\mathbf{x}'$, the theorem is proved. ■

   Corollary 2.2. (Reciprocity Theorem.) *Green's function is symmetric for the self-adjoint operators with respect to $\mathbf{x}$ and $\mathbf{x}'$, i.e., $G(\mathbf{x}, \mathbf{x}') = G(\mathbf{x}', \mathbf{x})$.*

   PROOF. The proof follows from Green's theorem (1.12). Substitute $G_1 = G(x, x')$, $G_2 = G(x, x')$ in the identity by dividing the interval of integration into three subintervals $x_0 \le x \le x'$, $x' \le x \le x''$, $x'' \le x \le x_1$, treating each separately. The proof is completed by taking into account both discontinuities at the points $x = x'$ and $x = x''$. ■

Courant and Hilbert [1968:355] note that "The symmetry of Green's function expresses a reciprocity frequently occurring in physics: If the force 1, applied at the point x', produces the result $G(x, x')$ at the point x, then the same force 1 acting at x produces the same result at x'."

Theorem 2.3. *If $f(x)$ is a continuous or piecewise continuous function of x, then the function*

$$u(x) = \int_{x_0}^{x_1} G(x, s) f(s) \, ds \qquad (2.45)$$

is a solution of the equation $L[u] = -f(x)$, which satisfies the prescribed boundary conditions.

PROOF. From (2.45) we have

$$u'(x) = \int_{x_0}^{x_1} G'(x, s) f(s) \, ds,$$

$$u''(x) = \int_{x_0}^{x_1} G''(x, s) f(s) \, ds + \int_{x_0}^{x_1} G''(x, s) f(s) \, ds$$
$$\qquad + G'(x, x - 0) f(x) - G'(x.x + 0) f(x)$$
$$= \int_{x_0}^{x_1} G''(x, s) f(s) \, ds + [G'(x, x + 0) - G'(x, x - 0)] \, f(x)$$
$$= \int_{x_0}^{x_1} G''(x, s) f(s) \, ds - \frac{f(x)}{p(x)}.$$

Thus, $pu'' + p'u' - qu = \int_{x_0}^{x_1} (pg'' + p'G' - qG) f(s) \, ds - f(x)$. This completes the proof since $L[G] = 0$. To prove the converse, we integrate by parts setting $u = G$ over two intervals of integration $x_0 \le x \le s$ and $s \le x \le x_1$. Thus, the general solution is

$$u(x) = \int_{x_0}^{x_1} G(x, s) f(s) \, ds + pG'u \Big|_{x_0}^{x_1}. \quad \blacksquare$$

An application of Theorem 2.2 will allow us to express the solution of the partial differential equation $L[u](x) = \delta(x, x')$ in terms of Green's function $G(x, x')$. Therefore, it is not required to use the adjoint operator L^* to find a Green's function.

Example 2.14. Consider the boundary value problem $u'' + 2u' + u = f(x)$, $u(0) = 0 = u(l)$. Its solution is given by

$$u(x) = -\int_0^s \frac{s}{l} (l - x) f(s) e^{-(x-s)} \, ds - \frac{x}{l} \int_0^s (l - s) f(s) e^{-(x-s)} \, ds, \quad (2.46)$$

which leads to Green's function

$$G(x, s) = \begin{cases} -\dfrac{x}{l} (l - s) e^{-(x-s)} & \text{for } 0 \le x < s, \\[2mm] -\dfrac{s}{l} (l - x) e^{-(x-s)} & \text{for } s < x \le l. \end{cases} \qquad (2.47)$$

Note that the same Green's function is obtained if we directly solve the problem $u'' + 2u' + u = \delta(x - s)$, $u(0) = 0 = u(l)$. In this case the operator $D^2 + 2D + 1$ is not self-adjoint, and the adjoint operator is given by $D^2 - 2D + 1$. Green's function $G^*(x, s)$ for this adjoint operator satisfies the boundary value problem $u'' - 2u' + u = \delta(x - s)$, $u(0) = 0 = u(l)$. But the solution $G^*(x, s)$ for this problem is such that $G^*(s, x) = G(x, s)$. The function $G^*(s, x)$ is given by (2.47). Green's function in this case is not symmetric. Also note that once Green's function (2.47) is known, the solution u of the problem, given by (2.46), can be easily determined. This is known as the *Green's function method* of finding the solution to boundary value problems. ∎

The Green's function is the response at the point x from an impulse at the point s. It is the influence function which depends on the geometry of the domain and the boundary conditions.

We will show that for every equation and boundary condition the Green's function gives a kind of 'inverse'. We have the following result:

Theorem 2.4. *The Green's function $G(x, s)$ for a linear differential equation $L[u] = f$ is the solution when f is a δ-function at s: $L[G] = \delta(x - s)$. For bounded f the solution for $L[u] = f$ is a superposition of the free-space Green's functions (or fundamental solutions) such that*

$$u(x) = \int_a^b G(x, s) f(s) \, ds. \tag{2.48}$$

PROOF. By applying L on both sides of (2.48) we find that the right side is $L[G] = \delta(x - s)$. Thus, (2.48) becomes an identity, and hence $L[u] = f$. ∎

The equation $L[G] = \delta(x - s)$ shows that G is the inverse of L. This becomes clear if we compare this situation with the finite-dimensional case where L is a matrix. In the language of algebra, when L multiplies a column of the matrix L^{-1}, it produces a column of the identity matrix I. Notice that a column of I is like a δ-function, where the nonzero part is concentrated at a single point. Thus, expressing $f(x)$ as an integral of the δ-function is like writing a vector f as $\sum_j f_j \delta_j$, where δ_j is the jth column of the identity matrix. This means that formula (2.6) corresponds to the identity $f = If$. Next, note that in the continuous case there is a 'column' for every point s. In the identity matrix this column is δ concentrated at a single point. In L^{-1} the column is G which is response to δ. Then (2.48) is precisely equivalent to $u = L^{-1}f$, where we have an integral operator instead of a sum over all columns. Thus, the response to a distributed source f is the sum or integral of the response to the point source δ, which is the kind of inverse operation provided by the Green's function.

The existence of Green's function is equivalent to that of a unique solution of the homogeneous boundary value problem for the equation $L[u] = f(x)$. Therefore, the following *alternative* exists: Under given homogeneous boundary conditions, either the equation $L[u] = f(x)$ has a uniquely determined solution $u(x)$ for every given x, or the homogeneous equation $L[u] = 0$ has a nontrivial solution.

It is obvious from the above examples that Green's function $G(x, s)$ does not depend on the input function $f(t)$. Let T is the linear (integral) operator defined by

$$T : f \longmapsto \int_{x_0}^{x} G(x, s) f(s) \, ds, \quad \text{or} \quad T[f](x) = \int_{x_0}^{x} G(x, s) f(s) \, ds,$$

where the operator T is equivalent to the inverse linear operator L^{-1} introduced in (2.35), and G is a continuous function on the square $I \times I = \{(x, s) \mid x, s \in I\}$. This formulation leads to inverse problems which are typically ill posed. Examples of ill-posed inverse problems include integral equations of the first kind, tomography, and inverse scattering. Since these topics are beyond the scope of this book, interested readers are advised to consult the following literature: Groetsch [1984], Porter and Sterling [1993], Kythe and Puri [2002], Kythe and Schäferkotter [2005] on integral equations of the first kind where the last two books also include useful Mathematica® codes.

In Chapter 3 we will study a generalization of the variation of parameters method, which leads to the Green's function method for initial and boundary value problems, while in Chapters 4 and 5 we study Bernoulli's separation method and integral transforms which will be useful in developing different methods for determining Green's functions for partial differential operators, including the Sturm-Liouville systems, parabolic, hyperbolic and elliptic equations, and fractional partial differential equations. This will be followed by the conformal mapping method and computational construction of Green's functions in different types of convex and starlike 2-D regions.

2.5. Fundamental Solutions

Consider the boundary value problem

$$L[u](\mathbf{x}) = f(\mathbf{x}) \quad \text{in} \quad D \subset \mathbb{R}^n, \tag{2.49a}$$
$$B[u] = 0 \quad \text{on} \quad \partial D = S, \tag{2.49b}$$

where $f \in C(D)$, and (2.49b) represents linear initial and boundary conditions. This problem has a unique solution in terms of Green's function $G(\mathbf{x}, \mathbf{x}')$:

$$u(\mathbf{x}) = \int_{D} G(\mathbf{x}, \mathbf{x}') f(\mathbf{x}') d\mathbf{x}', \tag{2.50}$$

where $d\mathbf{x}'$ denotes integration with respect to the variable $\mathbf{x}'$ (source point or singularity). Green's function $G(\mathbf{x}, \mathbf{x}')$ is singular at the fixed point $\mathbf{x}' \in D$, and the singular part of $G(\mathbf{x}, \mathbf{x}')$, denoted by $u^*(\mathbf{x}, \mathbf{x}')$, is known as the fundamental solution (or singular solution, or 'free-space' Green's function, or Green's function in the large) for the operator L such that

$$L[u^*](\mathbf{x})(\mathbf{x}, \mathbf{x}') = \delta(\mathbf{x}, \mathbf{x}'). \tag{2.51}$$

Note that Eq (2.51) is simply Eq (2.49a), where $u(\mathbf{x})$ is replaced by $u^*(\mathbf{x}, \mathbf{x}')$ and $f(\mathbf{x})$ by $\delta(\mathbf{x}, \mathbf{x}')$. In particular, for example, the fundamental solution for the Laplacian operator ∇^2 is unique up to a harmonic function. Some important properties of the fundamental solution u^* for the operator L are:

(i) $u^*(\mathbf{x}, \mathbf{x}')$ is defined everywhere except at $\mathbf{x} = \mathbf{x}'$ where it is singular.

(ii) $u^*(\mathbf{x}, \mathbf{x}')$ is unique up to a harmonic function of $\mathbf{x}$, i.e., any two fundamental solutions for L with the same pole $\mathbf{x}'$ differ by a solution of the homogeneous equation $Lu^* = 0$. In the absence of boundary conditions the homogeneous equation will have many solutions, and u^* will not be unique.

(iii) $u^*(\mathbf{x}, \mathbf{x}') = u^*(\mathbf{x}', \mathbf{x})$.

(iv) $u^*(\mathbf{x}, \mathbf{x}')$ depends only on the distance $r = |\mathbf{x} - \mathbf{x}'|$ for all $n \geq 1$.

(v) $\nabla^2_{\mathbf{x}} u^* = 0$ for all $\mathbf{x} \neq \mathbf{x}'$ and $n \geq 1$.

(vi) $\displaystyle \int_S \frac{\partial u^*}{\partial n} ds = 1$, where S denotes the surface of the sphere $|\mathbf{x} - \mathbf{x}'| = r$ for any $r > 0$, $\mathbf{n}$ is the outward normal to S, and $\dfrac{\partial}{\partial n} = \mathbf{n} \cdot \nabla_{\mathbf{x}}$.

Physically, the function u^* in (2.51) is the response to a concentrated unit source located at $\mathbf{x} = \mathbf{x}'$. An important physical application of the fundamental solution $u^*(\mathbf{x})$ is that it enables us to solve the nonhomogeneous equation

$$L[u](\mathbf{x}) = f(\mathbf{x}), \tag{2.52}$$

where $f \in C_0^\infty(\mathbb{R}^n)$. In fact, let $f(\mathbf{x})$ represent the source term in the form of a total point sources $f(\mathbf{x}')\delta(\mathbf{x} - \mathbf{x}')$, i.e.,

$$f(\mathbf{x}) = \int_{\mathbb{R}^n} f(\mathbf{x}')\delta(\mathbf{x} - \mathbf{x}')\, d\mathbf{x}'.$$

Then, in view of (2.51), there exists an influence function $f(\mathbf{x}')u^*(\mathbf{x} - \mathbf{x}')$ at each point. Note that if L does not have constant coefficients, then

$$u(\mathbf{x}) = \int_{\mathbb{R}^n} u^*(\mathbf{x}, \mathbf{x}')f(\mathbf{x}')\, d\mathbf{x}' \tag{2.53}$$

is the solution of (2.52).

Example 2.15. (Potential flow) The potential flow in a one-dimensional region is defined by the Sturm-Liouville equation $\dfrac{d}{dx}\left[a(x)\dfrac{du}{dx}\right] = f(x), \quad 0 \leq x \leq l$, where u denotes the potential. This equation is found in problems of transverse deflection of a cable, axial deformation of a bar, heat transfer along a fin in heat exchangers, flow though pipes, laminar incompressible flow through a channel under constant pressure gradient, linear flow through porous media, and electrostatics. Let a function $u^*(x, s)$, which is sufficiently continuous and differentiable as often as needed, be a solution of

$$\frac{d}{dx}\left[a\frac{du^*}{dx}\right] = \delta(x, s).$$

For $a = \text{const}$ (for a homogeneous isotropic medium), the fundamental solution for this equation is given by

$$u^*(x, s) = \frac{1}{2a}(l - r), \quad r = |x - s|. \ \blacksquare$$

Example 2.16. (Bending of an elastic beam) This problem is defined by

$$\frac{d^2}{dx^2}\left(b\frac{d^2u}{dx^2}\right) = f(x), \quad 0 \le x \le l,$$

where $b = EI$ is the flexural rigidity of the beam (E being the modulus of elasticity and I the moment of inertia). Let $u^*(x, s)$ be the fundamental solution of

$$b\frac{d^4u^*}{dx^4} = \delta(x, s).$$

The fundamental solution for this equation, subject to the boundary conditions

$$u^*\left(0, 0^+\right) = \frac{b}{2} = u^*\left(l, l^-\right), \qquad u^*\left(l, 0^+\right) = 0 = u^*\left(l, l^-\right),$$

$$u^{*\prime}\left(0, 0^+\right) = \frac{bl}{2} = u^{*\prime}\left(0, l^-\right), \quad u^{*\prime}\left(l, 0^+\right) = -\frac{bl^3}{12} = u^{*\prime}\left(l, l^-\right),$$

is given by

$$u^*(x, s) = \lambda l^3\left(2 + |\rho|^3 - 3|\rho|^2\right), \tag{2.54}$$

where $\rho = r/l$, $\lambda = b/12$, and $r = x - s$. $\ \blacksquare$

Example 2.17. The Heaviside function $H(x)$ defines the distribution $H'(x) = \delta(x)$. Note that since $H(x)$ is not differentiable at $x = 0$ in the classical sense, the distributional definition (2.17) for $H'(x)$ when formally integrated gives

$$H(x) = \int_0^\infty \delta(x)\,dx = \begin{cases} 0, & x < 0 \\ 1, & x > 0. \end{cases}$$

It is obvious from (2.17) that the Heaviside function $H(x)$ is a fundamental solution for the basic differential operator ∂ on $\mathbb{R}$ $\blacksquare$.

Example 2.18. In $\mathbb{R}^n$ the radial solutions $v(r)$ of the potential equation $\nabla^m u = 0$, which depends only on the distance $r(\ne 0)$ of the point $\mathbf{x}$ from a fixed point $\mathbf{x}'$ so that $r = |\mathbf{x} - \mathbf{x}'|$, are given by the equation

$$\frac{d^2v}{dr^2} + \frac{n-1}{r}\frac{dv}{dr} = 0.$$

This equation has the solutions

$$v(r) = \begin{cases} c_1 + c_2 \, r^{2-n}, & n > 2, \\ c_1 + c_2 \, \log r, & n = 2, \end{cases}$$

where c_1, c_2 are arbitrary constants.; These solutions exhibit the so-called *characteristic singularity* at $r = 0$. The function

$$s\,(\mathbf{x}, \mathbf{x}') = \begin{cases} \dfrac{1}{(n-2)\,S_n(1)} \, |\mathbf{x} - \mathbf{x}|^{2-n}, & n > 2, \\ -\dfrac{1}{2\pi} \, \log |\mathbf{x} - \mathbf{x}'|, & n = 2, \end{cases} \tag{2.55}$$

is called the *singularity function* for $\nabla^n u = 0$. The function $s(\mathbf{x}, \mathbf{x}')$ satisfies the equation $\nabla^n s = 0$ for $\mathbf{x} \neq \mathbf{x}'$ and has the singularity at $\mathbf{x} = \mathbf{x}'$. Every solution $u^*(\mathbf{x}, \mathbf{x}')$ of the potential equation $\nabla^n u = 0$ in D of the form

$$u^*(\mathbf{x}, \mathbf{x}') = s(\mathbf{x}, \mathbf{x}') + \phi(\mathbf{x}), \quad \mathbf{x}' \in D,$$

is called a *fundamental solution* in D for the operator ∇^n with a singularity at $\mathbf{x} = \mathbf{x}'$, where $\phi(\mathbf{x}) \in C^2(D), \phi(\mathbf{x}) \in C^1(D \cup \partial D)$, and $\nabla^n \phi = 0$ in D. ∎

We have provided in Appendix B a list of fundamental solutions for some differential operators. For an extensive study of fundamental solution for differential operators, see Kythe [1996]. Fundamental solutions play a significant role in the boundary element method. A detailed study of this subject can be found in Brebbia and Dominguez [1992], and Kythe [1995].

2.6. Exercises

2.1. Show that the following functions define the Dirac δ-function as a limiting case:

(a) $\delta(x) = \lim\limits_{a \to 0} \dfrac{1}{\sqrt{\pi a}} \, e^{-x^2/a}, \ |x| < a$;

(b) $\delta(x) = \lim\limits_{a \to 0} \dfrac{1}{\pi} \dfrac{a}{a^2 + x^2}, \ |x| < a$;

(c) $\delta(x) = \lim\limits_{a \to 0} \begin{cases} 0, & x < -\dfrac{a}{2}, \\ \dfrac{1}{a}, & -\dfrac{a}{2} < x < \dfrac{a}{2}, \\ 0, & x > \dfrac{a}{2}; \end{cases}$

(d) $\delta(x) = \lim\limits_{a \to \infty} \dfrac{\sin ax}{\pi x} = \lim\limits_{a \to \infty} \dfrac{1}{2\pi} \int_{-\infty}^{\infty} e^{i\,xt} \, dt.$

2.2. Let $\delta(x)$ be defined as in Exercise 2.1(c). Show that $\lim\limits_{a \to 0} \int_{-\infty}^{\infty} f(x)\delta(x)\,dx = f(0)$.

2.3. Show that $\int_{-\infty}^{\infty} \delta(t)\,dt = 1$.

HINT. Use property (2.9) (iii).

2.4. Show that $t\,\delta(t) = 0$. HINT. Set $t\phi(t) = \psi(t)$. Then $\int_{-\infty}^{\infty} \delta(t)\,t\phi(t)\,dt = \int_{-\infty}^{\infty} \psi(t)\delta(t)\,dt = \psi(0) = 0$.

2.5. Show that the integral $I = \int_{-\infty}^{\infty} \delta\left(f(t)\right)\phi(t)\,dt = \dfrac{\phi(t')}{|f'(t')|}$, where f is a monotone function that vanishes at $t = t'$, and ϕ is an arbitrary function.

SOLUTION. Set $y = f(t)$. Then $I = \int_{-\infty}^{\infty} \delta(y)\psi(y)\,dy$, where $\psi(y) = \phi(y)/|f'(y)|$.

2.6. Show that $\delta\left(f(x)\right) = \dfrac{\delta(x, s)}{|f'(s)|}$. HINT. Use $\int_{-\infty}^{\infty} \delta(x - s)\phi(x)\,dx = \phi(s)$.

2.7. Show that $\int_{-\infty}^{\infty} \delta(ax - b)\,\phi(x)\,dx = |a|^{-1}\,\phi\left(ba^{-1}\right)$. HINT. Set $ax - b = y$.

2.8. Show that $\delta(x) = \delta(-x)$.

2.9. Show that $\int_{-\infty}^{\infty} \delta'(x)\phi(x)\,dx = -\int_{-\infty}^{\infty} \delta(x)\phi'(x)\,dx = -\phi'(0)$,

$$\int_{-\infty}^{\infty} \delta''(x)\phi(x)\,dx = -\int_{-\infty}^{\infty} \delta'(x)\phi'(x)\,dx = \phi''(0).$$

HINT. Use $\int_{-\infty}^{\infty} f'(x)\phi(x)\,dx = -\int_{-\infty}^{\infty} f(x)\phi'(x)\,dx$. (*)

2.10. Show that $\int_{-\infty}^{\infty} H'(t)\phi(t)\,dt = \phi(0)$. HINT. Use formula (*) of Exercise 2.9.

2.11. Show that $\int_{-\infty}^{\infty} f'(x)\phi(x)\,dx = \int_{-\infty}^{\infty} \mathrm{sgn}(x)\,\phi(x)\,dx$, where $\mathrm{sgn}(x)$ is the signum function defined by $\mathrm{sgn}(x) = \begin{cases} 1, & x > 0, \\ 0, & x = 0, \\ -1, & x < 0. \end{cases}$

HINT. Use formula (*) of Exercise 2.9.); then left-side integral

$$= -\int_{-\infty}^{0} (-x)\phi'(x)\,dx - \int_{0}^{\infty} x\phi'(x)\,dx = -\int_{-\infty}^{0} \phi(x)\,dx + \int_{0}^{\infty} \phi(x)\,dx$$
$$= \text{right-side}.$$

2.12. Show that $f'(x) = |x|' = \mathrm{sgn}(x)$. HINT. Follows from Exercise 2.11.

2.13. Show that $\dfrac{d}{dx}\,\mathrm{sgn}(x) = 2\delta(x)$.

2.14. Evaluate $I = \int_{-1}^{1} |x|\,\psi''(x)\,dx$. ANS. $I = \psi'(-1) - \psi'(-1) - \psi(1) - \psi(-1) + 2\psi(0)$.

2.15. Solve $L[K(x,t)] = \delta(x - t)$, where $L \equiv \dfrac{d^2}{dx^2}$, and K is the kernel function.

SOLUTION. Integrating $\dfrac{d^2}{dx^2} K(x,t) = \delta(x - t)$ and using $H'(x) = \delta(x)$, we get $\dfrac{d}{dx} K(x,t) = H(x - t) + a(t)$, where $a(t)$ is an arbitrary function. Integrating again we find that $K(x,t) = \int_{-\infty}^{\infty} H(x - t)\, dx + x\, a(t) + b(t) = (x - t)H(x - t) + xa(t) + b(t)$. where $b(t)$ is an arbitrary function.

2.16. If $f(t)$ is an integrable function with compact support, show that $u(x) = \displaystyle\int_{-\infty}^{\infty} K(x,t)\, f(t)\, dt$ satisfies the equation $\dfrac{d^2u}{dx^2} = f(x)$, where K is the kernel function of Exercise 2.15.

2.17. Show that $\displaystyle\int_{-\infty}^{\infty} \delta'(x) f(x)\, dx = -f'(0)$, assuming that $f'(x)$ is continuous at $x = 0$.

2.18. Let the δ-function be defined by $\delta_a(x) = \begin{cases} 0, & x < 0, \\ a\, e^{-ax}, & x > 0. \end{cases}$

Find a representation for $\delta(x)$.

HINT. The function $\Delta_a(x)$ has the singularity at $x = 0^+$.

ANS. $\delta(x) = \lim\limits_{a \to \infty} a\, e^{-ax}$.

3

Sturm-Liouville Systems

The simple case of determining Green's functions for ordinary differential equations of second order is important because it brings out the methods involved with different types of boundary conditions and the shape of the boundary, and it provides a framework for constructing Green's functions for different ordinary and partial differential operators.

3.1. Ordinary Differential Equations

We will discuss the theory and methods of solving nonhomogeneous ordinary differential equations of order n, $n = 1, 2, \ldots$, of the form $P(D)[u](x) = f(x)$, where $P(D)$ is the linear ordinary differential operator defined by

$$P(D) = a_0(x) \, D^n + a_1(x) \, D^{n-1} + \cdots + a_{n-1}(x) \, D + a_n(x), \quad a_0 \neq 0, \quad (3.1)$$

which represents a polynomial of degree n, where $D \equiv d/dx$, $D^n \equiv d^n/dx^n$, and the coefficients $a_k(x)$, $k = 0, 1, 2, \ldots, n$, are continuous on some interval I on the x-axis, and the function $u(x) \in C^n(I)$.

3.1.1. Initial and Boundary Conditions. An ordinary (or partial) differential equation subject to prescribed conditions in the form of initial or boundary conditions is known as an initial value or a boundary value problem. The initial conditions, also known as *Cauchy conditions*, are the values of the unknown function u and an appropriate number of its derivatives at the initial point.

The boundary conditions fall into the following three categories:

(i) *Dirichlet boundary conditions* (also known as boundary conditions of the first kind), when the values of the unknown function u are prescribed at each point of the boundary ∂D of a given domain D.

(ii) *Neumann boundary conditions* (also known as boundary conditions of the second kind), when the values of the normal derivatives of the unknown function

u are prescribed at each point of the boundary ∂D.

(iii) *Robin boundary conditions* (also known as boundary conditions of the third kind, or mixed boundary conditions), when the values of a linear combination of the unknown function u and its normal derivative are prescribed at each point of the boundary ∂D.

The classical two-point boundary value problem of ordinary differential equations involves a second-order equation, an initial condition which is a boundary condition at the initial point, and a terminal condition which is another boundary condition at the end-point of the interval of definition of the problem. For example, the boundary conditions imposed on an equation $P(D)[u] = 0$ defined on an interval (a, b), can be one of the following forms: (i) $u(a) = u_0$, $u(b) = u_1$ (Dirichlet); (ii) $u'(a) = u_0$, $u'(b) = u_1$ (Neumann); or (iii) $a_1 u(a) + b_1 u(b) = c_1$, $b_1 u'(a) + b_2 u'(b) = c_2$ (Robin).

3.1.2. General Solution. The general solution of an ordinary differential equation of the form $P(D)[u](x) = f(x)$, where the linear differential operator $P(D)$ is defined by (3.1), is determined by first considering the associated homogeneous equation $P(D)[u](x) = 0$ and finding the set of its *primitives* which are linearly independent solutions of the given equation, denoted by $\{u_1(x), u_2(x), \ldots, u_n(x)\}$. The following result holds:

Theorem 3.1. *The set of primitives* $\{u_1(x), u_2(x), \ldots, u_n(x)\}$ *for the homogeneous equation* $P(D)[u](x) = 0$, *is linearly independent on the interval* $I = \{x : x_0 < x < +\infty\}$ *iff the Wronskian* $W(x_1, \ldots, x_n) \neq 0$ *for all* $x \in I$. *Moreover, every solution of the homogeneous equation* $P(D)[u](x) = 0$ *is a linear combination of the set of primitives for this equation.*

The proof of this theorem is easy and left as an exercise; it is also available in most textbooks on ordinary differential equations. The Wronskian is defined by

$$W(u_1, u_2, \ldots, u_n) = \begin{vmatrix} u_1(x) & u_2(x) & \cdots & u_n(x) \\ u_1'(x) & u_2'(x) & \cdots & u_n'(x) \\ u_1''(x) & u_2''(x) & \cdots & u_n''(x) \\ \cdots & \cdots & \cdots & \cdots \\ u_1^{(n-1)}(x) & u_2^{(n-1)}(x) & \cdots & u_n^{(n-1)}(x) \end{vmatrix} \neq 0.$$

!! For the sake of brevity we will henceforth denote the Wronskian by $W(x)$.

The complementary function for the nonhomogeneous equation $P(D)[u](x) = f(x)$ is $u_c(x) = c_1 u_1(x) + c_2 u_2(x) + \cdots + c_n u_n(x)$, where $c_1, c_2, \ldots, c_n$ are arbitrary constants. If $f(x) = 0$, then the function $u_c(x)$ provides the general solution of the given homogeneous equation. If $f(x) \neq 0$, then a particular solution (known as particular integral) u_p is usually determined by using the following methods: (i) The method of undetermined coefficients which is used if the given equation has constant coefficients a_k, $k = 0, 1, 2, \ldots, n$; (ii) the method of variation of parameters which is used if the given equation has variable coefficients $a_k(x)$, $k = 0, 1, 2, \ldots, n$; and sometimes (iii) the particular solution u_p can be determined by inspection. Then the general solution of the equation $P(D)[u](x) = f(x)$ is given by

$$u(x) = u_c(x) + u_p(x) = c_1 u_1(x) + c_2 u_2(x) + \cdots + c_n u_n(x) + u_p(x), \quad (3.2)$$

where the coefficients c_j, $j = 1, 2, \ldots, n$, are determined from the n prescribed initial or boundary conditions.

Example 3.1. Consider the ordinary differential equation $D^2 u - 2Du - 3u = 2e^x - 10 \sin x$. The set of primitives is $\{e^{3x}, e^{-x}\}$, and the Wronskian is

$$W(x) = \begin{vmatrix} e^{3x} & e^{-x} \\ 3e^{3x} & -e^{-x} \end{vmatrix} = -3e^{2x} \neq 0.$$

Thus, $u_c(x) = c_1 e^{3x} + c_2 e^{-x}$. Using the method of undetermined coefficients, we take the particular integral as $u_p(x) = Ae^x + B \sin x + C \cos x$, which after substituting into the given equation and solving the identity so obtained gives $A = -\frac{1}{2}$, $B = 2, C = -1$. Hence, the general solution is $u(x) = u_c(x) + u_p(x) = c_1 e^{3x} + c_2 e^{-x} - \frac{1}{2}e^x + 2 \sin x - \cos x$. If the initial conditions are prescribed as $u(0) = 2$ and $u'(0) = 4$, then we have an initial value problem. Using these conditions on $u(x)$, we find that $c_1 + c_2 = \frac{7}{2}$ and $3c_1 + c_2 = \frac{5}{2}$, which give $c_1 = \frac{3}{2}$ and $c_2 = 2$. Hence, the solution of this initial value problem is $u(x) = \frac{3}{2}e^{3x} + 2e^{-x} - \frac{1}{2}e^x + 2 \sin x - \cos x$. ■

3.1.3. Method of Variation of Parameters.

We will discuss this method in some detail since it leads to the development of the so-called *Green's function method* for initial value and boundary value problems involving ordinary differential equations. Consider the general second order equation

$$a_0(x)u'' + a_1(x)u' + a_2(x)u = f(x), \quad a_0 \neq 0, \quad (3.3)$$

where $u \in C^2$ on some interval I on the x-axis. Let the set of primitives be $\{u_1(x), u_2(x)\}$, where u_1 and u_2 are linearly independent so that the Wronskian $W(x) = W(u_1, u_2) \neq 0$. Thus, the complementary function is $u_c(x) = c_1 u_1(x) + c_2 u_2(x)$, where c_1, c_2 are arbitrary constants. Now, using the method of variation of parameters we take $u_p(x) = v_1(x)u_1(x) + v_2(x)u_2(x)$. Then $u_p' = v_1 u_1' + v_2 u_2' + v_1' u_1 + v_2' u_2$. Imposing the restriction that $v_1' u_1 + v_2' u_2 = 0$, which becomes the first condition, a further differentiation gives $u_p'' = v_1 u_1'' + v_2 u_2'' + v_1' u_1' + v_2' u_2'$. Then substituting these values of u_p, u_p' and u_p'' into Eq (3.3) we find that

$$a_0 \left[v_1 u_1'' + v_2 u_2'' + v_1' u_1' + v_2' u_2' \right] + a_1 \left[v_1 u_1' + v_2 u_2' \right] + a_2 \left[v_1 u_1 + v_2 u_2 \right] = f(x).$$

This equation, after using the fact that u_1 and u_2 satisfy the homogeneous equation $a_0(x)u'' + a_1(x)u' + a_2(x)u = 0$, simplifies to $v_1' u_1' + v_2' u_2' = f(x)/a_0$, which is the second condition. From these two conditions we have the system of two equations to determine v_1 and v_2:

$$u_1 v_1' + u_2 v_2' = 0, \quad u_1' v_1' + u_2' v_2' = \frac{f(x)}{a_0},$$

This system can be written as

$$a_0 \begin{vmatrix} u_1 & u_2 \\ u_1' & u_2' \end{vmatrix} \left\{ \begin{array}{c} v_1' \\ v_2' \end{array} \right\} = \left\{ \begin{array}{c} 0 \\ f(x) \end{array} \right\}.$$

The determinant on the left side is the Wronskian $W(x)$. Using Cramer's rule (§1.6) this system is solved to yield

$$v_1' = \frac{\begin{vmatrix} 0 & u_2 \\ f & u_2' \end{vmatrix}}{a_0 W(x)} = -\frac{f u_2}{a_0 W(x)}, \quad v_2' = \frac{\begin{vmatrix} u_1 & 0 \\ u_1' & f \end{vmatrix}}{a_0 W(x)} = \frac{f u_1}{a_0 W(x)},$$

which on integration with respect to x gives

$$v_1(x) = -\int^x \frac{f(s) u_2(s)}{a_0(s) W(s)}\, ds, \quad v_2(x) = \int^x \frac{f(s) u_1(s)}{a_0(s) W(s)}\, ds. \qquad (3.4)$$

Thus, the particular integral is given by

$$u_p(x) = \int^x \frac{\Delta(x, s)}{a_0(s) W(s)} f(s)\, ds = \int^x g(x, s) f(s)\, ds, \qquad (3.5)$$

where

$$\Delta(x, s) = \begin{vmatrix} u_1(s) & u_2(s) \\ u_1(x) & u_2(x) \end{vmatrix}, \quad g(x, s) = \frac{\Delta(x, s)}{a_0(s) W(s)},$$

and $g(x, s)$ is the kernel in the integral (3.5) for u_p. Then the general solution of Eq (3.3) is

$$u(x) = u_c(x) + u_p(x) = c_1 u_1(x) + c_2 u_2(x) + \int^x g(x, s) f(s)\, ds. \qquad (3.6)$$

This analysis provides another form for the variation of parameters method to determine the particular integral u_p.

Example 3.2. Consider the equation $u'' + u = \tan x$, for which the complementary function is $u_c(x) = c_1 \cos x + c_2 \sin x$, where c_1, c_2 are arbitrary constants. Using the variation of parameters method, we find that $v_1' = -\sin x \tan x = \cos x - \sec x$ and $v_2' = \cos x \tan x = \sin x$, which after integration give $v_1 = \sin x - \ln|\sec x + \tan x| + c_3$ and $v_2 = -\cos x + c_4$. Then

$$u_p(x) = A \cos x + B \sin x - \cos x \ln|\sec x + \tan x|,$$

where A and B are the arbitrarily assigned values of c_1 and c_2, respectively. Hence, the solution can be written as

$$u(x) = C_1 \cos x + C_2 \sin x - \cos x \ln|\sec x + \tan x|,$$

where $C_1 = c_1 + A$ and $C_2 = c_2 + B$. Alternatively, using formula (3.5) we have $a_0(s) = 1, W(s) = 1$, and

$$\Delta(x, s) = u_1(s) u_2(x) - u_1(x) u_2(s) = \sin x \cos s - \cos x \sin s = \sin(x - s).$$

Then $g(x, s) = \sin(x - s)$, and the particular integral is given by

$$
u_p(x) = \int^x \sin(x - s)\tan s\, ds = -\cos x \sin x + \left[\sin x - \ln|\sec x + \tan x|\right]\cos x
$$
$$
= -\cos x \ln|\sec x + \tan x|,
$$

which is the same as found above by the variation of parameters method. ∎

3.2. Initial Value Problems

We will develop the method for determining Green's function for the nth-order ordinary differential equation $P(D)[u] = f$ subject to prescribed n initial conditions. The existence and uniqueness theorem for the general second-order ordinary differential equation $F(x, u, u', u'') = 0$, or $u'' = f(x, u, u')$ states that if $f, f_u, f_{u'}$ are continuous in an open region R of the 3-D xuu'-space and if the point (x_0, u_0, u_0') is in R, then in some interval about x_0 there exists a unique solution $u = \phi(x)$ of the above general second-order differential equation that satisfies the prescribed initial conditions $u(x_0) = u_0$, $u'(x_0) = u_0'$. The following uniqueness theorem holds for the general nth-order initial value problem involving either the space variable x or time variable t (for proof see Coddington [1989], or Coddington and Levinson [1955]).

Theorem 3.2. *Let $\alpha_1, \ldots, \alpha_n$ be any n constants, and let x_0 be any real number. In any interval $I = (a, b)$ containing x_0 there exists at most one solution $\phi(x)$ of $P(D)u = 0$ satisfying $\phi(x_0) = \alpha_1, \phi'(x_0) = \alpha_2, \ldots, \phi^{(n-1)}(x_0) = \alpha_n$.*

3.2.1. One-Sided Green's Functions.
The following results are direct generalization of formula (3.6), which is used to find the particular integral for the nonhomogeneous equation $P(D)[u](x) = f(x)$ with variable coefficients.

Theorem 3.3. *(For initial value problems.) Let $\{u_1(t), \ldots, u_n(t)\}$ be the set of primitives for the homogeneous equation $P(D)[u](t) = 0$, and let $W(t)$ denote the Wronskian for this primitive. If $\Delta(t, s)$ denotes the determinant*

$$
\Delta(t, s) = \begin{vmatrix}
u_1(s) & \cdots & u_n(s) \\
u_1'(s) & \cdots & u_n'(s) \\
\cdots & \cdots & \cdots \\
u_1^{(n-2)}(s) & \cdots & u_n^{(n-2)}(s) \\
u_1(t) & \cdots & u_n(t)
\end{vmatrix},
$$

and if t_0 is the initial point of the interval $t > t_0$, then

$$
u_p(t) = \int_{t_0}^t \frac{\Delta(x, s)}{a_0(s)\, W(s)}\, f(s)\, ds \tag{3.7}
$$

is the unique particular integral of the initial value problem $P(D)[u](t) = f(t)$, subject to the n initial conditions $u^{(j)}(t_0) = u_j(t_0)$ for $j = 0, 1, \ldots, n - 1$.

PROOF. Let u_p be of the form $u_p = v_1(t)u_1(t) + \cdots + v_n(t)u_n(t)$. Impose the following $n - 1$ conditions on the derivatives $v_1'(t), \ldots, v_n'(t)$:

$$v_1' u_1^{(k)}(t) + \cdots + v_n'(t)u_n^{(k)}(t) = 0. \quad k = 0, 1, 2, \ldots, n - 2,$$

and substitute into the nonhomogeneous equation $P(D)[u] = f$, which provides the additional condition

$$v_1'(t)u_1^{(n-1)}(t) + \cdots + v_n'(t)u_n^{(n-1)}(t) = f(t).$$

Then using Cramer's rule (see §1.6) solve the n equations for $v_1', \ldots, v_n'$, integrate each with respect to t to determine $v_1(t), v_2(t), \ldots, v_n(t)$. The result follows after substituting the values of $v_k(t)$, $k = 1, , 2, \ldots, n$, into the assumed form of $u_p(t)$. ∎

Formula (3.7) is remarkable in that the kernel

$$g(t, s) = \frac{\Delta(t, s)}{a_0(s) W(s)}, \quad t > t_0, \tag{3.8}$$

which is known as the *one-sided Green's function* for the problem, can be determined as soon as the primitives $\{u_1(t), \ldots, u_n(t)\}$ of solutions are known. Sometimes the one-sided Green's function (3.8) is written as

$$g(t, s) = H(t - t_0) \frac{\Delta(t, s)}{a_0(s) W(s)},$$

where $H(t - t_0)$ is the Heaviside function (see §2.1.1). This method is also valid if the independent variable is the space variable x instead of the time variable t, in which case the factor $H(x - x_0)$ provides the non-zero solution for $x > x_0$.

For the non-homogeneous equation $P(D)u(t) = f(t)$ with prescribed initial conditions, note that the one-sided Green's function $g(t, s)$ is always obtained from the associated homogeneous equation. Thus, for a fixed s the one-sided Green's function $g(t, s)$ is the solution of the homogeneous initial value problem:

$$P(D) g(t, s) = 0, \quad \text{for } t > t_0,$$

$$g(t, s)\Big|_{t=s} = 0, \quad \frac{dg}{dt}\Big|_{t=s} = \frac{1}{p(s)}. \tag{3.9}$$

Example 3.3. Consider the initial value problem $(D^2 + k^2)x(t) = 0$ such that (a) $x(0) = x_0$, $x'(0) = 0$, $t > 0$, and (b) $x(0) = x_0$, $x'(0) = u_0$, $t > 0$. This problem describes the free motion of a mass m which is attracted by a force proportional to its distance from the origin and (a) starts from rest at $x = x_0$, and (b) starts with an initial velocity u_0 at $x = x_0$ away from the origin. The primitive set is $\{\cos kt, \sin kt\}$; thus, $x_c(t) = c_1 \cos kt + c_2 \sin kt$. In both cases we find from (3.6) that the one-sided Green's function is $g(t, s) = \dfrac{H(t)}{k} \sin k(t - s)$, which satisfies the properties (3.9).

Example 3.4. Consider the equation $u'' - \dfrac{1}{t}u' + \dfrac{u}{t^2} = t^3, t > 1$, subject to the initial conditions $u(1) = u_0$ and $u'(1) = v_0$. Since the primitive set is $\{t, t \ln t\}$, we have $u_1(t) = t, u_2(t) = t \ln t$. Also, the Wronskian is $W(t) = t \neq 0$, and $a_0(t) = 1$. The one-sided Green's function is given by

$$g(t, s) = \frac{H(t-1)}{s} \begin{vmatrix} s & s \ln s \\ t & t \ln t \end{vmatrix} = H(t-1)t(\ln t - \ln s),$$

which satisfies the properties (3.9). By formula (3.7)

$$u_p(t) = \int_1^t g(t, s) f(s)\, ds = \int_1^t s^3 t (\ln t - \ln s)\, ds = \frac{t^5}{16} - \frac{t}{16} - \frac{t}{4}\ln t,$$

and the solution is $u(t) = c_1 t + c_2 t \ln t + u_p(t)$, where the initial conditions yield $c_1 = u_0, c_2 = v_0 - u_0$, and thus, $u(t) = \dfrac{t^5}{16} + \left(u_0 - \frac{1}{16}\right)t + \left(v_0 - u_0 - \frac{1}{4}\right)t \ln t$, which satisfies the prescribed initial conditions. ∎

Note that formula (3.5) for ordinary differential equations of order greater than 2 always yields u_p, and hence the general solution, as the following example shows.

Example 3.5. To solve the initial value problem: $x^3 u''' - 3x^2 u'' + 6xu' - 6u = x^5, x > 0$, such that $u(1) = 1, u'(1) = 0, u''(1) = 2$, by one-sided Green's function method, note that the primitive set is $\{x, x^2, x^3\}$; thus, $u_1(x) = x, u_2(x) = x^2, u_3(x) = x^3$, and the Wronskian is $W(x) = \begin{vmatrix} x & x^2 & x^3 \\ 1 & 2x & 3x^2 \\ 0 & 2 & 6x \end{vmatrix} = 2x^3 \neq 0$. Also,

$$\Delta(x, s) = \begin{vmatrix} s & s^2 & s^3 \\ 1 & 2s & 3s^2 \\ x & x^2 & x^3 \end{vmatrix} = x^3 s^2 - 2x^2 s^3 + xs^4,$$ and $a_0(x) = x^3$. Hence, by formula (3.5)

$$g(x, s) = \frac{\Delta(x, s)}{a_0(s)W(s)} = \frac{x^3 s^2 - 2x^2 s^3 + xs^4}{2s^6}.$$

With $f(x) = x^5$, the particular integral $u_p(x)$ is given by

$$u_p(x) = \int_0^x g(x, s) f(s)\, ds = \int_0^x \left(x^3 s - 2x^2 s^2 + xs^3\right) ds = \frac{x^5}{24}.$$

The general solution of this initial value problem is $u(x) = c_1 u_1(x) + c_2 u_2(x) + c_3 u_3(x) + u_p(x) = c_1 x + c_2 x^2 + c_3 x^3 + \dfrac{x^5}{24}$. Applying the initial conditions we find that $c_1 = \frac{31}{8}, c_2 = -\frac{14}{3}, c_3 = \frac{7}{4}$, and the final solution of this initial value problem is $u(x) = \frac{31}{8}x - \frac{14}{3}x^2 + \frac{7}{4}x^3 + \frac{1}{24}x^5$. Note that this problem can also be solved by the method of variation of parameters, but it may sometimes require more work. However, this may not always be the case, as is shown below in Example 3.12. ∎

Theorem 3.4. *The one-sided Green's function for a constant-coefficient operator* $P(D) = a_0 D^n + a_1 D^{n-1} + \cdots + a_n$ *is given by* $g(t-s)$, $a_0 \neq 0$, *where* $g(u)$ *is a solution of*

$$P(D)[g(u)] = 0, \ \text{such that} \ g(0) = 0, \ldots, \ g^{n-2}(0) = 0, \ g^{n-1}(0) = 1.$$

Example 3.6. Consider the constant-coefficient $P(D) = D^3 + 4D$. Then the one-sided Green's function for the related initial vale problem is found by solving

$$g'''(u) + 4g'(u) = 0, \quad g(0) = g'(0) = 0, \quad g''(0) = 1,$$

which is $g(y) = \frac{1}{4}(1 - \cos 2y)$, $y = t - s$. Hence, one-sided Green's function is

$$g(t-s) = \frac{1}{4} H(t) [1 - \cos 2(t-s)].$$

It can verified that this Green's function satisfies the given initial conditions, as follows:

$$g(0) = g(t-s)\big|_{t=s} = 0; \ g'(0) = \frac{dg(t-s)}{dt}\Big|_{t=s} = \frac{1}{2}\sin 2(t-s)\big|_{t=s} = 0; \ \text{and}$$
$$g''(0) = \frac{d^2 g(t-s)}{dt^2}\Big|_{t=s} = \cos 2(t-s)\big|_{t=s} = 1. \ \blacksquare$$

3.2.2. Wronskian Method. An alternate effective method to determine the particular integral u_p for the initial value problems involving the equation $P(D)[u](t) = f(t)$ is provided by the following theorem (see Brauer and Nohel [1989]).

Theorem 3.5. *Let* $\{u_1(t), u_2(t), \ldots, u_n(t)\}$ *be the primitive set for the equation* $P(D)[u](t) = 0$. *Then the general solution for the nonhomogeneous equation* $P(D)[u](t) = f(t)$, $t \in I = \{t : t_0 < t <= +\infty\}$, *satisfying the initial conditions* $u(t_0) = \alpha_1, u'(t_0) = \alpha_2, \ldots, u^{(n-1)}(t_0) = \alpha_n$, *where* $\alpha_k, k = 1, \ldots, n$, *are real numbers, is given by*

$$u(t) = u_c(t) + u_p(t), \quad u_c(t) = c_1 u_1(t) + \ldots + c_n u_2(t)$$

is the complementary function for the associated homogeneous equation, and

$$u_p(t) = \sum_{j=1}^{n} u_j(t) \int_{t_0}^{t} \frac{W_j(s)}{a_0(s) W(s)} f(s)\, ds, \ t > t_0, \tag{3.10}$$

$W(t) = W(u_1, \ldots, u_n)$ *is the Wronskian, and* $W_j(t)$ *is obtained from* $W(t)$ *by replacing the* j*th column by* $[0 \ 0 \ \cdots \ 1]^T$.

It can be verified that formula (3.10) is the same as (3.7), and it is left as an exercise. Note that either formula (3.7) or formula (3.10) is useful for determining the particular integral $u_p(x)$, since both of these formulas are equivalent to each other.

Example 3.7. Consider $u'' + \frac{1}{t}u' - \frac{1}{t^2}u = f(t), t > 0$, where the function f is continuous for all $t \in I$. The primitive set for the associated homogeneous equation is $\{t, 1/t\}$, so the Wronskian is $W(t) = \begin{vmatrix} t & 1/t \\ 1 & -1/t^2 \end{vmatrix} = -\dfrac{2}{t}, t > 0$. Here,

$W_1(t) = \begin{vmatrix} 0 & 1/t \\ 1 & -1/t^2 \end{vmatrix} = -\dfrac{1}{t}$, and $W_2(t) = \begin{vmatrix} t & 0 \\ 1 & 1 \end{vmatrix} = t$, and thus,

$$u_p(t) = \int_{t_0}^{t} \frac{s}{2} \left(\frac{t}{s} - \frac{s}{t} \right) f(s)\, ds.$$

The one-sided Green's function is given by $g(s,t) = \dfrac{s}{2} \left(\dfrac{t}{s} - \dfrac{s}{t} \right), t > t_0$, which satisfies the properties (3.9) ∎

3.2.3. Systems of First-Order Differential Equations. One-sided Green's function for a system of n first-order linear ordinary differential equations can be determined by generalizing the above method, and is explained in Appendix F.

3.3. Boundary Value Problems

The method of variation of parameters is generally used to solve the second order ordinary differential equation

$$a_0(x)\, u'' + a_1(x)\, u' + a_2(x)\, u = F(x), \quad a < x < b, \tag{3.11}$$

where it is assumed that $a_0(x)$ is continuously differentiable, and $a_1(x), a_2(x)$ are continuous functions on the interval $I = (a, b)$. Further, it is assumed that $a_0(x) > 0$ for $x > a$, and it does not approach zero as $x \to a$. If we multiply the above equation by $\dfrac{1}{a_0(x)} \exp \left\{ \int_a^x \dfrac{a_1(s)}{a_0(s)}\, ds \right\}$, and set $p(x) = \exp \left\{ \int_a^x \dfrac{a_1(s)}{a_0(s)} \right\}$, $q(x) = \dfrac{a_2(x)}{a_0(x)} p(x)$, and $f(x) = \dfrac{F(x)}{a_0(x)} p(x)$, the above differential equation becomes

$$\frac{d}{dx} \left[p(x) \frac{du}{dx} \right] + q(x)\, u = f(x), \quad a < x < b, \tag{3.12}$$

where the function $p(x)$ is continuously differentiable and positive on the interval $I = (a, b)$, and $q(x)$ and $f(x)$ are continuous on I. Eq (3.12) is known as the *self-adjoint Sturm-Liouville equation.*

Theorem 3.6. *The Sturm-Liouville operator of Eq (3.12) is self-adjoint.*

PROOF. Consider a general linear, second-order, differential equation of the form

$$L[u](x) \equiv a_0(x) \frac{d^2 u(x)}{dx^2} + a_1(x) \frac{du(x)}{dx} + a_2(x)\, u(x), \quad a_0(x) \neq 0, \tag{3.13}$$

where a_0, a_1 and a_2 are real functions of x defined over the interval $a \leq x \leq b$, such that the first $(2 - j)$ derivatives of $a_j(x)$ are continuous, and the zeros of $a_0(x)$ are the singular points. This equation is a Sturm-Liouville equation if $a(x) = a_1(x)/a_0(x)$, and $q(x) = a_2(x)/a_0(x)$. An *adjoint operator*, denoted by L^*, is defined by

$$L^*[u] = \frac{d^2}{dx^2}[a_0 u] - \frac{d}{dx}[a_1 u] + a_2 u$$

$$= a_0 \frac{d^2 u}{dx^2} + (2a_0' - a_1)\frac{du}{dx} + (a_0'' - a_1' + a_2)\, u. \tag{3.14}$$

Comparing (3.13) and (3.14) we obtain the necessary and sufficient condition for $L = L^*$, which is $a_0'(x) = a_1(x)$, so that when this condition is satisfied the operators L and L^* are said to be *self-adjoint*. ∎

Self-adjoint operators are also called *Hermitian operators*. An important property is:

$$\int_a^b v\, L[u](x)\, dx = \int_a^b u\, L[v](x)\, dx, \tag{3.15}$$

which is a consequence of the above theorem. Further, if we integrate $vL[u]$ with respect to x over an interval $[a, x]$, where $v(x)$ is any function of x, we obtain

$$\int_a^x vL[u]\, dx = \Big[(va_0)u' - (va_0)'u + (va_2)u\Big]_a^x$$

$$+ \int_a^x \big[(va_)'' - (va_1)' + (va_2)\big]\, u. \tag{3.16}$$

If we differentiate both sides with respect to x , then

$$vL[u] - uL^*[v] = \frac{d}{dx}\big[a_0(u'v - uv') + (a_1 - a_0')uv\big], \tag{3.17}$$

which is known as *Lagrange's identity* for the Sturm-Liouville operator L. Again, if we integrate (3.16) from a to b, we get

$$\int_a^b \big(vL[u] - uL^*[v]\big)\, dx = \Big[a_0(u'v - uv') + (a_1 - a_0')uv\Big]_a^b. \tag{3.18}$$

For the self-adjoint Sturm-Liouville operator L formula (3.18) yields

$$\int_a^b \big(vL[u] - uL[v]\big)\, dx = \Big[a_0(u'v - uv')\Big]_a^b. \tag{3.19}$$

3.3.1. Sturm-Liouville Boundary Value Problems.
Consider the nonhomogeneous, second-order boundary value problem on $x \in (a, b)$:

$$\big[p(x)\, u'\big]' + q(x)\, u = f(x), \quad a < x < b, \tag{3.20a}$$

$$a_1\, u(a) + a_2\, u'(a) = 0, \quad b_1\, u(b) + b_2\, u'(b) = 0, \quad a \le x \le b. \tag{3.20b}$$

Before we solve this boundary value problem, we will first consider the associated homogeneous equation $[p(x) u']' + q(x) u = 0$ with the boundary conditions (3.20b). This equation has the primitives (linearly independent solutions) $u_1(x)$ and $u_2(x)$, such that their Wronskian

$$W(x) = \begin{vmatrix} u_1(x) & u_2(x) \\ u_1'(x) & u_2'(x) \end{vmatrix} = u_1(x)u_2'(x) - u_1'(x)u_2(x) \neq 0 \text{ on } [a, b].$$

The general solution of the homogeneous equation is $u(x) = c_1 u_1(x) + c_2 u_2(x)$. The first condition in (3.20b) is satisfied if

$$a_1 u(a) + a_2 u'(a) = a_1 c_1 u_1(a) + a_1 c_2 u_1(a) + a_2 c_1 u_1'(a) + a_2 c_2 u_2'(a) = 0,$$

which implies that

$$a_1 a_2 \begin{bmatrix} u_1(a) & u_2(a) \\ u_1'(a) & u_2'(a) \end{bmatrix} \begin{Bmatrix} c_1 \\ c_2 \end{Bmatrix} = \begin{Bmatrix} 0 \\ 0 \end{Bmatrix}.$$

The determinant of the matrix on the left side is $W(a) \neq 0$. Thus, this system has a unique solution unless a_1 or a_2 is zero. Suppose $a_1 = 0$ and $a_2 \neq 0$. Then, we solve

$$c_1 u_1'(a) + c_2 u_2'(a) = 0.$$

This equation has at least one solution. Let $u_1(x)$ be that solution of the homogeneous equation satisfying the first condition in (3.20b). Similarly, there will always be a solution of this equation, say $u_2(x)$, which satisfies the second condition in (3.20b). Let these two solutions $u_1(x)$ and $u_2(x)$ be such that they are not multiples of each other. Then there is no single function, except the trivial solution $u(x) \equiv 0$, that satisfies the homogeneous boundary value problem, and that 0 is not an eigenvalue of this boundary value problem (see §3.4).

To find the solution of the nonhomogeneous boundary value problem (3.20a,b), note that a particular solution of Eq (3.20a) is given by

$$u_p(x) = \int_a^x \frac{u_2(x)u_1(s)}{p(s)W(s)} f(s) \, ds - \int_a^x \frac{u_1(x)u_2(s)}{p(s)W(s)} f(s) \, ds. \tag{3.21}$$

Thus, the general solution of Eq (3.20a) is $u(x) = c_1 u_1(x) + c_2 u_2(x) + u_p(x)$. The first condition in (3.20b) gives

$$a_1 u(a) + a_2 u'(a)$$
$$= a_1 [c_1 u_1(a) + c_2 u_2(a)] + a_2 [c_1 u_1'(a) + c_2 u_2'(a)] + a_1 u_p(a) + a_2 u_p'(a)$$
$$= c_2 [a_1 u_1(a) + a_2 u_2'(a)] = 0.$$

If $a_1 u_1(a) + a_2 u_2'(a) = 0$, then $u_2(x)$ would satisfy the boundary value problem (3.20a,b), i.e., there would exist no function $u_2(x)$, contrary to our assumption. Hence, $a_1 u_1(a) + a_2 u_2'(a) \neq 0$, and so $c_2 = 0$. Similarly, the second condition in (3.20b) gives

$$b_1 u(1) + b_2 u'(b) = c_1 [b_1 u_1(b) + b_2 u_2'(b)] + c_2 [b_1 u_1(b) + b_2 u_2'(b)] +$$

$$+ b_2 \left\{ \int_a^x \frac{u_2'(b)u_1(s)}{p(s)W(s)} f(s)\, ds - \int_a^x \frac{u_1'(x)u_2(s)}{p(s)W(S)} f(s)\, ds \right\}$$

$$= c_1 \left[b_1 u_1(b) + b_2 u_1'(b) \right] - b_1 \int_a^b \frac{u_1(b)u_2(s)}{p(s)W(s)} f(s)\, ds$$

$$- b_2 \int_a^b \frac{u_1'(b)u_2(s)}{p(s)W(s)} f(s)\, ds$$

$$= c_1 \left[b_1 u_1(b) + b_2 u_1'(b) \right] - \left[b_1 u_1(b) + b_2 u_1'(b) \right] \int_a^b \frac{u_2(s)\, ds}{p(s)W(s)} = 0.$$

or $c_1 = \int_a^b \frac{u_2(s)\, ds}{p(s)W(s)} = 0$, and $c_2 = 0$. Hence, using (3.21) we find that solution of the problem as

$$u(x) = c_1 u_1(x) + u_p(x)$$

$$= \int_a^b \frac{u_1(x)u_2(s)}{p(s)W(s)} f(s)\, ds + \int_a^x \frac{u_2(x)u_1(s)}{p(s)W(s)} f(s)\, ds - \int_a^x \frac{u_1(x)u_2(s)}{p(s)W(s)} f(s)\, ds$$

$$= \int_a^x \frac{u_2(x)u_1(s)}{p(s)W(s)} f(s)\, ds + \int_x^b \frac{u_1(x)u_2(s)}{p(s)W(s)} f(s)\, ds = \int_a^b G(x,s)\, f(s)\, ds,$$

where

$$G(x,s) = \begin{cases} \dfrac{u_2(x)u_1(s)}{p(s)W(s)}, & a \le x < s, \\[2mm] \dfrac{u_1(x)u_2(s)}{p(s)W(s)}, & s < x \le b, \end{cases} \qquad (3.22)$$

is Green's function for the system (3.20a,b), and $\{u_1(x), u_2(x)\}$ is the set of primitives each of which satisfy the respective boundary conditions. Hence, the solution of the Sturm-Liouville system is given by

$$u(x) = \int_a^b G(x,s)\, f(s)\, ds.$$

3.3.2. Properties of Green's Functions. For each fixed s Green's function $G(x,s)$ satisfies the following five properties:

(i) $\dfrac{d}{dx}\left[p(x)\dfrac{dG}{dx} \right] + q(x)\, G = 0$, for $x \ne s$;

(ii) $G(a,s) = G(b,s) = 0$ (boundary conditions);

(iii) $G\big|_{x=s+0} - G\big|_{x=s-0} = 0$ (continuity condition);

(iv) $\dfrac{dG}{dx}\bigg|_{x=s+0} - \dfrac{dG}{dx}\bigg|_{x=s-0} = -\dfrac{1}{p(s)}$ (jump condition); and

(v) $G(x,s) = G(s,x)$ (symmetry condition).

3.3.3. Green's Function Method. We will summarize the method for constructing Green's function and finding solution for boundary value problems involving the Sturm-Liouville equation (3.20a), where $p(x) \neq 0$ and $a \leq x \leq b$, subject to the boundary conditions (3.20b), or generally for ordinary differential equations of the form $P(D)u(x) = f(x)$ subject to appropriate boundary conditions.

STEP 1. Find the primitive $\{u_1(x), u_2(x)\}$ of the associated homogeneous equation, so that the complementary function is $u_c(x) = c_1 u_1(x) + c_2 u_2(x)$.

STEP 2. Since Green's function satisfies the prescribed boundary conditions, let $g_1(x)$ be the value of $u_c(x)$ at the left boundary point $x = a$, and $g_2(x)$ the value of $u_c(x)$ at the other endpoint $x = b$. The set $\{g_1(x), g_2(x)\}$ is called the *basis* of solutions. Determine the Wronskian $W(x) = W(g_1(x), g_2(x))$.

STEP 3. Since Green's function must satisfy the continuity condition at $x = s$, and since the jump discontinuity in G_x at $x = s$ must be of magnitude $-1/p(x)$, take

$$G(x, s) = \begin{cases} g_1(x)\, g_2(s), & a \leq x < s, \\ g_1(s)\, g_2(x), & s < x \leq b. \end{cases}$$

STEP 4. Since the jump in the derivative G_x at $x = s$ is $g_1(s)\, g_2'(s) - g_1'(s)\, g_2(s) = W(x)$, divide the right side of $G(x, s)$ in Step 3 by $p(s)W(s)$. This gives the required Green's function as

$$G(x, s) = \begin{cases} \dfrac{g_1(x)\, g_2(s)}{p(s)W(s)}, & a \leq x < s, \\[2mm] \dfrac{g_1(s)\, g_2(x)}{p(s)W(s)}, & s < x \leq b. \end{cases} \tag{3.23}$$

The significance of determining the basis of solutions in Step 2 is evident from Step 4, where $G(x, s)$, split into two parts, is such that the first part contains $g_1(x)$ which satisfies the boundary condition at $x = a$, and the second part contains $g_2(x)$ which satisfies the boundary condition at $x = b$. Hence, Green's function constructed in Step 4 satisfies all five properties of §3.3.2.

This method does not apply to initial value problems. To confirm we give a counter-example by solving Example 3.4 using this method. The prescribed initial conditions yield $g_1(t) = u_0 t$, $g_2(t) = (v_0 - u_0)t \ln t$, so that the Wronskian is $W(t) = u_0(v_0 - u_0)t \neq 0$, and $p(t) = 1$. Then by (3.23), we have $G(t, s) = \begin{cases} st \ln t, & t < s, \\ st \ln s, & t > s \end{cases}$, which will give the solution as $u(t) = \frac{t^5}{16} - \frac{t}{16} - \frac{1}{4}t \ln t$, that does not satisfy the initial conditions.

Example 3.8. Consider the boundary value problem

$$u'' + u = f(x), \quad u(0) = 0 = u(1).$$

The homogeneous equation $u'' + u = 0$ has the general solution $u = c_1 \sin x + c_2 \cos x$. The boundary condition at $x = 0$ gives $c_2 = 0$, so $u = c_1 \sin x$. Choosing $c_1 = 1$ gives $u_1(x) = \sin x$. The condition at $x = 1$ gives $c_1 \sin(1) + c_2 \cos(1) = 0$, where by taking $c_1 = \cos(1)$ and $c_2 = \sin(1)$, we get $u_2(x) = \sin(x - 1)$. Thus, the basis of solutions is $\{\sin x, \sin(x-1)\}$. The Wronskian of the solutions is

$$W(x) = \begin{vmatrix} \sin x & \sin(x-1) \\ \cos x & \cos(x-1) \end{vmatrix} = \sin(1).$$

Hence, by (3.23) Green's function is

$$G(x,s) = \begin{cases} \dfrac{\sin(x-1)\sin s}{\sin(1)}, & 0 \le x < s, \\[3mm] \dfrac{\sin x \sin(s-1)}{\sin(1)}, & s < x \le 1. \end{cases}$$

In particular, if we take $f(x) = x$, then

$$u(x) = \int_0^1 G(x,s)f(s)\,ds = \int_0^x \frac{\sin(x-1)\sin s}{\sin 1}\,s\,ds + \int_x^1 \frac{\sin x \sin(s-1)}{\sin 1}\,s\,ds$$

$$= x - \frac{\sin x}{\sin 1}.$$

This result can be verified, for example, by the variation of parameters method, as follows: The Wronskin $W(x) = \begin{vmatrix} \sin x & \cos x \\ \cos x & -\sin x \end{vmatrix} = -1$, and $a_0(x) = 1$, so

$$\Delta(x,s) = \begin{vmatrix} \sin s & \cos s \\ \sin x & \cos x \end{vmatrix} = \sin s \cos x - \sin x \cos s = g(x,s),$$

and $u_p(x) = \int^x (\sin s \cos x - \sin x \cos s)\,s\,ds = x$, which leads to $u(x) = u_c(x) + u_p(x) = c_1 \sin x + c_2 \cos x + x$, and the boundary conditions yield $c_2 = 0$, $c_1 = -\dfrac{\sin x}{\sin 1}$; this finally gives the solution as $u(x) = -\dfrac{\sin x}{\sin 1} + x$, which is the same as given above. Compare this example with Example 2.9. ∎

Example 3.9. Consider the nonhomogeneous boundary value problem

$$\left(1+x^2\right)u'' + 2xu' - 2u = f(x), \quad u(0) = 0, \quad u(1) + u'(1) = 0.$$

The general solution of the homogeneous equation is $u = c_1 x + c_2 (1 + x \arctan x)$. The boundary condition at $x = 0$ gives $c_2 = 0$, and by choosing $c_1 = 1$, we have $g_1(x) = x$. The condition at $x = 1$ gives $c_1 + c_2(1 + \arctan 1) + c_1 + c_2 \left(\frac{1}{2} + \arctan 1\right) = 2c_1 + c_2 \left(\frac{3}{2} + \frac{\pi}{2}\right)$. Choosing $c_1 = \pi + 3$ and $c_2 = -4$, we find the other linearly independent solution $g_2(x) = (\pi + 3)x - 4(1 + x \arctan x)$. The Wronskian is given by

$$W(x) = \begin{vmatrix} x & (\pi+3)x - 4(1 + x \arctan x) \\ 1 & \pi+3 - 4\left(\dfrac{x}{1+x^2} + \arctan x\right) \end{vmatrix} = \frac{4}{1+x^2} \ne 0.$$

Hence, Green's function is

$$G(x, s) = \begin{cases} \dfrac{\pi+3}{4}\, xs - s\,(1 + x\arctan x), & 0 \le x < s, \\[2mm] \dfrac{\pi+3}{4}\, sx - x\,(1 + s\arctan s), & s < x \le 1. \end{cases}$$

Then the solution of this problem is given by $u(x) = \displaystyle\int_0^1 G(x, s) f(s)\, ds$. For instance, with $f(x) = 1$ we get

$$\begin{aligned} u(x) &= \int_0^1 G(x, s)\, ds \\ &= \int_0^x \frac{\pi+3}{4}\, xs - s\,(1 + x\arctan x)\, ds + \int_x^1 \frac{\pi+3}{4}\, sx - x\,(1 + s\arctan s)\, ds \\ &= \frac{x}{2}\arctan x - \frac{\pi+1}{8}\, x, \end{aligned}$$

where we have used the formula $\displaystyle\int x\arctan x\, dx = \frac{1}{2}(x^1 + 1)\arctan x - \frac{x}{2}$. This result is verified by the variation of parameters method or the Wronskin method, as follows: Since $W(x) = \begin{vmatrix} x & 1 + x\arctan x \\ 1 & \dfrac{x}{1+x^2} + \arctan x \end{vmatrix} = -\dfrac{1}{1+x^2}$, and $a_0(x) = 1 + x^2$, we obtain $\Delta(x, s) = -\begin{vmatrix} s & 1 + s\arctan s \\ x & 1 + x\arctan x \end{vmatrix} = x(1 + s\arctan x) - s(1 + x\arctan x)$. Thus,

$$u_p(x) = \int^x [x(1 + s\arctan s) - s(1 + x\arctan x)]\, ds = \frac{x}{2}\arctan x,$$

so that the general solution is $u(x) = u_c(x) + u_p(x) = c_1 x + c_2(1 + x\arctan x) + \dfrac{x}{2}\arctan x$. After applying the prescribed boundary conditions we find that $c_2 = 0$ and $c_1 - \dfrac{\pi+1}{8}$, which gives $u(x) = \dfrac{x}{2}\arctan x - \dfrac{\pi+1}{8}\, x$, same as above. ∎

Example 3.10. Consider the problem $u'' = f(x)$, $0 < x < 1$, such that $u(0) = u(1) = 0$. In view of the above properties (i) and (ii), Green's function is

$$G(x, s) = \begin{cases} a_1(s)\, x, & 0 \le x < s, \\ a_2(s)\,(1 - x), & s < x \le 1, \end{cases}$$

where a_1 and a_2 are constants which depend on s only. By the above condition (v), we have $a_1(s) = A(1 - s)$, $a_2(s) = A\, s$, where A is a constant independent of s. Then the continuity condition (iii) is automatically satisfied, and the jump condition (iv) gives

$$\frac{dG}{dx}\Big|_{x=s+0} = A\, s\,(-1), \qquad \frac{dG}{dx}\Big|_{x=s-0} = A\,(1 - s).$$

Thus, $A = 1$, and Green's function is

$$G(x,s) = \begin{cases} (1-s)\,x, & 0 \le x < s, \\ s\,(1-x), & s < x \le 1, \end{cases}$$

and the particular integral is

$$u_p(x) = \int_0^x x(1-s)\,f(s)\,ds + \int_x^1 (1-x)\,s\,f(s)\,ds. \;\blacksquare$$

Example 3.11. Consider the transverse displacement of a string of unit length, $0 < x < 1$, under tension at fixed $x = 0$, and connected to a spring with spring constant b at $x = 1$, and under a unit transverse force applied at $x = 1$ and a transverse force $f(x)$ per unit length applied along the string. This problem is defined by

$$u'' = f(x), \quad u(0) = 0, \quad b\,u(1) + u'(1) = 1.$$

Under the above properties (i) and (ii) Green's function is given by

$$G(x,s) = \begin{cases} a_1(s)\,x, & 0 \le x < s, \\ a_2(s)\,[1 + b\,(1-x)], & s < x \le 1. \end{cases}$$

By the symmetry condition (v), we have $a_1(s) = A\,[1 + b\,(1-s)]$, $a_2 = As$, where A is a constant not dependent on s. By condition (iv), we have $As(-b) - A\,[1 + b\,(1-s)] = -1$, which gives $A = \dfrac{1}{1+b}$. Hence,

$$G(x,s) = \begin{cases} \dfrac{x\,[1 + b\,(1-s)]}{1+b}, & 0 \le x < s, \\[2mm] \dfrac{s\,[1 + b\,(1-x)]}{1+b}, & s < x \le 1, \end{cases}$$

and

$$u(x) = \frac{x}{1+b} + \frac{1 + b\,(1-x)}{1+b} \int_0^x s\,f(s)\,ds$$
$$+ \frac{x}{1+b} \int_x^1 [1 + b\,(1-s)]\,f(s)\,ds.$$

If $b = 0$, we obtain Green's function for the boundary value problem $u'' = f$, $0 < x < 1$, such that $u(0) = 0$, $u'(1) = 1$. $\blacksquare$

Sometimes it becomes easier to find the solution of a boundary value problem if it is possible to get a particular solution simply by inspection. This method is useful if we need to find the solution $u(x)$ without determining Green's function. The following example illustrates this short cut.

Example 3.12. Consider the problem

$$u'' - u = 2x, \quad 0 < x < 1,$$
$$u'(0) = 0, \ u(1) = 0.$$

By inspection, a particular solution of the differential equation is $u = 2x$, and the basis of solutions is $\{e^x, e^{-x}\}$. Then the general solution is

$$u(x) = 2x + A e^x + B e^{-x}.$$

Using the boundary conditions, we find that $A = -\dfrac{2(1+e)}{1+e^2}$, $B = -\dfrac{2e\,(1-e)}{1+e^2}$. Hence,

$$u(x) = 2x - \frac{2(1+e)}{1+e^2} e^x - \frac{2e(1-e)}{1+e^2} e^{-x}.$$

Note that it will a lengthier process to find this solution through Green's function method. ∎

Example 3.13. (a) $L[u] = u'' = 0, 0 < x < 1$, such that $u(0) = 0 = u(1)$. Green's function is

$$G(x, s) = \begin{cases} (1-s)\,x, & \text{if } 0 \le x < s, \\ (1-x)\,s, & \text{if } s < x \le 1. \end{cases}$$

(b) $L[u] = u'' = 0, 0 < x < 1$, such that $u(0) = 0, u'(1) = 0$. Green's function is

$$G(x, s) = \begin{cases} x, & \text{if } 0 \le x < s, \\ s, & \text{if } s < x \le 1. \end{cases}$$

(c) $L[u] = u'' = 0, -1 \le x \le 1$, such that $u(-1) = 0 = u(1)$. Green's function is

$$G(x, s) = -\frac{1}{2}\Big\{|x - s| + xs - 1\Big\}.$$

(d) $L[u] = u'' = 0, 0 \le x \le 1$, such that $u(0) = u(1), u'(0) = -u'(1)$. Green's function is

$$G(x, s) = -\frac{1}{2}\,|x - s| + \tfrac{1}{4}. \quad \blacksquare$$

Example 3.14. $L[u] = xu'' + u' = 0, 0 \le x \le 1$, such that $u(1) = 0, u(0)$ finite. Green's function is

$$G(x, s) = \begin{cases} -\ln s, & \text{if } 0 \le x < s, \\ -\ln x, & \text{if } s < x \le 1. \end{cases}$$

This problem is associated with the Bessel function $J_0(x)$ of order zero. Note that Bessel's equation of order n is $x^2 y'' + xy' + (x^2 - n^2)\,y = 0, n \ge 0$, and thus Bessel's equation of order zero is $xy'' + y' + y = 0$. ∎

Example 3.15. $L[u] = (xu')' - \dfrac{n^2}{x} u = 0, 0 \le x \le 1$, such that $u(1) = 0$, $u(0)$ finite. Green's function is

$$G(x, s) = \begin{cases} \dfrac{1}{n}\left[\left(\dfrac{x}{s}\right)^n - (xs)^n\right], & \text{if } 0 \le x \le s, \\[3mm] \dfrac{1}{n}\left[\left(\dfrac{s}{x}\right)^n - (xs)^n\right], & \text{if } s < x \le 1. \end{cases}$$

This problem is associated with the Bessel function $J_n(x)$ of order n. Note that Bessel's equation of order n is $x^2 y'' + xy' + (x^2 - n^2) y = 0$, $n \ge 0$. $\blacksquare$

Example 3.16. (Linear Oscillator.) The linear oscillator equation for a vibrating string of length l is defined by

$$u''(x) + \lambda u(x) = 0, \quad u(0) = 0 = u(l).$$

The boundary conditions imply that the string is clamped at both end points. To determine Green's function, first we find the solutions of the corresponding homogeneous equation which is $L[u](x) = u''(x) = 0$ under the prescribed conditions: A solution is $u_1(x) = x$ which satisfies the condition at $x = 0$, while the other solution is $u_2(x) = 1 - x$ which satisfies the condition at $x = 1$. We say that the primitive set of solutions is $\{x, 1 - x\}$. Also, since the Wronskian $W(x) = u_1 u_2' - u_1' u_2 = -1$, and $p(x) = 1$, we get $A = -1$. Thus, Green's function is

$$G(x, s) = \begin{cases} x(1 - s), & 0 \le x < s, \\ s(1 - x), & s < x \le 1, \end{cases}$$

and the solution of of this problem is $u(x) = \lambda \displaystyle\int_0^1 G(x, s)\, ds$. Note that λ is not the wavelength of the linear oscillator. $\blacksquare$

3.4. Eigenvalue Problem for Sturm-Liouville Systems

Consider the Sturm-Liouville system:

$$\frac{d}{dx}\left[p(x)\frac{du}{dx}\right] + \lambda \rho(x)\, u = 0, \quad a < x < b,$$
$$a_1 u(a) + a_1 u'(a) = 0, \quad b_1 u(b) + b_2 u'(b) = 0, \tag{3.24}$$

where λ is real, $\rho(x)$ is the weight function, p and q are the functions as defined before, and a_i, b_i, $i = 1, 2$, are real constants such that a_1 and b_1, or a_2 and b_2 are both not zero and $a_1^2 + a_2^2 > 0$ and $b_1^2 + b_2^2 > 0$. It is obvious that the system (3.24) always has a trivial solution $u = 0$. The nontrivial solution of this problem consists of the *eigenfunctions* $\phi_n(x)$ and the corresponding values of λ which are called as the *eigenvalues* λ_n for the problem. The pair (λ_n, ϕ_n) is known as the *eigenpair*.

For example, the set $\{1, \cos x, \sin x, \cos 2x, \sin 2x, \dots\}$ of (orthogonal) functions consists of the eigenfunctions for the problem $u'' + \lambda u = 0$, $u(-\pi) = u(\pi)$, $u'(-\pi) = u'(\pi)$, with the corresponding eigenvalues $\lambda_n = n^2$ for $n = 1, 2, \dots$. Another example is as follows: For the eigenvalue problem $u'' + \lambda u = 0$, where $0 < x < L$, and (a) subject to the (Dirichlet) boundary conditions $u(0) = 0 = u(L)$, the eigenpair is

$$\lambda_n = \frac{n\pi}{L}, \quad \phi_n(x) = \sin \lambda_n x, \quad n = 1, 2, \dots;$$

and (b) subject to the (Neumann) boundary conditions $u'(0) = 0 = u'(L)$, the eigenpair is

$$\lambda_n = \frac{n\pi}{L}, \quad \phi_n(x) = \cos \lambda_n x, \quad n = 0, 1, 2, \dots.$$

Theorem 3.7. *Let the functions p, q, ρ, and p' in Eq (3.24) be real-valued and continuous on the interval $a \leq x \leq b$. Let $u_m(x)$ and $\phi_n(x)$ be the eigenfunctions of the Sturm-Liouville problem (3.24) with corresponding eigenvalues λ_m and λ_n, respectively, such that $\lambda_m \neq \lambda_n$. Then*

$$\int_a^b \phi_m(x)\, \phi_n(x)\, \rho(x)\, dx = 0, \quad m \neq n,$$

i.e., the eigenfunctions ϕ_m and ϕ_n are orthogonal with respect to the weight function $\rho(x)$ on the interval $a \leq x \leq b$.

PROOF. Let λ_n and μ_m be two different eigenvalues with the corresponding eigenfunctions ϕ_n and ψ_m, $n = 1, 2, \dots$. Then

$$(p\,\phi_n')' - q\,\phi_n + \lambda_n \rho \phi_n = 0, \quad (p\,\psi_n')' - q\,\psi_n + \mu_m \rho \psi_n = 0.$$

If we subtract ϕ_n times the second equation from ψ_m times the first equation and integrate from a to b, we get

$$\int_a^b \left[\psi_m\,(p\,\phi')' - \phi_n\,(p\,\psi_m') + (\lambda_n - \mu_m)\,\rho \phi_n \psi_m \right] dx = 0.$$

Since $\psi_m\,(p\,\phi_n')' - \phi_n\,(p\,\psi_m')' = (\psi_m p\,\phi_n' - \phi_n p\,\psi_m')$, and since ϕ_n and ψ_m are zero at both end points, then first two terms of the integrand are zero and we have

$$(\lambda_n - \mu_m) \int_a^b \rho \phi_n \psi_m\, dx = 0.$$

Thus, since $\lambda_n \neq \mu_m$, this yields $\int_a^b \rho \phi_n \psi_m\, dx = 0$, which means that the eigenfunctions corresponding to different eigenvalues are orthogonal with respect to the weight function $\rho(x)$ ∎

Example 3.17. Consider $u'' + \lambda u = 0$, $x \in (0, \pi)$, and $u(0) = 0 = u(\pi)$. For $\lambda = 0$, the problem has a trivial solution. For $\lambda < 0$ the primitive is $\{\cos\sqrt{\lambda}x, \sin\sqrt{\lambda}x\}$, and the solution of this problem is $u(x) = c_1\cos\sqrt{\lambda}x + c_2\sin\sqrt{\lambda}x$. The boundary conditions give $c_1 = 0$, and $c_2\sin\sqrt{\lambda}\pi = 0$, so $c_2 = 0$ would yield the trivial solution. Hence $c_2 \neq 0$, so $\sin\sqrt{\lambda}\pi = 0$, which gives the eigenvalues $\lambda_n = n^2$, $n = 1, 2, \ldots$, where $\lambda_n \to \infty$ as $n \to \infty$ and the corresponding eigenfunctions are $\phi_n(x) = \sin nx$. Green's function is given by

$$G(x,s) = \begin{cases} \dfrac{\sin\sqrt{\lambda}x\cos\sqrt{\lambda}s}{\sqrt{\lambda}}, & 0 \le x < s, \\[2mm] \dfrac{\sin\sqrt{\lambda}s\cos\sqrt{\lambda}x}{\sqrt{\lambda}}, & s < x \le \pi. \blacksquare \end{cases}$$

Example 3.18. (Cauchy-Euler Equation) Consider the problem

$$x^2 u'' + xu' + \lambda u = 0, \quad 1 \le x \le e; \quad u(1) = 0 = u(e).$$

This equation reduces to the Sturm-Liouville equation (3.11) which has the primitive $\{x^{i\sqrt{\lambda}}, x^{-i\sqrt{\lambda}}\} = \{\cos(\sqrt{\lambda}\ln x), \sin(\sqrt{\lambda}\ln x)\}$. Green's function for this problem can be easily determined. The solution of this problem is

$$u(x) = c_1\cos(\sqrt{\lambda}\ln x) + c_2\sin(\sqrt{\lambda}\ln x),$$

The first condition gives $c_1 = 0$, so taking $c_2 \neq 0$, we get $\sin\sqrt{\lambda} = 0$, which yields the eigenvalues $\lambda_n = n^2\pi^2$, $n = 1, 2, \ldots$; the corresponding eigenfunctions are $\phi_n(x) = \sin(n\pi\ln x)$, $n = 1, 2, \ldots$ $\blacksquare$

3.4.1. Eigenpairs. Consider the differential equation

$$L[u](x) + \lambda\rho(x)u(x) = 0, \tag{3.25}$$

where L is the Sturm-Liouville operator, λ a constant, and the weight function ρ is a known function of x such that $\rho(x) > 0$ except possibly at isolated points where $\rho(x) = 0$. For a given choice of λ, the eigenfunction $\phi_\lambda(x)$ satisfies Eq (3.25) and the prescribed boundary conditions and it corresponds to the eigenvalue λ. The existence of an eigenfunction $\phi_\lambda(x)$ does not necessarily depend on an arbitrary choice of the eigenvalue λ. This restricts the acceptable eigenvalues λ to a discrete set. A list of the values of $p(x), q(x), \rho(x)$, and λ for ordinary differential equations encountered in this book is given in Table 3.1.

Properties of Hermitian (self-adjoint) operators are as follows:

(i) Their eigenvalues are real;

(ii) Their eigenfunctions are orthogonal; and

(iii) Their eigenfunctions for a complete set. However, this property holds only for linear, second-order Sturm-Liouville differential operators. Proofs of these properties can be found in most textbooks on differential equations.

Table 3.1.

Equation	$p(x)$	$q(x)$	$\rho(x)$	λ
Legendre's	$1 - x^2$	0	1	$n(n+1)$
Associated Legendre's	$1 - x^2$	$-\dfrac{m^2}{1-x^2}$	1	$n(n+1)$
Bessel's	x	$-\dfrac{n^2}{x}$	x	a^2
Hermite's	e^{-x^2}	0	e^{-x^2}	α
Laguerre's	xe^{-x}	0	e^{-x}	α
Associated Laguerre's	$x^{k+1}e^{-x}$	0	$x^k e^{-x}$	$\alpha - k$

The eigenfunctions $\phi_n(x)$ of a self-adjoint (Hermitian) operator form a complete set. This means that any continuous or at least piecewise continuous function $f(x)$ can be approximated to any degree of accuracy by a series of the form

$$f(x) = \sum_{n=0}^{\infty} a_n \phi_n(x),$$ (3.26)

where the coefficients a_n are determined by

$$a_m = \int_a^b f(x)\phi_m(x)\rho(x)\, dx.$$ (3.27)

The orthogonality of the eigenfunctions $\phi_m(x)$ implies that only the mth term survives. Thus, the set $\{\phi_n(x)\}$ is a complete set if the limit of the mean square error is zero:

$$\lim_{m\to\infty} \int_a^b \left[f(x) - \sum_{n=0}^{m} a_n \phi_n(x) \right]^2 \rho(x)\, dx = 0.$$

3.4.2. Orthonormal Systems. A function f is said to be *normalized* with respect to the weight function ρ on the interval $a \le x \le b$ iff $\int_a^b |f(x)|^2 \rho(x)\, dx = 1$.

For example, the function $f(x) = \sqrt{\dfrac{2}{\pi}}\,\sin x$ is normalized with respect to $\rho(x) = 1$ on $[0, \pi]$ since $\int_0^\pi \left(\sqrt{\dfrac{2}{\pi}}\,\sin x \right) dx = 1$. Let $\{\phi_n\}, n = 1, 2, \ldots$, be an infinite set of functions defined on the interval $a \le x \le b$. The set $\{\phi_n\}$ forms an *orthonormal system* with respect to the weight function $\rho(x)$ on $a \le x \le b$ if (i) it is an orthogonal system with respect to the weight function $\rho(x)$ on $a \le x \le b$, and (ii) every function of this system is normalized with respect to $\rho(x)$ on $a \le x \le b$; i.e., the set $\{\phi_n\}$ is orthonormal with respect to $\rho(x)$ on $a \le x \le b$ if

$$\int_a^b \phi_m(x)\phi_n(x)\rho(x)\, dx = \begin{cases} 0, & m \neq n, \\ 1, & m = n. \end{cases}$$

For example, the set $\{\phi_n\} = \left\{\sqrt{\dfrac{2}{\pi}}\sin x\right\}$, $\rho = 1$, is an orthonormal system on $[0, \pi]$.

Consider the following boundary value problem involving the Sturm-Liouville equation:

$$[p(x)u']' + [q + \lambda\rho]\, u = 0, \quad a \leq x \leq b,$$
$$A_1 u(a) + A_2 u'(a) = 0, \quad B_1 u(b) + B_2 u'(b) = 0,$$

where A_1, A_2, B_1, B_2 are real constants such that A_1 and A_2 are not both zero and B_1 and B_2 are not both zero. Let $\{\lambda_n\}$ denote the infinite set of eigenvalues arranged in a monotone increasing sequence $\lambda_1 < \lambda_2 < \cdots$. If $\phi_n, n = 1, 2, \ldots,$ is one of the eigenfunctions corresponding to the eigenvalue λ_n for each $n = 1, 2, \ldots$, then the infinite set of eigenfunctions $\{\phi_1, \phi_2, \ldots\}$ is an orthogonal system with respect to $\rho(x)$ on $a \leq x \leq b$. If ϕ_n is an eigenfunction, then $\mu_n \phi_n$ is also an eigenfunction for arbitrary scalar $\mu_n \neq 0$ which corresponds to the eigenvalue λ_n. So we form a new set of functions $\{\mu_1\phi_1, \mu_2\phi_2, \ldots\}$ which is orthogonal with respect to $\rho(x)$ on $a \leq x \leq b$. Let this set be denoted by E. We can choose constants $\mu_1, \mu_2, \ldots$ such that every eigenfunction in the set E is also normalized with respect to $\rho(x)$ on $a \leq x \leq b$. Then the set E of eigenfunctions $\{\mu_1\phi_1, \mu_2\phi_2, \ldots\}$ will be an orthonormal system with respect to $\rho(x)$ on $a \leq x \leq b$. Now, to determine the constants $\mu_1, \mu_2, \ldots$ so that the set E of eigenfunctions becomes orthonormal with respect to $\rho(x)$ on $a \leq x \leq b$, note that no eigenfunction $\phi_n, n = 1, 2, \ldots,$ is identically zero on $a \leq x \leq b$. Let

$$\int_a^b [\phi_n(x)]^2\, \rho(x)\, dx = M_n > 0, \quad n = 1, 2, \ldots,$$

which implies that,

$$\int_a^b \left[\frac{1}{\sqrt{M_n}}\phi_n(x)\right]^2 \rho(x)\, dx = 1, \quad n = 1, 2, \ldots.$$

Hence, the set $\left\{\dfrac{1}{\sqrt{M_1}}\phi_1, \dfrac{1}{\sqrt{M_2}}\phi_2, \ldots\right\}$ is an orthonormal set with respect to $\rho(x)$ on $a \leq x \leq b$, i.e., the set $E = \{\mu_1\phi_1, \mu_2\phi_2, \ldots\}$ is orthonormal with respect to $\rho(x)$ on $a \leq x \leq b$, if μ_n is chosen as

$$\mu_n = \frac{1}{\sqrt{M_n}} = \frac{1}{\sqrt{\int_a^b [\phi_n(x)]^2\, \rho(x)\, dx}}, \quad n = 1, 2, \ldots. \tag{3.28}$$

Example 3.19. Let $\phi_n(x) = c_n \sin nx$, $n = 1, 2, \ldots$ on $0 \leq x \leq \pi$, where $c_n, n = 1, 2, \ldots,$ are nonzero constants. We will form the set of orthonormal eigenfunctions $E = \{\mu_n\phi_n\}$, where μ_n is given by (3.28). Here,

$$M_n = \int_0^\pi [c_n \sin nx]^2 (1)\, dx = \frac{c_n^2 \pi}{2},$$

$$\mu_n = \frac{1}{\sqrt{M_n}} = \frac{1}{c_n}\sqrt{\frac{2}{\pi}};$$

thus, $\mu_n \phi_n(x) = \left(\frac{1}{c_n}\sqrt{\frac{2}{\pi}}\right) c_n \sin nx = \sqrt{\frac{2}{\pi}} \sin nx,\ n = 1, 2, \dots$ ∎

3.4.3. Eigenfunction Expansion.

Assuming that an expansion of a function f in an infinite series of orthonormal eigenfunctions $\{\phi_1, \phi_2, \dots\}$ exists, let

$$f(x) = \sum_{n=1}^\infty c_n \phi_n(x), \tag{3.29}$$

for each $x \in [a, b]$. To determine the coefficients c_n, we proceed formally, and multiply both sides of (3.29) by $\phi_k(x)\rho(x)$, to obtain

$$f(x)\phi_k(x)\rho(x) = \sum_{n=1}^\infty c_n \phi_n(x)\phi_k(x)\rho(x),$$

Integrating both sides from a to b we get

$$\int_a^b f(x)\phi_k(x)\rho(x)\, dx = \int_a^b \left[\sum_{n=1}^\infty c_n \phi_n(x)\phi_k(x)\rho(x)\right] dx$$

$$= \sum_{n=1}^\infty \int_a^b c_n \phi_n(x)\phi_k(x)\rho(x)\, dx$$

$$= \sum_{n=1}^\infty c_n \int_a^b \phi_n(x)\phi_k(x)\rho(x)\, dx$$

$$= c_k,$$

since $\{\phi_n\}$ is an orthonormal set with respect to $\rho(x)$ on $a \le x \le b$. This result is true for the kth coefficient c_k for $k = 1, 2, \dots$. Hence

$$c_n = \int_a^b f(x)\phi_n(x)\rho(x)\, dx,\ n = 1, 2, \dots.$$

In fact it can be shown that the series $\sum_{n=1}^\infty c_n \phi_n(x)$ converges uniformly to $f(x)$ on $a \le x \le b$, and thus the above formal procedure is justified.

Example 3.20. Obtain the eigenfunction expansion for $f(x) = \pi x - x^2$, $0 \le x \le \pi$, in the series of orthonormal eigenfunctions $\{\phi_n\}$ of the Sturm-Liouville system $u'' + \lambda u = 0,\ u(0) = 0 = u(\pi)$. Since the set $\{\phi_n\} = \left\{\sqrt{\frac{2}{\pi}} \sin nx\right\}$, $n = 1, 2, \dots$, is orthonormal with respect to $\rho(x) = 1$ on $[0, \pi]$, let

$$f(x) = \sum_{n=1}^{\infty} c_n \phi_n(x), \tag{3.30}$$

where

$$
\begin{aligned}
c_n &= \int_0^{\pi} (\pi x - x^2) \left(\sqrt{\frac{2}{\pi}} \sin nx \right) (1)\, dx \\
&= \sqrt{\frac{2}{\pi}} \left[\pi \int_0^{\pi} x \sin nx\, dx - \int_0^{\pi} x^2 \sin nx\, dx \right] \\
&= \sqrt{\frac{2}{\pi}} \left\{ \left[\frac{\pi}{n^2} \sin nx - \frac{\pi x}{n} \cos nx \right]_0^{\pi} - \right. \\
&\qquad \left. - \left[\frac{2x}{n^2} \sin nx + \frac{2}{n^3} \cos nx - \frac{x^2}{n} \cos nx \right]_0^{\pi} \right\} \\
&= \sqrt{\frac{2}{\pi}} \left[\left(-\frac{\pi^2}{n} \cos nx \right) - \left(\frac{2}{n^3} \cos nx - \frac{\pi^2}{n} \cos nx - \frac{2}{n^3} \right) \right] \\
&= \sqrt{\frac{2}{\pi}} \frac{2}{n^3} (1 - \cos n\pi) =
\begin{cases}
\sqrt{\frac{2}{\pi}} \frac{4}{n^3} & \text{if } n \text{ is odd}, \\
0 & \text{if } n \text{ is even}.
\end{cases}
\end{aligned}
$$

Thus, the series (3.30) becomes

$$
\begin{aligned}
f(x) &= \sum_{n=1}^{\infty} \frac{8}{\pi n^3} \sin nx \ \text{ if } n \text{ is odd} \\
&= \frac{8}{\pi} \sum_{n=1}^{\infty} \frac{\sin(2n-1)x}{(2n-1)^3},
\end{aligned}
$$

and we have formally

$$\pi x - x^2 = \frac{8}{\pi} \sum_{n=1}^{\infty} \frac{\sin(2n-1)x}{(2n-1)^3}, \quad 0 \le x \le \pi.$$

It can be shown that this series converges uniformly and absolutely to $\pi x - x^2$ on the interval $0 \le x \le \pi$. ∎

The following theorem provides a bilinear eigenfunction expansion of a Green's function for the regular Sturm-Liouville system.

Theorem 3.8. *If $G(x,s)$ is Green's function for a regular Sturm-Liouville system, defined by*

$$
\begin{aligned}
L[u] &\equiv [p(x)u'(x)]' + q(x)u = f(x), \ a \le x \le b \\
a_1 u(a) &+ a_2 u'(a) = 0, \ b_1 u(b) + b_2 u'(b) = 0,
\end{aligned}
\tag{3.31}
$$

and the associated eigenvalue problem $L[\psi] = \lambda \psi$, $a \leq x \leq b$, subject to the boundary conditions $a_1\psi(a) + a_2\psi'(a) = 0$, $b_1\psi(b) + b_2\psi'(b) = 0$, has countably many eigenvalues λ_n with the corresponding eigenfunctions $\psi_n(x)$, then Green's function can be represented as an eigenfunction expansion

$$G(x,s) = \sum_{n=1}^{\infty} \frac{1}{\lambda_n} \psi_n(x)\psi_n(s). \qquad (3.32)$$

PROOF. Assume that the solution $u(x)$ of the regular Sturm-Liouville system with the forcing term $f(x)$ can be represented as an eigenfunction expansion

$$u(x) = \sum_{n=1}^{\infty} a_n\,\psi_n(x), \quad f(x) = \sum_{n=1}^{\infty} f_n\psi_n(x), \qquad (3.33)$$

where the coefficients a_n are to be determined, while the coefficients f_n are the Fourier coefficients determined from $f_n = \int_a^b \psi_n(s)\,ds$. Substituting (3.33) into the Sturm-Liouville equation (3.31), we get

$$L\left[\sum_{n=1}^{\infty} a_n\psi_n(x)\right] = \sum_{n=1}^{\infty} f_n\,\psi_n(x).$$

Since the left side of this equality is

$$L\left[\sum_{n=1}^{\infty} a_n\psi_n(x)\right] = \sum_{n=1}^{\infty} a_nL\left[\psi_n(x)\right] = \sum_{n=1}^{\infty} a_n\lambda_n\psi_n(x),$$

then equating the right side of these two equations we get

$$a_n = \frac{1}{\lambda_n}\,f_n = \frac{1}{\lambda_n}\int_a^b \psi_n(s)f(s)\,ds.$$

Hence, from (3.33) we find that

$$u(x) = \sum_{n=1}^{\infty} \frac{1}{\lambda_n}\left\{\int_a^b \psi_n(s)f(s)\,ds\right\}\psi_n(x)$$

$$= \int_a^b \left[\sum_{n=1}^{\infty} \frac{1}{\lambda_n}\psi_n(x)\psi(s)\right]f(s)\,ds = \int_a^b G(x,s)f(s)\,ds, \qquad (3.34)$$

which, in view of (3.8) with the upper limit $x = b$, yields (3.32). ∎

Example 3.21. Consider the problem $\left(x^2\,u'\right)' + \lambda u = 0$, $0 < x < 1$, such that $u(1) = 0$ and u is bounded. Green's function is

$$G(x,s) = \begin{cases} \dfrac{1}{x} & \text{for } s \leq x, \\[2mm] \dfrac{1}{s} & \text{for } s \geq x. \end{cases}$$

Since $\rho(x) = 1$, we find that $\int_0^1 \int_0^1 [G(x, s)]^2 \rho(s) \, dx \, ds = \infty$. The fact that the finiteness of the double integral in (3.42) does not hold shows that the condition (3.42) is sufficient but not necessary for completeness of the set of eigenfunctions. Moreover, the solution of the differential equation subject to the boundary condition at $x = 1$ is $u(x) = x^{-1/2} \sin\left(\sqrt{\lambda + 1/4} \, \ln x\right)$, which is unbounded, and $\int_0^1 \rho \, u^2 \, dx = +\infty$. Hence, this problem has no eigenfunctions. ∎

3.4.4. Data for Eigenvalue Problems. We provide some useful data for the solution of eigenvalue problems with the three types of boundary conditions. Consider $u'' + \lambda u = 0$, $0 < x < L$, such that $a_1 u(0) + b_1 u'(0) = 0$, $a_2 u(L) + b_2 u'(L) = 0$, and let the eigenfunction expansion of an arbitrary function $f(x)$ be

$$f(x) = \sum_{n=1}^{\infty} c_n \phi_n(x), \quad \text{where } c_n = \frac{1}{\|\phi_n\|^2} \int_a^b f(x)\phi_n(x) \, dx.$$

Set $h_i = a_i/b_i$, $i = 1, 2$. Then the nine cases for the three types of boundary conditions are as follows.

CASE 1. Dirichlet at $x = 0$ and at $x = L$: $a_1 \neq 0$, $b_1 = 0$ $a_2 \neq 0$, $b_2 = 0$. Then $\phi_n(x) = \sin \lambda_n x$; $\|\phi_n\|^2 = \dfrac{L}{2}$; and λ_n are the roots of $\sin nL = 0$, i.e., $\lambda_n = \dfrac{n\pi}{L}$, $n = 0, 1, 2, \ldots$.

CASE 2. Dirichlet at $x = 0$ and Neumann at $x = L$: $a_1 \neq 0$, $b_1 = 0$ $a_2 = 0$, $b_2 \neq 0$. Then $\phi_n(x) = \sin \lambda_n x$; $\|\phi_n\|^2 = \dfrac{L}{2}$; and λ_n are the roots of $\cos \lambda L = 0$, i.e., $\lambda_n = \dfrac{(2n - 1)\pi}{2L}$, $n = 1, 2, \ldots$.

CASE 3. Dirichlet at $x = a$ and Robin at $x = L$: $a_1 \neq 0$, $b_1 = 0$ $a_2 \neq 0$, $b_2 \neq 0$. Then $\phi_n(x) = \sin \lambda_n x$; $\|\phi_n\|^2 = \dfrac{\lambda_n L - \sin \lambda_n L \cos \lambda_n L}{2\lambda_n}$; and λ_n are the roots of $\lambda + h_2 \tan \lambda L = 0$. If $L = -\dfrac{b_2}{a_2} > 0$, then $(0, x)$ is the eigenpair.

CASE 4. Neumann at $x = 0$ and Dirichlet at $x = L$: $a_1 = 0$, $b_1 \neq 0$ $a_2 \neq 0$, $b_2 = 0$. Then $\phi_n(x) = \cos \lambda_n x$; $\|\phi_n\|^2 = \dfrac{L}{2}$; and λ_n are the roots of $\cos \lambda L = 0$, i.e., $\lambda_n = \dfrac{(2n - 1)\pi}{L}$, $n = 1, 2, \ldots$.

CASE 5. Neumann at $x = 0$ and Neumann at $x = L$: $a_1 = 0$, $b_1 \neq 0$ $a_2 = 0$, $b_2 \neq 0$. Then $\phi_n(x) = \cos \lambda_n x$; $\|\phi_n\|^2 = \dfrac{L}{2}$, where L is replaced by $2L$ for $n = 0$; and λ_n are the roots of $\sin \lambda L = 0$, i.e., $\lambda_n = \dfrac{n\pi}{L}$, $n = 0, 1, 2, \ldots$.

CASE 6. Neumann at $x = 0$ and Robin at $x = L$: $a_1 = 0$, $b_1 \neq 0$ $a_2 \neq 0$, $b_2 \neq 0$. Then $\phi_n(x) = \cos \lambda_n x$; $\|\phi_n\|^2 = \dfrac{\lambda_n L + \sin \lambda_n L \cos \lambda_n L}{2\lambda_n}$; and λ_n are the roots of $\lambda \tan \lambda L = h_2$.

CASE 7. Robin at $x = 0$ and Dirichlet at $x = L$: $a_1 \neq 0, b_1 \neq 0 \, a_2 \neq 0, b_2 = 0$.
Then $\phi_n(x) = \sin \lambda_n(L - x)$; $\|\phi_n\|^2 = \dfrac{\lambda_n L - \sin \lambda_n L \cos \lambda_n L}{2\lambda_n}$; and λ_n are the
roots of $\lambda \cot \lambda L = h_1$.

CASE 8. Robin at $x = 0$ and Neumann at $x = L$: $a_1 \neq 0, b_1 \neq 0 \, a_2 = 0$,
$b_2 \neq 0$. Then $\phi_n(x) = \cos \lambda_n(L - x)$;

$$\|\phi_n\|^2 = \frac{\lambda_n L + \sin \lambda_n L \cos \lambda_n L}{2\lambda_n};$$

and λ_n are the roots of $\lambda \tan \lambda L = -h_1$.

CASE 9. Robin at $x = 0$ and Robin at $x = L$: $a_1 \neq 0, b_1 \neq 0 \, a_2 \neq 0, b_2 \neq 0$.
Then $\phi_n(x) = \left(\lambda_n \cos \lambda_n x - h_1 \sin \lambda_n x\right)$;

$$\|\phi_n\|^2 = \frac{1}{2}\left[(\lambda_n^2 + h_1^2)\left(L + \frac{h_2}{\lambda_n^2 + h_2^2}\right) - h_1\right];$$

and λ_n are the roots of $\tan \lambda L = -\dfrac{\lambda(h_1 - h_2)}{\lambda^2 + h_1 h_2}$. If $L = \dfrac{1}{h_1} - \dfrac{1}{h_2} > 0$, then
$(0, x - 1/h_1)$ is the eigenpair.

3.5. Periodic Sturm-Liouville Systems

This system is defined by

$$\frac{d}{dx}\left[p(x)\frac{du}{dx}\right] + [q(x) + \lambda \rho(x)]\, u = 0, \quad a \leq x \leq b,$$
$$p(a) = p(b), \tag{3.35}$$
$$u(a) = u(b), \; u'(a) = u'(b).$$

In this case the solution is of period $(b - a)$.

Example 3.22. Consider the Sturm-Liouville system

$$u'' + \lambda u = 0, \quad -\pi < x < \pi; \quad u(= \pi) = u(\pi), \; u'(-\pi) = u'(\pi).$$

Here $p(x) = 1$, and so $p(-\pi) = p(\pi)$. The system is periodic, and the primitive of
solutions is $\{\cos \sqrt{\lambda} x, \sin \sqrt{\lambda} x\}$. For $\lambda > 0$, the solution is $u(x) = c_1 \cos \sqrt{\lambda} x +
c_2 \sin \sqrt{\lambda} x$, and the boundary conditions give $c_2 \sin \sqrt{\lambda}\pi = 0$, $c_1 \sin \sqrt{\lambda}\pi = 0$.
Since $c_1 \neq 0, c_2 \neq 0$, we must have $\sin \sqrt{\lambda}\pi = 0$, or $\lambda = n^2$, $n = 1, 2, \ldots$.
Thus, for $\lambda > 0$, and for every eigenvalue $\lambda_n = n^2$, there are two linearly independent eigenfunctions $u_n(x) = \cos nx$ and $\sin nx$. Further, for $\lambda = 00$, we get a
single eigenvalue $\lambda = 0$ and the corresponding eigenfunction $u(x) = 1$. Lastly, for
$\lambda < 0$, there are no eigenvalues. Hence, the eigenvalues are $\lambda_n = 0, n^2$, with the
corresponding eigenfunctions $\{1, \cos nx, \sin nx\}$. ∎

3.6. Singular Sturm-Liouville Systems

A Sturm-Liouville equation is said to be *singular* when the interval I is semi-infinite or infinite, or when the functions p and q vanish, or when one of the functions p and q become unbounded at one or both end points of the a finite interval I. A singular Sturm-Liouville equation with appropriate boundary conditions is called a *singular Sturm-Liouville system*. The boundary conditions in this system are not like those in a regular Sturm-Liouville system. The boundary conditions imposed on a singular Sturm-Liouville system differ from those imposed on the regular Sturm-Liouville systems in that they are not separated boundary conditions but are often prescribed as a bounded function $u(x)$ at the singular boundary point. For example, consider

$$\int_{a+\varepsilon}^{b} (vL[u] - uL[v]) \, dx = p(b)\big[u'(b)v(b) - u(b)v'(b)\big]$$
$$- p(a+\varepsilon)\big[u'(a+\varepsilon)v(a+\varepsilon) - u(a+\varepsilon)v'(a+\varepsilon)\big], \tag{3.36}$$

whre $\varepsilon > 0$ is a small number. When this equation is subjected to the boundary conditions

$$\lim_{x \to a^+} p(x)\big[u'(x)v(x) - u(x)v'(x)\big] = 0, \tag{3.37}$$

$$p(b)\big[u'(a+\varepsilon)v(a+\varepsilon) - u(a+\varepsilon)v'(a+\varepsilon)\big] = 0, \tag{3.38}$$

then (3.36) yields

$$\int_{a}^{b} (vL[u] - uL[v]) \, dx = 0. \tag{3.39}$$

In particular, if $p(a) = 0$, the boundary conditions (3.37) and (3.38) are replaced by the conditions: (i) $u(x)$ and $u'(x)$ are finite as $x \to a$, and (ii) $b_1 u(b) + b_2 u'(b) = 0$, where b_1, b_2 are constants. Thus, a singular Sturm-Liouville system is *self-adjoint* if the condition (3.39) holds such that the functions $u(x)$ and $v(x)$ satisfy the boundary conditions.

Consider the equation

$$\frac{d}{dx}\left[p(x)\frac{\partial u}{\partial x}\right] - q(x)\,u + \lambda\rho(x)u = 0, \quad 0 < x < 1, \tag{3.40}$$

where $p, q \in C$, $p \in C^\infty$; $p > 0, q > 0, q \geq 0$ for $0 \leq x \leq 1$. The set of eigenfunctions $\{u_n(x)\}$ is complete in the interval $[0, 1]$ in the following sense: The eigenfunctions $u_n(x)$ are orthonormal in $L_2(0, 1)$ if there is no nontrivial $u \in L_2(0, 1)$ orthogonal to all u_n, i.e., $\langle u, u_n \rangle = 0$ for $n = 0, 1, 2, \ldots$ implies that $u = 0$ almost everywhere. Thus, if $\{u_n\}$ is an orthonormal set in $L_2(0, 1)$, and if the expansion of $u \in L_2(0, 1)$ has the Fourier coefficients c_n, then $\|u - \sum_{n=0}^{N} c_n u_n\| \to 0$ as $N \to \infty$ iff $\{u_n\}$ is complete.

In many physical problems the conditions p, q, ρ and q are satisfied in the open interval $(0, 1)$ but not at one or both end points, where the functions p and ρ may approach zero or infinity, and q may become infinite (at $x = 0$). Since $u(0) = 0 = u(1)$, we require that u remain bounded at $x = 0$ and $xu' \to 0$. The integral $\int_0^1 \int_0^1 [G(x, s)]^2 \, \rho(x)\rho(s) \, dx \, ds$ is finite since Green's function $G(x, s)$ is continuous in both x and s for $0 \le x, s \le 1$.

Example 3.23. (Legendre's equation) Let the singular Sturm-Liouville system involving Legendre's equation

$$L[u] = \frac{d}{dx} \left[\left(1 - x^2\right) u' \right] + \lambda u = 0, \quad -1 < x < 1,$$

be subject to boundary conditions that both u and u' remain finite as $x \to \pm 1$. Since $p(x) = 1 - x^2$ vanishes at $x = \pm 1$, $q(x) = 0$, and $\rho(x) = 1$, this problem has the eigenvalues $\lambda_n = n(n + 1)$ and the eigenfunctions $\phi_n(x) = P_n(x)$ for $n = 0, 1, 2, \ldots$, which are the Legendre polynomials of order n, finite at $x = \pm 1$, and orthogonal to each other with respect to the weight function $\rho(x) = 1$ on $-1 \le x \le 1$;

that is. $\int_{-1}^1 P_n(x) p_m(x) \, dx = \begin{cases} 0, & n \ne m \\ \dfrac{2}{2n + 1}, & n = m \end{cases}$. The function $P_n(x)$ represents a surface zonal harmonic or a Legendre's coefficient, or simply a Legendrian, of order n. A few surface zonal harmonics are given in §C.1.1 and §C.1.2. In general, any surface zonal harmonic is obtained from the two of the preceding orders by using the formula $(n + 1)P_{n+1}(x) = (2n + 1)x P_n(x) - nP_{n-1}(x)$.

There are four different forms of surface zonal harmonics, namely,

(i) a polynomial in x, given in §C.1.1.;

(ii) an expression involving cosines of multiples of θ, where $x = \cos \theta$ (see §C.1.1);

(iii) a form involving derivative with respect to x: $P_n(x) = \dfrac{1}{2^n \, n!} \dfrac{d^n}{dx^n} \left(x^2 - 1\right)^n$; and

(iv) a form involving a definite integral: $P_n(x) = \dfrac{1}{\pi} \int_0^\pi \left(x + \sqrt{x^2 - 1} \cdot \cos s\right)^n ds$.

The problems involving surface zonal harmonics $P_n(x)$ start with the determination of the value of a potential function in terms of powers of x. Thus, a given function $f(x)$ can be developed into a Fourier series involving $P_n(x)$ of the form

$$f(x) = \sum_{n=0}^\infty A_n P_n(x), \quad -1 < x < 1, \text{ where } A_n = \frac{2n + 1}{2} \int_{-1}^1 f(x) P_n(x) \, dx$$

$$= \begin{cases} \dfrac{(2n + 1)m(m - 1)(m - 2) \cdots (m - n + 1)}{(m - n + 1)(m - n + 3) \cdots (m + n + 1)} \\ \qquad \text{if } n < m + 1 \text{ and } n + m \text{ is even,} \\ 0 \qquad \text{if } n > m \text{ or if } n + m \text{ is odd.} \end{cases}$$

The Sturm-Liouville boundary value problem involving the associated Legendre's equation of order n

$$\left[\left(1 - x^2\right) u'\right]' + \left[\lambda - \frac{m^2}{1 - x^2}\right] u = 0, \quad m = 0, 1, 2, \ldots, \quad 0 \le x \le 1,$$

has the eigenvalues $\lambda_n = n(n + 1)$ and the eigenfunctions $\phi_n(x) = P_n^m(x)$ for $n = 0, 1, 2, \ldots$, which are called associated Legendre functions of first kind and order n, defined by $P_n^m(x) = (1 - x^2)^{m/2} \dfrac{d^m}{dx^m} P_n(x)$ and known as *spherical harmonics* (see Chapter 9). If the above equation is subject to the boundary conditions that $u(0)$ and $u(1)$ are finite, Green's function for the interval $[0, 1]$ when $n = 0$ is given by

$$G(x, s) = \begin{cases} \dfrac{1}{2m} \left(\dfrac{1 + x}{1 - x} \dfrac{1 - s}{1 + s}\right)^{m/2}, & \text{if } x < s, \\[3mm] \dfrac{1}{2m} \left(\dfrac{1 + s}{1 - s} \dfrac{1 - x}{1 + x}\right)^{m/2}, & \text{if } s < x. \end{cases}$$

However, this result fails for $m = 0$, since the equation $L[u] \equiv \left[\left(1 - x^2\right) u'\right]' - \dfrac{m^2}{1 - x^2} u = 0$ has the solution $u = 1$ which is regular everywhere and satisfies the boundary conditions. ∎

Example 3.24. (Bessel's equation) Consider the singular Sturm-Liouville system involving Bessel's equation

$$\frac{d}{dx} \left(x \frac{du}{dx}\right) + \left(\lambda x - \frac{n^2}{x}\right) u = 0, \quad 0 < x < a, \quad \lambda \ge 0, \tag{3.41}$$

$$u(a) = 0, \quad \text{and } u, u' \text{ are finite as } x \to 0^+,$$

where $n \ge 0$ is constant. In this equation $p(x) = x$, $q(x) = -n^2/x$, and $\rho(x) = x$; also $p(0) = 0$, $q(x) \to -\infty$ as $x \to 0^+$, and $\rho(0) = 0$. Hence, it is a singular Sturm-Liouville system. If $\lambda = k^2$, $k = 0, 1, 2, \ldots$, the eigenfunctions are $\phi_n(x) = J_n(kx)$, $n = 1, 2, \ldots$, which are the Bessel functions of the first kind and order n; the eigenvalues are $\lambda_n = k_n^2$. Also, $J_n(0^+) < +\infty$, $J_n'(0^+) < +\infty$. The eigenfunctions are orthogonal with respect to the weight function $\rho(x) = x$. For $\lambda = 1$ Eq (3.41) is Bessel's equaion of order n. For $\lambda = 0$ Bessel's equation has two solutions: x^n and x^{-n} for $n > 0$. Green's function subject to the conditions that G is bounded and $x\, G' \to 0$ at $x = 0$, and $u(1) = 0$, is given by

$$G(x, s) = \begin{cases} \dfrac{x^n \left(s^{-n} - s^n\right)}{2n} & \text{for } x < s, \\[3mm] \dfrac{s^n \left(x^{-n} - x^n\right)}{2n} & \text{for } s < x. \end{cases} \tag{3.42}$$

This Green's function is not bounded and approaches infinity as $x \to 0$. Hence, the convergence is uniform in every interval $\varepsilon \le x \le 1$, where $\varepsilon > 0$, but not in the whole interval $[0,1]$. Note that Green's function $G(x,s)$ does not depend on $\rho(x)$. For $\lambda = 1$ and $n = 0$ Eq (3.41) reduces to Fourier's equation which is $\dfrac{d^2u}{dx^2} + \dfrac{1}{x}\dfrac{du}{dx} + u = 0$ with general solution $u = AJ_0(x) + BK_0(x)$, where the function $J_0(x)$ is called a *cylindrical harmonic* or Bessel function of order zero, and $K_0(x)$ another *cylindrical harmonic* or Bessel function of the second kind and of order zero. In general, the function $u = J_n(x)$ is a particular solution of Bessel's equation (3.41) and is called a *cylindrical harmonic* or Bessel function of order n, and unless n is an integer, we have $u = AJ_n(x) + BJ_{-n}(x)$ as the general solution of Bessel's equation (3.41). If n is an integer, then $J_n(x) = (-1)^n J_{-n}(x)$, and then $u = AJ_n(x) + BK_n(x)$ is the general solution of Bessel's equation (3.41). For $\lambda = 0$ and $n = 0$ Eq (3.41) reduces to $xu'' + u' = 0$, which has linearly independent solutions 1 and $-\ln x$, which with $p(x) = x$ and $W(x) = -1/x \neq 0$ for $x \neq 0$ gives Green's function in this case is the same as in Example 3.14. ∎

Example 3.25. (Hermite's equation) Consider Hermite's equation

$$u'' - 2xu' + \lambda u = 0, \quad -\infty < x < \infty, \tag{3.43}$$

which is not self-adjoint. If we set $v(x) = e^{-x^2/2} u(x)$, Eq (3.43) becomes

$$v'' + \left[(1 - x^2) + \lambda\right] v = 0, \quad -\infty < x < \infty, \tag{3.44}$$

which is self-adjoint. The eigenvalues of Eq (3.44) are $\lambda_n = 2n$ for nonnegative integers n, and the corresponding eigenfunctions are $\phi_n(x) = e^{-x^2/2} H_n(x)$, where $H_n(x)$ are the Hermite polynomials of order n, which are solutions of Eq (3.43). If we impose the boundary conditions $v \to 0$ as $x \to \pm\infty$, the the polynomials $H_n(x)$ satisfy these conditions because $x^n e^{-x^2/2} \to 0$ as $x \to \pm\infty$. Since $H_n(x)$ are square-integrable, we have the orthogonality relation

$$\int_{-\infty}^{\infty} H_n(x)H_k(x) e^{-x^2/2}\, dx = \begin{cases} 0, & n \neq k, \\ 2^n n!\, \sqrt{\pi}, & n = k. \end{cases}$$

From Eqs (3.43) and (3.34) it is obvious that the Hermite polynomials $u(x) = H_n(x)$ and the corresponding orthogonal function $v(x) = e^{-x^2/2} H_n(x)$ are characterized as solutions of the eigenvalue problem

$$\left(e^{-x^2/2} u'\right)' + \lambda e^{-x^2/2} u = 0, \tag{3.45}$$

and Eq (3.44) respectively, where the eigenvalues are $\lambda = 0, 2, 4, \ldots, -\infty < x < \infty$, and subject to the boundary condition that the eigenfunctions u must remain finite at both $x = \pm\infty$ of order that of a finite power of x. If the series solution of Eq (3.45) is taken as $u = \sum\limits_{n=1}^{\infty} a_n x^n$, then the recurrence relation is $a_{n+2} = \dfrac{2n - \lambda}{(n + 1)(n + 2)} a_n$. Hence, either the series breaks off in the case when $\lambda = 2n$ is a non-negative even integer and thus represents the Hermite polynomials $H_n(x)$, or the series has infinitely many nonvanishing coefficients and converges for all x. All the coefficients a_n which occur as soon as $(2n - \lambda)$ becomes positive have the same sign. In the second case the terms $a_n x^n$ occur for arbitrary large n. Thus, $u(\pm\infty)$ becomes infinite of an order greater than any finite power of x. Hence, u cannot be an eigenfunction for the problem and the Hermite polynomials are the only solutions of the above eigenvalue problem.

For $\lambda = 0$ the function $u(x) = e^{-x^2/2}$ is a solution of Eq (3.43) in u instead of v, i.e., of $L[u] \equiv u'' + (1 - x^2) u = 0$. Assuming a solution of the form $u = w e^{-x^2/2}$, we obtain $w'' + 2w'x = 0$, which has the solutions $w = \text{const}$ and $w = c_1 \int_{c_2}^{x} e^{-t^2} \, dt$. Thus, $u(x) = c_1 e^{-x^2/2} \int_{c_2}^{x} e^{-t^2} \, dt$. Hence, the particular solutions which vanish at $x = \pm\infty$ are given by $g_1(x) = a\, e^{-x^2/2} \int_{x}^{\infty} e^{-t^2} \, dt$ and

$$g_2(x) = b\, e^{-x^2/2} \int_{-\infty}^{x} e^{-t^2} \, dt,$$ respectively. Since $p(x) = 1$ and $W(x) = ab\sqrt{\pi}$, using (3.23) we find that Green's function is given by

$$
G(x,s) = \begin{cases} \dfrac{1}{\sqrt{\pi}} e^{-(x^2+s^2)/2} \displaystyle\int_{-\infty}^{x} e^{-t^2} \, dt \int_{s}^{\infty} e^{-t^2} \, dt, & \text{if } x \leq s, \\[3ex] \dfrac{1}{\sqrt{\pi}} e^{-(x^2+s^2)/2} \displaystyle\int_{-\infty}^{s} e^{-t^2} \, dt \int_{x}^{\infty} e^{-t^2} \, dt, & \text{if } x > s, \end{cases}
$$

where we have used $\int_{-\infty}^{\infty} e^{-t^2} \, dt = \sqrt{\pi}$ (Courant and Hilbert [1968:374]). ∎

Example 3.26. (Laguerre's equation) Laguerre's equation is

$$xu'' + (1 - x)u' + nu = 0, \quad n = 0, 1, 2, \ldots,$$

and its solutions are the Laguerre polynomials $u = L_n(x) = e^x \dfrac{d^n}{dx^n} (x^n e^{-x})$. The eigenvalue equation for the Laguerre polynomials $L_n(x)$ is

$$xu'' + (1 - x)u' + \lambda u = 0, \tag{3.46}$$

with eigenvalues $\lambda = n$, where n is a positive integer. In self-adjoint form this equation is

$$\left(x\, e^{-x}\, u'\right)' + \lambda e^{-x} u = 0, \tag{3.47}$$

subject to the boundary conditions that u remains finite at $x = 0$ and $u(\infty)$ does not become infinite of an order greater than a positive power of x. For the associated orthogonal functions $v = e^{-x/2} L_n(x)$, we have the Sturm-Liouville eigenvalue equation

$$(xv')' + \left(\frac{1}{2} - \frac{x}{4}\right) v + \lambda v = 0, \tag{3.48}$$

where we require regularity at $x = 0$. Note that $w = x^{-1/2}L_n$ satisfies the self-adjoint eigenvalue equation

$$\left(x^2 w'\right)' - \frac{x^2 - 2x - 1}{4} w + \lambda x w = 0,$$

where this solution must vanish at $x = 0$. The corresponding eigenvalues are the positive integers $\lambda = n$, and it leads to the associated Laguerre functions of the first kind $u = L_n^m(x) = \dfrac{d^m}{dx^m} L_n(x)$.

The Laguerre polynomials $e^{-x/2} L_n(x)$ are solutions of the equation (3.48) in u instead of v:

$$xu'' + u' + \left(\frac{1}{2} - \frac{x}{4}\right) + \lambda u = 0.$$

For a particular value $\lambda = -1$, we define $L[u] = xu'' + u' - \left(\dfrac{1}{2} + \dfrac{x}{4}\right) u$. Then the equation $L[u] = 0$ has the particular solution $u = e^{x/2}$. Assuming the general solution of the form $u = w\, e^{x/2}$, we get $w = c_1 \displaystyle\int_{c_2}^x \frac{e^{-t}}{t}\, dt$. The two particular solutions which are regular at $x = 0$ and vanish at $x = \pm\infty$ are $g_1(x) = be^{x/2} \displaystyle\int_x^\infty \frac{e^{-t}}{t}\, dt$ and $g_2(x) = ae^{x/2}$, respectively. Since $W(x) = ab/x$ and $p(x) = x$, using (3.23) Green's function for this problem is

$$G(x, s) = \begin{cases} e^{(x+s)/2} \displaystyle\int_s^\infty \dfrac{e^{-t}}{t}\, dt, & \text{if } x < s, \\[2ex] e^{(x+s)/2} \displaystyle\int_x^\infty \dfrac{e^{-t}}{t}\, dt, & \text{if } s < x. \blacksquare \end{cases}$$

Example 3.27. No Green's function exists for the equation $L[u] = u''$ in $-\infty < x < \infty$, subject to the condition that u remain finite for all x, because the homogeneous equation $u'' = 0$ has solution $u = \text{const}$, which is regular at $x = \pm\infty$. $\blacksquare$

3.7. Exercises

3.1. Find one-sided Green's function for the following initial value problems:

(a) $u^{(n)} = 0$; (b) $t^2 u'' + 2tu' - 2u = 0$, $t > 0$; (c) $u''' + 2u'' - u' - 2u = 0$.

(d) $t^2 u'' + 4t\, u' + \left(2 + t^2\right) u = f(t), t > 1$.

ANS. (a) $g(t, s) = \dfrac{(t - s)^{n-1}}{(n - 1)!}$; (b) $g(t.s) = \dfrac{t^3 - s^3}{3t^2}$;

(c) $g(t - s) = \frac{1}{6}\, e^{t-s} - \frac{1}{2}\, e^{-}(t - s) + \frac{1}{3}\, e^{-2(t-s)}$;

(d) HINT. Primitive: $\{t^{-2}\cos t, t^{-2}\sin t\}$. ANS. $g(t,s) = H(t-1)\dfrac{\sin(t-s)}{t^2}$.

3.2. Determine Green's Function and the solution for the following initial value problems:

(1) $u'' + \dfrac{1}{x}u' - \dfrac{1}{x^2}u = f(x)$, $u(1) = 0$, $u'(1) = 1$;

(2) $u'' + \dfrac{1}{4x^2}u = f(x)$, $u(0) = u'(0) = 1/2$;

(3) $x^2 u'' - 7xu' + 15u = f(x)$, $u(1) = 0 = u'(1)$;

(4) $x^2 u'' - xu' + u = 0$, $u(0) = 0$, $u'(0) = 1$;

(5) $xu'' - (x+1)u' + u = f(x), u(1) = -1 = u'(1)$;

(6) $x^2\left(1-x^2\right)u'' - 2xu' + 2u = 0$, $u(1/2) = 0, u'(1/2) = 1$;

(7) $u'' + u' - 2u = e^x$, $u(0) = 1, u'(0) = 0$;

(8) $x^2 u'' + 4xu' + \left(2+x^2\right)u = f(x)$, $x > 0$, $u(\pi/2) = 1 = u'(\pi/2)$;

(9) $\left(3x^2 - 1\right)u'' + 3(3x - 1)u' - 9u = f(x)$, $x > 1/3$, $u(1) = 0 = u'(1)$;

(10) $u'' - \dfrac{2}{x^2}u = f(x)$, $0 < x < \infty$;

HINT. Primitives are: (1) $\{x, x^{-1}\}$; (2) $\{x^{1/2}, x^{1/2}\ln x\}$; (3) $\{x^3, x^5\}$;

(4) $\{x, x\ln x\}$; (5) $\{e^x, x+1\}$; (6) $\left\{x, \dfrac{x}{2}\ln\dfrac{1+x}{1-x}\right\}$; (7) $\{e^x, e^{-2x}\}$;

(8) $\{\cos x/x^2, \sin x/x^2\}$; (9) $\{3x - 1, (3x-1)^{-1}\}$; (10) $\{x^2, x^{-1}\}$.

3.3. Solve the initial value problem $u'' - \dfrac{2}{x^2}u = \dfrac{1}{x}$, $u(1) = 0, u'(1) = 1$.

HINT. Primitive: $\{x^2, x^{-1}\}$; $W(x) = -3$; $g(x,s) = \frac{1}{3}\left(x^2 s^{-1} - x^{-1}s^2\right)$.

ANS. $u_p(x) = -\dfrac{x}{2}$; $u(x) = \frac{2}{3}x^2 - \frac{1}{6}x^{-1} - \dfrac{x}{2}$.

3.4. Obtain one-sided Green's functions for the initial value problem $\ddot{x}(t) + k^2 x(t) = 0$ such that (a) $x(0) = x_0$, $\dot{x}(0) = 0$, $t > 0$, and (b) $x(0) = x_0$, $\dot{x}(0) = u_0$, $t > 0$. This problem describes the free motion of a mass m which is attracted by a force proportional to its distance from the origin and (a) starts from rest at $x = x_0$, and (b) starts with an initial velocity u_0 at $x = x_0$ away from the origin.

HINT. The primitive is $\{\cos kt, \sin kt\}$, and in both cases one-sided Green's function is $g(t,s) = H(t)\sin k(t-s)$.

3.5. Consider the problem of vibration of a spring which hangs vertically with its upper end fixed, and a mass of 16 lb is attached to its lower end. After coming to rest the mass is pulled down 2 inches and released. Assuming that the medium offers a resistance of $v/64$ lb, where v is expressed in ft/sec, the equation of motion is $\dfrac{16}{g} D^2 x(t) + \dfrac{1}{64} Dx(t) + 48x(t) = 0$, and the initial conditions are: $x(0) = 0$ and $Dx(1/16) = 0$. Using $g = 32$, show that Green's function is

$$g(t,s) \approx \frac{H(t)}{a_0(s)W(s)}\left[v_1(t)\,v_2(s) - v_1(s)\,v_2(t)\right],$$

where $v_1(t) = \frac{1}{6}e^{-0.0156t}\cos 9.8t$ and $v_2(t) = 0.000265e^{-0.0156t}\sin 9.8t$.

HINT. $a_0(t) = 0.5,\ W(v_1, v_2) = \dfrac{0.000265(9.8)}{6}\,e^{-19.8t}$.

3.6. Show that if $u_1(x)$ and $u_2(x)$ are linearly independent solutions of the homogeneous equation $[p(x)\,u']' + q(x)\,u = 0$, then the Wronskian $W\,(u_1(x), u_2(x))$ is a constant multiple of $1/p(x)$.

HINT. Differentiate $p(x)\,W\,(u_1(x), u_2(x))$ with respect to x and show that $p(x)\,W\,(u_1(x), u_2(x)) = \text{const}$.

3.7. Consider the Sturm-Liouville equation $L[u] = u'' = 0, 0 < x < 1$, subject to the boundary conditions: (a) $u(0) = 0 = u(1)$; (b) $u(0) = 0,\ u'(1) = 0$; (c) $u(-1) = 0 = u(1)$; and (d) $u(0) = u(1),\ u'(0) = -u'(1)$. Determine Green's function in each case.

ANS. (a) $G(x, s) = \begin{cases} (1-s)\,x, & \text{if } x \le s, \\ (1-x)\,s, & \text{if } x > s. \end{cases}$; (b) $G(x, s) = \begin{cases} x, & \text{if } x < s, \\ s, & \text{if } s < x. \end{cases}$

(c) $G(x, s) = -\frac{1}{2}\left\{|x - s| + xs - 1\right\}$; (d) $G(x, s) = -\frac{1}{2}\,|x - s| + \frac{1}{4}$.

3.8. Find the eigenpairs of the following Sturm-Liouville problems:

(a) $u'' + \lambda u = 0,\ 0 < x < 1;\ u(0) = 0, u(1) + u'(1) = 0$;

(b) $u'' + \lambda u = 0,\ 0 < x < 1;\ u'(0) + u'(1) = 0$;

(c) $u'' + \lambda u = 0,\ 0 < x < 1;\ u(0) = u(1), u'(0) + u'(1) = 0$;

(d) $u'' + 2u' + 4\lambda u = 0,\ 0 < x < a;\ u(0) = 0 = u(a)$.

ANS. (a) For $\lambda > 0$: $\lambda = k_n^2 \approx \frac{1}{4}\,(2n+1)^2\pi^2;\ u_n(x) = \sin k_n x,\ n = 1, 2, \dots$;

(b) For $\lambda_n = n^2\pi^2;\ u_n(x) = \cos n\pi x,\ n = 0, 1, 2, \dots$;

(c) For $\lambda_n = 4n^2\pi^2;\ u_n(x) = \{\sin 2n\pi x, \cos 2n\pi x\},\ n = 0, 1, 2, \dots$;

(d) For $1 - 4\lambda = -k_n^2 < 0$: $\lambda_n = \frac{1}{4}\,(1 + n^2\pi^2/a^2);\ u_n(x) = e^{-x}\sin(n\pi x/a)$, ds $n = 1, 2, \dots$.

3.9. Show that Green's function for 1-D potential equation[1], defined by

$$\frac{d}{dx}\left[a\,\frac{du}{dx}\right] = f(x), \quad 0 \le x \le l,$$

---

[1] This equation is found in problems of transverse deflection of a cable, axial deformation of a bar, heat transfer along a fin in heat exchangers, flow though pipes, laminar incompressible flow through a channel under constant pressure gradient, linear flow through porous media, and electrostatics.

in a homogeneous isotropic medium ($a = $ const), is given by

$$G(x,s) = \frac{1}{2a}(l-r), \quad r = |x-s|.$$

HINT. Solve $a\dfrac{d^2G}{dx^2} = \delta(x,s)$. Green's function is also the fundamental solution for this potential operator.

3.10. Solve: $\left[(x+1)^2 u'\right]' - u = f(x), \quad 0 < x < 1,$ subject to the boundary conditions $u(0) = 0 = u'(1)$.

ANS. $u(x) = \dfrac{1}{\sqrt{5}}\displaystyle\int_0^x (1+x)^{-1/2}(1+s)^{-1/2}\left\{\left(\dfrac{1+x}{1+s}\right)^{\sqrt{5}/2} - \left(\dfrac{1+s}{1+x}\right)^{\sqrt{5}/2}\right\}$

$$\cdot f(s)\,ds + \frac{1}{\sqrt{5(1+x)}}\left[(1+x)^{\sqrt{5}/2} - (1+x)^{-\sqrt{5}/2}\right].$$

3.11. Solve $x^2 u'' + xu' + u = \ln x, \quad x > 1,$ subject to the conditions $u(1) = 0,\ u'(1) = 0.$ HINT. $x^{\pm i} = \cos(\ln x) \pm i\,\sin(\ln x).$ Find a particular solution by inspection.   ANS. $u(x) = \ln x.$

3.12. Solve $\sqrt{1+x^2}\,u' + u = x, \ x > 0,$ subject to the condition $u(0) = 0.$

HINT. Reduce the equation to the form $[p(x)\,u]' = f(x).$

ANS. $u(x) = \dfrac{1}{2}\left[x - \left(x + \sqrt{1+x^2}\right)^{-1}\ln\left(x + \sqrt{1+x^2}\right)\right].$

3.13. Solve $u'' - u = e^x,\ 0 < x < 1,$ such that $u(0) = 1,\ u'(1) = 0.$

ANS. $u(x) = \dfrac{1}{2}\,x\,e^x - \dfrac{e\,\sinh x}{\cosh 1} + \dfrac{\cosh(1-x)}{\cosh 1}.$

HINT. The primitive of solutions is $\{e^x, e^{-x}\}$, and Green's function is

$$G(x,s) = \begin{cases} \dfrac{\sinh(x-1)\,\sinh s}{\cosh 1}, & x < s, \\[2mm] \dfrac{\sinh x\,\sinh(s-1)}{\cosh 1}, & s < x. \end{cases}$$

3.14. Show that the initial value problem $\dot{x} = x^{1/3},\ x(0) = 0$ has infinitely many solutions passing through the origin of the (x,t)-plane,

ANS. the solutions are given by $x_\nu(t) = \begin{cases} 0, & 0 \le t \le \nu, \\[1mm] \left[\frac{2}{3}(t-\nu)\right]^{3/2}, & \nu < t \le 1, \end{cases}$

where $\nu \in [0,1]$ is a real number.

3.15. The homogeneous Sturm-Liouville equation $\dfrac{d}{x}\left[p(x)\dfrac{du}{dx}\right] = 0$ is satisfied by $\dfrac{du}{dx} = \dfrac{1}{p(x)}.$ Use this to obtain a second solution for Legendre's equation.

ANS. $u_2(x) = \dfrac{1}{2} \ln \dfrac{1+x}{1-x}$.

3.16. Determine Green's function for the Sturm-Liouville system $\dfrac{d^2 u}{dx^2} = 0, u(0) =$
$0 = u'(1), 0 < x < 1$. ANS. $G(x,s) = \begin{cases} x, & x < s, \\ s, & s < x \end{cases}$.

3.17. Find Green's function for the system

$$\dfrac{d^2 u(x)}{dx^2} - k^2 u(x) + \lambda \dfrac{e^{-x}}{x} u(x) = 0, \ u(0) = 0 = u(\infty), \ 0 < x < \infty.$$

ANS. $G(x,s) = \begin{cases} \dfrac{1}{k} e^{-ks} \sinh kx, & x < s, \\ \dfrac{1}{k} e^{-ks} \sinh kx, & s < x. \end{cases}$

3.18. Obtain the expansion for the function $f(x) = 1, 1 \le x \le e^{\pi}$, in a series of eigenfunctions $\{\phi_n\}$ of the Sturm-Liouville system $[xu']' + \dfrac{\lambda}{x} u = 0, \ u(1) = 0 = u(e^{\pi})$.

ANS. $1 = \dfrac{2}{\pi} \displaystyle\sum_{n=1}^{\infty} \dfrac{1 - (-1)^n}{n} \sin(n \ln |x|) = \dfrac{4}{\pi} \displaystyle\sum_{n=1}^{\infty} \dfrac{\sin[(2n-1)\ln|x|]}{2n-1}$.

4

Bernoulli's Separation Method

In the study of linear partial differential equations and determination of Green's functions there are two methods that are mostly used. They are: Bernoulli's separation method, commonly known as the method of separation of variables, and the integral transform method. In this chapter we will present Bernoulli's separation method by means of some useful examples, which will explain the working of the method in detail.

4.1. Coordinate Systems

Let a point P be located by the polar cylindrical coordinates (r, θ, z) as well as rectangular cartesian coordinates (x, y, z) (see Fig. 4.1).

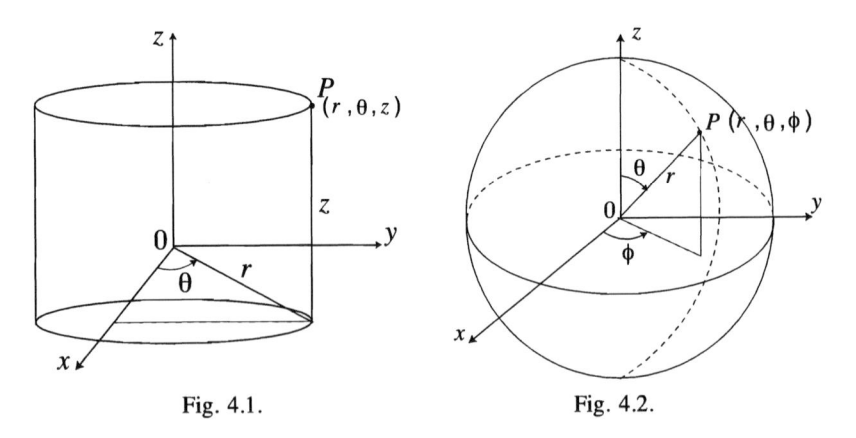

Fig. 4.1.　　　　　　　　　　Fig. 4.2.

The transformation between these two coordinate systems is defined by

$$
\begin{array}{ll}
x = r\cos\theta & r = \sqrt{x^2 + y^2} \\
y = r\sin\theta \quad \text{or} & \theta = \arctan(y/x) \\
z = z & z = z.
\end{array}
$$

Similarly, the point $P(x, y, z)$ is defined in the spherical coordinate system (r, θ, ϕ) by the transformations (see Fig. 4.2)

$$
\begin{aligned}
x &= r \sin \theta \cos \phi \\
y &= r \sin \theta \sin \phi \\
z &= r \cos \theta
\end{aligned}
\qquad \text{or} \qquad
\begin{aligned}
r &= \sqrt{x^2 + y^2 + z^2} \\
\theta &= \arccos(z/r) \\
\phi &= \arctan(y/x).
\end{aligned}
$$

We will study linear partial differential equations of the form

$$
\nabla^2 u + k^2 u = 0, \tag{4.1}
$$

where the 3-D Laplacian ∇^2 is expressed in different coordinate systems as follows:

In rectangular cartesian : $\dfrac{\partial^2}{\partial x^2} + \dfrac{\partial^2}{\partial y^2} + \dfrac{\partial^2}{\partial z^2}$;

In polar cylindrical : $\dfrac{\partial^2}{\partial r^2} + \dfrac{1}{r}\dfrac{\partial}{\partial r} + \dfrac{1}{r^2}\dfrac{\partial^2}{\partial \theta^2} + \dfrac{\partial^2}{\partial z^2}$;

In spherical : $\dfrac{1}{r^2}\dfrac{\partial}{\partial r}\left(r^2 \dfrac{\partial}{\partial r}\right) + \dfrac{1}{\sin \theta}\dfrac{\partial}{\partial \theta}\left(\sin \theta \dfrac{\partial}{\partial \theta}\right) + \dfrac{1}{\sin^2 \theta}\dfrac{\partial^2}{\partial \phi^2}$.

We will analyze the cases in which Eq (4.1) is separable. Thus,

if $k^2 = 0$, then Eq (4.1) is Laplace's equation;

if $k^2 = $ a positive constant, then Eq (4.1) is Helmholtz's equation;

if $k^2 = $ a negative constant, then Eq (4.1) is diffusion equation;

if $k^2 = $ constant $\times$ kinetic energy, then Eq (4.1) is Schrödinger's wave equation.

4.2. Partial Differential Equations

We will discuss the classification of second-order nonhomogeneous partial differential equations of the form

$$
a_{11}\, u_{xx} + 2a_{12}\, u_{xy} + a_{22}\, u_{yy} + b_1\, u_x + b_2\, u_y + c_0\, u = f(x, y). \tag{4.2}
$$

If $f = 0$ in Eq (4.2), the most general form of a second-order homogeneous equation is

$$
a_{11}\, u_{xx} + 2a_{12}\, u_{xy} + a_{22}\, u_{yy} + b_1\, u_x + b_2\, u_y + c_0\, u = 0. \tag{4.3}
$$

To show a correspondence of this equation with an algebraic quadratic equation, we replace u_x by α, u_y by β, u_{xx} by α^2, u_{xy} by $\alpha\beta$, and u_{yy} by β^2. Then the left side of Eq (4.3) reduces to a second degree polynomial in α and β:

$$
P(\alpha, \beta) = a_{11}\alpha^2 + 2a_{12}\alpha\beta + a_{22}\beta^2 + b_1\alpha + b_2\beta + c_0.
$$

It is known from analytical geometry and algebra that the polynomial equation $P(\alpha, \beta) = 0$ represents a *hyperbola, parabola* or *ellipse* according as its discriminant $a_{12}^2 - a_{11}a_{22}$ is positive, zero or negative. Thus, Eq (4.3) is classified as hyperbolic, parabolic or elliptic according as the quantity $a_{12}^2 - a_{11}a_{22}$ is positive, zero or negative.

An alternative approach to classify the types of Eq (4.3) is based on the following theorem:

Theorem 4.1. *The relation $\phi(x, y) = C$ is a general integral of the ordinary differential equation*

$$a_{11}\, dy^2 - 2a_{12}\, dx\, dy + a_{22}\, dx^2 = 0 \qquad (4.4)$$

iff $u = \phi(x, y)$ is a particular solution of the equation

$$a_{11}\, u_x^2 + 2a_{12}\, u_x\, u_y + a_{22}\, u_y^2 = 0. \qquad (4.5)$$

Eq (4.4), or (4.5), is called the *characteristic equation* of the partial differential equation (4.2) or (4.3); the related integrals are called the *characteristics*. Eq (4.4), regarded as a quadratic equation in dy/dx, yields two solutions:

$$\frac{dy}{dx} = \frac{a_{12} \pm \sqrt{a_{12}^2 - a_{11}\, a_{22}}}{a_{11}}.$$

The expression under the radical sign determines the type of the differential equation (4.2) or (4.3). Thus, as before, Eq (4.2) or (4.3) is of the hyperbolic, parabolic or elliptic type according as the quantity $a_{12}^2 - a_{11}a_{22} \gtreqless 0$.

Most frequently encountered partial differential equations of mathematical physics are:

1. Laplalce's equation: $\nabla^2 u = 0$. This equation occurs in the study of (i) electromagnetics including electrostatics, dielectrics, steady currents, and magnetostatics; (ii) hydrodynamics (irrotational flow of perfect fluids and surface waves); (iii) heat flow; and (iv) gravitation.

2. Poisson's equation: $\nabla^2 u = -\rho/\varepsilon_0$.

3. Helmholtz's (wave) and time-dependent diffusion equations: $\nabla^2 u \pm k^2 u = 0$. These equations occur in (i) elastic waves in solids including vibrating strings, bars, membranes; (ii) acoustics; (iii) electromagnetic waves; and (iv) nuclear reactors.

4. The time-dependent diffusion equation: $\nabla^2 u = \dfrac{1}{a^2}\dfrac{\partial u}{\partial t}$, and the corresponding 4-D analog of the Laplacian in Minkowski space: $\Box^2 = \nabla^2 + \dfrac{\partial^2}{\partial x_4^2} = \dfrac{\partial^2}{\partial x^2} + \dfrac{\partial^2}{\partial y^2} + \dfrac{\partial^2}{\partial z^2} + \dfrac{\partial^2}{(ic)^2 \partial t^2}$.

5. The time-dependent wave equation: $\Box^2 u = 0$, where $\Box = \dfrac{\partial^2}{\partial t^2} - c^2 \nabla^2$.

6. The scalar potential equation: $\Box^2 u = -\rho/\varepsilon_0$.

7. The Klein-Gordon equation: $\Box^2 u = \mu^2 u$, and the associated vector equation in which the scalar function u is replaced by a vector function.

8. Schrödinger's wave equation: $-\dfrac{\hbar^2}{2m} \nabla^2 u + Vu = i\,\hbar \dfrac{\partial u}{\partial t}$, and the time-independent equation: $-\dfrac{\hbar^2}{2m} \nabla^2 u + Vu = Eu$.

9. Equations for elastic waves and for viscous fluids, and telegrapher's equation.

10. Maxwell's coupled equations for electric and magnetic fields, and Dirac's equations for relativistic electron wave equations.

All these equations can be written in the form $L[u] = F$, where L is the linear differential operator $L\left(\dfrac{\partial}{\partial x}, \dfrac{\partial}{\partial y}, \dfrac{\partial}{\partial z}, \dfrac{\partial}{\partial t}, x, y, z \right)$, F is a known function, and u is the unknown scalar (or vector) function. Also, all these equations are second-order partial differential equations except Maxwell's and Dirac's equations which are first-order but involve two unknown functions. Sometimes higher order equations are found, as in the slow motion of a viscous fluid and the theory of an elastic body which are governed by the fourth-order equation $(\nabla^2)^2 u = \left(\dfrac{\partial^4}{\partial x^4} + 2\dfrac{\partial^4}{\partial x^2 \partial y^2} + \dfrac{\partial^4}{\partial y^4} \right) u = 0$. The techniques mostly used to solve these equations are: (i) Bernoulli's separation method which separates the equations into ordinary differential equations; (ii) integral transform methods; (iii) integral solutions using Green's functions; and (iv) numerical computation. We will henceforth study different methods based on these techniques.

A important aspect of Green's functions in the case of partial differential equation is similar to the case of ordinary differential equations where Green's function method is used to determine the solution of a given linear partial differential equation $L[u](x) = f(x)$, as proved in the following theorem.

Theorem 4.2. *If $f(x)$ is a continuous or piecewise continuous function of x, $x_0 \le x \le x_1$, then the function*

$$u(x) = \int_{x_0}^{x_1} G(x, x') f(x')\, dx' \tag{4.6}$$

is a solution of the equation $L[u] = -f(x)$, which satisfies the prescribed boundary conditions.

PROOF. Let $u(x) = \displaystyle\int_{x_0}^{x_1} G(x, x') f(x')\, dx'$. Then

$$u'(x) = \int_{x_0}^{x_1} G'(x, x') f(x')\, dx',$$

$$u''(x) = \int_{x_0}^{x_1} G''(x, x') f(x')\, dx' + \int_{x_0}^{x_1} G''(x, x') f(x')\, dx'$$

$$+ G'(x, x - 0) f(x) - G'(x, x + 0) f(x)$$

$$= \int_{x_0}^{x_1} G''(x, x') f(x')\, dx' + [G'(x, x + 0) - G'(x, x - 0)]\, f(x)$$

$$= \int_{x_0}^{x_1} G''(x, x') f(x')\, dx' - \frac{f(x)}{p(x)}.$$

Thus, $pu'' + p'u' - qu = \int_{x_0}^{x_1} (pg'' + p'G' - qG)\, f(x')\, dx' - f(x)$. Since Green's function $G(x, x')$ satisfies the equation $L[G(x, x')] = \delta(x - x')$, we find by using shifting property (2.10) of the δ-function that $L[u] = \int_{x_0}^{x_1} f(s) L[G(x, s)]\, ds = \int_{x_0}^{x_1} f(x) \delta(x - s)\, ds = f(x)$.

To prove the converse in 2-D case, let D denote a domain with a rectifiable boundary ∂D, and let $L[u]$ be any of the partial differential operators listed in §4.2. In particular, we consider the boundary value problem: $L[u] \equiv \nabla^2 u = f(x, y)$ subject to the boundary conditions $u\big|_{(x,y)\in\partial D} = 0$. If B represents the region $(x - x')^2 + (y - y')^2 \leq \varepsilon^2$, then $\lim\limits_{\varepsilon \to 0} \iint_B \nabla^2 G(x, y; x', y')\, dx\, dy = -1$. We use Green's second identity (1.16) with $f = 1$ and $g = G(x, x')$, and obtain

$$\lim_{\varepsilon \to 0} \iint_B \nabla^2 G(x, y; x', y')\, dx\, dy = \oint_{\partial B} \frac{\partial G}{\partial n}\, ds,$$

where n is the direction of the outward normal to the boundary ∂B. Obviously, n can be replaced by $r = \sqrt{(x - x')^2 + (y - y')^2}$, and we have

$$\lim_{\varepsilon \to 0} \oint_{C_\varepsilon} \frac{\partial G(x, y; x', y')}{\partial r}\, ds = -1,$$

where C_ε is the circle with radius ε and center at (x', y'). In terms of the polar cylindrical coordinates: $x = x' + r\cos\theta$, $y = y' + r\sin\theta$, with $ds = \sqrt{r^2 + r'^2}\, d\theta$, we find that $\oint_{C_\varepsilon} \frac{\partial G}{\partial r}\, ds = \int_0^{2\pi} \frac{\varepsilon\, d\theta}{\varepsilon} = 2\pi$. Thus, if we choose for $G(x, y; x', y')$ a function such that its derivative has a singularity of the type $-\dfrac{1}{2\pi r}$, then we can take $G(x, y; x', y') = \dfrac{1}{2\pi} \log r + g(x, y; x', y')$, where $g(x, y; x', y') \in C^2(D)$. Hence,

$$\lim_{\varepsilon \to 0} \oint_{C_\varepsilon} \frac{\partial}{\partial r}\left[\frac{1}{2\pi} \log r + g(x, y; x', y')\right] ds = 1.$$

Now, if we substitute $u(x)$ as defined by (4.6) in the equation $L[u] \equiv \nabla^2 u = f(x)$, we get

$$\nabla^2 u = \nabla \iint_D G(x,y;x',y')f(x',y')\,dx'\,dy' = \iint_D \nabla G(x,y;x',y')f(x',y')\,dx'\,dy'$$

$$= \frac{1}{2\pi}\iint_D \nabla^2(\log r)\,f(x',y')\,dx'\,dy', \quad \text{since } \nabla^2 g = 0 \text{ everywhere}$$

$$= \frac{1}{2\pi}\lim_{\varepsilon\to 0} f(\bar{x},\bar{y})\oint_{C_\varepsilon}\frac{ds}{r} = f(x,y).$$

The 3-D case can be similarly completed. ∎

4.3. Bernoulli's Separation Method

We will present this method by using certain examples which involve simple partial differential equations of mathematical physics.

4.3.1. Laplace's Equation in a Cube.
Consider Laplace's equation in the cube $D = \{0 < x, y, z < \pi\}$:

$$\nabla^2 u \equiv \frac{\partial^2 u}{\partial x^2} + \frac{\partial^2 u}{\partial y^2} + \frac{\partial^2 u}{\partial z^2} = 0, \quad \text{in } D, \tag{4.7}$$

such that $u(0,0,0) = 0 = u(\pi,\pi,\pi)$, and $u(x,y,0) = g(x,y)$. This problem arises in electrostatics where u is the potential which has the prescribed value $g(x,y)$ on the face $z = 0$ of the cube D, while the other faces are kept at zero potential. The solution also defines an equilibrium temperature distribution when the faces of the cube are kept at temperature zero and $g(x,y)$ as defined above. Since the maximum principle[1] holds for Laplace's equation in 3-D as well as 2-D, it implies that a 3-D boundary value problem for Laplace's equation has at most one solution, which vanishes continuously with the prescribed boundary values. We will find this solution by Bernoulli's separation method: Let $u = X(x)\,Y(y)\,Z(z)$, and substituting it in Eq (4.7) we get $\dfrac{X''}{X} + \dfrac{Y''}{Y} = -\dfrac{Z''}{Z}$. Since the left-hand side is independent of z and the right-hand side depends only on z, both sides must be equal to a constant, i.e.,

$$\frac{X''}{X} + \frac{Y''}{Y} = -\frac{Z''}{Z} = k_1, \quad \frac{X''}{X} = k_1 - \frac{Y''}{Y} = k_2.$$

Thus, we have $X'' - k_2 X = 0$, $Y'' - (k_1 - k_2)\,Y = 0$, $Z'' + k_1 Z = 0$, with the respective boundary conditions: $X(0) = X(\pi) = 0$, $Y(0) = Y(\pi) = 0$, $Z(\pi) = 0$. The boundary value problem for X is an eigenvalue problem, for which we must have

---

[1] This principle states that a nonconstant function which is harmonic inside a bounded domain D with boundary ∂D and continuous on the closed region $\overline{D} = D \cup \partial D$ attains its maximum and minimum values only on the boundary ∂D. Note that u is harmonic iff $\nabla^2 u = 0$ (see Theorem 8.1).

$k_2 = -n^2$, where $n \in \mathbb{Z}^+$, and the corresponding eigenfunction is $X = \sin nx$. The boundary value problem for Y is also an eigenvalue problem, for which $k_1 - k_2 = -m^2$, $m \in \mathbb{Z}^+$, and thus, $Y = \sin my$. But $k_1 = -m^2 - n^2$, so Z is a multiple of $\sinh \sqrt{m^2 + n^2}\, (\pi - z)$. Hence, we have a solution of the form

$$u(x, y, z) = \sum_{n=1}^{\infty} \sum_{m=1}^{\infty} c_{nm} \sin nx \sin my \sinh \sqrt{m^2 + n^2}\, (\pi - z).$$

The boundary condition at $z = 0$ gives

$$g(x, y) = \sum_{n=1}^{\infty} \sum_{m=1}^{\infty} c_{nm} \sin nx \sin my \sinh \sqrt{m^2 + n^2}\, \pi.$$

If we set

$$c_{nm} \sinh \sqrt{m^2 + n^2}\, \pi = d_{nm} = \frac{4}{\pi^2} \int_0^{\pi} \int_0^{\pi} g(x, y) \sin nx \sin my \, dx \, dy,$$

then the solution is given by

$$g(x, y) = \sum_{n=1}^{\infty} \sum_{m=1}^{\infty} d_{nm} \sin nx \sin my \, \frac{\sinh \sqrt{m^2 + n^2}\, (\pi - z)}{\sinh \sqrt{m^2 + n^2}\, \pi}, \qquad (4.8)$$

where the Fourier series of $g(x, y)$ converges uniformly for $z \geq 0$.

4.3.2. Laplace's Equation in a Cylinder. Consider the problem

$$\frac{\partial^2 u}{\partial r^2} + \frac{1}{r}\frac{\partial u}{\partial r} + \frac{1}{r}^2 \frac{\partial^2 u}{\partial \theta^2} + \frac{\partial^2 u}{\partial z^2} = 0 \qquad (4.9)$$

in the cylinder $0 < r < a$, $0 < \theta < 2\pi$, $0 < z < \pi$, such that $u(r, \theta, 0) = u(r, \theta, \pi) = 0$, $u(a, \theta, z) = g(\theta, z)$. Using Bernoulli's separation method, we take $u = R(r)\Theta(\theta)Z(z)$, and substituting it in Eq (4.9) we get

$$\frac{R'' + \dfrac{1}{r} R'}{R} + \frac{1}{r^2}\frac{\Theta''}{\Theta} = -\frac{Z''}{Z} = k_1,$$

or

$$\frac{r^2 R'' + r R'}{R} - r^2 k_1 = -\frac{\Theta''}{\Theta} = k_2.$$

Thus, solving $Z'' + k_1 Z = 0$, $Z(0) = Z(\pi) = 0$, we get $k_1 = n^2$, $n \in \mathbb{Z}^+$, and $Z_n = \sin nz$. Again, $\Theta'' + k_2 \Theta = 0$, where Θ is periodic with period 2π, i.e., $\Theta(0) = \Theta(2\pi)$. Then $k_2 = m^2$, $m = 0, 1, 2, \ldots$, with the corresponding eigenfunctions $\Theta_0 = 1$, $\Theta_m = \{\sin m\theta, \cos m\theta\}$, $m = 1, 2, \ldots$. Finally, for R we have the differential equation

$$R''_{mn} + \frac{1}{r} R'_{mn} - \left(\frac{m^2}{r^2} + n^2\right) R_{mn} = 0,$$

which has the singularity at $r = 0$. The solution at $r = a$ is

$$R_{mn}(r) = \frac{I_m(nr)}{I_m(na)},$$

where I_m is the Bessel function with imaginary argument, which converges for all r. Hence, the solution is given by

$$u(r, \theta, z) = \frac{1}{2} \sum_{n=1}^{\infty} c_{n0} \frac{I_0(nr)}{I_0(na)} \sin nz$$

$$+ \sum_{n=1}^{\infty} \sum_{m=1}^{\infty} \frac{I_m(nr)}{I_m(na)} (c_{nm} \sin m\theta + d_{nm} \cos m\theta) \sin nz, \quad (4.10)$$

where

$$c_{nm} = \frac{2}{\pi^2} \int_0^{\pi} \int_0^{2\pi} g(\theta, z) \cos m\theta \sin nz \, d\theta \, dz,$$

$$d_{nm} = \frac{2}{\pi^2} \int_0^{\pi} \int_0^{2\pi} g(\theta, z) \sin m\theta \sin nz \, d\theta \, dz,$$

and the series in (4.10) converges uniformly for $r < a$, and $0 < z < \pi$. Also see Example 4.10 below.

4.3.3. Laplace's Equation in a Sphere. Consider the boundary value problem in a sphere of radius a:

$$\frac{\partial^2 u}{\partial r^2} + \frac{2}{r} \frac{\partial u}{\partial r} + \frac{1}{r^2 \sin \theta} \frac{\partial}{\partial \theta} \left(\sin \theta \frac{\partial u}{\partial \theta}\right) + \frac{1}{r^2 \sin^2 \theta} \frac{\partial^2 u}{\partial \phi^2} = 0, \quad r < a,$$

$$u(a, \theta, \phi) = f(\theta, \phi). \quad (4.11)$$

Using Bernoulli's separation method, this Laplace's equation has a solution of the form $u = R(r)\Theta(\theta)\Phi(\phi)$ if

$$\frac{R'' + 2R'/r}{R} + \frac{(\sin \theta \, \Theta')'}{\Theta r^2 \sin \theta} + \frac{\Phi''}{\Phi r^2 \sin^2 \theta} = 0.$$

Multiplying this equation by $r^2 \sin^2 \theta$ and transposing the last term, we see that Φ''/Φ must be a constant. Since Φ is a periodic function of period of 2π, we have $\Phi = \cos m\phi$ or $\Phi = \sin m\phi$, $m = 0, 1, 2, \ldots$. Thus,

$$\frac{R'' + 2R'/r}{R} + \frac{(\sin \theta \, \Theta')'}{\Theta r^2 \sin \theta} - \frac{m^2}{r^2 \sin^2 \theta} = 0.$$

Multiplying by r^2 and transposing the first term, we obtain two equations:

$$r^2 \left(R'' + \frac{2}{r} R' \right) - \lambda R = 0, \tag{4.12}$$

$$(\sin \theta\, \Theta')' = \frac{m^2}{\sin \theta} \Theta + \lambda \sin \theta\, \Theta = 0, \tag{4.13}$$

where λ is a constant. The equation for Θ is singular at both end points $\theta = 0, \pi$. If we impose the condition that Θ and Θ' remain bounded at these end points, this becomes an eigenvalue problem in Θ with two singular points. If we use the new variable $t = \cos \theta$ and let $\Theta(\theta) = P(\cos \theta)$, then Eq (4.13) becomes

$$\frac{d}{dt}\left[(1 - t^2) \frac{dP}{dt} \right] - \frac{m^2}{1 - t^2} P + \lambda P = 0, \quad -1 < t < 1. \tag{4.14}$$

where $\lambda = n(n + 1)$ yields the associated Legendre's equation

$$(1 - t^2)\, \Theta'' - 2t\, \Theta' + \left[n(n + 1) - \frac{m^2}{1 - t^2} \right] \Theta = 0, \tag{4.15}$$

which has the solutions $\Theta(t) = P_n^m(t)$, or $\Theta(\theta) = P_n^m(\cos \theta)$, where $P_n^m(\cos \theta)$ are the associated Legendre polynomials of order $m = 0, \pm 1, \pm 2, \dots , \pm n$, and degree $n = 0, 1, 2, \dots$. Also see Example 4.12 below.

4.3.4. Helmholtz's Equation in Cartesian Coordinates. We rewrite Eq (4.1) as

$$\frac{\partial^2 u}{\partial x^2} + \frac{\partial^2 u}{\partial y^2} + \frac{\partial^2 u}{\partial z^2} + k^2 u = 0, \tag{4.16}$$

and solve it by using Bernoulli's separation method: Set $u(x, y, z) = X(x)Y(y)Z(z)$ and substitute in Eq (4.16). Then we obtain

$$YZ\frac{d^2 X}{dx^2} + XZ\frac{d^2 Y}{dy^2} + XY\frac{d^2 Z}{dz^2} + k^2 XYZ = 0,$$

or dividing by XYZ and rearranging terms we have

$$\frac{1}{X}\frac{d^2 X}{dx^2} = -k^2 - \frac{1}{Y}\frac{d^2 Y}{dy^2} - \frac{1}{Z}\frac{d^2 Z}{dz^2}.$$

At this stage we have only one separation of variables: The left side is a function of x only, while the right side depends on y and z. This means that we are equating a function of x to a function of y and z, although x, y and z are all independent variables. Thus, the behavior of x as an independent variable cannot be determined by y and z. To resolve this situation we introduce another arbitrary constant l^2 and choose

$$\frac{1}{X}\frac{d^2 X}{dx^2} = -l^2, \tag{4.17}$$

$$\frac{1}{Y}\frac{d^2 Y}{dy^2} + \frac{1}{Z}\frac{d^2 Z}{dz^2} = l^2 - k^2, \tag{4.18}$$

where the sign of l^2 will be fixed in specific problems subject to prescribed boundary conditions. Now, we rewrite Eq (4.18) as

$$\frac{1}{Y}\frac{d^Y}{dy^2} = l^2 - k^2 - \frac{1}{Z}\frac{d^2 Z}{dz^2},$$

where we arrive at a second separation of variables by equating a function of y to a function of z, which is accomplished by equating each side of the above equation to another constant of separation $-m^2$, i.e.,

$$\frac{1}{Y}\frac{d^2 Y}{dy^2} = -m^2, \tag{4.19}$$

$$\frac{1}{Z}\frac{d^2 Z}{dz^2} = l^2 - k^2 + m^2 = -n^2. \tag{4.20}$$

As a result of this separation method we obtain a set of three ordinary differential equations (4.17), (4.19) and (4.20), which can be solved by known methods. Hence, the solution of Eq (4.16) is given by

$$u_{lmn}(x, y, z) = X_l(x)\, Y_m(y)\, Z_n(z), \tag{4.21}$$

subject to prescribed boundary conditions; the constants l, m, n are chosen such that the condition $k^2 = l^2 + m^2 + n^2$ holds. Then, the general solution of Eq (4.16) is given by

$$u(x, y, z) = \sum_{l,m,n} c_{lmn}\, u_{lmn}(x, y, z), \tag{4.22}$$

where the coefficients c_{lmn} are chosen so that u satisfies prescribed boundary conditions of the problem.

4.3.5. Helmholtz's Equation in Spherical Coordinates. This equation with positive k^2 in spherical coordinates is

$$\frac{1}{r^2}\frac{\partial}{\partial r}\left(r^2\frac{\partial}{\partial r}\right) + \frac{1}{\sin\theta}\frac{\partial}{\partial\theta}\left(\sin\theta\frac{\partial}{\partial\theta}\right) + \frac{1}{\sin^2\theta}\frac{\partial^2}{\partial\phi^2} + k^2\, u = 0. \tag{4.23}$$

As before, we set $u(r, \theta, \phi) = R(r)\Theta(\theta)\Phi(\phi)$, substitute it in Eq (4.23), and divide throughout by $R\Theta\Phi$, to obtain

$$\frac{1}{Rr^2}\frac{d}{dr}\left(r^2\frac{dR}{dr}\right) + \frac{1}{\Theta r^2 \sin\theta}\frac{d}{d\theta}\left(\sin\theta\frac{d\Theta}{d\theta}\right) + \frac{1}{\Phi r^2 \sin^2\theta}\frac{d^2\Phi}{d\phi^2} = -k^2.$$

Multiplying this equation by $r^2 \sin^2 \theta$, we separate the Φ part and get

$$\frac{1}{\Phi}\frac{d^2\Phi}{d\phi^2} = -r^2 \sin^2 \theta \left[k^2 + \frac{1}{r^2 R}\frac{d}{dr}\left(r^2 \frac{dR}{dr} \right) + \frac{1}{r^2 \sin^2 \theta \Theta}\frac{d}{d\theta}\left(\sin \theta \frac{d\Theta}{d\theta} \right) \right].$$
$$(4.24)$$

This equation equates a function of ϕ to a function of r and θ. But since r, θ, and ϕ are independent variables, each side of Eq (4.24) must be equal to a constant. Again, since in all physical problems ϕ denotes the azimuth angle, we assume a periodic solution, and take the separation constant as $-m^2$. Then after separation we obtain

$$\frac{1}{\Phi}\frac{d^2\Phi}{d\phi^2} = -m^2, \qquad (4.25)$$

$$\frac{1}{r^2 R}\frac{d}{dr}\left(r^2 \frac{dR}{dr} \right) + \frac{1}{r^2 \sin^2 \theta \Theta}\frac{d}{d\theta}\left(\sin \theta \frac{d\Theta}{d\theta} \right) - \frac{m^2}{r^2 \sin^2 \theta} = -k^2. \qquad (4.26)$$

Multiplying Eq (4.26) by r^2 and rearranging the terms we get

$$\frac{1}{R}\frac{d}{dr}\left(r^2 \frac{dR}{dr} \right) + k^2 r^2 = -\frac{1}{\sin \theta \Theta}\frac{d}{d\theta}\left(\sin \theta \frac{d\Theta}{d\theta} \right) + \frac{m^2}{\sin^2 \theta}. \qquad (4.27)$$

Next, separating the remaining two variables and equating each side of Eq (4.27) to a constant l, we obtain

$$\frac{1}{r^2}\frac{d}{dr}\left(r^2 \frac{dR}{dr} \right) + \left(k^2 - \frac{l}{r^2} \right) R = 0, \qquad (4.28)$$

$$\frac{1}{\sin \theta}\frac{d}{d\theta}\left(\sin \theta \frac{d\Theta}{d\theta} \right) + \left(l - \frac{m^2}{\sin^2 \theta} \right) \Theta = 0. \qquad (4.29)$$

The three equations (4.25), (4.28) and (4.29) are ordinary differential equations, which can be solved by known methods. Note that Eq (4.29) is Legendre's equation with the constant $l = n(n+1)$. Thus, the general solution of Eq (4.23) is given by

$$u(r, \theta, \phi) = \sum_{l,m} R_l(r)\Theta_{lm}(\theta)\Phi_m(\phi). \qquad (4.30)$$

The constant k^2 is a function of r in Schrödinger's wave equation, and a function of r, θ and ϕ in the problem of the hydrogen atom (see §7.8).

4.3.6. Wave Equation. Consider the wave equation for the space-dependent amplitude of an electromagnetic wave in a cylindrical wave guide: $\nabla^2 u + k^2 u = 0$. Taking the central axis of the cylindrical wave guide as the z-axis we find that in a wave guide with perfectly conducting walls there is no attenuation and hence, no dependence on z. Then the equation to be solved by Bernoulli's separation method for this problem is

$$\frac{\partial^2 u}{\partial r^2} + \frac{1}{r}\frac{\partial u}{\partial r} + \frac{1}{r^2}\frac{\partial^2 u}{\partial \theta^2} + k^2 u = 0. \qquad (4.31)$$

Set $u(r, \theta) = R(r)\Theta(\theta)$ in this equation, which gives

$$\frac{1}{Rr}\frac{d}{dr}\left(r\frac{dR}{dr}\right) + \frac{1}{\Theta r^2}\frac{d^2\Theta}{d\theta} + k^2 = 0.$$

Multiplying by r^2 we separate the θ variable and get

$$\frac{1}{\Theta}\frac{d^2\Theta}{d\theta} = -m^2, \tag{4.32}$$

where θ is the azimuth angle in this case; thus, the solution of E (4.32) will be periodic of the form $\sin m\theta$, $\cos m\theta$. The equation in r is given by

$$\frac{1}{r}\frac{d}{dr}\left(r\frac{dR}{dr}\right) + \left(k^2 - \frac{m^2}{r^2}\right)R = 0, \tag{4.33}$$

which is Bessel's equation, for which see Example 3.24.

4.4. Examples

Some useful examples follow.

Example 4.1. (Hyperbolic equation.) The problem of a vibrating string is governed by the one-dimensional wave equation. Now, consider the boundary value problem

$$\frac{\partial^2 u}{\partial t^2} = c^2\frac{\partial^2 u}{\partial x^2}, \quad 0 < x < l, \tag{4.34}$$

$$u(0, t) = 0 = u(l, t), \quad t > 0, \tag{4.35}$$

$$u(x, 0) = f(x), \quad u_t(x, 0) = h(x), \quad 0 < x < l, \tag{4.36}$$

where $f \in C^1(0, l)$ is a given function. We seek a solution of the form $u(x, t) = X(x)T(t)$, where X is a function of x only and T a function of t only, and assume that a solution of this form exists. We will carry out the details to see if the method works for this problem. Note that $\frac{\partial^2 u}{\partial t^2} = X T''$, and $\frac{\partial^2 u}{\partial x^2} = X'' T$, where the primes denote the derivative with respect to its corresponding independent variable. Then Eq (4.34) reduces to $X T'' = c^2 X'' T$, or, after separating the variables, it becomes

$$\frac{T''}{T} = c^2\frac{X''}{X}. \tag{4.37}$$

Since the left side of Eq (4.37) is a function of t only and the right side a function of x only, the only situation where $X(x)$ and $T(t)$ have solutions for all x and all t is when $\dfrac{c^2 X''(x)}{X} = \dfrac{T''(t)}{T} = \text{const.}$ Hence, from (4.37) we can write

$$\frac{1}{c^2}\frac{T''}{T} = \frac{X''}{X} = k, \quad k = \text{const.} \tag{4.38}$$

The set of equations (4.38) is equivalent to two ordinary differential equations:

$$T'' - k c^2 T = 0, \tag{4.39}$$

$$X'' - k X = 0. \tag{4.40}$$

It is apparent from Eq (4.38) that the constant k in Eqs (4.39) and (4.40) has the same value. The general solution of Eq (4.39) is

$$T(t) = \begin{cases} c_1 e^{c\sqrt{k}t} + c_2 e^{-c\sqrt{k}t} & \text{for } k > 0 \\ c_1 t + c_2 & \text{for } k = 0 \\ c_1 \cos c\sqrt{-k}t + c_2 \sin c\sqrt{-k}t & \text{for } k < 0, \end{cases} \tag{4.41}$$

and of Eq (4.40) is

$$X(x) = \begin{cases} d_1 e^{\sqrt{k}x} + d_2 e^{-\sqrt{k}x} & \text{for } k > 0 \\ d_1 x + d_2 & \text{for } k = 0 \\ d_1 \cos \sqrt{-k}x + d_2 \sin \sqrt{-k}x & \text{for } k < 0. \end{cases} \tag{4.42}$$

In view of the boundary conditions (4.35) we must have

$$X(0)T(t) = 0 = X(l)T(t), \quad \text{for all } t \geq 0. \tag{4.43}$$

Using these conditions in (4.42) for $k > 0$ we get the system of equations

$$\begin{aligned} X(0) &= d_1 + d_2 = 0, \\ X(l) &= d_1 e^{\sqrt{k}l} + d_2 e^{-\sqrt{k}l} = 0. \end{aligned} \tag{4.44}$$

The system (4.44) is has a nontrivial solution iff the determinant of its coefficients vanishes. But since

$$\begin{vmatrix} 1 & 1 \\ e^{\sqrt{k}l} & e^{-\sqrt{k}l} \end{vmatrix} = e^{-\sqrt{k}l} - e^{\sqrt{k}l} \neq 0,$$

a nonzero solution for $X(x)$ in (4.42) for $k > 0$ is not possible. Next, for $k = 0$, the boundary conditions (4.43) imply that $d_1 = 0$ and $d_2 = 0$. Hence, there is no nonzero solution for $k = 0$. Finally, for $k < 0$, let us set $k = -\lambda^2$. Then the general solution (4.42) in this case becomes $X(x) = d_1 \cos \lambda x + d_2 \sin \lambda x$, which, under the boundary conditions (4.43), yields $X(0) = d_1 = 0$ and $X(l) = d_2 \sin \lambda l = 0$. To avoid a trivial solution in this case, we choose λ such that λl is a positive multiple of π, i.e., $\lambda l = n\pi$, or $\lambda = \dfrac{n\pi}{l}$. The positive values of λ are chosen because the negative multiples give the same eigenfunctions as the positive ones. This result leads to an infinite set of solutions which are denoted by $X_n(x) = d_{2,n} \sin \dfrac{n\pi x}{l}$, where each eigenfunction $\sin \dfrac{n\pi x}{l}$ corresponds to the eigenvalue

$$k = -\frac{n^2 \pi^2}{l^2}. \tag{4.45}$$

The solutions for $T(t)$ for the choice of $k < 0$, as in (4.45), are then obtained from (4.41) as

$$T_n(t) = c_{1,n} \cos \frac{n\pi ct}{l} + c_{2,n} \sin \frac{n\pi ct}{l}.$$

Then the infinite set of solutions is

$$u_n(x,t) = X_n(x)T_n(t) = \left[A_n \cos \frac{n\pi ct}{l} + B_n \sin \frac{n\pi ct}{l} \right] \sin \frac{n\pi x}{l}, \qquad (4.46)$$

where the constants A_n and B_n are determined from the initial conditions. The eigenfunctions are contained in the solution (4.46), whereas the eigenvalues for this boundary value problem are given by (4.45).

The next step is to obtain the particular solution which satisfies the initial conditions (4.36). At this point it may so happen that none of the solutions (4.46) will satisfy (4.36). In view of the superposition principle, any finite sum of the solutions (4.46) is also a solution of this boundary value problem. We should, therefore, find a linear combination of those solutions, which also satisfies the initial conditions (4.36). Even if this technique fails, we can always try an infinite series of solutions (4.46); that is,

$$u(x,t) = \sum_{n=1}^{\infty} X_n(x)T_n(t) = \sum_{n=1}^{\infty} \left[A_n \cos \frac{n\pi ct}{l} + B_n \sin \frac{n\pi ct}{l} \right] \sin \frac{n\pi x}{l}. \quad (4.47)$$

Now, we can take this series expansion formally and verify that the boundary conditions (4.35) are still satisfied. We will use the initial conditions to find the constants A_n and B_n. Using the first of the initial conditions (4.36) we get

$$u(x,0) = \sum_{n=1}^{\infty} A_n \sin \frac{n\pi x}{l} = f(x). \qquad (4.48)$$

This infinite series is a Fourier sine series. Hence, $f(x)$ can be regarded as an odd function with period $2l$. Thus, we expand this function $f(x)$ on the interval $0 \le x \le l$ such that $f(-x) = -f(x)$ on the interval $-l \le x \le 0$, and $f(x+2l) = f(x)$ for all x. Then the coefficients A_n for $n = 1, 2, \ldots$ are given by

$$A_n = \frac{2}{l} \int_0^l f(x) \sin \frac{n\pi x}{l} \, dx.$$

Taking the time-derivative of (4.47), we get

$$u_t = \frac{\pi c}{l} \sum_{n=1}^{\infty} n \left[B_n \cos \frac{n\pi ct}{l} - A_n \sin \frac{n\pi ct}{l} \right] \sin \frac{n\pi x}{l},$$

which, in view of the second of the initial conditions (4.36), gives

$$u_t(x,0) = \frac{\pi c}{l} \sum_{n=1}^{\infty} n B_n \sin \frac{n\pi x}{l} = h(x),$$

where

$$B_n = \frac{2}{n\pi c} \int_0^l h(x) \sin \frac{n\pi x}{l} \, dx, \quad n = 1, 2, \ldots.$$

Hence, the solution (4.47) is completely determined.

D'ALEMBERT FORM OF THE SOLUTION. We will now derive the d'Alembert solution for this problem. From (4.47) we have

$$u = \frac{1}{2} \sum_{n=1}^{\infty} A_n \left\{ \sin \frac{n\pi(x+ct)}{l} + \sin \frac{n\pi(x-ct)}{l} \right\}$$

$$+ \frac{1}{2} \sum_{n=1}^{\infty} B_n \left\{ \cos \frac{n\pi(x-ct)}{l} - \cos \frac{n\pi(x+ct)}{l} \right\} \tag{4.49}$$

$$= \frac{1}{2} \left[f(x+ct) + f(x-ct) \right] + \frac{1}{2} \left[-g(x+ct) + g(x-ct) \right],$$

where $f(z) = \sum_{n=1}^{\infty} A_n \sin \frac{n\pi z}{l}$, as in (4.48). Let $g(z) = \sum_{n=1}^{\infty} B_n \cos \frac{n\pi z}{l}$.
Then $g'(z) = -\frac{1}{c} \frac{\pi c}{l} \sum_{n=1}^{\infty} n B_n \sin \frac{n\pi z}{l} = -\frac{1}{c} h(z)$, and from (4.49) we obtain the formal solution, known as the *d'Alembert solution* for this problem, as

$$u(x,t) = \phi(x-ct) + \psi(x+ct), \tag{4.50}$$

where c is the wave velocity, and

$$\phi(x-ct) = \frac{1}{2} \left[f(x-ct) + g(x-ct) \right]$$
$$\psi(x+ct) = \frac{1}{2} \left[f(x+ct) - g(x+ct) \right].$$

An interpretation of the solution of this problem is as follows: At each point x of the string we have $u(x,t) = \sum_{n=1}^{\infty} a_n \cos \frac{n\pi c}{l}(t+\delta_n) \sin \frac{n\pi x}{l}$. This equation describes a harmonic motion with amplitudes $a_n \sin \frac{n\pi x}{l}$. Each such motion of the string is called a standing wave, which has its nodes at the points where $\sin(n\pi x/l) = 0$; these points remain fixed during the entire process of vibration. But the string vibrates with maximum amplitudes a_n at the points where $\sin(n\pi x/l) = \pm 1$. For any t the structure of the standing wave is described by $u(x,t) = \sum_{n=1}^{\infty} C_n(t) \sin \frac{n\pi x}{l}$, where

$C_n(t) = a_n \cos \omega_n(t + \delta_n), \omega_n = \dfrac{n\pi c}{l}$. At times t when $\cos \omega_n(t + \delta_n) = \pm 1$, the velocity becomes zero and the displacement reaches its maximum value. ∎

Example 4.2. (Parabolic Equation) Consider the one-dimensional heat conduction equation

$$\frac{\partial u}{\partial t} = k\frac{\partial^2 u}{\partial x^2}, \quad 0 < x < l, \tag{4.51}$$

subject to the boundary and initial conditions

$$u(0, t) = 0 = u(l, t), \quad t > 0, \tag{4.52a}$$
$$u(x, 0) = f(x), \quad 0 < x < l, \tag{4.52b}$$

where $f \in C^1$ is a prescribed function. In physical terms, this problem represents the heat conduction in a rod when its ends are maintained at zero temperature while the initial temperature u at any point of the rod is prescribed as $f(x)$. We assume the solution in the form $u(x, t) = X(x)T(t)$, which after substitution into Eq (4.51) yields the set of equations

$$\frac{1}{k}\frac{T'}{T} = \frac{X''}{X}. \tag{4.53}$$

As in Example 4.1, the only situation where these two expressions can be equal is for each of them to be constant, say each equal to m. Then Eq (4.53) yields two ordinary differential equations

$$T' - mk\,T = 0, \tag{4.54}$$
$$X'' - m\,X = 0, \tag{4.55}$$

where the boundary conditions (4.52a) reduce to

$$X(0)\,T(t) = 0 = X(l)\,T(t), \quad \text{or} \quad X(0) = 0 = X(l), \tag{4.56}$$

except for the case when the rod has zero initial temperature at every point. This situation, being uninteresting, can be neglected. We notice that for a nonzero solution of the problem (4.55)-(4.56) we must choose negative values of m. Hence we set $m = -\lambda^2$, and find that the eigenvalues $m = -n^2\pi^2/l^2$ have the corresponding eigenfunctions $X_n(x) = A_n \sin \dfrac{n\pi x}{l}$. Eq (4.54) then becomes $T' + \dfrac{kn^2\pi^2}{l^2}T = 0$ whose general solution for each n is given by $T_n(t) = B_n e^{-kn^2\pi^2 t/l^2}$. Hence, we consider an infinite series of the form

$$u(x, t) = \sum_{n=1}^{\infty} X_n(x)T_n(t) = \sum_{n=1}^{\infty} C_n \sin \frac{n\pi x}{l} e^{-kn^2\pi^2 t/l^2}. \tag{4.57}$$

Now, we use the initial condition (4.52b) in (4.57) and obtain

$$u(x, 0) = \sum_{n=1}^{\infty} C_n \sin \frac{n\pi x}{l} = f(x), \tag{4.58}$$

which shows that $f(x)$ can be represented as a Fourier sine series, by extending f as an odd, piecewise continuous function of period $2l$ with piecewise continuous derivatives. Equation (4.58) gives the coefficients C_n as

$$C_n = \frac{2}{l} \int_0^l f(x) \sin \frac{n\pi x}{l} \, dx, \ n = 1, 2, \ldots . \tag{4.59}$$

Hence, the solution (4.57) is completely determined for this problem. Note that the series in (4.58) converges since $u(x, 0)$ does, and the exponential expression in (4.57) is less than 1 for each n and all $t > 0$ and approaches zero as $t \to \infty$. ∎

An interesting situation, considered in the next example, arises if the function $f(x)$ is zero in the initial condition (4.52b), but the boundary conditions are nonhomogeneous.

Example 4.3. Consider the dimensionless partial differential equation governing the plane wall transient heat conduction

$$u_t = u_{xx}, \quad 0 < x < 1, \tag{4.60}$$

with the boundary and initial conditions

$$u(0, t) = 1, \quad u(1, t) = 0, \quad t > 0, \tag{4.61}$$
$$u(x, 0) = 0, \quad 0 < x < 1. \tag{4.62}$$

Since the nonhomogeneous boundary condition in (4.61) does not allow us to compute the eigenfunctions, we proceed as follows: First, we find a particular solution of the problem, which satisfies only the boundary conditions. Although there is more than one way to determine the particular solution, we can, for example, take the steady-state case, where the equation becomes $u_{xx} = 0$, which is independent of t, and after integrating twice, has the general solution $u(x) = c_1 x + c_2$, with the boundary conditions $u(0) = 1$, $u(1) = 0$. Thus, $c_1 = -1$, $c_2 = 1$, and the steady-state solution is $u(x) = 1 - x$. Next, we formulate a homogeneous problem by writing $u(x, t)$ as a sum of the steady-state solution $u(x)$ and a transient term $v(x, t)$, i.e., $u(x, t) = u(x) + v(x, t)$, or

$$v(x, t) = u(x, t) - u(x). \tag{4.63}$$

Thus, the problem reduces to finding $v(x, t)$. If we substitute v from (4.63) into (4.60), we get

$$v_t = v_{xx}, \tag{4.64}$$

where the boundary conditions (4.61) and the initial condition (4.62) reduce to

$$v(0, t) = u(0, t) - u(0) = 0,$$
$$v(1, t) = u(1, t) - u(1) = 0, \tag{4.65}$$

and

$$v(x,0) = u(x,0) - u(x) = x - 1. \tag{4.66}$$

Notice that the problem (4.64)-(4.66) is the same as in Example 4.2 with $k = 1$, $l = 1$, $f(x) = x - 1$, and u replaced by v. Hence, its general solution from (4.57) is given by

$$v(x,t) = \sum_{n=1}^{\infty} C_n e^{-n^2 \pi^2 t} \sin n\pi x,$$

and the coefficients C_n are determined from (4.59) as

$$C_n = 2 \int_0^1 (x - 1) \sin n\pi x \, dx = -\frac{2}{n\pi}.$$

Thus,

$$v(x,t) = -\frac{2}{\pi} \sum_{n=1}^{\infty} \frac{1}{n} e^{-n^2 \pi^2 t} \sin n\pi x,$$

and finally from (4.63)

$$u(x,t) = 1 - x - \frac{2}{\pi} \sum_{n=1}^{\infty} \frac{1}{n} e^{-n^2 \pi^2 t} \sin n\pi x.$$

In general, if the thickness of the plate is l, the solution is

$$u(x,t) = 1 - \frac{x}{l} - \frac{2}{\pi} \sum_{n=1}^{\infty} \frac{1}{n} e^{-n^2 \pi^2 t / l^2} \sin \frac{n\pi x}{l}. \tag{4.67}$$

The solution for the half-space is derived by letting $l \to \infty$. Since

$$\lim_{l \to \infty} u(x,t) = 1 - \frac{2}{\pi} \sum_{n=1}^{\infty} \frac{l}{n\pi} e^{-n^2 \pi^2 t / l^2} \sin \left(\frac{n\pi x}{l} \right) \cdot \frac{\pi}{l},$$

let $n\pi/l = \xi$ and $\pi/l = d\xi$. Then

$$\lim_{l \to \infty} u(x,t) = 1 - \frac{2}{\pi} \int_0^{\infty} \frac{1}{\xi} e^{-\xi^2 t} \sin \xi t \, d\xi = 1 - \mathrm{erf}\left(\frac{x}{2\sqrt{t}} \right) = \mathrm{erfc}\left(\frac{x}{2\sqrt{t}} \right). \ \blacksquare$$

Example 4.4. (Elliptic equation.) Consider the steady-state heat conduction or potential problem for the rectangle $R = \{0 < x < a, \ 0 < y < b\}$:

$$u_{xx} + u_{yy} = 0, \quad x, y \in R, \tag{4.68}$$

subject to the Dirichlet boundary conditions

$$u(0, y) = 0 = u(a, y), \ u(x, 0) = 0, \ u(x, b) = f(x). \tag{4.69}$$

Physically, this problem arises if three edges of a thin isotropic rectangular plate are insulated and maintained at zero temperature, while the fourth edge is subjected to a variable temperature $f(x)$ until the steady-state conditions are attained throughout R. Then the steady-state value of $u(x, y)$ represents the distribution of temperature in the interior of the plate. As before, we seek a solution of the form $u(x, y) = X(x)Y(y)$, which, after substitution into Eq (4.68), leads to the set of two ordinary differential equations :

$$X'' - cX = 0, \tag{4.70}$$

$$Y'' + cY = 0, \tag{4.71}$$

where c is a constant, as in Example 4.2. Since the first three boundary conditions in (4.69) are homogeneous, they become

$$X(0) = 0, \quad X(a) = 0, \quad Y(0) = 0, \tag{4.72}$$

but the fourth boundary condition which is nonhomogeneous must be used separately. Now, taking $c = -\lambda^2$, as before, the solution of (4.70) subject to the first two boundary conditions in (4.72) leads to the eigenpairs as

$$\lambda_n^2 = \frac{n^2 \pi^2}{a^2}, \quad X_n(x) = \sin \frac{n \pi x}{a}, \quad n = 1, 2, \dots,$$

while for these eigenvalues the solutions of (4.68) satisfying the third boundary condition in (4.72) are

$$Y_n(y) = \sinh \frac{n \pi y}{a}, \quad n = 1, 2, \dots . \tag{4.73}$$

Hence, for arbitrary constants $C_n, n = 1, 2, \dots$, we get

$$u(x, y) = \sum_{n=1}^{\infty} C_n \sin \frac{n \pi x}{a} \sinh \frac{n \pi y}{a}, \tag{4.74}$$

where the coefficients C_n are determined by using the fourth boundary condition in (4.69). Thus,

$$u(x, b) = f(x) = \sum_{n=1}^{\infty} C_n \sin \frac{n \pi x}{a} \sinh \frac{n \pi b}{a}, \quad 0 < x < a,$$

which, in view of formula (4.59), yields

$$C_n \sinh \frac{n \pi b}{a} = \frac{2}{a} \int_0^a f(x) \sin \frac{n \pi x}{a} \, dx, \quad n = 1, 2, \dots . \tag{4.75}$$

This solves the problem completely. In particular, if $f(x) = f_0 = \text{const}$, then

$$C_n \sinh \frac{n\pi b}{a} = \frac{2f_0[1 - (-1)^n]}{n\pi}.$$

Thus, from (4.74), we have

$$u(x, y) = \frac{2f_0}{\pi} \sum_{n=1}^{\infty} \frac{1 - (-1)^n}{n} \frac{\sin(n\pi x/a) \sinh(n\pi y/a)}{\sinh(n\pi b/a)}. \ \blacksquare \qquad (4.76)$$

Example 4.5. Consider the problem of transverse vibrations of a stretched string of length l fastened at both ends, initially distorted into some given curve and then allowed to vibrate. Let the equilibrium state of the string be along the x-axis and one end at the origin, and suppose that the string is initially distorted into a given curve defined by $u = f(x)$. The problem is to find an expression for u which will be the solution of the equation

$$\frac{\partial^2 u}{\partial t^2} = c^2 \frac{\partial^2 u}{\partial x^2}, \qquad (4.77)$$

subject to the following conditions: (i) $u(0, t) = 0$, (ii) $u(l, t) = 0$, (iii) $u(x, 0) = f(x)$, and (iv) $\frac{du}{dt}(x, 0) = 0$. The condition (iv) signifies the fact that the string starts from rest. To solve this problem, we assume $u = e^{ax+bt}$, $b > 0$, and substituting it into Eq (4.77) we find that $b^2 = a^2 c^2$, which gives the general solution of Eq (4.77) for any value of a as $u = e^{ax+act}$. To convert this solution into a trigonometric form we replace a by $\pm ai$. The $u = e^{(x+ct)ai}$ and $u = e^{-(x+ct)ai}$ become two solutions of Eq (4.77). Adding these values and dividing by 2 we have $\cos a(x \pm ct)$, while subtracting them and dividing by $2i$ we have $\sin a(x + ct)$. This means that $u = \cos a(x+ct)$, $u = \cos a(x-ct)$, $u = \sin a(x+ct)$, and $u = \sin a(x-ct)$ each is a solution of Eq (4.77). Again, using trigonometric identities these four solutions can be rewritten as $u = \cos ax \cos act$, $u = \sin ax \sin act$, $u = \sin ax \cos act$, and $u = \cos ax \sin act$. Notice that the third form satisfies conditions (i) and (iv) for any value of a and it will satisfy condition (ii) if $a = \frac{k\pi}{l}$, where k is an integer. Then, if we take

$$u = \sum_{k=1}^{\infty} A_k \sin \frac{k\pi x}{l} \cos \frac{k\pi ct}{l}, \qquad (4.78)$$

where A_k, $k = 1, 2, \ldots$, are undetermined coefficients, we have the solution of Eq (4.77) which satisfies conditions (i), (ii), and (iv). Using condition (iii), we find from (4.78) that

$$y = \sum_{k=1}^{\infty} A_k \sin \frac{k\pi x}{l} = f(x), \text{ where } A_k = \frac{2}{l} \int_0^l f(x) \sin \frac{k\pi x}{l} \, dx. \ \blacksquare$$

Example 4.6. To find the potential function u due to a circular wire ring of radius a of small cross section, first notice that the potential satisfies the relation

$u = \dfrac{M}{\sqrt{a^2 + x^2}}$, where M is the mass of the ring, and x is the distance from the center of the ring. Using spherical coordinates with the center of the ring as the origin of the coordinates and the axis of the ring as the polar axis, and using symmetry we find that the potential, being independent of ϕ, is the solution of the equation

$$r\frac{\partial^2 (ru)}{\partial r^2} + \frac{1}{\sin\theta}\frac{\partial}{\partial\theta}\left(\sin\theta\frac{\partial u}{\partial\theta}\right) = 0, \tag{4.79}$$

subject to the condition that $u = \dfrac{M}{\sqrt{a^2 + x^2}}$ when $\theta = 0$. We assume $u(r,\theta) = r^n P(\theta)$, where $n > 0$ is an integer and P is a function of θ, and substituting it in (4.79) we find that

$$n(n+1)r^n P + \frac{r^n}{\sin\theta}\frac{\partial}{\partial\theta}\left(\sin\theta\frac{\partial P}{\partial\theta}\right) = 0,$$

or dividing by r^n,

$$n(n+1)P + \frac{1}{\sin\theta}\frac{\partial}{\partial\theta}\left(\sin\theta\frac{\partial P}{\partial\theta}\right) = 0.$$

To obtain a solution for this equation we set $x = \cos\theta$. Then this equation becomes

$$\frac{d}{dx}\left[(1 - x^2)\frac{dP}{dx}\right] + n(n+1)P = 0, \tag{4.80}$$

where P is now a function of x. Assuming a series solution $P(x) = \displaystyle\sum_{k=0}^{\infty} a_k x^k$ and substituting in Eq (4.80) we get

$$\sum_{k=0}^{\infty}\left[k(k-1)a_k x^{k-2} - k(k+1)a_k x^k + n(n+1)a_k x^k\right] = 0.$$

Since this relation is true for any x, the coefficients of any given power of x, for example, of x^k, must vanish and thus,

$$(k+2)(k+1)a_{k+2} - [k(k+1) - n(n+1)]\,a_k = 0,$$

which gives the recurrence relation

$$a_{k+2} = -\frac{n(n+1) - k(k+1)}{(k+1)(k+2)}\,a_k.$$

Now, if $k = n$, then $a_{k+2} = 0$, $a_{k+4} = 0$, and so on. Rewriting this relation as

$$a_k = -\frac{(k+2)(k+1)}{(n-k)(n+k+1)}\,a_{k+2},$$

we begin with $k = n - 2$ and obtain

$$a_{n-2} = -\frac{n(n-1)}{2 \cdot (2n-1)} a_n,$$

$$a_{n-4} = \frac{n(n-1)(n-2)(n-3)}{2 \cdot 4 \cdot (2n-1)(2n-3)} a_n,$$

$$a_{n-6} = -\frac{n(n-1)(n-2)(n-3)(n-4)(n-5)}{2 \cdot 4 \cdot 6 \cdot (2n-1)(2n-3)(2n-5)} a_n, \text{ and so on.}$$

If n is even, then we will end with a_0; if n is odd, then we will end with a_1. Thus, a solution of Eq (4.80) is

$$P(x) = a_n \left[x^n - \frac{n(n-1)}{2 \cdot (2n-1)} x^{n-2} + \frac{n(n-1)(n-2)(n-3)}{2 \cdot 4 \cdot (2n-1)(2n-3)} x^{m-4} - \cdots \right],$$

where a_n is arbitrary. It is convenient to take $a_n = \dfrac{(2n-1)((2n-3)\cdots 1}{n!}$, which
acts like a normalizing factor to give $P(1) = 1$. This function P which is a particular
solution of Eq (4.80), is usually denoted by $P_n(x)$ or $P_n(\cos\theta)$ and is called the
Legendre polynomials of order n, or *Legendre's coefficients* or *surface zonal harmonics*. A particular solution of Eq (4.79) is $u(r,\theta) = r^n P_n(\cos\theta)$ and is sometimes
called a *solid zonal harmonic*. This solution can be written in the form of an infinite
series as

$$u(r,\theta) = \sum_{k=0}^{\infty} A_k r^k P_k(\cos\theta), \tag{4.81}$$

where A_k are arbitrary constants to be determined from the condition at $\theta = 0$, i.e.,
$\sum_{k=0}^{\infty} A_k r^k = \dfrac{M}{\sqrt{a^2 + r^2}}$. Since by binomial theorem

$$\frac{M}{\sqrt{a^2 + r^2}} = \frac{M}{a} \left[1 - \frac{1}{2}\frac{r^2}{a^2} + \frac{1 \cdot 3}{2 \cdot 4}\frac{r^4}{a^4} - \frac{1 \cdot 3 \cdot 5}{2 \cdot 4 \cdot 6}\frac{r^6}{a^6} + \cdots \right], \quad r < a,$$

the required solution for $r < a$ is

$$u(r,\theta) = \frac{M}{a} \left[P_0(\cos\theta) - \frac{1}{2}\frac{r^2}{a^2}P_2(\cos\theta) + \frac{1 \cdot 3}{2 \cdot 4}\frac{r^4}{a^4}P_4(\cos\theta) \right.$$
$$\left. - \frac{1 \cdot 3 \cdot 5}{2 \cdot 4 \cdot 6}\frac{r^6}{a^6}P_6(\cos\theta) + \cdots \right].$$

The formula to compute a few values of $P_n(x)$ is

$$P_n(x) = \frac{(2n-1)((2n-3)\cdots 1}{n!} \left[x^n - \frac{n(n-1)}{2 \cdot (2n-1)} x^{n-2} \right.$$
$$+ \frac{n(n-1)(n-2)(n-3)}{2 \cdot 4 \cdot (2n-1)(2n-3)} x^{n-4}$$
$$\left. - \frac{n(n-1)(n-2)(n-3)(n-4)(n-5)}{2 \cdot 4 \cdot 6 \cdot (2n-1)(2n-3)(2n-5)} x^{n-6} + \cdots \right]. \tag{4.82}$$

The values of $P_n(x)$ and $P_n(\cos\theta)$ for $n = 1,\ldots,5$ are given in §C.1.1 and §C.1.2. ∎

We will now justify the value of the coefficient a_n which we have taken above as equal to $\dfrac{(2n-1)((2n-3)\cdots 1}{n!}$ in the definition (4.82) for $P_n(x)$. Consider the potential $u = 1/r$, where $r = |\mathbf{x} = \mathbf{x}_1|$, $\mathbf{x} = (x,y,z)$, $\mathbf{x}_1 = (x_1,y_1,z_1)$. In the spherical coordinates $x = r\sin\theta\cos\phi$, $y = r\sin\theta\sin\phi$, $z = r\cos\theta$, we choose $\mathbf{x} = (r,\theta,\phi)$ and $\mathbf{x}_1 = (r_1,\theta_1,\phi)$, where the point $\mathbf{x}_1$ is chosen so as to be independent of ϕ, and $\theta_1 = 0$. Then we find that

$$u = \frac{1}{r} = \frac{1}{\sqrt{r^2 - 2rr_1\cos\theta + r_1^2}},$$

is a solution of Eq (4.79). This solution can be written as

$$\text{(a)}: \quad u = \frac{1}{r}\left[1 - 2\frac{r_1}{r}\cos\theta + \left(\frac{r_1}{r}\right)\right]^{-1/2};$$

$$\text{or} \quad \text{(b)}: \quad u = \frac{1}{r_1}\left[1 - 2\frac{r}{r_1}\cos\theta + \left(\frac{r}{r_1}\right)\right]^{-1/2}.$$

We set $t = \dfrac{r_1}{r}$ in (a), and $t = \dfrac{r}{r_1}$ in (b). Then, since $x = \cos\theta$, we will now discuss in these two cases the expression $\left[1 - 2t\cos\theta + t^2\right]^{-1/2}$, which is known as the *generating function* for the Legendre polynomials $P_n(x)$.

In case (a), this generating function can be developed into a convergent series involving positive integral powers of t for $t < 1$:

$$\left[1 - 2t\cos\theta + t^2\right]^{-1/2} = \sum_{n=0}^{\infty} p_n(\cos\theta)t^n,$$

where the coefficients $p_n(\cos\theta)$ are yet to be determined. Thus, in this case we have $u = \dfrac{1}{r}\displaystyle\sum_{n=0}^{\infty} p_n(\cos\theta)\left(\dfrac{r_1}{r}\right)^n$. Substituting it in Legendre's equation (4.15) yields

$$\sum_{n=0}^{\infty} \frac{r_1^n}{r^{n+1}}\left[n(n+1)p_n(\cos\theta) + \frac{1}{\sin\theta}\frac{d}{d\theta}\left(\sin\theta\frac{dp_n}{d\theta}\right)\right] = 0,\ r > r_1.$$

Since the coefficients of each power of r must be zero, we obtain

$$\frac{1}{\sin\theta}\frac{d}{d\theta}\left(\sin\theta\frac{dp_n}{d\theta}\right) + n(n+1)\,p_n = 0.$$

With $x = \cos\theta$, this equation reduces to

$$(1 - x^2)\frac{d^2p_n}{dx^2} - 2x\frac{dp_n}{dx} + n(n+1)\,p_n = 0.$$

Hence, $t = p_n(x) = p_n(\cos\theta)$ is a solution of Legendre's equation, and so $p_n(x) \equiv P_n(x)$. Similarly, in case (b) the generating function can again be developed into a convergent series involving positive integral powers of t for $t < 1$. Then $u = \dfrac{1}{r_1} \displaystyle\sum_{n=0}^{\infty} p_n(\cos\theta)\left(\dfrac{r}{r_1}\right)^n$, which leads to

$$\sum_{n=0}^{\infty} \frac{r^n}{r_1^{n+1}}\left[n(n+1)p_n(\cos\theta) + \frac{1}{\sin\theta}\frac{d}{d\theta}\left(\sin\theta\frac{dp_n}{d\theta}\right)\right] = 0, \ r < r_1,$$

and this again shows that $t = p_n(x) = p_n(\cos\theta)$ is a solution of Legendre's equation, and so $p_n(x) \equiv P_n(x)$. But since p_n is the coefficient of the t^n in the development of the above generating function, we have $\left[1 - 2tx + t^2\right]^{-1/2} = \left[1 - t(2x - t)\right]^{-1/2}$, which when expanded by the binomial theorem gives

$$\frac{(2n-1)((2n-3)\cdots 1}{n!}\left[x^n - \frac{n(n-1)}{2(2n-1)}x^{n-2}\right.$$
$$\left. + \frac{n(n-1)(n-2)(n-3)}{2\cdot 4\cdot(2n-1)(2n-3)}x^{n-4} - \cdots\right],$$

and this is precisely (4.82) which defines $P_n(x)$.

The constant in formula (4.82) for $P_n(x)$ can be found as follows: It is obviously the value of $P_n(x)$ when $x = 0$. Thus, it is the coefficient of t^n in the development of $\left(1 + t^2\right)^{-1/2}$, which is

$$\left(1 + t^2\right)^{-1/2} = 1 - \frac{1}{2}t^2 + \frac{1\cdot 3}{2\cdot 4}t^2 - \frac{1\cdot 3\cdot 5}{2\cdot 4\cdot 6}t^6 - \cdots,$$

and the coefficient of t^n is

$$\begin{cases} (-1)^{n/2}\dfrac{1\cdot 3\cdot 5\cdots(n-1)}{2\cdot 4\cdot 6\cdots n}, & \text{if } n \text{ is an even number,} \\[2mm] (-1)^{(n-1)/2}\dfrac{1\cdot 3\cdot 5\cdots n}{2\cdot 4\cdot 6\cdots(n-1)}, & \text{if } n \text{ is an odd number. } \blacksquare \end{cases}$$

Example 4.7. Consider the problem of temperature distribution within a solid sphere of radius 1, one half of the surface of which is kept at the constant temperature zero and the other half at a constant temperature unity. Taking the diameter perpendicular to the plane separating the two halves as the axis of spherical coordinates, we solve the equation

$$\frac{1}{r^2}\frac{\partial}{\partial r}\left(r^2\frac{\partial u}{\partial r}\right) + \frac{1}{\sin\theta}\frac{\partial}{\partial\theta}\left(\sin\theta\frac{\partial u}{\partial\theta}\right) + \frac{1}{\sin^2\theta}\frac{\partial^2 u}{\partial\phi^2} = 0,$$

which, because of symmetry, is independent of ϕ, and reduces to Eq (4.79), subject to the conditions that $u = 1$ from $\theta = 0$ to $\theta = \pi/2$, and $u = 0$ from $\theta = \pi/2$ to

$\theta = \pi$. Then a particular solution of this equation is $u = r^n P_n(\cos\theta)$, where $n > 0$ is an integer, and $u(r, \theta) = \sum\limits_{k=0}^{\infty} A_k r^k P_k(\cos\theta)$, where A_k are yet undetermined coefficients. For $r = 1$, we have $u = \sum\limits_{k=0}^{\infty} A_k P_k(\cos\theta)$, or equivalently, $u = \sum\limits_{k=0}^{\infty} A_k P_k(x)$, where $u = 0$ from $x = -1$ to $x = 0$ and $u = 1$ from $x - 0$ to $x = 1$. Thus, since $P_0(x) = 1$ (see §C.1), we have $A_0 = \frac{1}{2}\int_0^1 P_0(x)\,dx = \frac{1}{2}$, and $A_n = \dfrac{2n+1}{2}\int_0^1 P_n(x)\,dx$ for $n \geq 1$, or since

$$\int_0^1 P_n(x)\,dx = \begin{cases} 0 & \text{if } n \text{ is even,} \\[2mm] (-1)^{(n-1)/2}\dfrac{1}{n(n+1)}\dfrac{1\cdot3\cdot5\cdots n}{2\cdot4\cdot6\cdots(n-1)} & \text{if } n \text{ is odd,} \end{cases}$$

we obtain the desired solution as

$$\begin{aligned} u &= \frac{1}{2} + \frac{3}{4}P_1(x) - \frac{7}{8}\frac{1}{2}P_3(x) + \frac{11}{12}\frac{1\cdot3}{2\cdot4}P_5(x) + \cdots \\ &= \frac{1}{2} + \frac{3}{4}P_1(\cos\theta) - \frac{7}{8}\frac{1}{2}P_3(\cos\theta) + \frac{11}{12}\frac{1\cdot3}{2\cdot4}P_5(\cos\theta) + \cdots . \blacksquare \end{aligned}$$

Example 4.8. The 2-D Laplace's equation is $\dfrac{\partial^2 u}{\partial x^2} + \dfrac{\partial^2 u}{\partial y^2} = 0$. Using Bernoulli's separation method, let $u = X(x)Y(y)$, which after substitution in the above equation, leads to $\dfrac{Y''}{Y} = -\dfrac{X''}{X} = m^2$, or equivalently, to

$$Y'' - m^2 Y = 0, \quad X'' + m^2 X = 0.$$

The general solution of the first equation is $Y = A e^{my} + B e^{-my}$, and that of the second $X = C \sin mx + D \cos mx$. Thus, there are four particular solutions of the 2-D Laplace's equation, namely, $u(x,y) = e^{my}\sin mx$, $u(x,y) = e^{my}\cos mx$, $u(x,y) = e^{-my}\sin mx$, $u(x,y) = e^{-my}\cos mx$. $\blacksquare$

Example 4.9. The 1-D wave equation is $\dfrac{\partial^2 u}{\partial t^2} = c^2\dfrac{\partial^2 u}{\partial x^2}$. Using Bernoulli's separation method, let $u = T(t)X(x)$, which after substitution in the above equation, leads to $\dfrac{T''}{c^2 T} = \dfrac{X''}{X} = -m^2$, or equivalently, to

$$X'' + m^2 X = 0, \quad T'' + c^2 m^2 T = 0.$$

The general solution of the first equation is $X = A \sin mx + B \cos mx$, and that of the second $T = C \sin mct + D \cos mct$. Hence, there are four particular solutions of the

1-D wave equation: $u = \sin mx \cos mct$, $u = \sin mx \sin mct$, $u = \cos mx \cos mct$, and $u = \cos mx \sin mct$. ∎

Example 4.10. Laplace's equation in cylindrical coordinates is given by

$$\frac{\partial^2 u}{\partial r^2} + \frac{1}{r}\frac{\partial u}{\partial r} + \frac{1}{r^2}\frac{\partial^2 u}{\partial \theta^2} + \frac{\partial^2 u}{\partial z^2} = 0. \tag{4.83}$$

Consider the following problems:

(a) The convex surface and one base of a cylinder of radius a and length b are kept at the constant temperature zero, and the temperature at each point of the other base is a preassigned function $f(r)$ of the distance from the center of the base. It is required to determine the temperature at any point of the cylinder after the steady state is attained. Thus, we must solve Eq (4.83) subject to the following conditions: $u(r, \theta, 0) = 0$, $u(a, \theta, z) = 0$, and $u(r, \theta, b) = f(r)$. Since, because of the symmetry $\frac{\partial^2 u}{\partial \theta^2} = 0$, we use Bernoulli's separation method by taking $u = R(r)Z(z)$, which breaks up Eq (4.83) into two equations

$$\frac{d^2 R}{dr^2} + \frac{1}{r}\frac{dR}{dr} + m^2 R = 0, \qquad \frac{d^2 Z}{dz^2} - m^2 Z = 0.$$

Solving these equations, we find the particular solutions of the problem as

$$u(r, z) = J_0(mr)\sinh(mz), \qquad u(r, z) = J_0(mr)\cosh(mz).$$

If m_k is a zero of $J_0(ma)$, then $u(r, z) = J_0(m_k r)\sinh(m_k z)$ satisfies Laplace's equation and the first two conditions. If $f(r) = \sum_{k=1}^{\infty} A_k J_0(m_k r)$, where m_k are zeros of $J_0(ma)$, then the required solution is

$$u(r, z) = \sum_{k=1}^{\infty} A_k \frac{\sinh(m_k z)}{\sin(m_k b)} J_0(m_k r). \tag{4.84}$$

(b) If instead of keeping the convex surface of the cylinder at temperature zero, we surround it it by an insulation impervious to heat, the condition $u(a, z) = 0$ is then replaced by $u_r(a, z) = 0$, or by $\left.\frac{dJ_0(mr)}{dr}\right|_{r=a} = 0$, i.e., by $m\,J_0'(ma) = 0$, or by $J_1(ma) = 0$. Then (4.84) will be the solution of this problem where m_k are now the zeros of $J_1(ma)$.

(c) If instead of keeping the convex surface of the cylinder at temperature zero, we allow it to cool in air at the temperature zero, the condition $u(a, z) = 0$ will be replaced by $u_r(a, z) + hu(a, z) = 0$, i.e., $u(r, z) = J_0(mr)\sinh(mz)$ will be replaced by $m\,J_0(ma) + h\,J_0(ma) = 0$, that is, by $ma\,J_0(ma) + ah\,J_0(ma) = 0$

or by $ma\,J_1(ma) - ha\,J_0(ma) = 0$. Then (4.84) will the solution of this problem where m_k are now the zeros of $ma\,J_1(ma) - ha\,J_0(ma) = 0$. ∎

Example 4.11. Consider Laplace's equation (4.11) in spherical coordinates, independent of ϕ:

$$\frac{\partial^2 u}{\partial r^2} + \frac{2}{r}\frac{\partial u}{\partial r} + \frac{1}{r^2 \sin\theta}\left(\sin\theta\frac{\partial u}{\partial\theta}\right) = 0. \tag{4.85}$$

Using Bernoulli's separation method, let $u = R(r)\Theta(\theta)$, which after substitution in Eq (4.85) gives

$$\frac{r}{R}\frac{d^2(rR)}{dr^2} = -\frac{1}{\sin\theta}\frac{d}{d\theta}\left(\sin\theta\frac{d\Theta}{d\theta}\right). \tag{4.86}$$

Since each side of Eq (4.86) must be equal to a constant, say α^2, this equation is equivalent to the following two equations:

$$r\frac{d^2(rR)}{dr^2} - \alpha^2 R = 0, \tag{4.87}$$

$$\frac{1}{\sin\theta}\frac{d}{d\theta}\left(\sin\theta\frac{d\Theta}{d\theta}\right) + \alpha^2\Theta = 0. \tag{4.88}$$

Eq (4.87) can be written as

$$r^2\frac{d^2 R}{dr^2} + 2r\frac{dR}{dr} - \alpha^2 R = 0,$$

and has the general solution as $R(r) = Ar^n + Br^m$, where $n = -\frac{1}{2} + \sqrt{\alpha^2 + \frac{1}{4}}$ and $m = -\frac{1}{2} - \sqrt{\alpha^2 + \frac{1}{4}}$. Thus, $m = -n - 1$, and we can write $\alpha^2 = n(n+1)$, where n is arbitrary. Then $R(r) = Ar^n + Br^{-n-1}$. Thus, $R = r^n$ and $R = r^{-n-1}$ are particular solutions of Eq (4.87) with $\alpha^2 = n(n+1)$, i.e., of the equation

$$r^2\frac{d^2 R}{dr^2} + 2r\frac{dR}{dr} - n(n+1)R = 0. \tag{4.89}$$

Again, with this value of α^2 Eq (4.88) becomes

$$\frac{1}{\sin\theta}\frac{d}{d\theta}\left(\sin\theta\frac{d\Theta}{d\theta}\right) + n(n+1)\Theta = 0,$$

where $n > 0$ is an integer (see Example 4.6), so that its particular solution is $\Theta(\theta) = P_n(\cos\theta)$. Hence, $u(r,\theta) = r^n P_n(\cos\theta)$ and $u(r,\theta) = \frac{1}{r^{n+1}}P_n(\cos\theta)$, where n is a positive integer, are particular solutions of Eq (4.85), where the first of these solutions is known as a *solid zonal harmonic*.

We will now determine a series solution of the equation

$$x^2 \frac{d^2 u}{dx^2} + 2x \frac{du}{dx} - n(n+1)u = 0, \tag{4.90}$$

where n is a positive integer. This equation is same as Eq (4.89) with r replaced by x and R by u. Assume that $u = \sum_{k=0}^{\infty} a_k x^k$. Then substituting it in Eq (4.90) we get $\sum_{k=0}^{\infty} [k(k+1) - n(n+1)] a_k x^k = 0$. This, being an identity equation, gives $[k(k+1) - n(n+1)] a_k \equiv 0$. Hence, $a_k = 0$ for all values of n except $n = k$ and $n = -k - 1$. Then, $u = Ax^n + Bx^{-n-1}$ is the general solution of Eq (4.90), and $u = x^n$ and $u = \frac{1}{x^{n+1}}$ are its particular solutions. If n is not a positive integer, this method fails and does not yield any result. ∎

Example 4.12. (*Circular drum*) Consider the problem of vibrations of a stretched circular membrane of radius a fastened on its circumference, which can be referred to as a circular drum, subject to the conditions that the membrane is initially distorted into a given form which has circular symmetry about the axis through the center perpendicular to the plane of the boundary, and then allowed to vibrate. Recall that a function of the coordinates of a point has circular symmetry about an axis when its value is not affected by rotating the point through an angle about the axis, and a surface has circular symmetry about an axis when it is a surface of revolution about the axis. Using the spherical coordinates the distortion u due to the symmetry is independent of ϕ, and is governed by the equation

$$\frac{\partial^2 u}{\partial t^2} = c^2 \left(\frac{\partial^2 u}{\partial r^2} + \frac{1}{r} \frac{\partial u}{\partial r} \right), \tag{4.91}$$

subject to the conditions $u(r, 0) = f(r)$, $u_t(r, 0) = 0$, and $u(a, t) = 0$. Using Bernoulli's separation method, let $u(r, t) = R(r)T(t)$. Substituting it in Eq (4.91) we get $RT'' = c^2 T \left(R'' + \frac{1}{r} R' \right)$. Since the left side of this equation does not involve r while the right side does not contain t, each side must be a constant, say $-m^2$, where where m is an yet undetermined constant. Then this equation can be written as

$$\frac{T''}{c^2 T} = \frac{1}{R} \left(R'' + \frac{R'}{r} \right) = -m^2,$$

which is equivalent to the following two equations:

$$T'' + m^2 c^2 T = 0, \tag{4.92}$$

$$R'' + \frac{1}{r} R' + m^2 R = 0. \tag{4.93}$$

The particular solutions of Eq (4.92) are $T = \cos mct$ and $T = \sin mct$. To solve Eq (4.93), set $r = x/m$. Then it becomes

$$\frac{d^2 R}{dx^2} + \frac{1}{x}\frac{dR}{dx} + R = 0. \tag{4.94}$$

This equation was first studied by Fourier while considering the cooling of a cylinder, and therefore, it is sometimes known as Fourier's equation. We will find a series solution for Eq (4.94): Assume that R can be expressed in the form of an infinite series in powers of x, i.e., take $R = \sum_{k=0}^{\infty} a_k\, x^k$. Then, substituting it in Eq (4.94), we obtain

$$\sum_{k=0}^{\infty} \left[k(k-1)a_k\, x^{k-2} + ka_k\, x^{k-2} + a_k\, x^k\right] = 0.$$

This equation must be true for all values of x. The coefficient of any power of x, say of x^{k-2}, must vanish, which gives $k(k-1)a_k + ka_k + a_{k-2} = 0$, or $k^2 a_k + a_{k-2} = 0$, or

$$a_{k-2} = -k^2 a_k. \tag{4.95}$$

If $k = 0$, then $a_{k-2} = 0$, $a_{k-4} = 0$, and so on. Thus, we start with $k = 0$, and (4.95) gives the recurrence relation

$$a_k = -\frac{a_{k-2}}{k^2},$$

whence we get $a_2 = -\dfrac{a_0}{2^2}$, $a_4 = \dfrac{a_0}{2^2 \cdot 4^2}$, $a_6 = -\dfrac{a_0}{2^2 \cdot 4^2 \cdot 6^2}$, and so on. Hence,

$$R = a_0 \left[1 - \frac{x^2}{2^2} + \frac{x^4}{2^2 \cdot 4^2} - \frac{x^6}{2^2 \cdot 4^2 \cdot 6^2} + \cdots\right],$$

where a_0 is arbitrary, provided the series is convergent. If we take $a_0 = 1$, then $R = J_0(x)$, where $J_0(x)$ is the Bessel function of order zero, and the corresponding series is convergent for all real or purely imaginary values of x, since the series

$$J_0(r) = 1 - \frac{r^2}{2^2} + \frac{r^4}{2^2 \cdot 4^2} - \frac{r^6}{2^2 \cdot 4^2 \cdot 6^2} + \cdots$$

composed of the moduli of x, is convergent for all $r = |x|$. Using the ratio test for this series, we find that the ratio $\left|\dfrac{a_{n+1}}{a_n}\right| = \dfrac{r^2}{4n^2} \to 0$ as $n \to \infty$. Hence, $J_0(x)$ is absolutely convergent. The function $J_0(x)$ is called a *cylindrical harmonic* or the Bessel function of order zero.

Note that Eq (4.94) was obtained by substituting $x = mr$; thus,

$$R = J_0(mr) = 1 - \frac{(mr)^2}{2^2} + \frac{(mr)^4}{2^2 \cdot 4^2} - \frac{(mr)^6}{2^2 \cdot 4^2 \cdot 6^2} + \cdots.$$

Hence, $u(r,t) = J_0(mr) \cos mct$ or $u(r,t) = J_0(mr) \sin mct$ is a particular solution of Eq (4.91). But since $u(r,t) = J_0(mr) \cos mct$ satisfies the prescribed second condition for any value of m, in order to satisfy the first condition, m must be taken such that $J_0(ma) = 0$, i.e., m must be a zero of $J_0(ma)$. It is known that $J_0(x)$ has an infinite number of real positive zeros (see Abramowitz and Stegun [1965]), any one of which can be obtained to any required degree of approximation. Let these zeros be $x_1, x_2, x_3, \ldots$. Then, if $\dfrac{x_1}{a} = m_1$, $\dfrac{x_2}{a} = m_2$, $\dfrac{x_3}{a} = m_3$, and so on, we get

$$u(r,t) = \sum_{k=1}^{\infty} A_k J_0(m_k r) \cos m_k ct, \qquad (4.96)$$

where A_k are constants, as a solution of Eq (4.91) which satisfies second and third prescribed conditions. Now, to satisfy the first prescribed condition, for $t = 0$ the solution (4.96) reduces to $u(r,0) = \sum_{k=1}^{\infty} A_k J_0(m_k r)$. If $f(r)$ can be expressed as a series of this form, the solution of the problem can be obtained by substituting the coefficients of that series for A_k, $k = 1, 2, \ldots$. ∎

Example 4.13. (*Cooling ball*) Consider the boundary value problem

$$u_t = \nabla^2 u, \quad 0 \le r < 1, \quad 0 \le \phi \le \pi, \quad t > 0,$$

subject to the conditions $u(r, \phi, 0) = f(r, \phi)$, and $u(1, \phi, t) = 0$, where the function $u = u(r, \phi, t)$ is independent of θ. The problem describes the temperature distribution in the interior of the unit ball dropped in cold water. The first condition implies that the temperature u is not uniform but depends only on r, ϕ and t. Thus, the solution is assumed formally as

$$u(r, \phi, t) = R(r)\, \Phi(\phi)\, T(t), \qquad (4.97)$$

which, after separating the variables, gives

$$\frac{R'' + \dfrac{2}{r} R'}{R} + \frac{\Phi'' + \cot\phi\, \Phi'}{r^2\, \Phi} = -\alpha^2 = \frac{T'}{T} = \lambda. \qquad (4.98)$$

The left side of Eq (4.98) can be expressed as

$$\left(\frac{R'' + \dfrac{2}{r} R'}{R} + \alpha^2 \right) r^2 = -\frac{\Phi'' + \cot\phi\, \Phi'}{\Phi}.$$

In order that this equation be satisfied, the terms on each side must be constant. Thus,

$$\left(\frac{R'' + \dfrac{2}{r} R'}{R} + \alpha^2 \right) r^2 = \mu = -\frac{\Phi'' + \cot\phi\, \Phi'}{\Phi}. \qquad (4.99)$$

It is known that $\mu = n(n + 1)$ (Courant and Hilbert [1968]). Then Eq (4.99) yields

$$\Phi'' + \cot\phi\,\Phi' + n(n+1)\,\Phi = 0,$$
$$r^2\,R'' + 2r\,R' + \alpha^2 r^2\,R - n(n+1)\,R = 0. \tag{4.100}$$

The first equation reduces to Legendre's equation by using the substitution $z = \cos\phi$ (see Example 4.8), and its bounded solution is given by $\Phi = P_n(\cos\phi)$. If we set $x = \alpha r$ in the second equation, it becomes

$$x^2 R'' + 2x\,R' + \left[x^2 - n(n+1)\right]\,R = 0,$$

which under the transformation $w = \sqrt{x}\,R$ reduces to Bessel's equation

$$x^2 w'' + x\,w' + \left[x^2 - (n + \tfrac{1}{2})^2\right]\,w = 0,$$

and has a bounded solution $w = J_{n+1/2}(x)$. Hence, $R(r) = \dfrac{J_{n+1/2}(\alpha r)}{\sqrt{\alpha r}}$. The eigenfunctions $\psi_{mn} = \dfrac{J_{n+1/2}(\alpha_{mn} r)}{\sqrt{\alpha_{mn} r}}\,P_n(\cos\phi)$ form an orthogonal set. Using the boundary condition at $r = 1$, we get $J_{n+1/2}(\alpha) = 0$. Let α_{mn} denote the positive zeros of $J_{n+1/2}(\alpha)$. Then the solution for the temperature distribution in the unit ball is given by

$$u(r, \phi, t) = \sum_{\substack{m=1 \\ n=0}}^{\infty} C_{mn} \frac{e^{-\alpha_{mn}^2 t}\,J_{n+1/2}(\alpha_{mn} r)}{\sqrt{\alpha_{mn} r}}\,P_n(\cos\phi),$$

where, after using $\displaystyle\int_{-1}^{1} P_n^2(x)\,dx = \frac{2}{2n+1}$,

$$C_{mn} = \frac{(2n+1)\sqrt{\alpha_{mn}}}{\beta(\alpha_{mn})} \int_0^\pi \int_0^1 r^{3/2} J_{n+1/2}(\alpha_{mn} r) P_n(\cos\theta) f(r,\phi)\,\sin\phi\,dr\,d\phi,$$

and

$$\beta(\alpha_{mn}) = 2 \int_0^1 r\,J_{n+1/2}^2(\alpha_{mn} r)\,dr = \left[J_{n+1/2}'(\alpha_{mn})\right]^2. \quad\blacksquare$$

Example 4.14. Consider the steady-state heat conduction or potential problem

$$u_{xx} + u_{yy} = 0, \quad 0 < x < \pi, \quad 0 < y < 1,$$

subject to the mixed boundary conditions

$$u(x, 0) = u_0 \cos x, \quad u(x, 1) = u_0 \sin^2 x, \quad u_x(0, y) = 0 = u_x(\pi, y). \tag{4.101}$$

The separation of variables technique leads to the following set of ordinary differential equations:

$$X'' + \lambda^2 X = 0, \quad X'(0) = 0 = X'(\pi), \tag{4.102}$$

$$Y'' - \lambda^2 Y = 0. \tag{4.103}$$

The eigenvalues and the corresponding eigenfunctions for (4.102) are

$$\lambda_0 = 0, \quad X_0(x) = 1,$$
$$\lambda_n = n^2, \quad X_n(x) = \cos nx, \quad n = 1, 2, \ldots,$$

and subsequently the solutions of (4.103) are

$$Y_n(y) = \begin{cases} A_0 + B_0 y, & n = 0, \\ A_n \cosh ny + B_n \sinh ny, & n = 1, 2, \ldots. \end{cases} \tag{4.104}$$

Hence, using the superposition principle, we get

$$u(x, y) = A_0 + B_0 y + \sum_{n=1}^{\infty} [A_n \cosh ny + B_n \sinh ny] \cos nx. \tag{4.105}$$

Now, the first boundary condition in (4.101) leads to

$$u(x, 0) = A_0 + \sum_{n=1}^{\infty} A_n \cos nx = u_0 \cos x. \tag{4.106}$$

By comparing the coefficients of similar terms on both sides of (4.106), we find that $A_0 = 0$, $A_1 = u_0$, and $A_n = 0$ for $n \geq 2$. Hence, the solution becomes

$$u(x, y) = B_0 y + u_0 \cosh y \cos x + \sum_{n=1}^{\infty} B_n \sinh ny \cos nx. \tag{4.107}$$

Similarly, using the second boundary condition in (4.101) we find from (4.107) that

$$u(x, 1) = B_0 + u_0 \cosh 1 \cos x + \sum_{n=1}^{\infty} B_n \sinh n \cos nx$$
$$= u_0 \sin^2 x = u_0 \frac{1 - \cos 2x}{2},$$

from which, after comparing the coefficients of similar terms on both sides, we get

$$B_0 = \frac{u_0}{2}, B_1 = -\frac{u_0 \cosh 1}{\sinh 1}, B_2 = -\frac{u_0}{2 \sinh 2}, B_n = 0 \quad \text{for } n \geq 3. \text{ Hence,}$$

from (4.107) the general solution is given by

$$u(x,y) = \frac{u_0}{2}y + u_0 \left[\cosh y - \frac{\cosh 1 \sinh y}{\sinh 1}\right]\cos x - u_0\frac{\sinh 2y}{2\sinh 2}\cos 2x$$

$$= u_0 \left[\frac{1}{2}y + \frac{\sinh(1-y)}{\sinh 1}\cos x - \frac{\sinh 2y}{2\sinh 2}\cos 2x\right]. \blacksquare \tag{4.108}$$

4.5. Exercises

4.1. Show that the general equation of the hydrogen atom

$$\nabla^2 u(r,\theta,\phi) + \left[k^2 + f(r) + \frac{1}{r}\frac{2}{} g(\theta) + \frac{1}{r^2\sin^2\theta}h(\phi)\right]u(r,\theta,\phi) = 0$$

is separable, where the functions f, g and h are functions of the variable shown, and k^2 is a constant.

HINT. Use Bernoulli's separation method: take $u = R(r)\Theta(\theta)\Phi(\phi)$, and separate the variables.

4.2. Show that if $x = 1$, then $P_n(x) = 1$.

HINT. The generating function is $(1 - 2t + t^2)^{-1/2} = (1-t)^{-1} = 1+t+t^2+ t^3 + \cdots$. The coefficients of each power being 1, we get $P_n(1) = 1$.

4.3. Compute the value of potential $u(r,\theta)$ for the circular ring (see Example 4.6), using the following data: $M = 1$, $a = 1$, and

(a) $r = 0.2$, $\theta = 0$,   (b) $r = 0.2$, $\theta = \pi/4$,   (c) $r = 0.2$, $\theta = \pi/2$,

(d) $r = 0.6$, $\theta = 0$,   (e) $r = 0.6$, $\theta = \pi/6$,   (f) $r = 0.6$, $\theta = \pi/3$,

(g) $r = 0.6$, $\theta = \pi/2$.

ANS. (a) 0.9806, (b) 0.9948, (c) 1.0102, (d) 0.8540, (e) 0.8941,
       (f) 1.0037, (g) 1.0763.

4.4. Show that Laplace's equation is not completely separable in toroidal coordinates.

NOTE. The toroidal coordinates are defined by

$$x = \frac{a\sinh v \cos\phi}{\cosh v - \cos u}, \quad y = \frac{a\sinh v \sin\phi}{\cosh v - \cos u}, \quad z = \frac{a\sinh u}{\cosh v - \cos u},$$

$$h_1^2 = h_2^2 = \frac{a^2}{(\cosh v - \cos u)^2}, \quad h_3^2 = \frac{a^2\sinh^2 v}{(\cosh v - \cos u)^2},$$

and Laplace's equation is

$$\nabla^2\Phi = \frac{(\cosh v - \cos u)^3}{a^2}\frac{\partial}{\partial u}\left(\frac{1}{\cosh v - \cos u}\frac{\partial\Phi}{\partial u}\right)$$

$$+ \frac{(\cosh v - \cos u)^3}{a^2\sinh v}\frac{\partial}{\partial v}\left(\frac{\sinh v}{\cosh v - \cos u}\frac{\partial\Phi}{\partial v}\right) + \frac{(\cosh v - \cos u)^2}{a^2\sinh^2 v}\frac{\partial\Phi}{\partial\phi}.$$

This coordinate system is seldom used although it has some very rare physical applications, such as vortex rings.

4.5. Show that Bernoulli's separation method applied to Laplace's equation in parabolic cylindrical coordinates leads to Hermite's equation $y'' - 2xy' + 2ny = 0$.

HINT. Parabolic cylindrical coordinates are defined by $x = \frac{1}{2}(u^2 - v^2)$, $y = uv$, $z = z$, and Laplace's equation is

$$\nabla^2 w = \frac{1}{u^2 + v^2}\left(\frac{\partial^2 w}{\partial u^2} + \frac{\partial^2 w}{\partial v^2}\right) + \frac{\partial^2 w}{\partial z^2}.$$

4.6. If $J_n(x)$ is a solution of Bessel's equation, verify that $u(x) = x^a J_n(bx^c)$ satisfies the equation

$$u'' + \frac{1 - 2a}{x}u' + \left[(bcx^{c-1})^2 + \frac{a^2 - n^2c^2}{x^2}\right]u = 0.$$

4.7. Solve $u_t = a^2 u_{xx} + f(x,t)$, $0 < x < l$, subject to the boundary conditions $u(0,t) = 0 = u(l,t)$ for $0 \le x \le l$, and the initial condition $u(x,0) = 0$ for $t > 0$.

ANS. $u(x,t) = \sum_{n=1}^{\infty} u_n(t) \sin\frac{n\pi x}{l}$, where

$$u_n(t) = \int_0^t e^{n^2\pi^2 a^2(\tau-t)/l^2} f_n(\tau)\,d\tau, \quad f_n(t) = \frac{2}{l}\int_0^l f(\xi,t)\sin\frac{n\pi\xi}{l}\,d\xi.$$

4.8. Find the interior temperature of the cooling ball of Example 4.13, if

$$f(r,\phi) = \begin{cases} 1, & 0 \le \phi < \pi/2 \\ 0, & \pi/2 \le \phi < \pi. \end{cases}$$

ANS. $u(r,\phi,t) = \sum_{m,n=1}^{\infty} C_{mn}\frac{J_{n+1/2}(\alpha_{mn}r)P_n(\cos\phi)}{\sqrt{\alpha_{mn}r}}e^{-\alpha_{mn}^2 t}$,

$$C_{mn} = \frac{(2n+1)P_n'(0)\sqrt{\alpha_{mn}}\int_0^1 r^{3/2}J_{n+1/2}(\alpha_{mn}r)\,dr}{2n(n+1)\int_0^1 rJ_{n+1/2}^2(\alpha_{mn}r)\,dr}$$

$$= \frac{(2n+1)P_n'(0)\sqrt{\alpha_{mn}}\int_0^1 r^{3/2}J_{n+1/2}(\alpha_{mn}r)\,dr}{n(n+1)J_{n-1/2}^2(\alpha_{mn})},$$

where α_{mn} are the consecutive positive roots of $J_{n+1/2}(\alpha) = 0$.

4.9. Solve $u_t = u_{xx}$, $-\pi < x < \pi$, subject to the conditions $u(x,0) = f(x)$, $u(-\pi,t) = u(\pi,t)$, and $u_x(-\pi,t) = u_x(\pi,t)$, where $f(x)$ is a periodic function of period 2π. This problem describes the heat flow inside a rod of length 2π, which is shaped in the form of a closed circular ring.

HINT: Assume $X(x) = A \cos \omega x + B \sin \omega x$.

ANS. $\omega_n = n$; $u(x,t) = \sum_{n=0}^{\infty} e^{-n^2 t} (a_n \cos nx + b_n \sin nx)$, where

$$a_n = \frac{1}{\pi} \int_{-\pi}^{\pi} f(x) \cos nx \, dx, \ b_n = \frac{1}{\pi} \int_{-\pi}^{\pi} f(x) \sin nx \, dx, \ a_0 = \frac{1}{2\pi} \int_{-\pi}^{\pi} f(x) \, dx.$$

4.10. Solve the problem $u_t = \nabla^2 u$, $r < 1$, $0 < z < 1$, such that $u(r, z, 0) = 1$, $u(1, z, t) = 0$, and $u(r, 0, t) = 0 = u(r, 1, t)$. This problem describes the temperature distribution inside a homogeneous isotropic solid circular cylinder.

ANS. $u(r, z, t) = \sum_{m,n=1}^{\infty} C_{mn} e^{-(\lambda_m^2 + n^2 \pi^2)t} J_0(\lambda_m r) \sin n\pi z$, where λ_m are

the zeros of J_0, and $C_{mn} = \dfrac{4(1 - (-1)^n)}{n\pi \lambda_m J_1(\lambda_m)}$.

4.11. Solve the steady-state problem of temperature distribution in a half-cylinder $0 \le r \le 1$, $0 \le \theta \le \pi$, $0 \le z \le 1$, where the flat faces are kept at $0°$ and the curved surface at $1°$.

ANS. $u(r, \theta, z) = \dfrac{16}{\pi^2} \sum_{\substack{m,n=1 \\ m,n \text{ odd}}}^{\infty} \dfrac{I_m(n\pi r)}{I_m(n\pi)} \dfrac{\sin n\pi z}{mn} \sin m\theta$.

4.12. Solve $u_{tt} = c^2 (u_{xx} + u_{yy})$ in the rectangle $R = \{(x, y) : 0 < x < a, \ 0 < y < b\}$, subject to the condition $u = 0$ on the boundary of R for $t > 0$, and the initial conditions $u(x, y, 0) = f(x, y)$, $u_t(x, y, 0) = g(x, y)$. This problem describes a vibrating rectangular membrane.

ANS. $u(x, y, t) = \sum_{m,n=1}^{\infty} (A_{mn} \cos \lambda_{mn} t + B_{mn} \sin \lambda_{mn} t) \sin \dfrac{m\pi x}{a} \sin \dfrac{n\pi y}{b}$,

where

$$A_{mn} = \frac{4}{ab} \int_0^b \int_0^a f(x, y) \sin \frac{m\pi x}{a} \sin \frac{n\pi y}{b} \, dx \, dy,$$

$$B_{mn} = \frac{4}{ab\lambda_{mn}} \int_0^b \int_0^a g(x, y) \sin \frac{m\pi x}{a} \sin \frac{n\pi y}{b} \, dx \, dy,$$

for $m, n = 1, 2, \ldots$; the eigenvalues are $\lambda_{mn} = c\pi \sqrt{\dfrac{m^2}{a^2} + \dfrac{n^2}{b^2}}$.

4.13. Solve $u_{xx} - u_{tt} = e^{-a^2 \pi^2 t} \sin a\pi x$, subject to the conditions $u(x, 0) = 0$, $u_t(x, 0) = 0$, and $u(0, t) = u(1, t) = 0$, where a is an integer.

ANS. $\dfrac{1}{a^2 \pi^2 (1 + a^2 \pi^2)} \left[\cos a\pi t - e^{-a^2 \pi^2 t} - a\pi \sin a\pi t \right] \sin a\pi x$.

4.14. Solve $u_{xx} + u_{yy} = 0$, under the conditions $u(x, 0) = 0 = u(x, \pi)$, $u(0, y) = 0$, and $u(\pi, y) = \cos^2 y$.

ANS. $u = \sum_{n=1}^{\infty} C_n \sinh nx \sin ny$, where $C_n = \dfrac{2}{\pi \sinh n\pi} \displaystyle\int_0^\pi \cos^2 y \sin ny \, dy =$

$\dfrac{1}{\pi \sinh n\pi} \left[1 - (-1)^n \right] \left[\dfrac{1}{n} - \dfrac{n}{n^2 - 4} \right]$, $n \neq 2$, and $C_2 = 0$.

4.15. Solve Poisson's equation $u_{xx} + u_{yy} = -1$, $0 < x, y < 1$, subject to the Dirichlet boundary conditions $u(0, y) = 0 = u(1, y) = u(x, 0) = u(x, 1)$.

ANS. $u(x, y) = \dfrac{16}{\pi^4} \displaystyle\sum_{\substack{j,k=1 \\ j,k \text{ odd}}} \dfrac{\sin j\pi x \sin k\pi y}{jk \, (j^2 + k^2)}$.

4.16. Solve $u_{rr} + \dfrac{1}{r} u_r = u_{tt}$, subject to the conditions $u(r, 0) = u_0 r$, $u_t(r, 0) = 0$, and $u(a, t) = au_0$, $\lim_{r \to 0} |u(r, t)| < \infty$, where u_0 is a constant.

ANS. $u = \displaystyle\sum_{i=1}^{\infty} A_i J_0(\alpha_i r) \cos \alpha_i t + u_0 a$, where α_i are the consecutive positive roots of $J_0(\alpha a) = 0$, and

$$A_i = u_0 \dfrac{\int_0^a r(r - a) J_0(\alpha_i r) \, dr}{\int_0^a r J_0^2(\alpha_i r) \, dr} = 2u_0 \dfrac{\int_0^a r(r - a) J_0(\alpha_i r) \, dr}{a^2 J_1^2(\alpha_i a)}.$$

4.17. Solve $u_{rr} + \dfrac{1}{r} u_r + \dfrac{1}{r^2} u_{\theta\theta} = 0$, subject to the conditions $u = 0$ for $\theta = 0$ or $\pi/2$, and $u_r = \sin \theta$ at $r = a$.

ANS. $u = \displaystyle\sum_{n=1}^{\infty} C_n r^{2n} \sin 2n\theta$, where $C_n = \dfrac{4(-1)^{n+1}}{\pi(4n^2 - 1)a^{2n}}$.

4.18. Solve the problem of vibrations of a plucked string of length l, stretched and fixed at both ends. The string is raised to a height b at $x = a$ and then released to vibrate freely. The problem to be solved consists of the wave equation $u_{tt} = c^2 u_{xx}$, $0 < x < l$, $t > 0$, subject to the boundary conditions $u(0, t) = 0 = u(l, t)$ and the initial condition $u(x, 0) = f(x) = \begin{cases} \dfrac{bx}{a}, & 0 \leq x \leq a, \\ \dfrac{b(l - x)}{l - a}, & a \leq x \leq l. \end{cases}$

HINT. See Example 4.1; note that $h(x) = 0$, so $B_n = 0$, and

$$A_n = \dfrac{2}{l} \int_0^l f(x) \sin \dfrac{n\pi x}{l} \, dx$$

$$= \dfrac{2}{l} \left[\int_0^a \dfrac{bx}{a} \sin \dfrac{n\pi x}{l} \, dx + \int_a^l \dfrac{b(l - x)}{l - a} \sin \dfrac{n\pi x}{l} \, dx \right] = \dfrac{2bl^2}{a(l - a)\pi^2 n^2} \sin \dfrac{n\pi a}{l}.$$

ANS. $u(x, t) = \dfrac{2bl^2}{a(l - a)\pi^2} \displaystyle\sum_{n=1}^{\infty} \dfrac{1}{n^2} \sin \dfrac{n\pi a}{l} \sin \dfrac{n\pi x}{l} \cos \dfrac{n\pi ct}{l}$.

4.19. Show that the temperature distribution at any time t in a long cylinder of radius a, which is kept inside at temperature 1 throughout and whose convex surface outside is kept at temperature zero, is given by $u(r,t) = \sum\limits_{k=1}^{\infty} A_k\, e^{-c^2 m_k^2 t}\, J_0\,(m_k r)$, where m_k are the zeros of $J_0(ma)$, and $1 = \sum\limits_{k=1}^{\infty} A_k\, J_0\,(m_k r)$.

4.20. Use Bernoulli's separation method to solve the problem of transverse vibrations of a beam governed by $u_{tt} + a^2 u_{xxxx} = 0$, subject to the conditions $u(0,t) = u(L,t) = u_{xx}(0,t) = u_{xx}(L,t) = u_t(x,0) = 0$, and $u(x,0) = f(x)$.

SOLUTION. Let $u = X(x)T(t)$, then we have $\dfrac{X^{(4)}}{X} = -\dfrac{T''}{a^2 T} = \lambda$, where λ is a parameter. By standard arguments it can be shown that the relevant values of λ are positive. Let $\lambda = \alpha^4$. Then the solutions for X and T are given by $X = A\cos\alpha x + B\sin\alpha x + C\cosh\alpha x + D\sin\alpha x$, and $T = E\cos\alpha^2 t + F\sin\alpha^2 t$. The condition $X(0) = 0$ means $A + C = 0$, and $X(L) = 0$ yields $A\cos\alpha L + B\sin\alpha L + C\cosh\alpha L + D\sin\alpha L = 0$, and $X_{xx}(0) = 0$ implies $2\alpha^2 A = 0$, which gives $A = C = 0$, and $X_{xx}(L) = 0$, which yields $\alpha^2(B\sinh\alpha L - D\sin\alpha L) = 0$. Thus, we have a pair of two homogeneous equations:

$$B\sinh\alpha L - D\sin\alpha L = 0, \quad B\sinh\alpha L + D\sin\alpha L = 0.$$

For B and D to have nontrivial values, we must have $\sinh\alpha L \sin\alpha L = 0$, i.e., $\alpha L = n\pi$, and $B = 0$ and $T(0) = 0$ are equivalent to $F = 0$. Absorbing E in D and using the initial condition, we get:

$$u = \sum_{n=1}^{\infty} D_n \sin\frac{n\pi x}{L}\cos\frac{a n^2\pi^2 t}{L^2}, \text{ where } D_n = \frac{2}{L}\int_0^L f(x)\sin\frac{n\pi x}{L}dx.$$

5

Integral Transforms

The integral transform method is very useful in the study of linear partial differential equations and determination of Green's functions. A function $f(x)$ may be transformed by the formula $F(s) = \int_a^b k(s,x)f(x)\,dx$, provided $F(s)$ exists, where s denotes the variable of the transform, $F(s)$ is the integral transform of $f(x)$, and $k(s,x)$ is the *kernel* of the transform. An integral transform is a linear transformation, which, when applied to a linear initial or boundary value problem, reduces the number of independent variables by one for each of its application. Thus, a partial differential equation can be reduced to an algebraic equation by repeated application of integral transforms. The algebraic problem is generally easy to solve for the function $F(s)$, and the solution of the problem is found by obtaining the function $f(x)$ from $F(s)$ by inversion formula of the transform applied to the problem. In this chapter we will discuss basic definitions and some important result of the Laplace, Fourier, finite Fourier, multiple Laplace and Fourier, and Hankel transforms, which are used in this book. Tables containing certain important pairs for these transforms are provided in Appendix D.

5.1. Integral Transform Pairs

Definitions of some transforms and their inverses are given below.

1. Laplace transform: $\mathcal{L}\{f(t)\} \equiv F(s) = \overline{f}(s) = \displaystyle\int_0^\infty f(t)\,e^{-st}\,dt,\ \Re\{s\} > 0,$

   and its inverse: $\mathcal{L}^{-1}\{F(s)\} \equiv f(t) = \dfrac{1}{2\pi i}\displaystyle\int_{c-i\infty}^{c+i\infty} F(s)\,e^{st}\,ds,\ c > 0.$

2. Fourier cosine transform: $\mathcal{F}_c\{f(x)\} \equiv \widetilde{f}_c(\alpha) = \sqrt{\dfrac{2}{\pi}}\displaystyle\int_0^\infty f(x)\cos(x\alpha)\,dx,$

   and its inverse: $\mathcal{F}^{-1}\{\widetilde{f}_c(\alpha)\} \equiv f(x) = \sqrt{\dfrac{2}{\pi}}\displaystyle\int_0^\infty \widetilde{f}_c(\alpha)\cos(x\alpha)\,d\alpha.$

3. Fourier sine transform: $\mathcal{F}_s\{f(x)\} \equiv \widetilde{f}_s(\alpha) = \sqrt{\dfrac{2}{\pi}}\displaystyle\int_0^\infty f(x)\sin(x\alpha)\,dx,$

and its inverse: $\mathcal{F}^{-1}\{\tilde{f}_s(\alpha)\} \equiv f(x) = \sqrt{\dfrac{2}{\pi}} \displaystyle\int_0^\infty \tilde{f}_s(\alpha) \sin(x\alpha)\, d\alpha.$

4. Fourier complex transform: $\mathcal{F}\{f(x)\} = \tilde{f}(\alpha) = \dfrac{1}{\sqrt{2\pi}} \displaystyle\int_{-\infty}^\infty f(x)\, e^{ix\alpha}\, dx,$

and its inverse: $\mathcal{F}^{-1}\{\tilde{f}(\alpha)\} \equiv f(x) = \dfrac{1}{\sqrt{2\pi}} \displaystyle\int_{-\infty}^\infty \tilde{f}(\alpha) e^{-ix\alpha}\, d\alpha.$

5. Hankel transform of order n:

$$\mathcal{H}_n\{f(x)\} \equiv F_n(\sigma) = \hat{f}_n(\sigma) = \int_0^\infty x f(x) J_n(\sigma x)\, dx,$$

and its inverse: $\mathcal{H}_n^{-1}\{F_n(\sigma)\} \equiv f(x) = \displaystyle\int_0^\infty \sigma F_n(\sigma)\, J_n(\sigma x)\, d\sigma.$

These definitions are not unique, particularly in the case of Fourier transforms.

5.2. Laplace Transform

Some basic properties of the Laplace transform are:

(i) $\mathcal{L}\left\{e^{at} f(t)\right\} = F(s-a)$, and $\mathcal{L}^{-1}\left\{F(s-a)\right\} = e^{at} f(t)$.

(ii) $\mathcal{L}\left\{H(t-a)f(t-a)\right\} = e^{-as} F(s)$, and $\mathcal{L}^{-1}\left\{e^{-as} F(s)\right\} = H(t-a)f(t-a)$.

(iii) Convolution Theorem: $\mathcal{L}^{-1}\left\{G(s)F(s)\right\} \equiv F \star G$

$$= \int_0^t f(t-u)g(u)\, du = \int_0^t f(u)g(t-u)\, du.$$

(iv) $\mathcal{L}\left\{\dfrac{d^n f(t)}{dt^n}\right\} = s^n F(s) - s^{n-1}f(0) - s^{n-2}f'(0) - \cdots - sf^{(n-2)}(0) - f^{(n-1)}(0),$

and $\mathcal{L}^{-1}\left\{s^n F(s)\right\} = \dfrac{d^n}{dt^n} f(t).$

(v) $\mathcal{L}^{-1}\left\{\dfrac{1}{s}F(s)\right\} = \displaystyle\int_0^t f(u)\, du.$

(vi) $\mathcal{L}\left\{t^n f(t)\right\} = (-1)^n \dfrac{d^n F}{ds^n}$, and $\mathcal{L}^{-1}\left\{(-1)^n \dfrac{d^n F}{ds^n}\right\} = t^n f(t).$

(vii) If $\mathcal{L}\{f(x,t)\} = F(x,s)$, then

$$\mathcal{L}\left\{\dfrac{\partial f(x,t)}{\partial x}\right\} = \dfrac{\partial F(x,s)}{\partial x}, \quad \text{and} \quad \mathcal{L}^{-1}\left\{\dfrac{\partial F(x,s)}{\partial x}\right\} = \dfrac{\partial f(x,t)}{\partial x}.$$

The last two results, based on the Leibniz theorem, are extremely effective. This theorem states that if $g(x,t)$ is an integrable function of t for each value of x, and the partial derivative $\dfrac{\partial g(x,t)}{\partial x}$ exists and is continuous in the region under consideration, and if $f(x) = \displaystyle\int_a^b g(x,t)\, dt$, then $f'(x) = \displaystyle\int_a^b \dfrac{\partial g(x,t)}{\partial x}\, dt.$

Some frequently used Laplace transforms are as follows.

$$\mathcal{L}\{e^{at}\} = \frac{1}{s-a}, \quad \mathcal{L}\{te^{at}\} = \frac{1}{(s-a)^2},$$

which by repeated differentiation n-times gives $\mathcal{L}\{t^n e^{at}\} = \frac{n!}{(s-a)^{n+1}}$. After replacing a by ib, choosing an appropriate n, and comparing the real and imaginary parts on both sides, we get the Laplace transforms of functions $t^n \cos bt$ and $t^n \sin bt$, and then combining with Property (i), we obtain the Laplace transforms of functions $t^n e^{at} \cos bt$ and $t^n e^{at} \sin bt$. For example, if we choose $n = 2$, then $\mathcal{L}\{t^2 e^{at}\} = \frac{2!}{(s-a)^3}$, which after setting letting $a = ib$ gives $\mathcal{L}\{t^2 e^{ibt}\} = \frac{2!}{(s-ib)^3}$, and $\mathcal{L}\{t^2(\cos bt + i\sin bt)\} = \frac{2(s+ib)^3}{(s^2+b^2)^3}$. Expanding the numerator on the right side we get $\mathcal{L}\{t^2(\cos bt + i\sin bt)\} = \frac{2(s^3 + 3is^2b - 3sb^2 - ib^3)}{(s^2+b^2)^3}$.

Equating the real and imaginary parts in the above formula we have

$$\mathcal{L}\{t^2 \cos bt\} = \frac{2(s^3 - 3sb^2)}{(s^2+b^2)^3}, \quad \mathcal{L}\{t^2 \sin bt\} = \frac{2(3s^2b - b^3)}{(s^2+b^2)^3},$$

and transforms of $\mathcal{L}\{e^{at}t^2 \cos bt\}$ and $\mathcal{L}e^{at}\{t^2 \sin bt\}$ can now be easily obtained. An important Laplace inverse is

$$\mathcal{L}^{-1}\left\{\frac{e^{-a\sqrt{s}}}{s}\right\} = \operatorname{erfc}\frac{a}{2\sqrt{t}}, \tag{5.1}$$

where $\operatorname{erf}(x) = \frac{2}{\sqrt{\pi}}\int_0^x e^{-u^2}\,du$, and $\operatorname{erfc}(x) = 1 - \operatorname{erf}(x) = \frac{2}{\sqrt{\pi}}\int_x^\infty e^{-u^2}\,du$.

Example 5.1. $\mathcal{L}^{-1}\left\{\frac{e^{-a\sqrt{s}}}{\sqrt{s}}\right\}$ is obtained by differentiating formula (5.1) with respect to a. Again, $\mathcal{L}^{-1}\left\{\frac{e^{-a\sqrt{s}}}{\sqrt{s}}\right\} = \frac{1}{\sqrt{\pi t}}e^{-a^2/4t}$ is obtained after differentiating (5.1) with respect to a and canceling out the negative sign on both sides. The classical method of deriving $\mathcal{L}^{-1}\left\{\frac{e^{-a\sqrt{s}}}{\sqrt{s}}\right\}$ is by contour integration, or an interesting method is given in Churchill [1972], which is explained in the next example. ■

Example 5.2. Define $\frac{e^{-a\sqrt{s}}}{\sqrt{s}} = y$ and $e^{-a\sqrt{s}} = z$. Then $y' = \frac{dy}{ds} = -\frac{1}{2s^{3/2}}e^{-a\sqrt{s}} - \frac{a}{2s}e^{-a\sqrt{s}}$, which yields $2sy' + y + az = 0$. Similarly, $z' = -\frac{a}{2\sqrt{s}}e^{-a\sqrt{s}}$ yields $2z' + ay = 0$. Taking the inverse transform of these equations, we obtain $aG - F - 2tF' = 0$, and $aF - 2tG = 0$, where $\mathcal{L}^{-1}\{y\} = F(t)$ and $\mathcal{L}^{-1}\{z\} = G(t)$. From these two equations in F and G, we get $F' = \frac{1}{2t}\left(\frac{a^2F}{2t} - F\right)$, whose solution is $F = \dfrac{A}{\sqrt{t}} e^{-a^2/4t}$, which gives $G = \dfrac{aA}{2\sqrt{t^3}} e^{-a^2/4t}$. Note that if

$a = 0$, then $y = \dfrac{1}{\sqrt{s}}$, and $F(t) = \dfrac{1}{\sqrt{\pi t}}$ implies that $A = \dfrac{1}{\sqrt{\pi}}$. Hence,

$$F(t) = \frac{1}{\sqrt{\pi t}} e^{-a^2/4t}, \quad G = \frac{a}{\sqrt{\pi t^3}} e^{-a^2/4t}.$$

Then we integrate $\mathcal{L}^{-1}\left\{\dfrac{e^{-a\sqrt{s}}}{\sqrt{s}}\right\} = \dfrac{1}{\sqrt{\pi t}} e^{-a^2/4t}$ with respect to a from 0 to a and

obtain $\mathcal{L}^{-1}\left\{\dfrac{e^{-a\sqrt{s}}}{s}\right\}$. In this problem we have assumed that $\mathcal{L}\left\{\dfrac{1}{\sqrt{t}}\right\} = \sqrt{\dfrac{\pi}{s}}$. ∎

Example 5.3. $\mathcal{L}^{-1}\left\{e^{-a\sqrt{s}}\right\} = \dfrac{a}{2\sqrt{\pi t^3}} e^{-a^2/4t}$ is obtained by differentiating

with respect to a the formula for $\mathcal{L}^{-1}\left\{\dfrac{e^{-a\sqrt{s}}}{\sqrt{s}}\right\}$ in the previous example and canceling

out the negative sign. ∎

Example 5.4. If we integrate the formula $\mathcal{L}^{-1}\left\{\dfrac{e^{-a\sqrt{s}}}{s}\right\} = \operatorname{erfc}\dfrac{a}{2\sqrt{t}}$ with

respect to a from 0 to a, we get $\displaystyle\int_0^a \mathcal{L}^{-1}\left\{\dfrac{e^{-x\sqrt{s}}}{s}\right\} dx = \int_0^a \operatorname{erfc}\dfrac{x}{2\sqrt{t}} dx$. Now,
after changing the order of integration and the Laplace inversion and carrying out the
integration on the left side, we get

$$\int_0^a \mathcal{L}^{-1}\left\{\frac{e^{-x\sqrt{s}}}{s}\right\} dx = \mathcal{L}^{-1}(s^{-3/2} - s^{-3/2}e^{-a\sqrt{s}}), \tag{5.2}$$

and the right side yields

$$\int_0^a \operatorname{erfc}\frac{x}{2\sqrt{t}} dx = \left[x \operatorname{erfc}\frac{x}{2\sqrt{t}}\right]_0^a + \frac{1}{\sqrt{\pi t}}\int_0^a x\, e^{-x^2/4t}\, dx$$

$$= a \operatorname{erfc}\frac{a}{2\sqrt{t}} - 2\sqrt{\frac{t}{\pi}}e^{-a^2/4t} + 2\sqrt{\frac{t}{\pi}}.$$

Since $\mathcal{L}^{-1}\left\{s^{-3/2}\right\} = 2\sqrt{\dfrac{t}{\pi}}$, we get

$$\mathcal{L}^{-1}\left\{s^{-3/2}e^{-a\sqrt{s}}\right\} = 2\sqrt{\frac{t}{\pi}}e^{-a^2/4t} - a \operatorname{erfc}\frac{a}{2\sqrt{t}}. \;\blacksquare$$

We state a very useful theorem without proof.

Theorem 5.1. *If* $G(s) = \displaystyle\sum_1^\infty G_k(s)$ *is uniformly convergent, then*

$$\mathcal{L}^{-1}G(s) = g(t) = \sum_{k=1}^\infty g_k(t), \tag{5.3}$$

where $\mathcal{L}^{-1}G_k(s) = g_k(t)$.

Example 5.5. Since

$$\mathcal{L}^{-1}\left\{s^{-3/2}e^{-1/s}\right\}$$

$$= \mathcal{L}^{-1}\left\{\frac{1}{s^{3/2}}\left[1 - \frac{1}{s} + \frac{1}{2!s^2} - \frac{1}{3!s^3} + \cdots + (-1)^n\frac{1}{n!s^n} + \cdots\right]\right\}$$

$$= \mathcal{L}^{-1}\sum_0^\infty \frac{(-1)^n}{n!\,s^{n+3/2}} = \frac{1}{\sqrt{\pi}}\sum_0^\infty \frac{(-1)^n(2\sqrt{t})^{2n+1}}{(2n+1)!} = \frac{1}{\sqrt{\pi}}\sin(2\sqrt{t}),$$

this result and second part of Property (iv) give

$$\mathcal{L}^{-1}\left\{s^{-1/2}e^{-1/s}\right\} = \frac{1}{\sqrt{\pi t}}\cos(2\sqrt{t}). \blacksquare \qquad (5.4)$$

Example 5.6. Solve the heat conduction equation $k\,T_{xx} = T_t$ in the semi-infinite medium $0 \le x < \infty, -\infty < y, z < \infty$, which has an initial zero temperature, while its face $x = 0$ is maintained at a time-dependent temperature $f(t)$, and find the temperature T for $t > 0$. By applying the Laplace transform to the equation we get $\overline{T}_{xx} = \frac{s}{k}\overline{T}$, where $\overline{T} = \mathcal{L}\{T\}$. The solution of this equation is $\overline{T} = Ae^{mx} + Be^{-mx}$, where $m = \sqrt{s/k}$. Since $\overline{T}$ remains bounded as $x \to \infty$, we find that $A = 0$. The boundary condition at $x = 0$ in the transform domain yields $B = \bar{f}(s)$, where $\bar{f}(s)$ is the Laplace transform of $f(t)$. Thus, the solution in the transform domain is $\overline{T} = \bar{f}(s)\,e^{-mx}$. To carry out Laplace inversion, we use the convolution property and Example 5.2 and get

$$T = \int_0^t \frac{x\,e^{-x^2/4k\tau}}{2\tau\sqrt{\pi k\tau}}\,f(t-\tau)\,d\tau.$$

If $\bar{f}(s) = 1$, then the solution for T reduces to $T = \dfrac{x\,e^{-x^2/4kt}}{2t\sqrt{\pi kt}}$. This solution is the fundamental solution for the heat conduction equation for the half-space. In the special case when $f(t) = T_0$, the solution is given by

$$T = T_0\,\mathrm{erfc}\left(\frac{x}{2\sqrt{kt}}\right). \blacksquare$$

5.2.1. Definition of Dirac Delta Function. It is known that $\mathcal{L}\{\delta(t)\} = 1$. We will explain the existence of the δ-function heuristically. Consider the Heaviside unit step function $H(t)$ which is defined in §2.1.1. Since the Laplace transform of $H(t)$ is $\overline{H}(s) = \dfrac{1}{s}$, then by Property (iv) of the Laplace transforms, $\mathcal{L}H'(t) = s\overline{H}(s) = 1$. Obviously, $H'(t)$ vanishes for $|t| > 0$ and does not exist for $t = 0$. From the graph of $H(t)$ (Fig. 2.3), it is clear that there is a vertical jump at $t = 0$. Therefore, it seems reasonable to assume that $\lim_{t\to 0} H'(t) \to \infty$. But since $\int_{-\varepsilon}^{\varepsilon} H'(t)\,dt = 1$, it is obvious that a function like $H'(t)$ exists only in the generalized sense. The function $H'(t)$ is a generally denoted by $\delta(t)$ such that

$$\delta(t) = \begin{cases} 0 & \text{for } |t| > 0, \\ \infty & \text{for } t = 0, \end{cases} \quad \text{and} \quad \int_{-\varepsilon}^{\varepsilon} \delta(t)\, dt = 1. \tag{5.5}$$

For other properties of δ-function, see §2.1.

Example 5.7. Solve the wave equation $u_{tt} = c^2 u_{xx}$ subject to the initial condition $u = u_t = 0$ for $t \le 0$, and the boundary conditions $u(0) = 0$ and $u_x(l) = T$. By applying the Laplace transform to the wave equation, we get $\bar{u}_{xx} = c^{-2} s^2 \bar{u}$, which has the solution: $\bar{u} = A\, e^{-sx/c} + B\, e^{sx/c}$. Applying the boundary conditions in the transform domain, we get $A + B = 0$ and $-A\, e^{-sl/c} + B\, e^{sl/c} = \dfrac{cT}{s^2}$. Solving for A and B and substituting their values in the solution for $\bar{u}$, we get

$$\begin{aligned}
\bar{u}(x,s) &= \frac{cT}{s^2} \frac{\sinh(sx/c)}{\cosh(sl/c)} = \frac{cT}{s^2} \left(\frac{e^{sx/c} - e^{-sx/c}}{e^{sl/c} + e^{-sl/c}} \right) \\
&= \frac{T}{s^2} \sum_{0}^{\infty} (-1)^n \left(e^{-s[(2n+1)l-x]/c} - e^{-s[(2n+1)l+x]/c} \right),
\end{aligned}$$

which, after inversion, gives

$$\begin{aligned}
u(x,t) = T \sum_{0}^{\infty} (-1)^n \Bigg[&\left(t - \frac{(2n+1)l + x}{c} \right) H \left(t - \frac{(2n+1)l + x}{c} \right) \\
&- \left(t - \frac{(2n+1)l - x}{c} \right) H \left(t - \frac{(2n+1)l - x}{c} \right) \Bigg]. \ \blacksquare
\end{aligned}$$

5.3. Fourier Integral Theorems

The following three theorems are the basis of Fourier transforms:

Theorem 5.2. (Fourier integral theorem) *If $f(x)$ satisfies the Dirichlet conditions on the entire real line and is absolutely integrable on $(-\infty, \infty)$, then*

$$\frac{1}{2} [f(x+0) + f(x-0)] = \frac{1}{2\pi} \int_{-\infty}^{\infty} e^{-i\alpha x}\, d\alpha \int_{-\infty}^{\infty} f(u) e^{i\alpha u}\, du. \tag{5.6}$$

Theorem 5.3. (Fourier cosine theorem) *If $f(x)$ satisfies the Dirichlet conditions on the non-negative real line and is absolutely integrable on $(0, \infty)$, then*

$$\frac{1}{2} [f(x+0) + f(x-0)] = \frac{2}{\pi} \int_{0}^{\infty} d\alpha \int_{0}^{\infty} f(u) \cos(\alpha u) \cos(\alpha x)\, du. \tag{5.7}$$

Theorem 5.4. (Fourier sine theorem) *If $f(x)$ satisfies the Dirichlet conditions on the non-negative real line and is absolutely integrable on $(0, \infty)$, then*

$$\frac{1}{2}[f(x+0) + f(x-0)] = \frac{2}{\pi} \int_0^\infty d\alpha \int_0^\infty f(u)\sin(\alpha u)\sin(\alpha x)\, du. \qquad (5.8)$$

If $f(x)$ is continuous, then $\frac{1}{2}[f(x+0) + f(x-0)] = f(x)$.

These three integrals form the basis of the Fourier transforms. For proof, see Sneddon [1957].

5.3.1. Properties of Fourier Transforms. Let $\mathcal{F}\{f(x)\} = \widetilde{f}(\alpha)$. Then

(1) $\mathcal{F}\{f(x-a)\} = e^{i\alpha a}\,\widetilde{f}(\alpha)$.

(2) $\mathcal{F}\{f(ax)\} = \dfrac{1}{|a|}\widetilde{f}(\alpha/a)$.

(3) $\mathcal{F}\{e^{iax}f(x)\} = \widetilde{f}(\alpha + a)$.

(4) $\mathcal{F}\{\widetilde{f}(x)\} = f(-\alpha)$.

(5) $\mathcal{F}\{x^n f(x)\} = (-i)^n \dfrac{d^n}{d\alpha^n}\,\widetilde{f}(\alpha)$.

(6) $\mathcal{F}\{f(ax)e^{ibx}\} = \dfrac{1}{|a|}\widetilde{f}\left(\dfrac{\alpha+b}{a}\right)$.

5.3.2. Fourier Transforms of the Derivatives of a Function. We assume that $f(x)$ is differentiable n times and the function and its derivatives approach zero as $|x| \to \infty$. Then

$$\widetilde{f}^{(p)}(\alpha) = (-i\alpha)\widetilde{f}^{(p-1)},$$

where $\widetilde{f}^{(p)}$ is the Fourier transform of $f^{(p)}(x)$, which is the p-th derivative of $f(x)$ for $0 \le p \le n$. If $\lim\limits_{x\to\infty} f^{(p)}(x) = 0$, and $\lim\limits_{x\to 0} f^{(p)}(x) = \sqrt{\dfrac{\pi}{2}}\, c_p$, then

$$\widetilde{f}_c^{(p)} = -c_{p-1} + \alpha \widetilde{f}_s^{(p-1)}, \qquad (5.9)$$

and

$$\widetilde{f}_s^{(p)} = -\alpha \widetilde{f}_c^{(p-1)}. \qquad (5.10)$$

5.3.3. Convolution Theorems for Fourier Transform. The convolution (or Faltung) of $f(t)$ and $g(t)$ over $(-\infty, \infty)$ is defined by

$$f \star g = \frac{1}{\sqrt{2\pi}} \int_{-\infty}^\infty f(\eta)g(x-\eta)\, d\eta = \frac{1}{\sqrt{2\pi}} \int_{-\infty}^\infty f(x-\eta)g(\eta)\, d\eta. \qquad (5.11)$$

Theorem 5.5. Let $\widetilde{f}(\alpha)$ and $\widetilde{g}(\alpha)$ be the Fourier transforms of $f(x)$ and $g(x)$, respectively. Then the inverse Fourier transform of $\widetilde{f}(\alpha)\,\widetilde{g}(\alpha)$ is

$$\mathcal{F}^{-1}\left\{\widetilde{f}(\alpha)\widetilde{g}(\alpha)\right\} = \frac{1}{\sqrt{2\pi}}\int_{-\infty}^{\infty} f(\eta)g(x-\eta)\,d\eta.$$

Theorem 5.6. Let $\widetilde{f}(\alpha)$ and $\widetilde{g}(\alpha)$ be the Fourier transforms of $f(x)$ and $g(x)$, respectively. Then

$$\int_{-\infty}^{\infty} \widetilde{f}(\alpha)\widetilde{g}(\alpha)\,d\alpha = \int_{-\infty}^{\infty} f(-\eta)g(\eta)\,d\eta. \tag{5.12}$$

Example 5.8. Find the Fourier transform of $f(x) = e^{-k|x|}$, $k > 0$.

$$\mathcal{F}\{f(x)\} = \widetilde{f}(\alpha) = \frac{1}{\sqrt{2\pi}}\int_{-\infty}^{\infty} e^{-k|x|}e^{ix\alpha}\,dx$$

$$= \frac{1}{\sqrt{2\pi}}\left[\int_{-\infty}^{0} e^{kx}e^{ix\alpha}\,dx + \int_{0}^{\infty} e^{-kx}e^{ix\alpha}\,dx\right]$$

$$= \frac{1}{\sqrt{2\pi}}\left(\frac{1}{k+i\alpha} - \frac{1}{-k+i\alpha}\right) = \frac{k\sqrt{2}}{\sqrt{\pi}(k^2+\alpha^2)}.$$

Note that by Property 4, $\mathcal{F}\{f(x)\} = \dfrac{k\sqrt{2}}{\sqrt{\pi}(k^2+x^2)}$ yields $\widetilde{f}(\alpha) = e^{-k|\alpha|}$. ∎

Example 5.9. Find the Fourier transform of $f(x) = e^{-kx^2}$, $k > 0$.

$$\widetilde{f}(\alpha) = \frac{1}{\sqrt{2\pi}}\int_{-\infty}^{\infty} e^{-kx^2}e^{ix\alpha}\,dx$$

$$= \frac{1}{\sqrt{2\pi}}\int_{-\infty}^{\infty} e^{-k\left(x^2 - ix\alpha/k - \alpha^2/(4k^2) + \alpha^2/(4k^2)\right)}\,dx$$

$$= \frac{1}{\sqrt{2\pi}}\int_{-\infty}^{\infty} e^{-k\left((x-i\alpha/k)^2 - \alpha^2/(4k^2)\right)}\,dx = \frac{1}{\sqrt{2\pi k}}e^{-\alpha^2/(4k)}\int_{-\infty}^{\infty} e^{-u^2}\,du$$

$$= \frac{1}{\sqrt{2\pi k}}e^{-\alpha^2/(4k)}\sqrt{\pi} = \frac{1}{\sqrt{2k}}e^{-\alpha^2/(4k)}. \ \blacksquare$$

Example 5.10. Find the Fourier transform of

$$f(x) = \begin{cases} 0 & \text{for } x < 0, \\ xe^{-ax} & \text{for } x > 0,\ a > 0. \end{cases}$$

$$\mathcal{F}\{f(x)\} = \frac{1}{\sqrt{2\pi}}\int_{-\infty}^{\infty} f(x)e^{i\alpha x}\,dx = \frac{1}{\sqrt{2\pi}}\int_{0}^{\infty} xe^{-ax}e^{i\alpha x}\,dx$$

$$= \frac{xe^{-ax+i\alpha x}}{\sqrt{2\pi}(i\alpha - a)}\Big|_0^\infty - \frac{1}{\sqrt{2\pi}(i\alpha - a)} \int_0^\infty e^{-ax}e^{i\alpha x}\, dx$$

$$= -\frac{1}{\sqrt{2\pi}(i\alpha - a)} \int_0^\infty e^{-ax}e^{i\alpha x}\, dx = \frac{1}{\sqrt{2\pi}(i\alpha - a)^2} \cdot \blacksquare$$

Example 5.11. Find the Fourier transform of $f(x) = 0$ if $x < b$ and $f(x) = e^{-a^2 x^2}$ if $0 < b < x$. The solution is

$$\tilde{f}(\alpha) = \frac{1}{\sqrt{2\pi}} \int_b^\infty e^{-a^2 x^2}e^{i x \alpha}\, dx = \frac{1}{\sqrt{2\pi}} \int_b^\infty e^{-a^2\left(x - i\alpha/(2a^2)\right)^2 - \alpha^2/(4a^2)}\, dx$$

$$= \frac{e^{-\alpha^2/(4a^2)}}{a\sqrt{2\pi}} \int\limits_{(ab-i\alpha/2a)}^\infty e^{-u^2}\, du = \frac{1}{2a\sqrt{2}}e^{-\alpha^2/(4a^2)} \operatorname{erfc}\left(ab - \frac{i\alpha}{2a}\right). \blacksquare$$

Example 5.12. Find the solution of Laplace's equation $u_{xx} + u_{yy} = 0$ in the domain $|x| < \infty$ and $y \geq 0$, subject to the conditions that $u \to 0$ as $|x| \to \infty$ or as $y \to \infty$, and $u(x,0) = \delta(x)$. After applying the Fourier transform to the given differential equation with respect to x, we get $\tilde{u}_{yy} - \alpha^2 \tilde{u} = 0$, whose solution is $\tilde{u} = Ae^{-|\alpha|y}$. Applying the boundary condition at $y = 0$ in the transform domain, we get $\tilde{u}(\alpha, 0) = A = \frac{1}{\sqrt{2\pi}}$. Hence, $\tilde{u} = \frac{1}{\sqrt{2\pi}} e^{-|\alpha|y}$. On inverting, we obtain

$$u(x,y) = \frac{1}{\pi}\frac{y}{x^2 + y^2}.$$

Now, we use the convolution theorem to obtain the solution to the problem with arbitrary condition $u(x,0) = f(x)$. Then the solution is

$$u(x,y) = \frac{1}{\pi} \int_{-\infty}^\infty \frac{y f(\eta)}{(x - \eta)^2 + y^2}\, d\eta, \tag{5.13}$$

which is known as the Poisson integral representation for the Dirichlet problem in the half-plane. $\blacksquare$

Example 5.13. In this example we use both Laplace and Fourier transforms to solve the boundary value problem

$$u_{tt} - c^2 u_{xx} = e^{-|x|}\sin t, \quad u(x,0) = 0, \quad u_t(x,0) = e^{-|x|}.$$

Note that this equation in the Laplace transform domain becomes

$$s^2 \overline{u}(x,s) - c^2 \overline{u}_{xx}(x,s) = \frac{2 + s^2}{1 + s^2}e^{-|x|}.$$

Applying the complex Fourier transform, we get

$$(c^2\alpha^2 + s^2)\,\widetilde{\overline{u}}(x,s) = \frac{2 + s^2}{1 + s^2}\sqrt{\frac{2}{\pi}}\frac{1}{1 + \alpha^2}.$$

Thus, $\quad \widetilde{\overline{u}}(x,s) = \dfrac{2+s^2}{1+s^2} \sqrt{\dfrac{2}{\pi}} \dfrac{1}{(1+\alpha^2)(c^2\alpha^2+s^2)}$

$$= \dfrac{2+s^2}{(1+s^2)(s^2-c^2)} \sqrt{\dfrac{2}{\pi}} \left[\dfrac{1}{1+\alpha^2} - \dfrac{c^2}{c^2\alpha^2+s^2}\right].$$

On inverting the Fourier transform, we have

$$\overline{u}(x,s) = \dfrac{2+s^2}{(1+s^2)(s^2-c^2)} \left(e^{-|x|} - \dfrac{c}{s}e^{-s|x|/c}\right)$$

$$= \left\{ B\left(\dfrac{1}{s-c} - \dfrac{1}{s+c}\right) - \dfrac{1}{(1+c^2)(1+s^2)}\right\} e^{-|x|}$$

$$+ \left\{ \dfrac{2}{cs} - B\left(\dfrac{1}{s-c} + \dfrac{1}{s+c}\right) - \dfrac{cs}{(1+c^2)(1+s^2)}\right\} e^{-s|x|/c},$$

where $B = \dfrac{2+c^2}{2c(1+c^2)}$. After taking the Laplace inverse, we find that

$$u(x,t) = \left[B\left(e^{ct} - e^{-ct}\right) - \dfrac{1}{1+c^2}\sin t\right] e^{-|x|}$$

$$+ H(ct - |x|) \left[\dfrac{2}{c} - B\left(e^{ct-|x|} + e^{-ct+|x|}\right) - \dfrac{c}{1+c^2}\cos\left(t - |x|/c\right)\right]. \ \blacksquare$$

5.4. Fourier Sine and Cosine Transforms

Some important properties and theorems are given below.

5.4.1. Properties of Fourier Sine and Cosine Transforms.

(a) $\mathcal{F}_c\{f(x)\} = \widetilde{f}_c(\alpha), \quad \mathcal{F}_s\{f(x)\} = \widetilde{f}_s(\alpha),$

(b) $\mathcal{F}_c\{\widetilde{f}_c(x)\} = f(\alpha), \quad \mathcal{F}_s\{\widetilde{f}_s(x)\} = f(\alpha),$

(c) $\mathcal{F}_c\{f(kx)\} = \dfrac{1}{k}\widetilde{f}_c(\dfrac{\alpha}{k}), \quad k > 0,$

(d) $\mathcal{F}_s\{f(kx)\} = \dfrac{1}{k}\widetilde{f}_s(\dfrac{\alpha}{k}), \quad k > 0,$

(e) $\mathcal{F}_c\{f(kx)\cos bx\} = \dfrac{1}{2k}\left[\widetilde{f}_c\left(\dfrac{\alpha+b}{k}\right) + \widetilde{f}_c\left(\dfrac{\alpha-b}{k}\right)\right], \quad k > 0,$

(f) $\mathcal{F}_c\{f(kx)\sin bx\} = \dfrac{1}{2k}\left[\widetilde{f}_s\left(\dfrac{\alpha+b}{k}\right) - \widetilde{f}_s\left(\dfrac{\alpha-b}{k}\right)\right], \quad k > 0,$

(g) $\mathcal{F}_s\{f(kx)\cos bx\} = \dfrac{1}{2k}\left[\widetilde{f}_s\left(\dfrac{\alpha+b}{k}\right) + \widetilde{f}_s\left(\dfrac{\alpha-b}{k}\right)\right], \quad k > 0,$

(h) $\mathcal{F}_s\{f(kx)\sin bx\} = \dfrac{1}{2k}\left[\widetilde{f}_c\left(\dfrac{\alpha-b}{k}\right) - \widetilde{f}_c\left(\dfrac{\alpha+b}{k}\right)\right], \quad k > 0,$

(i) $\mathcal{F}_c\{x^{2n}f(x)\} = (-1)^n \dfrac{d^{2n}\tilde{f}_c(\alpha)}{d\alpha^{2n}}$,

(j) $\mathcal{F}_c\{x^{2n+1}f(x)\} = (-1)^n \dfrac{d^{2n+1}\tilde{f}_s(\alpha)}{d\alpha^{2n+1}}$,

(k) $\mathcal{F}_s\{x^{2n}f(x)\} = (-1)^n \dfrac{d^{2n}\tilde{f}_s(\alpha)}{d\alpha^{2n}}$,

(l) $\mathcal{F}_s\{x^{2n+1}f(x)\} = (-1)^{n+1} \dfrac{d^{2n+1}\tilde{f}_c(\alpha)}{d\alpha^{2n+1}}$.

5.4.2. Convolution Theorems for Fourier Sine and Cosine Transforms. These theorems are as follows:

Theorem 5.7. *Let $\tilde{f}_c(\alpha)$ and $\tilde{g}_c(\alpha)$ be the Fourier cosine transforms of $f(x)$ and $g(x)$, respectively, and let $\tilde{f}_s(\alpha)$ and $\tilde{g}_s(\alpha)$ be the Fourier sine transforms of $f(x)$ and $g(x)$, respectively. Then*

$$\mathcal{F}_c^{-1}\{\tilde{f}_c(\alpha)\tilde{g}_c(\alpha)\} = \frac{1}{\sqrt{2\pi}} \int_0^\infty g(s)\left[f(|x-s|) + f(x+s)\right] ds. \qquad (5.14)$$

Theorem 5.8. *If $\tilde{f}_s(\alpha)$, $\tilde{f}_c(\alpha)$, and $\tilde{g}_s(\alpha)$, $\tilde{g}_s(\alpha)$ are the Fourier sine and cosine transforms of $f(x)$ and $g(x)$, respectively, then the following results hold:*

(i) $\displaystyle\int_0^\infty \tilde{f}_c(\alpha)\,\tilde{g}_s(\alpha)\sin\alpha x\,d\alpha = \frac{1}{2}\int_0^\infty g(s)[f(|x-s|) - f(x+s)]\,ds,$ $\qquad (5.15)$

(ii) $\displaystyle\int_0^\infty \tilde{f}_s(\alpha)\,\tilde{g}_c(\alpha)\sin\alpha x\,d\alpha = \frac{1}{2}\int_0^\infty f(s)[g(|x-s|) - g(x+s)]\,ds,$ $\qquad (5.16)$

(iii) $\displaystyle\int_0^\infty \tilde{f}_s(\alpha)\,\tilde{g}_s(\alpha)\cos\alpha x\,d\alpha$

$$= \frac{1}{2}\int_0^\infty g(t)\left[H(t+x)\,f(t+x) + H(t-x)\,f(t-x)\right] dt$$

$$= \frac{1}{2}\int_0^\infty f(t)\left[H(t+x)\,g(t+x) + H(t-x)\,g(t-x)\right] dt, \qquad (5.17)$$

(iv) $\displaystyle\int_0^\infty \tilde{f}_c(\alpha)\,\tilde{g}_c(\alpha)\,d\alpha = \int_0^\infty f(s)g(s)\,ds = \int_0^\infty \tilde{f}_s(\alpha)\tilde{g}_s(\alpha)\,d\alpha.$ $\qquad (5.18)$

Proofs of all above theorems can be found in Sneddon [1978] or Kythe et al. [2003].

Example 5.14. Find the Fourier cosine transform of $f(x)$, where

$$f(x) = \begin{cases} x & \text{for } 0 < x < 1, \\ 2 - x & \text{for } 1 < x < 2, \\ 0 & \text{for } 2 < x < \infty. \end{cases}$$

Here $\mathcal{F}_c[f(x)] = \sqrt{\dfrac{2}{\pi}}\left[\displaystyle\int_0^1 x\cos x\,dx + \int_1^2 (2-x)\cos x\,dx\right]$

$$= \sqrt{\frac{2}{\pi}} \left[\frac{x \sin \alpha x}{\alpha} \Big|_0^1 - \frac{1}{\alpha} \int_0^1 \sin \alpha x \, dx + \frac{(2-x)\sin \alpha x}{\alpha} \Big|_1^2 + \frac{1}{\alpha} \int_1^2 \sin \alpha x \, dx \right]$$

$$= \sqrt{\frac{2}{\pi}} \left[\frac{2\cos \alpha - 1 - \cos 2\alpha}{\alpha^2} \right]. \ \blacksquare$$

Example 5.15. We will use the Fourier transform to solve Example 5.6, which was earlier solved by the Laplace transform. The partial differential equation and the boundary and initial conditions are:

$$k\, u_{xx} = u_t,$$

$$u = 0 \quad \text{for } t \le 0; \quad u \to 0 \text{ as } x \to \infty; \quad \text{and} \quad u = T_0 \text{ at } x = 0 \text{ for } t > 0.$$

Applying the Fourier sine transform to the partial differential equation, we get

$$\frac{\partial \tilde{u}_s}{\partial t} + k\alpha^2 \tilde{u}_s = \sqrt{\frac{2}{\pi}} \, k\alpha T_0,$$

where $\tilde{u}_s$ is the Fourier sine transform of u. Its solution is given by

$$\tilde{u}_s = A e^{-k\alpha^2 t} + \sqrt{\frac{2}{\pi}} \frac{T_0}{\alpha}.$$

After using the initial condition at $t = 0$, we get $A + \sqrt{\dfrac{2}{\pi}} \dfrac{T_0}{\alpha} = 0$. This yields

$$\tilde{u}_s = \sqrt{\frac{2}{\pi}} \frac{T_0}{\alpha} \left(1 - e^{-k\alpha^2 t} \right). \text{ Thus,}$$

$$u(x,t) = \frac{2}{\pi} T_0 \int_0^\infty \frac{1}{\alpha} \left(1 - e^{-k\alpha^2 t} \right) \sin \alpha x \, d\alpha$$

$$= \frac{2}{\pi} T_0 \left[\int_0^\infty \frac{\sin \alpha x}{\alpha} d\alpha - \int_0^\infty \frac{\sin \alpha x}{\alpha} e^{-k\alpha^2 t} \, d\alpha \right]$$

$$= \frac{2}{\pi} T_0 \left[\frac{\pi}{2} - \frac{\pi}{2} \operatorname{erf} \frac{x}{2\sqrt{kt}} \right] = T_0 \operatorname{erfc} \frac{x}{2\sqrt{kt}}. \ \blacksquare$$

5.5. Finite Fourier Transforms

When the domain of the physical problem is finite, it is generally not convenient to use the transforms with an infinite range of integration. In many cases finite Fourier transform can be used with advantage. We define

$$\mathcal{F}_s\{f(x)\} = \tilde{f}_s(n) = \int_0^a f(x) \sin \left(\frac{n\pi x}{a} \right) dx \tag{5.19}$$

as the finite Fourier sine transform of $f(x)$ for $n = 1, 2, \ldots$. The function $f(x)$ is then given by

$$\mathcal{F}_s^{-1}\{\tilde{f}_s(n)\} = f(x) = \frac{2}{a} \sum_{n=1}^\infty \tilde{f}_s(n) \sin \left(\frac{n\pi x}{a} \right). \tag{5.20}$$

Similarly, the finite Fourier cosine transform is defined for $n = 0, 1, 2, \ldots$ by

$$\mathcal{F}_c\{f(x)\} = \tilde{f}_c(n) = \int_0^a f(x) \cos\left(\frac{n\pi x}{a}\right)\, dx, \tag{5.21}$$

and its inverse by

$$\mathcal{F}_c^{-1}\{\tilde{f}_s(n)\} = f(x) = \frac{\tilde{f}_c(0)}{a} + \frac{2}{a}\sum_{n=1}^{\infty} \tilde{f}_c(n) \cos\left(\frac{n\pi x}{a}\right). \tag{5.22}$$

From the theory of Fourier series it is known that the Fourier cosine series for $f(x)$ on the interval $0 < x < a$, which is $\frac{1}{a}f(0) + \sum_{n=1}^{\infty} a_n \cos\frac{n\pi x}{a}$, converges to the value $f(x)$ at each point of continuity in the above interval to the value $\frac{1}{2}\left[f(x+0) + f(x-0)\right]$ at each point of x of finite discontinuity (and also at all points of continuity) in this interval. Similarly, the Fourier sine series for a function $f(x)$ on the interval $0 < x < 0$, which is $\frac{2}{a}\sum_{n=1}^{\infty} b_n \sin\frac{n\pi x}{a}$, converges to the value $f(x)$ at each point of continuity in the above interval to the value $\frac{1}{2}\left[f(x+0) + f(x-0)\right]$ at each point of x of finite discontinuity (and also at all points of continuity) in this interval. Thus, we can alternatively define the finite Fourier sine and cosine transforms $\tilde{f}_c$ and $\tilde{f}_c$ by a_n and b_n, respectively, as is done in Weinberger [1965]. An example of this approach is given in §6.8, 8.7, 8.8.1 and 8.11. The finite Fourier sine transform is the general choice because of its simplicity of use.

For $a = \pi$ set $\frac{\pi x}{a} = z$. Then the above definitions become

$$\tilde{f}_s(n) = \int_0^a f(x) \sin\left(\frac{n\pi x}{a}\right)\, dx = \frac{a}{\pi}\int_0^{\pi} \sin(nz) f\left(\frac{az}{\pi}\right)\, dz = \frac{a}{\pi}\mathcal{F}_s\left\{f\left(\frac{ax}{\pi}\right)\right\},$$

$$\tilde{f}_c(n) = \int_0^a f(x) \cos\left(\frac{n\pi x}{a}\right)\, dx = \frac{a}{\pi}\int_0^{\pi} \cos(nz) f\left(\frac{az}{\pi}\right)\, dz = \frac{a}{\pi}\mathcal{F}_c\left\{f\left(\frac{ax}{\pi}\right)\right\}.$$

Example 5.16. (a) For $f(x) = 1$ we have

$$\mathcal{F}_s\{1\} = \tilde{f}_s(n) = \int_0^a \sin\frac{n\pi x}{a}\, dx = \frac{a}{n\pi}\left[1 - (-1)^n\right]; \text{ and}$$

$$\mathcal{F}_c\{1\} = \tilde{f}_c(n) = \int_0^a \cos\frac{n\pi x}{a}\, dx = \begin{cases} a, & n = 0, \\ 0, & n \neq 0. \end{cases}$$

(b) For $f(x) = x$, we have

$$\mathcal{F}_s\{x\} = \tilde{f}_s(n) = \int_0^a x \sin\frac{n\pi x}{a}\, dx = \frac{(-1)^{n+1}a^2}{n\pi}; \text{ and}$$

$$\mathcal{F}_c\{x\} = \tilde{f}_c(n) = \int_0^a x \cos\frac{n\pi x}{a}\, dx = \begin{cases} \dfrac{a^2}{2}, & n = 0, \\[2mm] \left(\dfrac{a}{n\pi}\right)^2 [1 - (-1)^n], & n \neq 0. \end{cases}$$

5.5.1. Properties. The following properties are useful:

(i) $\mathcal{F}_s\{f'(x)\} = -\left(\dfrac{n\pi}{a}\right)\tilde{f}_c(n),$

(ii) $\mathcal{F}_s\{f''(x)\} = -\left(\dfrac{n\pi}{a}\right)^2 \tilde{f}_s(n) + \left(\dfrac{n\pi}{a}\right)[f(0) + (-1)^{n+1}f(a)],$

(iii) $\mathcal{F}_c\{f'(x)\} = \left(\dfrac{n\pi}{a}\right)\tilde{f}_s(n) + (-1)^n f(a) - f(0),$ and

(iv) $\mathcal{F}_c\{f''(x)\} = -\left(\dfrac{n\pi}{a}\right)^2 \tilde{f}_c(n) + (-1)^n f'(a) - f'(0).$

5.5.2. Periodic Extensions. A function $g(x)$ is said to be an *odd periodic extension* of a periodic function $f(x)$ of period 2π if

$$g(x) = \begin{cases} f(x) & \text{for } 0 < x < \pi, \\ -f(-x) & \text{for } -\pi < x < 0. \end{cases}$$

That is, $g(x) = f(x)$ for $0 < x < \pi$, and $g(-x) = -g(x)$ and $g(x + 2\pi) = g(x)$ for $-\infty < x < \infty$.

A function $h(x)$ is said to be an *even periodic extension* of a periodic function $f(x)$ of period 2π if

$$h(x) = \begin{cases} f(x) & \text{for } 0 < x < \pi, \\ f(-x) & \text{for } -\pi < x < 0. \end{cases}$$

That is, $h(x) = f(x)$ for $0 < x < \pi$, and $h(-x) = h(x)$ and $g(x + 2\pi) = h(x)$ for $-\infty < x < \infty$.

A theorem on odd and even extensions of a periodic function f is given below without proof (Chruchill [1972]).

Theorem 5.9. *(i) If $g(x)$ is the odd periodic extension of $f(x)$ of period 2π, then for any constant α*

$$\mathcal{F}_s\{g(x-\alpha) + g(x+\alpha)\} = 2\cos n\alpha\, \mathcal{F}_s\{f(x)\}, \tag{5.23}$$

$$\mathcal{F}_c\{g(x+\alpha) + g(x-\alpha)\} = 2\sin n\alpha\, \mathcal{F}_s\{f(x)\}. \tag{5.24}$$

(ii) If $h(x)$ is an even periodic extension of $f(x)$ of period 2π, then for any constant α

$$\mathcal{F}_c c\{h(x-\alpha) + h(x+\alpha)\} = 2\cos n\alpha\, \mathcal{F}_c\{f(x)\}, \tag{5.25}$$

$$\mathcal{F}_c\{h(x-\alpha) + h(x+\alpha)\} = 2\sin n\alpha\, \mathcal{F}_c\{f(x)\}. \tag{5.26}$$

5.5.3. Convolution. Let $f(x)$ and $g(x)$ be two piecewise continuous periodic functions defined on $-\pi < x < \pi$. Then their *convolution* is defined by

$$f(x) \star g(x) = \int_{-\pi}^{\pi} f(x-u)\, g(u)\, du, \qquad (5.27)$$

which is a continuous periodic function of period 2π. The following result hold (Churchill [1972]):

Theorem 5.10. *Let $f_1(x)$ and $g_1(x)$ be the odd periodic extensions of $f(x)$ and $g(x)$, respectively on $0 < x < \pi$, and $f_2(x)$ and $g_2(x)$ their even periodic extensions on $0 < x < \pi$. Then*

$$\mathcal{F}_s\{f_1(x)\star g_2(x)\} = 2\widetilde{f}_s(n)\,\widetilde{g}_c(n), \quad \mathcal{F}_s\{f_2(x)\star g_1(x)\} = 2\widetilde{f}_c(n)\,\widetilde{g}_s(n),$$

$$\mathcal{F}_c\{f_1(x)\star g_1(x)\} = -2\widetilde{f}_s(n)\,\widetilde{g}_s(n), \quad \mathcal{F}_c\{f_2(x)\star g_2(x)\} = 2\widetilde{f}_c(n)\,\widetilde{g}_c(n),$$

or inversely,

$$\mathcal{F}_s^{-1}\{\widetilde{f}_s(n)\,\widetilde{g}_c(n)\} = \frac{1}{2}\{f_1(x)\star g_2(x)\},$$

$$\mathcal{F}_s^{-1}\{\widetilde{f}_c(n)\,\widetilde{g}_s(n)\} = 2\{f_2(x)\star g_1(x)\}, \qquad (5.28)$$

$$\mathcal{F}_s^{-1}\{\widetilde{f}_s(n)\,\widetilde{g}_s(n)\} = -\frac{1}{2}\{f_1(x)\star g_1(x)\},$$

$$\mathcal{F}_c^{-1}\{\widetilde{f}_c(n)\,\widetilde{g}_c(n)\} = \frac{1}{2}\{f_2(x)\star g_2(x)\}.$$

Example 5.17. To determine $\mathcal{F}^{-1}\left(n^2-a^2\right)^{-1}$, $n \neq 0$, let $\dfrac{1}{n_2-a^2} = \widetilde{f}_s(n)\,\widetilde{g}_s(n)$, where $\widetilde{f}_s(n) = \dfrac{n(-1)^{n+1}}{n^2-a^2}$ and $\widetilde{g}_s(n) = \dfrac{(-1)^{n+1}}{n}$, which gives $f(x) = \dfrac{\sin ax}{\sin a\pi}$, and $g(x) = \dfrac{x}{\pi}$, so that

$$\frac{1}{n^2-a^2} = \widetilde{f}_s(n)\widetilde{g}_s(n) = \mathcal{F}_s\left\{\frac{\sin ax}{\sin a\pi}\right\}\mathcal{F}_s\left\{\frac{x}{\pi}\right\}$$

$$= \mathcal{F}^{-1}\left\{\frac{1}{n^2-a^2}\right\} = \mathcal{F}^{-1}\{\widetilde{f}_s(n)\}\widetilde{g}_s(n)\} = -\frac{1}{2}\,f_1(x)\star g_1(x),$$

where $f_1(x)$ is the periodic extension of the odd function $f(x)$ and $g_1(x) = \dfrac{x}{\pi}$. Thus, using (5.28)

$$\mathcal{F}^{-1}\left\{\frac{1}{n^2-a^2}\right\} = -\frac{1}{2\pi}\int_{-\pi}^{\pi} f_1(x-u)\,g_1(u)\,du$$

$$= -\frac{1}{2\pi}\int_{-\pi}^{\pi} f_1(x-u)\,u\,du = -\frac{\cos[a(\pi-x)]}{a\sin a\pi}. \ \blacksquare$$

In the case of a function of two variables, $f(x, y)$, $0 < x < a$, $0 < y < b$, these finite transforms are denoted by $\widetilde{f}_s(n, y)$ and $\widetilde{f}_c(n, y)$, respectively, as the following example shows.

Example 5.18. Consider Laplace's equation in the rectangle $\{0 < x < a,\ 0 < y < b\}$

$$u_{xx} + u_{yy} = 0, \tag{5.29}$$

with the boundary conditions $u(0, y) = u(a, y) = u(x, b) = 0$, and $u(x, 0) = f(x)$. After applying the finite Fourier sine transform to $u(x, y)$ with respect to x from 0 to a, we have

$$(\tilde{u}_s)_{xx}(n) = \int_0^a u_{xx} \sin \frac{n\pi x}{a}\, dx$$

$$= \frac{n\pi}{a}\left[u(0, y) - (-1)^n u(a, y)\right] - \frac{n^2\pi^2}{a^2}\tilde{u}_s(n, y).$$

Then Equation (5.29) becomes $\left[\dfrac{d^2}{dy^2} - \dfrac{n^2\pi^2}{a^2}\right]\tilde{u}_s(n, y) = 0$. Solving for $\tilde{u}_s(n, y)$, we get $\tilde{u}_s(n, y) = Ae^{n\pi y/a} + Be^{-n\pi y/a}$. Since $\tilde{u}_s(n, b) = 0$, we can express $\tilde{u}_s(n, y)$ as $\tilde{u}_s(n, y) = A_n\left(e^{n\pi(y-b)/a} - e^{-n\pi(y-b)/a}\right)$. After applying the boundary condition at $y = 0$, we get

$$A_n\left(e^{-n\pi b/a} - e^{n\pi b/a}\right) = \overline{f}_s(n),$$

which, after solving for A_n and substituting its value into $\tilde{u}_s(n, y)$, yields

$$\tilde{u}_s(n, y) = -\frac{\sinh[n\pi(y - b)/a]}{\sinh(n\pi b/a)}\overline{f}_s(n).$$

Hence,

$$u(x, y) = \frac{2}{a}\sum_1^\infty \frac{\sinh[n\pi(b - y)/a]}{\sinh(n\pi b/a)}\overline{f}_s(n)\sin(n\pi x/a),$$

where $\overline{f}_s(n) = \displaystyle\int_0^a f(t)\sin\frac{n\pi t}{a}\, dt.$ ∎

Other examples are available in §6.8.

5.6. Multiple Transforms

Multiple transforms used in this book are defined below.

5.6.1. Multiple Laplace Transform.
Double Laplace transform is defined by

$$\mathcal{L}\{f(x, t)\} = \overline{F}(\kappa, s) = \overline{\overline{f}}(\kappa, s) = \int_0^\infty \int_{c-i\infty}^{c+i\infty} f(x, t)\, e^{-(\kappa+s)t}\, d\kappa\, dt, \tag{5.30}$$

and its inverse by

$$\mathcal{L}^{-1}\{\overline{F}(\kappa, s)\} = f(x, t) = \frac{1}{2\pi i} = \int_0^\infty \int_{c-i\infty}^{c+i\infty} \overline{F}(\kappa, s)\, e^{(\kappa+s)t}\, dt\, d\kappa, \quad (5.31)$$

where $\Re\{z\} > 0, \Re\{\kappa\} > 0$ and $c > 0$. For an application see §6.9.

5.6.2. Multiple Fourier Transform. Let $\mathbf{x} = (x_1, \dots, x_n) \in \mathbb{R}^n$. Then under the same assumptions on $f(\mathbf{x})$ as on $f(x) \in \mathbb{R}$, the multi-dimensional Fourier transform of $f(\mathbf{x})$ is defined by

$$\mathcal{F}\{f(\mathbf{x}) = \widetilde{f}(\varsigma) = \frac{1}{(2\pi)^{n/2}} \int_{-\infty}^\infty \cdots \int_{-\infty}^\infty \exp\{-i\,(\varsigma\cdot\mathbf{x})\}\, f(\mathbf{x})\, d\mathbf{x}, \quad (5.32)$$

where $\varsigma = (\varsigma_1, \dots, \varsigma_n)$ denotes the n-dimensional transform variable and $\varsigma \cdot \mathbf{x} = (\varsigma_1 x_1 + \cdots + \varsigma_n x_n)$.

The inverse Fourier transform is defined by

$$\mathcal{F}^{-1}\{\widetilde{f}(\varsigma)\} = f(\mathbf{x}) = \frac{1}{(2\pi)^{n/2}} \int_{-\infty}^\infty \cdots \int_{-\infty}^\infty \exp\{-i\,(\varsigma\cdot\mathbf{x})\}\, f(\mathbf{x})\, d\mathbf{x}, \quad (5.33)$$

In 2-D and 3-D cases we will use $\varsigma = (\alpha, \beta)$ and $\varsigma = (\alpha, \beta, \gamma)$, respectively. Thus, the *double Fourier transform* is defined by

$$\mathcal{F}\{f(x, y)\} = \frac{1}{2\pi} \int_{-\infty}^\infty \int_{-\infty}^\infty e^{-i\,(\alpha x+\beta y)}\, f(x, y)\, dx\, dy, \quad (5.34)$$

and its inverse by

$$\mathcal{F}^{-1}\{\widetilde{f}(\alpha, \beta)\} = \frac{1}{2\pi} \int_{-\infty}^\infty \cdots \int_{-\infty}^\infty e^{-i\,(\alpha x+\beta y)}\, f(x, y)\, d\alpha\, d\beta, \quad (5.35)$$

and the 3-D Fourier transform by

$$\mathcal{F}\{f(x, y, z)\} = \frac{1}{2\pi} \int_{-\infty}^\infty \int_{-\infty}^\infty e^{-i\,(\alpha x+\beta y+\gamma z)}\, f(x, y, z)\, dx\, dy\, dz, \quad (5.36)$$

and its inverse by

$$\mathcal{F}^{-1}\{\widetilde{f}(\alpha, \beta, \gamma)\} = \frac{1}{2\pi} \int_{-\infty}^\infty \cdots \int_{-\infty}^\infty e^{-i\,(\alpha x+\beta y+\gamma z)}\, f(x, y, z)\, d\alpha\, d\beta\, d\gamma, \quad (5.37)$$

Example 5.19. For $f(x, y, z) = \delta(x)\delta(y)\delta(z)$ the 3-D Fourier transform is $\widetilde{f}(\varsigma) = (2\pi)^{-3/2}$. ∎

Other examples are available in §6.3 and §7.10.1.

5.7. Hankel Transforms

The nth order Hankel transform of a function $f(x)$ is defined by

$$\mathcal{H}_n\{f(x)\} = \widehat{f}_n(\sigma) = \int_0^\infty x J_n(\sigma x) f(x)\, dx, \quad (5.38)$$

where $J_n(\sigma x)$ is the Bessel function of order n, and the integral on the right side is convergent. The inverse Hankel transform of order n is given by

$$\mathcal{H}^{-1}\{\widehat{f}_n(\sigma)\} = f(x) = \int_0^\infty \sigma J_n(\sigma x)\widehat{f}_n(\sigma)\,d\sigma, \qquad (5.39)$$

provided the integral exists.

Some basic properties of this transform are:

(i) (Scaling) If $\mathcal{H}_n\{f(x)\} = \widehat{f}_n(\sigma)$, then $\mathcal{H}_n\{f(ax)\} = \dfrac{1}{a^2}\,\widehat{f}_n\left(\dfrac{\sigma}{a}\right)$.

(ii) (Parseval's Relation) If $\mathcal{H}_n\{f(x)\} = \widehat{f}_n(\sigma)$ and if $\mathcal{H}_n\{g(x)\} = \widehat{g}_n(\sigma)$, then

$$\int_0^\infty x f(x)\,g(x)\,dx = \int_0^\infty \sigma\,\widehat{f}_n(\sigma)\,\widehat{g}_n(\sigma)\,d\sigma. \qquad (5.40)$$

(iii) (Transform of derivatives) If $\mathcal{H}_n\{f(x)\} = \widehat{f}(\sigma)$, then

$$\mathcal{H}_n\{f'(x)\} = \frac{\sigma}{2n}\left[(n-1)\widehat{f}_{n+1}(\sigma) - (n+1)\widehat{f}_{n-1}(\sigma)\right], \quad n \geq 1. \qquad (5.41)$$

$$\mathcal{H}_1\{f'(x)\} = -\sigma\,\widehat{f}_0(\sigma), \qquad (5.42)$$

provided $x f(x) \to 0$ as $x \to 0$ and $x \to \infty$.

(iv) If $\mathcal{H}_n\{f(x)\} = \widehat{f}(\sigma)$, then

$$\mathcal{H}_n\{f''(x)\} = \mathcal{H}_n\left\{\frac{1}{x}\frac{d}{dx}\left(x\frac{df}{dx}\right) - \frac{n^2}{x^2}f(x)\right\} = -\sigma^2\,\widehat{f}_n(\sigma), \qquad (5.43)$$

provided both $x f'(x)$ and $x f(x)$ vanish as $x \to 0$ and $x \to \infty$.

(v) The zero-order Hankel transform ($n = 0$) is defined by

$$\mathcal{H}_0\{f(x)\} = \widehat{f}(x) = \int_0^\infty x J_0(\sigma x) f(x)\,dx. \qquad (5.44)$$

Example 5.20. The following results hold:

(a) $\mathcal{H}_0\left\{\dfrac{\delta(x)}{x}\right\} = 1$;

(b) $\mathcal{H}_0\{H(a-x)\} = \displaystyle\int_0^a x J_0(\sigma x)\,dx = \dfrac{a}{\sigma}\,J_1(a\sigma)$;

(c) $\mathcal{H}_0\left\{\dfrac{e^{-ax}}{x}\right\} = \displaystyle\int_0^a e^{-ax}\,J_0(\sigma x)\,dx = \dfrac{1}{\sqrt{a^2+\sigma^2}}$;

(d) $\mathcal{H}_1\{e^{-ax}\} = \displaystyle\int_0^a x\,e^{-ax}\,J_0(\sigma x)\,dx = \dfrac{\sigma}{(a^2+\sigma^2)^{3/2}}$;

(e) $\mathcal{H}_1\left\{\dfrac{e^{-ax}}{x}\right\} = \displaystyle\int_0^a e^{-ax}\,J_1(\sigma x)\,dx = \dfrac{1}{\sigma}\left[1 - a\left(a^2+\sigma^2\right)^{-1/2}\right]$;

(f) $\mathcal{H}_n\{x^n H(a-x)\} = \displaystyle\int_0^a x^{n+1} J_n(\sigma x)\,dx = \dfrac{a^{n+1}}{\sigma} J_{n+1}(a\sigma);$

(g) $\mathcal{H}_n\{x^n a^{-ax^2}\} = \displaystyle\int_0^\infty x^{n+1} J_n(\sigma x)\,e^{-ax^2}\,dx = \dfrac{\sigma^n}{(2a)^{n+1}}\,e^{-\sigma^2/(4a)}.$

Other examples are available in §6.7, §7.4 and §7.10.3.

5.8. Summary: Variables of Transforms

Laplace transform: $\mathcal{L}\{f(t)\} = \overline{f}(s)$; variable of transform: s.

Double Laplace transform: $\mathcal{L}\{f(x,t)\} = \overline{\overline{f}}(\kappa,s)$; variable of transform: κ, s

Fourier transform: $\mathcal{F}\{f(x)\} = \hat{f}(\alpha)$; variable of transform: α.

Fourier sine transform: $\mathcal{F}_s\{f(x)\} = \tilde{f}_s(\alpha)$; variable of transform: α.

Fourier cosine transform: $\mathcal{F}_c\{f(x)\} = \tilde{f}_c(\alpha)$; variable of transform: α.

Finite Fourier transforms: n.

Fourier 2-D transform: $\mathcal{F}\{f(x)\} = \hat{f}(\varsigma)$; variable of transform: $\varsigma = (\alpha, \beta)$.

Fourier 3-D transform: $\mathcal{F}\{f(x)\} = \hat{f}(\zeta)$; variable of transform: $\zeta = (\alpha, \beta, \gamma)$.

Zero-order Hankel transform: $\mathcal{H}_0\{f(x)\} = \tilde{f}(\sigma)$; variable of transform: σ.

5.9. Exercises

5.1. Using the properties of §5.2 and $\mathcal{L}\left\{e^{at}\right\} = \dfrac{1}{s-a}$, derive the Laplace transform of $\sin at$, $\cos at$, $e^{bt}\sin at$, $e^{bt}\cos at$, $t^n\,e^{at}$, $\sinh bt$, and $\cosh bt$.

5.2. Show that $\mathcal{L}^{-1}\left\{e^{-a\sqrt{s}}\right\} = \dfrac{a}{2\sqrt{\pi t^3}}\,e^{-a^2/4t}$.

5.3. Show that (a) $\mathcal{L}^{-1}\left\{s^{-3/2}e^{-1/s}\right\} = \dfrac{1}{\sqrt{\pi}}\sin(2\sqrt{t})$; and

(b) $\mathcal{L}^{-1}\left\{s^{-1/2}e^{-1/s}\right\} = \dfrac{1}{\sqrt{\pi t}}\cos(2\sqrt{t})$.

5.4. Find (a) $\mathcal{L}^{-1}\left\{\dfrac{\cosh a\sqrt{s}}{s\,\cosh b\sqrt{s}}\right\}$, and (b) $\mathcal{L}^{-1}\left\{\dfrac{\sinh a\sqrt{s}}{\sinh b\sqrt{s}}\right\}$, $b > a > 0$.

HINT. Use $\cosh x = \left(e^x + e^{-x}\right)/2$, $\sinh x = \left(e^x - e^{-x}\right)/2$, and $(1+z)^{-1} = \displaystyle\sum_{n=0}^\infty (-1)^n z^n$.

ANS. (a) $\displaystyle\sum_{n=0}^\infty (-1)^n \left\{\operatorname{erfc}\left(\dfrac{(2n+1)b - a}{2\sqrt{t}}\right) + \operatorname{erfc}\left(\dfrac{(2n+1)b + a}{2\sqrt{t}}\right)\right\},$

(b) $\displaystyle\sum_{n=0}^{\infty}\left[\frac{(2n+1)b-a}{\sqrt{4\pi t^3}}e^{-[(2n+1)b-a]^2/(4t)}-\frac{(2n+1)b+a}{\sqrt{4\pi t^3}}e^{-[(2n+1)b+a]^2/(4t)}\right].$

5.5. Using the Laplace transform method, solve the partial differential equation $u_{tt}=u_{xx}$, with the initial conditions $u(x,0)=-\dfrac{(1-x)^2}{2}$, $u_t(x,0)=0$, and the boundary conditions $u_x(0,t)=1$ and $u_x(1,t)=0$.

ANS. $u=-\dfrac{1}{2}t^2-\dfrac{(1-x)^2}{2}.$

5.6. Using the Laplace transform method, solve the partial differential equation $u_t=u_{xx}$, with the initial condition $u(x,0)=0$ and the boundary conditions $u_x(0,t)=0$ and $u(1,t)=1$.

HINT. The solution in the transform domain is $\bar{u}=\dfrac{\cosh x\sqrt{s}}{s\cosh\sqrt{s}}.$

ANS. $u=\displaystyle\sum_{0}^{\infty}(-1)^n\left[\operatorname{erfc}\frac{2n+1-x}{2\sqrt{t}}+\operatorname{erfc}\frac{2n+1+x}{2\sqrt{t}}\right],$ or

$u=1-\displaystyle\sum_{0}^{\infty}(-1)^n\frac{4\cos(2n+1)\pi x/2}{(2n+1)\pi}e^{-(2n+1)^2\pi^2 t/4}.$

5.7. Use Laplace transform to solve the wave equation $u_{tt}=c^2 u_{rr}$, subject to the initial conditions $u(r,0)=u_t(r,0)=0$ and the boundary conditions $\dfrac{\partial u(\rho,t)}{\partial\rho}\Big|_{\rho=a}=f(t)$ and $u\to 0$ as $r\to\infty.$

HINT. Use $\mathcal{L}^{-1}\left\{\dfrac{F(s)}{s+k}\right\}=\int_0^t e^{-k(t-x)}f(x)\,dx\equiv\phi(t)$, $k=\dfrac{c}{a}$. Then $u=\mathcal{L}^{-1}\left\{-\dfrac{ac}{s+c/a}\dfrac{F(s)}{\rho}e^{-s(\rho-a)/c}\right\}$, then use Property (ii). ANS. $u(r,t)=-\dfrac{ca}{\rho}H\left(t-\dfrac{\rho-a}{c}\right)\phi\left(t-\dfrac{\rho-a}{c}\right).$

5.8. Show that $\mathcal{L}\{t^p\}=\dfrac{\Gamma(p+1)}{s^{p+1}}$, where $\Gamma(x)$ is the gamma function defined by $\Gamma(x)=\displaystyle\int_0^{\infty}t^{x-1}e^{-t}\,dt$, $x>0$. HINT. Set $st=x$ in $\mathcal{L}\{t^p\}=\int_0^{\infty}t^p e^{-st}\,dt.$ Then $\int_0^{\infty}t^p e^{-st}\,dt=\dfrac{1}{s^{p+1}}\int_0^{\infty}x^p e^{-x}\,dx=\dfrac{\Gamma(p+1)}{s^{p+1}}.$

5.9. Find the complex Fourier transform of the following functions:

(a) $f(x)=\begin{cases}0 & \text{if } x<0\\ e^{-ax} & \text{if } x>0.\end{cases}$ ANS. $\dfrac{1}{\sqrt{2\pi}(a-i\alpha)}.$

(b) $f(x)=\begin{cases}0 & \text{if } x<0,\\ \dfrac{1}{x}[e^{-ax}-e^{-bx}] & \text{if } x>0 \text{ and } b>a>0.\end{cases}$

HINT. Use part (a) and integrate with respect to a. ANS. $\dfrac{1}{\sqrt{2\pi}} \ln \dfrac{b - i\alpha}{a - i\alpha}$.

(c) $f(x) = \begin{cases} 0 & \text{if } |x| > a, \\ 1 - \dfrac{|x|}{a} & \text{if } |x| < a. \end{cases}$ ANS. $\sqrt{\dfrac{2}{\pi}} \dfrac{1}{a\alpha^2}(1 - \cos \alpha a)$.

(d) $f(x) = \cos ax^2$ and $f(x) = \sin ax^2$.

HINT: Use Example 5.9 and define k as ia.

ANS. $\dfrac{1}{2\sqrt{a}} \left(\sin \dfrac{\alpha^2}{4a} + \cos \dfrac{\alpha^2}{4a} \right)$ and $\dfrac{1}{2\sqrt{a}} \left(\sin \dfrac{\alpha^2}{4a} - \cos \dfrac{\alpha^2}{4a} \right)$.

(e) $f(x) = \begin{cases} 1 & \text{if } |x| < 1 \\ 0 & \text{if } |x| > 1. \end{cases}$ ANS. $\dfrac{\sqrt{2}}{\alpha\sqrt{\pi}} \sin \alpha$.

(f) $f(x) = \begin{cases} \sin kx & \text{if } |x| < 1 \\ 0 & \text{if } |x| > 1. \end{cases}$ ANS. $\dfrac{i}{\sqrt{2\pi}} \left[\dfrac{\sin(k - \alpha)}{k - \alpha} - \dfrac{\sin(k + \alpha)}{k + \alpha} \right]$.

(g) $f(x) = \begin{cases} \dfrac{1}{\sqrt{x}} & \text{if } x > 0 \\ = 0 & \text{if } x < 0 \end{cases}$.

HINT: Substitute $\alpha x = v^2 e^{i\pi/2}$. ANS. $\dfrac{e^{i\pi/4}}{\sqrt{2\alpha}} \operatorname{sgn} \alpha$.

(h) $f(x) = \begin{cases} \dfrac{1}{\sqrt{|x|}} & \text{if } x < 0 \\ 0 & \text{if } x > 0. \end{cases}$ ANS. $\dfrac{e^{-i\pi/4}}{\sqrt{2\alpha}} \operatorname{sgn} \alpha$.

(i) $f(x) = \dfrac{1}{\sqrt{|x|}}$. ANS. $\dfrac{1}{\sqrt{|\alpha|}}$.

(j) $f(x) = e^{-kx^2}, k > 0$. ANS. $\dfrac{1}{\sqrt{2k}} e^{-\alpha^2/(4k)}$.

(k) $f(x) = \begin{cases} 0 & \text{for } x < 0, \\ xe^{-ax} & \text{for } x > 0,\ a > 0. \end{cases}$ ANS. $\dfrac{1}{\sqrt{2\pi}(i\alpha - a)^2}$.

5.10. Prove the following properties of the Fourier convolution:

(i) $\dfrac{d}{dx}\{f(x) \star g(x)\} = f'(x) \star g(x) = f(x) \star g'(x)$,

(i) $\dfrac{d^2}{dx^2}\{f(x) \star g(x)\} = f'(x) \star g'(x) = f''(x) \star g(x)$,

(iii) $(f(x) \star g(x))^{(m+n)} = \left(f^{(m)}(x) \star g^{(n)}(x) \right)$,

(iv) $\displaystyle\int_{-\infty}^{\infty} f(x) \star g(x)\, dx = \int_{-\infty}^{\infty} f(u)\, du \star \int_{-\infty}^{\infty} g(v)\, dv$,

(v) If $g(x) = \dfrac{1}{2a}H(a - x)$, then $f(x) \star g(x)$ is the average value of $f(x)$ in $[x - a, x + a]$,

(vi) If $g_t(x) = \dfrac{1}{4\pi kt} e^{-x^2/(4kt)}$, then $g_t(x) \star g_s(x) = g_{t+s}(x)$.

5.11. Compute the Fourier cosine transform of $f(t)$, where

$$f(x) = \begin{cases} x & \text{if } 0 < x < 1, \\ 2 - x & \text{if } 1 < x < 2, \\ 0 & \text{if } x \geq 2. \end{cases} \quad \text{ANS. } \widetilde{f}(\alpha) = \frac{4}{\alpha^2}\sqrt{\frac{2}{\pi}} \cos(\alpha) \sin^2(\alpha/2).$$

5.12. Solve the partial differential equation $u_{tt} - c^2 u_{xx} = 0$, subject to the conditions $u(x,0) = f(x) + g(x)$ and $u_t(x,0) = c\,(f'(x) - g'(x))$, where $u(x,t)$, $f(x)$, $g(x)$, $u'(x,t)$, $f'(x)$, and $g'(x)$ all go to zero as $|x| \to \infty$.

ANS. $u = f(x + ct) + g(x - ct)$.

5.13. Find the Fourier sine and cosine transforms of $f(x) = \dfrac{1}{\sqrt{x}} e^{-ax}$.

ANS. $\mathcal{F}_c f(x) = \dfrac{\sqrt{\sqrt{a^2 + \alpha^2} + a}}{\sqrt{a^2 + \alpha^2}}, \mathcal{F}_s f(x) = \dfrac{\sqrt{\sqrt{a^2 + \alpha^2} - a}}{\sqrt{a^2 + \alpha^2}}.$

5.14. Find the Fourier cosine transform of $\dfrac{1}{x}\left[e^{-ax} - e^{-bx}\right]$, $\Re\{a\}, \Re\{b\} > 0$.

ANS. $\dfrac{1}{\sqrt{2\pi}} \ln\left(\dfrac{b^2 + \alpha^2}{a^2 + \alpha^2}\right)$.

5.15. Find the Fourier sine transform of $\dfrac{1}{x}e^{-ax}$, $\Re\{a\} > 0$. ANS. $\sqrt{\dfrac{2}{\pi}} \tan^{-1}\dfrac{\alpha}{a}$.

5.16. Find Fourier sine and cosine transforms of $f(x) = \sqrt{x}e^{-ax}$, $a > 0$.

HINT: $\widetilde{f}_c(\alpha) + i\widetilde{f}_s(\alpha) = \dfrac{1}{\sqrt{2(a - i\alpha)^3}}$; then express $(a - i\alpha)$ in the polar form.

ANS. $\widetilde{f}_c(\alpha) + i\widetilde{f}_c(\alpha) = \dfrac{e^{(3i/2)\,\tan^{-1}(\alpha/a)}}{\sqrt{2}(a^2 + \alpha^2)^{3/4}}$.

5.17. Find the Fourier sine transforms of the following functions: (i) $f(x) = x\,e^{-ax}$, $a > 0$; (ii) $f(x) = e^{-ax} \cos x$, $a > 0$.

HINT. Differentiate both sides of the integral $\int_0^\infty e^{-ax} \sin kx\, dx = \dfrac{a}{a^2 + \alpha^2}$

with respect to a.   ANS. (ii) $\{_s(\alpha) = \sqrt{\dfrac{a}{\pi}} \dfrac{2a\alpha}{(a^2 + \alpha^2)^2}$.

5.18. Show that $\mathcal{H}_0\{(a^2 - x^2)H(a - x)\} = \dfrac{4a}{\sigma^3} J_1(a\sigma) - \dfrac{2a^2}{\sigma^2} J_0(a\sigma)$.

6

Parabolic Equations

Diffusion is one of the basic processes by which material moves. It is a result of continuous motion of atoms, molecules and particles moving from regions of high to low concentration. Besides, there are other processes that produce inhomogeneity, and the combined result of diffusion and other processes is generally not precisely defined. Diffusion depends on three main parameters: (i) temperature, (ii) mass (size) of the diffusing particles and (iii) viscosity of the system. While the temperature is a measure of the average kinetic energy, which is due to the movement of particles in the system, in most systems the energy is equal to a constant times the temperature, and a higher kinetic energy means a higher viscosity, yet the speed of diffusion always increases with temperature. Since heavy particles have a lower velocity for a given kinetic energy, they diffuse more slowly than lighter, smaller ones. Diffusion is most rapid in a gas, slower in a liquid, and very slow or almost zero in a solid.

The flow of a quantity, such as heat, is described by the diffusion equation, where the total amount of the quantity does not change. Thus, both heat and mass are conserved, although their distributions may undergo slight changes. The first-order time derivative in the diffusion equation signifies a complete determination of the time development of the solution subject to prescribed initial temperature or mass distribution, and presence of any sources or sinks. It also means the irreversibility of the equation in the sense that the solution is always in the direction of time; thus, a change in the sign of time also changes the behavior of the solution. Green's functions and fundamental solutions for the parabolic operator play an important role in problems of heat conduction and neutron diffusion.

In this chapter we will derive Green's functions for the transient operator $L\left[D, \dfrac{\partial}{\partial t}\right]$, defined in the space $\mathbb{R}^3 \times \mathbb{R}^+$ of the independent variables $(\mathbf{x}, t) = (x, y, z, t)$, where $L(D)$ is the 3-D Laplacian operator in the space variables. This transient operator is also known as the 3-D diffusion operator or the heat conduction operator. It is often denoted simply by $\dfrac{\partial}{\partial t} - k\nabla^2$, where k denotes the thermal diffusivity. The problem of finding Green's functions for these operators is solved by different methods, but mostly by the integral transform method.

6.1. 1-D Diffusion Equation

We will consider boundary value problem for 1-D diffusion equation, which are solved by two different methods.

6.1.1. Sturm-Liouville System for 1-D Diffusion Equation. Consider the 1-D diffusion system

$$
\begin{aligned}
&\frac{\partial u}{\partial t} = \frac{\partial}{\partial x}\left[p(x)\frac{\partial u}{\partial x}\right] + q(x)u + F(x,t), \quad a \le x \le b,\ t > 0; \\
&a_1 u(a,t) + a_2 u_x(a,t) = 0,\ b_1 u(b,t) + b_2 u_x(b,t) = 0, \\
&u(x,0) = f(x), \quad a < x < b,
\end{aligned}
\tag{6.1}
$$

where p, q, F are continuous functions on the interval $I = [a,b]$, and $a_{1,2}$ and $b_{1,2}$ are real positive constants such that $a_1^2 + a_2^2 > 0$ and $b_1^2 + b_2^2 > 0$. Using the Sturm-Liouville operator L, Eq (6.1) becomes

$$
u_t = L[u] + F.
\tag{6.2}
$$

First, we take $F = 0$ and use Bernoulli's separation method: Let the solution of Eq (6.2) be of the form $u(x,t) = X(x)T(t) \ne 0$. Then this equation reduces to

$$
\begin{aligned}
&L[X] = \lambda X, \quad a \le x \le b, \\
&\frac{dT}{dt} = \lambda T, \quad t > 0,
\end{aligned}
\tag{6.3}
$$

where λ is the separation constant. The boundary conditions and the initial condition in (6.1) become

$$
\begin{aligned}
&a_1 X_1(a) + a_2 X'(a) = 0,\ b_1 X_1(b) + b_1 X'(b) = 0, \\
&T(0) = f(x), \quad a < x < b.
\end{aligned}
\tag{6.4}
$$

The system (6.3)–(6.4) is called the associated Sturm-Liouville system for the 1-D diffusion equation. The solution $u(x,t)$ of this system can be found in terms of the eigenpair $(\lambda_n, X_n(x))$ as

$$
u(x,t) = \sum_{n=1}^{\infty} X_n(x)T_n(t),
\tag{6.5}
$$

where the functions $X_n(t)$ are determined as follows: Assume that the forcing term $F(x,t)$ is expandable in terms of the eigenfunctions as

$$F(x,t) = \sum_{n=1}^{\infty} X_n(t)\, f_n(t), \tag{6.6}$$

where $f_n(t)$ are Fourier coefficients given by

$$f_n(t) = \int_a^b F(x,t)\, X_n(x)\, dx. \tag{6.7}$$

Substituting (6.5) and (6.6) into Eq (6.2), we get

$$\sum_{n=1}^{\infty} \dot{T}_n(t) X_n(t) = L\left[\sum_{n=1}^{\infty} T_n(t) X_n(x)\right] + \sum_{n=1}^{\infty} f_n(t) X_n(x)$$

$$= \sum_{n=1}^{\infty} \left[\lambda_n T_n(t) + f_n(t)\right] X_n(x),$$

which yields the ordinary differential equation

$$\dot{T}_n(t) - \lambda_n T_n(t) = f_n(t),$$

where the dot denotes derivative with respect to t. To solve this equation, we apply Laplace transform, which after inverting gives the solution as

$$T_n(t) = X_n(0)\, e^{\lambda_n t} + \int_0^t e^{\lambda_n(t-\tau)}\, f_n(\tau)\, d\tau, \tag{6.8}$$

where $u(x,0) = f(x) = \sum_{n=1}^{\infty} X_n(x) X_n(0) = \sum_{n=1}^{\infty} X_n(x) \int_a^b X_n(x') f(x')\, dx'$.
Substituting (6.8) into (6.5) we have

$$u(x,t) = \sum_{n=1}^{\infty} \left[\left\{\int_a^b X_n(x') f(x')\, dx'\right\} e^{\lambda_n t} + \int_0^t e^{\lambda_n(t-\tau)}\, f_n(\tau)\, d\tau\right] X_n(x)$$

$$= \int_a^b \left[\sum_{n=1}^{\infty} X_n(x)\, X_n(x')\, e^{\lambda_n t}\right] f(x')\, dx'$$

$$+ \int_0^t \int_a^b \sum_{n=1}^{\infty} \left[X_n(x) X_n(x')\, e^{\lambda_n(t-\tau)}\right] F(x',\tau)\, d\tau, \tag{6.9}$$

where the integration and summation are interchanged. If we define Green's function for the operator L by

$$G(x,x';t) = \sum_{n=1}^{\infty} X_n(x)\, X_n(x')\, e^{\lambda_n t}, \tag{6.10}$$

then the solution (6.9) can be written as

$$u(x,t) = \int_a^b G(x,x';t)f(x')\,dx' + \int_0^t \int_a^b G(x,x';t-\tau)F(x',\tau)\,dx'\,d\tau. \quad \blacksquare \quad (6.11)$$

6.1.2. Green's Function for 1-D Diffusion Equation.

Consider the boundary value problem defined by the nonhomogeneous 1-D diffusion equation.

$$u_t - k\,u_{xx} = f(x), \quad x \in \mathbb{R},\ t > 0, \tag{6.12}$$

subject to the boundary and the initial conditions[1]

$$u(x,t) \to 0 \text{ as } |x| \to \infty, \quad \text{and } u(x,0) = 0 \text{ for } x \in \mathbb{R}. \tag{6.13}$$

Note that the diffusion operator is not self-adjoint in $(-\infty, \infty) \times (0, \infty)$. However, in the Laplace transform domain it is self-adjoint. The equation and the boundary and initial conditions to be satisfied by Green's function $G(x, x'; t, t')$ are

$$\frac{\partial G}{\partial t} - k\,\frac{\partial^2 G}{\partial x^2} = \delta(x - x')\,\delta(t - t'), \tag{6.14a}$$

$$G(x, x'; 0, t') = 0, \quad \lim_{|x|\to\infty} G(x, x'; t, t') = 0. \tag{6.14b}$$

Applying Laplace transform to Eq (6.14a), we have

$$s\overline{G} - k\,\frac{d^2\overline{G}}{dx^2} = \delta(x - x')\,e^{-st'},$$

where the boundary condition (6.14b) becomes $\lim_{|x|\to\infty} \overline{G}(x, x'; s, t') = 0$. By property (iv) §2.3.2, the discontinuity condition at x' is

$$\frac{d\overline{G}(x'_+, x'; s, t')}{dx} - \frac{d\overline{G}(x'_-, x'; s, t')}{dx} = -\frac{1}{k}\,e^{-st'}. \tag{6.15}$$

The solution in the transform domain is

$$\overline{G}(x, s) = \begin{cases} A_1\,e^{cx} + B_1\,e^{-cx} & \text{if } x < x', \\ A_2\,e^{cx} + B_2\,e^{-cx} & \text{if } x > x', \end{cases}$$

where $c = \sqrt{s/k}$ and s is the variable of the transform. The condition at $x \to -\infty$ yields $B_1 = 0$ and at $x \to \infty$ yields $A_2 = 0$. The continuity of $\overline{G}(x, s)$ at x'

---

[1] If the variable t denotes time in the defining equation, then an initial condition must be prescribed.

gives $A_1 e^{cx'} = B_2 e^{-cx'}$, and the jump condition (6.15) on $\overline{G}_x(x,s)$ at x' gives $B_2 e^{-cx'} + A_1 e^{cx'} = \dfrac{1}{kc} e^{-st'}$; thus,

$$A_1 = \frac{1}{2kc} e^{-st'-cx'}, \qquad B_2 = \frac{1}{2kc} e^{-st'+cx'}.$$

Hence,

$$\overline{G}(x,s) = \begin{cases} \dfrac{1}{2kc} e^{-st'-c(x'-x)}, & x < x', \\[2mm] \dfrac{1}{2kc} e^{-st'-c(x-x')}, & x > x', \end{cases}$$

that is, for all x we have $\overline{G}(x,s) = \dfrac{1}{2kc} e^{-st'-c|x-x'|}$, which on inversion gives

$$G(x,x';t,t') = \frac{H(t-t')}{2\sqrt{\pi k(t-t')}} e^{-(x-x')^2/(4k(t-t'))}. \tag{6.16}$$

This Green's function is also the fundamental solution for the diffusion operator.

METHOD 2. Apply Fourier transform followed by Laplace transform to Eq (6.14a). This gives

$$s\,\overline{\widetilde{G}} + k\alpha^2\,\overline{\widetilde{G}} = -\frac{1}{\sqrt{2\pi}} e^{-st'+i\alpha x'},$$

where the tilde and the bar over G denote its Fourier and Laplace transforms, respectively. The solution for $\overline{\widetilde{G}}$ is

$$\overline{\widetilde{G}} = \frac{1}{\sqrt{2\pi}\,(s+k\alpha^2)} e^{-st'+i\alpha x'}.$$

Inversion of Laplace transform yields

$$\widetilde{G}(x,\alpha) = \frac{H(t-t')}{\sqrt{2\pi}} e^{-k\alpha^2(t-t')+i\alpha x'},$$

which, after inverting Fourier transform, gives the Green's function G which is the same as (6.16).

METHOD 3. If we apply Laplace transform followed by Fourier transform to the problem (6.12), we get

$$\overline{\widetilde{u}}(\alpha,s) = \frac{\widetilde{f}(\alpha)}{s+k\alpha^2},$$

such that $\overline{\widetilde{u}}(\alpha,s) \to 0$ as $|\alpha| \to \infty$. Then first inverting Laplace transform we have

$$\widetilde{u}(\alpha,t) = \widetilde{f}(\alpha)\, e^{-k\alpha^2 t}.$$

Note that if $\tilde{g}(\alpha) = e^{-k\alpha^2 t}$, then by inversion of Fourier transform we have $g(x) = \dfrac{1}{\sqrt{2\pi kt}}\, e^{-x^2/(4kt))}$. Then applying inverse Fourier transform we find that

$$u(x,t) = \frac{1}{\sqrt{2\pi}} \int_{-\infty}^{\infty} e^{i\alpha x}\, \tilde{f}(\alpha)\, e^{-k\alpha^2 t}\, d\alpha$$

$$= \frac{1}{2\sqrt{k\pi t}} \int_{-\infty}^{\infty} \tilde{f}(\xi)\, e^{-(x-\xi)^4/(4kt)}\, d\xi = \frac{1}{2\pi} \int_{-\infty}^{\infty} \tilde{f}(\alpha)\, G(x,t;\xi)\, d\xi,$$

where Green's function $G(x - x', t)$ is given by

$$G(x,t;x') = \frac{1}{2\sqrt{k\pi t}}\, e^{-(x-x')^2/(4kt)},$$

which is the same as (6.16).

The graphs for this Green's function $G(x,t)$ for some values of $t > 0$ are shown in Fig. 6.1.

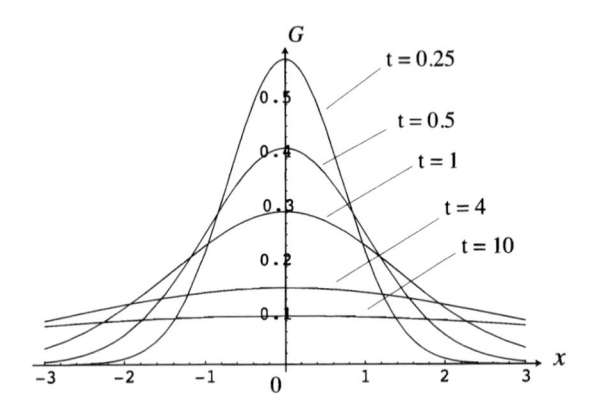

Fig. 6.1. Green's Function for 1-D Diffusion Operator.

6.2. 2-D Diffusion Equation

To find Green's function for the operator $\dfrac{\partial}{\partial t} - k\,\nabla^2$ in $\mathbb{R}^2$, $t > 0$, we consider the function $u(\mathbf{x}; t)$ which satisfies the following equation and the boundary conditions:

$$u_t - k\nabla^2 u = f(x,y)\,\delta(t), \quad -\infty < x, y < \infty, \ t > 0,$$
$$u(x,y,0) = 0 \quad \text{for all } (x,y) \in \mathbb{R}^2, \tag{6.17}$$
$$u(x,y,t) \to 0 \quad \text{as } r = \sqrt{x^2 + y^2} \to \infty.$$

Applying Laplace transform, followed by a double Fourier transform to the above system, we get

$$\widetilde{\widetilde{u}}(\alpha, s) = \frac{\widetilde{\widetilde{f}}(\alpha, \beta)}{s + k\gamma}, \quad \gamma = (\alpha, \beta),$$

which after the Laplace transform inversion gives

$$\widetilde{\widetilde{u}}(\alpha, \beta) = \widetilde{\widetilde{f}}(\alpha, \beta)\, e^{-k\gamma^2 t} \equiv \widetilde{\widetilde{f}}(\alpha, \beta)\, \widetilde{\widetilde{g}}(\alpha, \beta),$$

where $\widetilde{\widetilde{g}}(\alpha, \beta) = e^{-k\gamma^2 t}$, which after inverting yields

$$g(x, y) = \mathcal{F}^{-1}\left\{ e^{-k\gamma^2 t} \right\} = \frac{1}{2kt}\, e^{-(x^2 + y^2)/(4kt)}. \tag{6.18}$$

Then, by using the convolution theorem for Fourier transform (§5.3.3), we get

$$
\begin{aligned}
u(x, y, t) &= \frac{1}{4\pi kt} \int_{-\infty}^{\infty} \int_{-\infty}^{\infty} f(\xi, \eta)\, \exp\left[-\frac{(x - \xi)^2 + (y - \eta)^2}{4kt} \right] d\xi\, d\eta \\
&= \int_{-\infty}^{\infty} \int_{-\infty}^{\infty} f(\xi)\, G(r, \zeta)\, d\xi,
\end{aligned}
$$

where $r = |x - y|$, $\zeta = |\xi - \eta|$, and

$$G(r, \zeta) = \frac{H(t)}{4\pi kt}\, e^{-|r - \zeta|^2/(4kt)}, \tag{6.19}$$

which is the required Green's function (and the fundamental solution) for the 2-D diffusion operator.

6.2.1. Dirichlet Problem for the General Parabolic Equation in a Square. Consider

$$y^2 \frac{\partial^2 u}{\partial x^2} - 2xy \frac{\partial^2 u}{\partial x\, \partial y} + x^2 \frac{\partial^2 u}{\partial y^2} - x \frac{\partial u}{\partial x} - y \frac{\partial u}{\partial y} - u = 0, \tag{6.20}$$

subject to the boundary conditions $u(x, \pm 1) = x^2$, $u(\pm 1, y) = y^2$, on the square bounded by the lines $x = \pm 1, y = \pm 1$ (Fig. 6.2). In view of $\$\,4.2$ and Theorem 4.1, Eq (6.20) is parabolic since $a_{12}^2 - a_{11}a_{22} = 0$, and its characteristics are given by $y^2(dy)^2 + 2xy\, dx\, dy + x^2(dx)^2 = 0$, with the solution $\dfrac{dy}{dx} = -\dfrac{x}{y}$, or $x^2 + y^2 = c$, which is a single parameter family of concentric circles with center at the origin. If we transform Eq (6.20) into polar cylindrical coordinates by taking $x = r\cos\theta$, $y = r\sin\theta$, we obtain

$$\frac{\partial^2 u}{\partial \theta^2} - u = 0, \tag{6.21}$$

which is independent of the radial direction r. Since the solution on any circular path inside the square is required to be single-valued, it must return to the same values as θ changes by 2π. Both solutions of Eq (6.21) are real exponential functions, but they will be periodic only if $u = 0$ on this circular path. That is, $u = 0$ within and on the largest circle C inside the square, which is defined by $x^2 + y^2 = 1$ (Fig. 6.2). For the circular paths in the corners we solve a simple two-point boundary value problem.

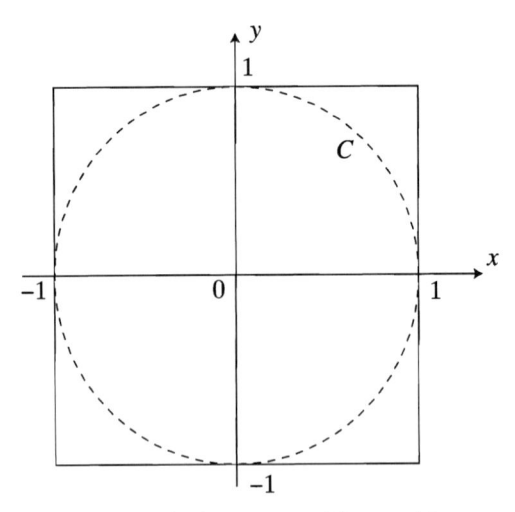

Fig. 6.2. The Square $x = \pm1, y = \pm1$.

Hence, the solution is given by

$$
u(x, y) = \begin{cases} 0, & \text{if } x^2 + y^2 \leq 1, \\ \\ (x^2 + y^2 - 1)\, \dfrac{\exp\left\{ \arcsin \dfrac{|x|}{r} \right\} + \exp\left\{ \arccos \dfrac{|y|}{r} \right\}}{\exp\left\{ \arcsin \dfrac{1}{r} \right\} + \exp\left\{ \arccos \dfrac{1}{r} \right\}}, \\ \quad \text{if } x^2 + y^2 > 1, \ -1 < x < 1, \ -1 < y < 1, \end{cases}
\tag{6.22}
$$

where $r = \sqrt{x^2 + y^2}$. This solution is identically zero within and on the largest circle C inside the square but nonzero in the corners only. In fact, the boundary conditions and the solution at the points $(\pm1, 0)$, $(0, \pm1)$ merge smoothly into the solution $u = 0$ at the core which is the region on and interior to the circle C defined by $x^2 + y^2 \leq 1$. This is because the prescribed boundary conditions tend to zero as the points $(\pm1, 0)$, $(0, \pm1)$ are approached since at these points the boundary of the square and the largest inscribed circle C meet. However, if the boundary conditions were prescribed arbitrarily on the square such that they (these boundary conditions) do not approach zero at the above four points, then the solution would break down in the neighborhood of these points, and we would no longer have a general solution. This shows that this Dirichlet problem in not well posed. ∎

6.3. 3-D Diffusion Equation

To find Green's function for the operator $\dfrac{\partial}{\partial t} - k\,\nabla^2$ in $\mathbb{R}^3$, $t > 0$, we first assume for the sake of simplicity that the singularity is at the origin and at $t = 0$. Green's function $G(\mathbf{x}; t)$ satisfies the following equation and the boundary conditions:

$$\frac{\partial G}{\partial t} - k\,\nabla^2 G = \delta(\mathbf{x})\,\delta(t),$$

$$G(\mathbf{x}; t) \to 0 \quad \text{as } |\mathbf{x}| \to \infty \text{ and for } t > 0. \tag{6.23}$$

Applying the triple Fourier transform with respect to the space variables, followed by Laplace transform with respect to t, to Eq (6.23), we get

$$s\,\widetilde{\overline{\overline{\overline{G}}}}(\alpha, \beta, \gamma, s) - k\,(\alpha^2 + \beta^2 + \gamma^2)\,\widetilde{\overline{\overline{\overline{G}}}}(\alpha, \beta, \gamma, s) = \frac{1}{(2\pi)^{3/2}},$$

where α, β, and γ are the variables of Fourier transform with respect to x, y, and z, respectively. Hence,

$$\widetilde{\overline{\overline{\overline{G}}}}(\alpha, \beta, \gamma, s) = \frac{1}{(2\pi)^{3/2}\left[s - k\,(\alpha^2 + \beta^2 + \gamma^2)\right]}.$$

Inverting first with respect to Laplace transform, we have

$$\widetilde{\overline{\overline{G}}}(\alpha, \beta, \gamma, s) = \frac{H(t)}{(2\pi)^{3/2}}\,e^{-k(\alpha^2 + \beta^2 + \gamma^2)t}.$$

Then inverting the triple Fourier transform, we get

$$G(\mathbf{x}; t) = \frac{H(t)}{(2\pi)^{3/2}\,(2kt)^{3/2}}\,e^{-(x^2 + y^2 + z^2)/4kt} = \frac{H(t)}{8(\pi k t)^{3/2}}\,e^{-|\mathbf{x}|^2/(4kt)}.$$

By translating the singularity to $\mathbf{x}'$ and t', Green's function for this operator with singularity at (x', t') is

$$G(\mathbf{x} - \mathbf{x}'; t - t') = \frac{H(t - t')}{8[\pi k(t - t')]^{3/2}}\,e^{-|\mathbf{x} - \mathbf{x}'|^2/(4k(t - t'))}. \ \blacksquare \tag{6.24}$$

6.3.1. Electrostatic Analog. Gauss's law for an electric charge q applied at the origin that produces an electric field $\mathbf{E}$, given by $\mathbf{E} = \dfrac{q\mathbf{r}_0}{4\pi\varepsilon_0 r^2}$, has two parts: One, defined by the surface integral

$$\int_S \mathbf{E}\cdot d\boldsymbol{\sigma} = \begin{cases} \dfrac{q}{\varepsilon_0}, & \text{if } S \text{ the closed surface includes the origin,} \\ 0, & \text{otherwise,} \end{cases}$$

which by the divergence theorem (see §1.7) gives the second part

$$\frac{q}{4\pi\varepsilon_0} \int_S \frac{\mathbf{r}_0 \cdot d\boldsymbol{\sigma}}{r^2} = \frac{q}{4\pi\varepsilon_0} \int_V \nabla \cdot \left(\frac{\mathbf{r}_0}{r^2}\right) dV = 0.$$

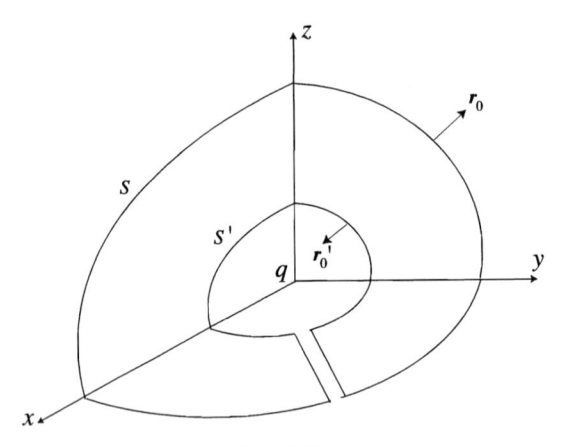

Fig. 6.3.

The first part can be resolved by indenting the origin by a small sphere S' of radius δ. Let the volume outside the outer surface S and the volume inside the surface S' be connected by a small hole, which joins the two surfaces S and S' and combines them into one simply-connected closed surface (Fig. 6.3). Since the radius of this (imaginary) hole may be made arbitrarily small which can vanish in the limit, the contribution by this hole to the surface integral is zero, and we obtain

$$\frac{q}{4\pi\varepsilon_0} \left(\int_S \frac{\mathbf{r}_0 \cdot d\boldsymbol{\sigma}}{r^2} + \int_{S'} \frac{\mathbf{r}_0 \cdot d\boldsymbol{\sigma}'}{\delta^2} \right) = 0.$$

In the second integral the element $d\boldsymbol{\sigma}' = -\mathbf{r}_0 \delta^2 \, d\omega$, where $d\omega$ is the volume element and the minus sign is taken because the positive normal $\mathbf{r}_0$ is directed outward from the volume. Since the outward $\mathbf{r}_0$ is in the negative radial direction, we have $\mathbf{r}_0' = -\mathbf{r}_0$. Thus, the second integral becomes

$$\int_{S'} \frac{\mathbf{r}_0 \cdot d\boldsymbol{\sigma}'}{\delta^2} = -\int_{S'} \frac{\mathbf{r}_0 \cdot d\mathbf{r}_0 \delta^2 \, d\omega}{\delta^2} = -4\pi.$$

Hence, Gauss's law becomes

$$\int_S \mathbf{E} \cdot d\boldsymbol{\sigma} = \frac{q}{4\pi\varepsilon_0} 4\pi = \frac{q}{\varepsilon_0}. \tag{6.25}$$

Also, if q denotes the total distributed charge enclosed by the surface S so that $q = \int_V \varrho \, dV$, where ϱ denotes the charge density, then $\int_S \mathbf{E} \cdot d\boldsymbol{\sigma} = \int_V \frac{\varrho}{\varepsilon_0} \, dV$, which by Gauss's theorem gives $\int_V \nabla \cdot \mathbf{E}\, dV = \int_V \frac{\varrho}{\varepsilon_0}\, dV$. Since the volume V is arbitrary, the two integrands must be equal, and thus

$$\nabla \cdot \mathbf{E} = \frac{\varrho}{\varepsilon_0}, \tag{6.26}$$

which is one of Maxwell's equations. Form (6.26) we can derive Poisson's equation by replacing $\mathbf{E}$ by $-\nabla u$; thus, Poisson's equation is

$$\nabla \cdot \nabla u = \nabla^2 u = -\frac{\varrho}{\varepsilon_0}, \tag{6.27}$$

which for $\varrho = 0$ (absence of electric charge) reduces to Laplace's equation $\nabla^2 u = 0$.

If the charges are point charges q_i, then the solution of Poisson's equation is

$$u = \frac{1}{4\pi\varepsilon_0} \sum_i \frac{q_i}{r_i}, \tag{6.28}$$

which is a superposition of single-point charge solutions.[2] Then

$$u\Big|_{r=0} = \frac{1}{4\pi\varepsilon_0 r^2} \int \frac{r(\mathbf{r})}{r}\, d\mathbf{r},$$

or for the potential at $\mathbf{r} = \mathbf{r}_1$ away from the origin and the charge $\mathbf{r} = \mathbf{r}_2$, we have

$$u(\mathbf{r}_1) = \frac{1}{4\pi\varepsilon_0 r^2} \int \frac{\varrho(\mathbf{r}_2)}{|\mathbf{r}_1 - \mathbf{r}_2|}\, d\mathbf{r}_2. \tag{6.29}$$

Since by Gauss's law,

$$\int_V \nabla^2(1/r)\, dV = \begin{cases} 0 & \text{if the volume } V \text{ does not include origin,} \\ -4\pi & \text{if the origin is included,} \end{cases}$$

we can write $\nabla^2\left(\dfrac{1}{4\pi r}\right) = -\delta(\mathbf{r})$, or if there is a translation of the charge from origin to to the position $\mathbf{r} = \mathbf{r}_2$, we have

$$\nabla^2\left(\frac{1}{4\pi r_{12}}\right) = -\delta(\mathbf{r}_1 - \mathbf{r}_2),$$

where $r_{12} = |\mathbf{r}_1 - \mathbf{r}_2|$, and the Dirac δ-function $\delta(\mathbf{r}_1 - \mathbf{r}_2) = 0$ for $\mathbf{r}_1 \neq \mathbf{r}_2$. Thus, in view of (6.29) the solution of the equation is

$$G(\mathbf{r}_1, \mathbf{r}_2) = \frac{1}{4\pi|\mathbf{r}_1 - \mathbf{r}_2|}, \tag{6.30}$$

---

[2] This follows from Coulomb's law for the force between two point charges q_1 and q_2, which is $\mathbf{E} = \dfrac{q_1 q_2 \mathbf{r}_0}{4\pi\varepsilon_0 r^2}$.

where $G(\mathbf{r}_1, \mathbf{r}_2)$ is known as *Green's function* for Poisson's equation (6.27). Green's function behaves like an influence function which enlarges or reduces the effect of the charge element $\varrho(\mathbf{r}_2)\,d\mathbf{r}_2$ depending on its distance from the field point $\mathbf{r}_1$. Thus, it represents the effect of a unit point source at $\mathbf{r}_2$ in producing a potential at $\mathbf{r}_1$. Also, the solution (6.29) can be written as

$$u\,(\mathbf{r}_1) = \frac{1}{\varepsilon_0 r^2} \int G(\mathbf{r}_1, \mathbf{r}_2)\,\varrho(\mathbf{r}_2)\,d\mathbf{r}_2. \qquad (6.31)$$

Using the 'cause and effect' terminology of physicists, the charge density $\varrho(\mathbf{r}_2)$ might be labeled as 'cause' that produces the 'effect' $u(\mathbf{r}_1)$, which means that a potential field is produced by the charge distribution, although the 'effectiveness' of the charge distribution to produce this potential depends on the distance between the element of charge $\varrho(\mathbf{r}_2)\,d\mathbf{r}_2$ and the point at $\mathbf{r}_1$. Thus, the 'influence' of the charge is represented by the quantity $\{4\pi|\mathbf{r}_1 - \mathbf{r}_2|\}^{-1}$, and for this reason this quantity, although designated as Green's function, is sometimes called an *influence function*.

6.4. Schrödinger Diffusion Operator

If we consider the nonhomogeneous case, then the Fourier heat equation

$$\left(\frac{1}{k}\frac{\partial}{\partial t} - \nabla^2\right) u(x,t) = f(x,t) \qquad (6.32)$$

has two interpretations:

(i) If $k > 0$ is a real constant, depending on the specific heat and thermal conductivity of the medium, then $u(x,t)$ represents a temperature distribution. The source function $f(x,t)$ on the right side describes local heat production minus absorption.

(ii) The function $u(x,t)$ represents a particle density and k is the diffusion coefficient. If k is purely imaginary such that $k = \dfrac{i\hbar}{2m}$, where m is the mass of the quantum particle, and $\hbar = 1.054 \times 10^{-34}$ joules-sec (or 6.625×10^{-27} erg-sec) is the Planck's constant, then Eq (6.32) for the 1-D steady-state Schrödinger equation for a free particle becomes

$$i\hbar\,\frac{\partial u(x,t)}{\partial t} + \frac{\hbar^2}{2m}\nabla^2 u(x,t) = 0. \qquad (6.33)$$

In order to solve this equation subject to the boundary and initial conditions

$$u(x,t) \to 0 \quad \text{as } |x| \to \infty,\ t > 0, \quad \text{and} \quad u(x,0) = f(x), \quad -\infty < x < \infty,$$

we apply Fourier transform to this system and get

$$\tilde{u}_t = -\frac{i\hbar\alpha^2}{2m}\,\tilde{u}, \quad \tilde{u}(\alpha,0) = f(\alpha).$$

The solution of this system in the transform domain is

$$\tilde{u}(\alpha,t) = f(\alpha)\,e^{-i\,b\alpha^2 t}, \quad b = \frac{\hbar}{2m}$$

which after inverting gives the solution as

$$
\begin{aligned}
u(x,t) &= \frac{1}{\sqrt{2\pi}} \int_{-\infty}^{\infty} f(\alpha)\,e^{i\,\alpha(x-b\alpha t)}\,d\alpha \\
&= \frac{1}{\sqrt{2\pi}} \int_{-\infty}^{\infty} f(y)\,e^{-i\,\alpha y}\,dy \int_{-\infty}^{\infty} e^{i\,\alpha(x-\alpha bt)}\,d\alpha \qquad (6.34) \\
&= \frac{1}{\sqrt{2\pi}} \int_{-\infty}^{\infty} f(y)\,dy \int_{-\infty}^{\infty} e^{i\,\alpha(x-y-b\alpha t)}\,d\alpha.
\end{aligned}
$$

Since

$$
\begin{aligned}
e^{i\,\alpha(x-y-b\alpha t)} &= \exp\left[-i\,bt\left\{\alpha^2 - 2\alpha\,\frac{x-y}{2bt} + \left(\frac{x-y}{2bt}\right)^2 - \left(\frac{x-y}{2bt}\right)^2\right\}\right] \\
&\quad \exp\left[-i\,bt\left(\alpha - \frac{x-y}{2bt}\right)^2\right] \exp\left[\frac{i\,(x-y)^2}{4bt}\right] \\
&= e^{i\,(x-y)^2/(4\,bt)}\,e^{-i\,bq^2 t}, \quad \text{where } q = \alpha - \frac{x-y}{2bt},
\end{aligned}
$$

the function $u(x,t)$ in (6.34) becomes

$$
\begin{aligned}
u(x,t) &= \frac{1}{2\pi} \int_{-\infty}^{\infty} f(y)\,e^{i\,(x-y)^2/(4\,bt)}\,dy \int_{-\infty}^{\infty} e^{-i\,bq^2 t}\,dq \\
&= \frac{1-i}{2\sqrt{2\pi bt}} \int_{-\infty}^{\infty} f(y)\,e^{i\,(x-y)^2/(4\,bt)}\,dy, \qquad (6.35)
\end{aligned}
$$

which, in view of (4.6), gives Green's function for the 1-D steady-state Schrödinger operator as

$$G(x;t) = \frac{(1-i)\,H(t)}{2\sqrt{2\pi bt}}\,e^{i\,(x-y)^2/(4\,bt)}. \qquad (6.36)$$

Note that in view of (6.16), Green's function in this case is

$$G(x;t) = H(t)\sqrt{\frac{m}{2i\pi\hbar\,t}}\,e^{imx^2/(2\hbar t)},$$

which reduces to (6.36). We have assumed in this problem that the source is of positive unit strength. Sometimes the notation h is used for the Plank's constant; it is related to $\hbar$ by $h = 2\pi\,\hbar$.

The argument in the notation $G(\mathbf{x},t;\mathbf{x}',t')$ is composed of two parts: The first part $\mathbf{x}, t$ denotes the field point where the *effect* of the impulsive heat source located at the source point signifies the temperature at the point $\mathbf{x}$ at time t. The second part $\mathbf{x}', t'$ denotes the *cause* which is the impulsive heat source situated at the point $\mathbf{x}'$ generating an instantaneous (impulsive) heat at an earlier time t'. The combined notation has the physical significance of an entire space-time process which can be visualized as $G(\textit{effect}; \textit{cause}) \equiv G(\mathbf{x}, t; \mathbf{x}', t')$.

6.5. Min-Max Principle

Let $R = \{(x, t) : 0 \le x \le l, 0 \le t \le T\}$ denote a closed rectangle in the (x, t)-plane. Then we have the following result:

Theorem 6.1. *If $u(x, t)$ is continuous on R and satisfies the 1-D diffusion equation in R, then $u(x, t)$ attains its maximum and minimum values on the sides $x = 0$ and $x = l$ or at time $t = 0$.*

PROOF. First, we will prove the maximum principle by contradiction. Let M be the maximum value of u in R, and, contrary to the hypothesis, we will assume that the maximum value on the vertical sides or the at time $t = 0$ is $M - \varepsilon$, where $\varepsilon > 0$. Let us further assume that the maximum value is attained at a point $(x_0, t_0) \in R$, i.e., $u(x_0, t_0) = M$, where $0 < x_0 < l$, $t_0 > 0$. Now, consider a function

$$v(x, t) = u(x, t) + \frac{\varepsilon}{4l^2} (x - x_0)^2,$$

which approaches u as $\varepsilon \to 0$. Then on the vertical sides of R and at $t = 0$ we have $v(x, t) \le M - \varepsilon + \dfrac{\varepsilon}{4} = M - \dfrac{3\varepsilon}{M}$, since $v(x_0, t_0) = M$. Thus, the maximum value is not attained on the vertical sides of R or at $t = 0$. Again, let (x_1, t_1), $0 < x_1 < l$, $0 < t_1 < T$ be a point in R where $v(x, t)$ attains its maximum value. Then, for the function v to satisfy the necessary condition for a maximum at (x_1, t_1) we must have $v_t = 0$ if $t_1 < T$ or $v_t = 0$ if $t_1 = T$, and $v_{xx} \le 0$, i.e., we must have $v_t - v_{xx} \ge 0$ at (x_1, t_1). But, since $v_t - v_{xx} = u_t - u_{xx} - \dfrac{\varepsilon}{2l^2} < 0$, we arrive at a contradiction, which proves the maximum part of the theorem. The minimum part is then proved because $-u$ satisfies the maximum principle. ∎

6.6. Diffusion Equation in a Finite Medium

Consider the 1-D diffusion equation $u_t = k u_{xx}$, $0 < x < a$, $t > 0$, subject to the initial and boundary conditions $u(x, 0) = 0$, $0 < x < a$; $u(0, t) = U_0$, $t > 0$, and $u_x(a, t) = 0$, $t > 0$. Applying Laplace transform with respect to t, we obtain the system:

$$\frac{d^2 \overline{u}}{dx^2} - \frac{s}{k} \overline{u} = 0, \quad 0 < x < a,$$

$$\overline{u}(0, s) = \frac{U_0}{s}, \quad \left[\frac{d\overline{u}}{dx}\right]_{x=a} = 0,$$

where s is the variable of the transform. The solution of this system in the transform domain is

$$\bar{u}(x,s) = U_0 \frac{\cosh\left((a-x)\sqrt{s/k}\right)}{s \cosh\left(a\sqrt{s/k}\right)}.$$

Using the Table D.1, the inverse Laplace transform gives the solution as

$$u(x,t) = U_0\left[1 + \frac{4}{\pi}\sum_{n=1}^{\infty}\frac{(-1)^n}{2n-1}\cos\frac{(2n-1)(a-x)\pi}{2a}e^{-(2n-1)^2\pi^2 kt/(4a^2)}\right]$$

$$= U_0\left[1 - \frac{4}{\pi}\sum_{n=1}^{\infty}\frac{1}{2n-1}\sin\frac{(2n-1)\pi x}{2a}e^{-(2n-1)^2\pi^2 kt/(4a^2)}\right]. \tag{6.37}$$

Note that the same result is obtained if this problem is solved by the Bernoulli's separation method.

6.7. Axisymmetric Diffusion Equation

Consider the axisymmetric diffusion equation in the polar cylindrical coordinate system

$$u_t = k\left(u_{rr} + \frac{1}{r}u_r\right), \quad 0 \le r \le \infty, \ t > 0,$$

which is subject to the initial condition $u(r,0) = f(r)$ for $0 < r < \infty$. Applying the zero-order Hankel transform we get

$$\frac{d\hat{u}}{dt} + k\sigma^2\hat{u} = 0, \quad \hat{u}(\sigma,0) = \hat{f}(\sigma),$$

where σ is the variable of the Hankel transform. The solution of this problem in the transform domain is

$$\hat{u}(\sigma,t) = \hat{f}(\sigma)e^{-k\sigma^2 t},$$

which after inversion gives the solution as

$$u(r,t) = \int_0^\infty \sigma\,\hat{f}(\sigma)\,J_0(\sigma r)\,e^{-k\sigma^2 t}\,d\sigma$$

$$= \int_0^\infty \sigma\left[\int_0^\infty z J_0(\sigma z)\,f(z)\,dz\right]e^{-k\sigma^2 t}\,J_0(\sigma r)\,d\sigma$$

$$= \int_0^\infty z\,f(z)\,dz\int_0^\infty \sigma\,J_0(\sigma z)J_0(\sigma r)\,e^{-k\sigma^2 t}\,d\sigma,$$

where we have interchanged the order of integration. Since

$$\int_0^\infty \sigma J_0(\sigma z)\,J_0(\sigma r)\,e^{-k\sigma^2 t}\,dy = \frac{1}{2kt}e^{-(r^2+z^2)/(4kt)}\,I_0\left(\frac{rz}{2kt}\right),$$

(see Gradsteyn and Rhyzik [2007; §6.626]), where I_0 is the modified Bessel function, the above solution becomes

$$u(r,t) = \frac{1}{2kt} \int_0^\infty e^{-(r^2+z^2)/(4kt)} \, z \, f(z) \, I_0\left(\frac{rz}{2kt}\right) dz. \qquad (6.38)$$

Let $f(r)$ represent a heat source concentrated at $r=0$. This situation can be visualized by considering the heat source concentrated in a circle of radius ε and letting $\varepsilon \to 0$, and assuming that $\lim\limits_{\varepsilon \to 0} \int_0^\varepsilon r \, f(r) \, dr = \dfrac{1}{2\pi}$, or equivalently, $f(r) = \dfrac{1}{2\pi}\dfrac{\delta(r)}{r}$. Then the solution of this problem due to the heat source concentrated at $r=0$ is

$$u(r,t) = \frac{1}{4\pi kt} \int_0^\infty \delta(r) \, I_0\left(\frac{rz}{2kt}\right) e^{-(r^2+z^2)/(4kt)} = \frac{1}{4\pi kt} e^{-r^2/(4kt)}. \qquad (6.39)$$

6.8. 1-D Heat Conduction Problem

This example illustrates an application of finite Fourier sine transform directly by using Fourier sine series as mentioned in §5.5.[3] Consider the nonhomogeneous heat conduction problem with Dirichlet boundary conditions:

$$\begin{aligned} u_t - u_{xx} &= f(x,t), \quad 0 < x < \pi, \ t > 0, \\ u(0,t) &= u(\pi,t) = 0, \quad u(x,0) = 0. \end{aligned} \qquad (6.40)$$

The solution of this problem, if any, can be formally written in a Fourier series for each fixed t as

$$u(x,t) = \sum_{n=1}^\infty b_n(t) \sin nx,$$

where $b_n(t) = \displaystyle\int_0^\pi u(x,t) \sin nx \, dx$ for an integer n . To determine the coefficients b_n, we take the finite Fourier sine transform of both sides of (6.40), which gives

$$b_n'(t) + n^2 b_n(t) = B_n(t), \qquad (6.41)$$

where

$$B_n(t) = \int_0^\pi f(x,t) \sin nx \, dx. \qquad (6.42)$$

Since the initial condition gives $b_n(0) = 0$, so, solving Eq (6.41) subject to this condition, we get $b_n(t) = \displaystyle\int_0^t e^{-n^2(t-t')} B_n(t') \, dt'$. Now, if the problem (6.40) has a solution $u \in C^2$, then we can formally write

$$u(x,t) = \frac{2}{\pi} \sum_{n=1}^\infty \left[\int_0^t e^{-n^2(t-t')} B_n(t') \, dt' \right] \sin nx.$$

---

[3] To use the finite Fourier sine transform, simply replace $b_n(t)$ and $B_n(t)$ by $\tilde{u}_s(n,t)$ and $\tilde{f}_s(n,t)$ in the solution provided in below for this boundary value problem.

By Schwarz's inequality, we have

$$|b_n(t)|^2 \leq \left\{ \int_0^t e^{-n^2(t-t')} \, dt' \right\} \left\{ \int_0^t B_n^2(t') \, dt' \right\}$$

$$= \frac{1}{2n^2} \left(1 - e^{-2n^2 t} \right) \int_0^t B_n^2(t') \, dt' \leq \frac{1}{2n^2} \int_0^t B_n^2 \, dt'.$$

Then, using the Schwarz's inequality for sums and the Parseval's equality (§1.9) we get

$$\left| \sum_{n=M+1}^N b_n(t) \sin nx \right|^2 \leq \frac{1}{2} \left\{ \sum_{n=M+1}^N \frac{1}{n^2} \right\} \left\{ \int_0^t \sum_{n=M+1}^N B_n^2 \, dt' \right\}$$

$$\leq \frac{1}{\pi} \sum_{n=M+1}^N \frac{1}{n^2} \int_0^t \int_0^\pi [f(x,t')]^2 \, dx \, dt'.$$

Since $\int_0^{t_0} \int_0^\pi [f]^2 \, dx \, dt$ converges for some $t_0 > 0$, the series $\sum b_n(t) \sin nx$ converges uniformly for $0 \leq x \leq \pi$, $0 \leq t \leq t_0$. Hence, we can write the solution as

$$u(x,t) = \frac{2}{\pi} \sum_{n=1}^\infty \left[\int_0^t e^{-n^2(t-t')} B_n(t') \, dt' \right] \sin nx, \tag{6.43}$$

where $u \in C^2$ and satisfies the given conditions in (6.40). Under suitable conditions on f, like $f(0,t) = 0 = f(\pi,t)$ and $\int_0^t f_{xx}^2 \, dx$ being uniformly bounded, the series in (6.43) can be differentiated term-by-term, and hence, u defined by (6.43) is the required solution.

Note that (i) Under some less restrictions on f, we use (6.42) and by formally interchanging integration and summation, obtain

$$u(x,t) = \frac{2}{\pi} \int_0^t \int_0^\pi G(x,x';t,t') \, f(x',t') \, dx' \, dt', \tag{6.44}$$

where

$$G(x,x',t) = \sum_{n=1}^\infty e^{-n^2 t} \sin nx \sin nx' \tag{6.45}$$

is Green's function for this problem.

(ii) If the initial condition in (6.40) is changed to $u(x,0) = g(x)$, then the above initial condition $b_n(0) = 0$ must be replaced by $b_n^*(0) = \int_0^\pi g(x) \sin nx \, dx$. Then

$$b_n(t) = \int_0^t e^{-n^2(t-t')} B_n(t') \, dt' + b_n^*(0) \, e^{-n^2 t},$$

and thus, the solution becomes

$$u(x,t) = \frac{2}{\pi} \sum_{n=1}^{\infty} \left[\int_0^t e^{-n^2(t-t')} B_n(t') \, dt' \right] \sin nx + \sum_{n=1}^{\infty} b_n^*(0) \, e^{-n^2 t} \sin nx.$$

$$(6.46)$$

Green's function in this case, however, remains the same as (6.45).

___

6.9. Stefan Problem

We will consider the one-phase Stefan problem with constant applied heat flux, which can be stated as follows: A solid, initially at the melting temperature, is heated by prescribing either the temperature or the heat flux at a boundary. Let the melting interface be denoted by $X(t)$. Then the problem is to determine the interface $X(t)$ as well as the temperature $u(x)$ in the liquid. The governing equation and the boundary conditions for the case of a prescribed heat flux at a boundary are

$$u_t = u_{xx}, \quad 0 < x < X(t), \quad t > 0,$$

$$u(x,t) = 0 \text{ for } x \geq X(t); \quad -u_x(X,t) = \frac{dX}{dt}, \tag{6.47}$$

$$u(0,t) = f(t), \quad \text{or} \quad -u_x(0,t) = g(t); \quad \text{and} \quad X(0) = 0.$$

This problem is solved by applying a double Laplace transform $\mathcal{L}\{u(x,t)\} = \overline{\overline{u}}\,(\kappa, s)$, defined by

$$\overline{\overline{u}}\,(\kappa, s) = \int_0^{\infty} e^{-st} \int_0^{X(t)} e^{-\kappa t} \, u(x,t) \, dx \, dt = \mathcal{L}\{u(x,t)\}, \tag{6.48}$$

where κ and s are the variables of Laplace transform, $\Re\{s\} > 0, \Re\{\kappa\} > 0$. Assuming that $X(t)$ is an increasing function and $0 < X(t) < t$ so that X has an inverse, which we will denote by ξ, i.e., $\xi\,(X(t)) = t$, we find that

$$\mathcal{L}\{u_t(x,t)\} = s\overline{\overline{u}},$$

$$\mathcal{L}\{u_{xx}(x,t)\} = \int_0^{\infty} e^{-st} \left[\int_0^X e^{-\kappa x} u_{xx} \, dx \right] dt$$

$$= \int_0^{\infty} e^{-st} \left[-\frac{dX}{dt} e^{-\kappa X} + g(t) + \kappa \int_0^X e^{-\kappa x} u_x \, dx \right] dt \tag{6.49}$$

$$= \overline{g}(s) + \kappa^2 \overline{\overline{u}} - \kappa \overline{f}(s) - \int_0^{\infty} e^{-s\,\xi(y)-\kappa y} \, dy,$$

where $\overline{f}(s)$ and $\overline{g}(s)$ are Laplace transforms of f and g, respectively. Thus,

$$\overline{\overline{u}}(\kappa, s) = \frac{\kappa \overline{f}(s) - \overline{g}(s) + \displaystyle\int_0^{\infty} e^{s\,\xi(y)-\kappa y} \, dy}{\kappa^2 - s}.$$

Notice that $\overline{\overline{u}}$ is defined for $\kappa = \sqrt{s}$ as well as for $\kappa = -\sqrt{s}$, which implies that the numerator in the above expression must vanish in either case, i.e., we must have

$$\overline{g}(s) = \sqrt{s}\,\overline{f}(s) + \int_0^\infty e^{-s\xi(y)-y\sqrt{s}}\,dy,$$

$$\overline{g}(s) = -\sqrt{s}\,\overline{f}(s) + \int_0^\infty e^{-s\xi(y)+y\sqrt{s}}\,dy,$$

Hence,

$$\overline{f}(s) = \int_0^\infty e^{-s\xi(y)}\,\frac{\sinh\left(y\sqrt{s}\right)}{\sqrt{s}}\,dy,$$

$$\overline{g}(s) = \int_0^\infty e^{-s\xi(y)}\cosh\left(y\sqrt{s}\right)\,dy. \tag{6.50}$$

Inverting $\overline{\overline{u}}$ with respect to κ, we get

$$\overline{u}(x,s) = \overline{f}(s)\cosh\left(x\sqrt{s}\right) - \frac{\overline{g}(s)\,\sinh\left(x\sqrt{s}\right)}{\sqrt{s}}$$

$$+ \int_0^\infty e^{s\xi(y)}\,\frac{\sinh\left((x-y)\sqrt{s}\,\right)}{\sqrt{s}}\,dy \tag{6.51}$$

$$= \int_0^\infty e^{s\xi(y)}\,\frac{\sinh\left((y-x)\sqrt{s}\,\right)}{\sqrt{s}}\,dy.$$

The above result, after integrating by parts and inverting, gives the nontrivial solution as

$$u(x,t) = \sum_{n=1}^\infty \frac{1}{(2n)!}\,\frac{d^n}{dt^n}\left[X(t)-x\right]^{2n}, \tag{6.52}$$

and

$$f(t) = \sum_{n=1}^\infty \frac{1}{(2n)!}\,\frac{d^n}{dt^n}\left[X(t)\right]^{2n},$$

$$g(t) = \sum_{n=1}^\infty \frac{1}{(2n-1)!}\,\frac{d^n}{dt^n}\left[X(t)\right]^{2n-1}. \tag{6.53}$$

Next, we will find an approximate solution of the integral equation for the prescribed flux $g(t)$ by considering the second integral equation in (6.50):

$$\overline{g}(s) = \int_0^\infty e^{-s\xi(y)}\cosh\left(y\sqrt{s}\right)\,dy.$$

Assuming that $\xi(y)$ is quadratic, i.e., taking $\xi(y) = A\,y + B y^2$, we get

$$\overline{g}(s) = \int_0^\infty e^{-s\left(A\,y+By^2\right)}\cosh\left(y\sqrt{s}\right)\,dy$$

$$= \int_0^\infty \frac{e^{-\sqrt{s}\left(A\,y+By^2\right)}}{\sqrt{s}}\cosh y\,dy,$$

which on inversion gives

$$g(t) = \frac{1}{\sqrt{\pi t}} \int_0^\infty e^{-(By^2 + A^2 y^2)/(4t)} \cosh y \, dy = \frac{1}{\sqrt{A^2 + 4Bt}} e^{t/(A^2 + 4Bt)}.$$

(6.54)

Since the time derivative of $t = AX + BX^2$ is $1 = \dot{X}(A + 2BX)$, and $4tB + A^2 = (A + 2BX)^2 = 1/\dot{X}^2$, we find for the case when t is a quadratic function of X that

$$g(t) = \dot{X} \exp\left[t \dot{X}^2\right].$$

(6.55)

Note that $g(t) = e^t$ for $X = t$; $g(t) = \dfrac{a \, e^{a^2}}{\sqrt{t}}$ for $X = 2a\sqrt{t}$; and $g = 1$ when $\dot{X} \exp\left[t \dot{X}^2\right] = 1$, so that the precise value of X can be determined as follows: Let $c = t\dot{X}^2$, so $\dot{X} = e^{-c}$, $t = ce^{2c}$, which gives $\dfrac{dX}{dc} = (2c + 1)e^c$. Integrating this equation by parts with respect to c and using the initial condition $X(0) = 0$ we find the parametric solution of this differential equation as $X(t) = 1 + (2c - 1)e^c$, $t = ce^{2c}$.

The algorithm to compute values of X uses the following series for $X(t)$ when $g = 1$:

$$X(t) = t - \frac{1}{2!}t^2 + \frac{5}{3!}t^3 - \frac{51}{4!}t^4 + \frac{827}{5!}t^5 - \frac{18961}{6!}t^6$$
$$+ \frac{574357}{7!}t^7 - \frac{21995899}{8!}t^8 + \frac{1032666859}{9!}t^9 - \cdots,$$

where the remainder after n terms is bounded and approaches ≈ 0.04 as $n \to \infty$. By using parametric solutions Ruehr [2002] has computed the values of X at $t = 10^k$, $k = 0, 1, 2, \ldots, 7$, as follows:

$X(1) = 0.772443774949597096137987895609995394$,

$X(10) = 4.56549276368834105587472$,

$X(100) = 21.29139510280473148$,

$X(1000) = 87.161250425814$, $\quad X(10^4) = 332.089233225$,

$X(10^5) = 1212.124436$, $\quad X(10^6) = 4302.9380$, $\quad X(10^7) = 14984.1$.

MATLAB® solution of the equation $\dot{X} \exp\left[t \dot{X}^2\right] = 1$ such that $X(0) = 0$ is given by

$$X = -1 + W(2t) \exp\left\{\tfrac{1}{2} W(2t) + 1\right\},$$

where $W(t)$ is the Lambert W-function, and it is presented in Fig. 6.4.

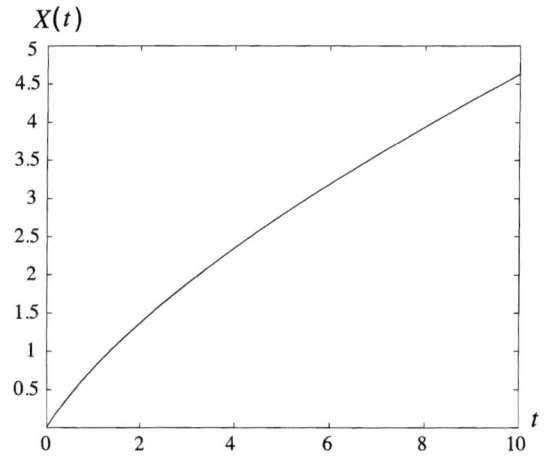

Fig. 6.4. Solution of Equation $\dot{X} \exp\left[t\,\dot{X}^2\right] = 1, X(0) = 0$.

6.10. 1-D Fractional Diffusion Equation

Consider the general boundary value problem of the fractional diffusion equation

$$D_t^p\, u - \nabla^2 u = 0 \quad \text{in } \mathbb{R}^n \times \mathbb{R}^+ \tag{6.56}$$

where $D_t^p \equiv \dfrac{\partial^p}{\partial t^p}$, $0 < p \le 1$, is the Caputo time derivative of the fractional order p (see Appendix E). For $p = 1$, we get the classical diffusion equation (see §6.1 for $n = 1$), and $p = 0$ gives Helmholtz's equation (see §4.3.4, 4.3.5, 8.5, 8.5.1, 8.6, and 8.12.4). We will solve the boundary value problem involving Eq (6.56) for $n = 1$, which is defined by

$$\frac{\partial^p u}{\partial t^p} = k\,\frac{\partial^2 u}{\partial x^2}, \quad x \in \mathbb{R}, \ t > 0,$$
$$u(x,t) \to 0 \quad \text{as } |x| \to \infty, \tag{6.57}$$
$$\left[D_t^{p-1} u(x,t)\right]_{t=0} = f(x), \quad x \in \mathbb{R},$$

where k is the diffusivity constant. Applying Fourier transform to (6.57) with respect to x and subject to the given boundary condition, we get

$$D_t^p\, \tilde{u}(\alpha,t) = -k\,\alpha^2\, \tilde{u},$$
$$\left[D_t^{p-1}\, \tilde{u}(\alpha,t)\right]_{t=0} = \tilde{f}(\alpha), \tag{6.58}$$

where $\widetilde{u}(\alpha, t)$ is Fourier transform of $u(x, t)$ and α is the variable of Fourier transform. Now, taking Laplace transform of (6.58), we have

$$\overline{\widetilde{u}}(\alpha, s) = \frac{\widetilde{f}(\alpha)}{s^p + k \alpha^2},$$

whose inverse Laplace transform gives

$$\widetilde{u}(\alpha, t) = \widetilde{f}(\alpha) \, t^{p-1} \, E_{p,p} \left(-k \, \alpha^2 t^p \right), \tag{6.59}$$

where $E_{p,q}$ is the Mittag-Leffler function defined by

$$E_{p,q}(z) = \sum_{m=0}^{\infty} \frac{z^m}{\Gamma(pm + q)}, \quad p > 0, \ q > 0. \tag{6.60}$$

Again, taking the inverse Fourier transform of (6.59) we obtain

$$u(x, t) = \int_{-\infty}^{\infty} G(x - x', t) \, f\left(x' \right) \, dx', \tag{6.61}$$

where

$$G(x, t) = \frac{1}{\pi} \int_{-\infty}^{\infty} t^{p-1} \, E_{p,p} \left(-k \, \alpha^2 t^p \right) \, \cos \alpha x \, d\alpha. \tag{6.62}$$

To evaluate this integral we take Laplace transform of $G(x, t)$. Using the formula [Costabel, 2004]

$$\mathcal{L}\left\{ t^{mp+q-1} E_{p,q}^{(m)} \left(\pm a t^p \right) \right\} = \frac{m! \, s^{p-q}}{(s^p \mp a)^{m+1}}, \tag{6.63}$$

where $E_{p,q}^{(m)} = \dfrac{d^m}{dz^m} E_{p,q}(z)$, we obtain

$$\overline{G}(x, s) = \frac{1}{\pi} \int_{-\infty}^{\infty} \frac{\cos \alpha x}{s^p + k \alpha^2} \, d\alpha = \frac{1}{2\sqrt{k}} s^{-p/2} \exp\left\{ -\frac{|x|}{\sqrt{k}} s^{p/2} \right\}.$$

Taking the inverse Laplace transform of this result finally we get

$$G(x, t) = \frac{1}{2\sqrt{k}} t^{p/2-1} \, W \left(-x', -\frac{p}{2}, \frac{p}{2} \right), \tag{6.64}$$

where $x' = \dfrac{|x|}{t^{p/2} \sqrt{k}}$ and $W(z, p, q)$ is the Wright function defined by

$$W(z, p, q) = \sum_{n=0}^{\infty} \frac{z^n}{n! \, \Gamma(pn + q)}. \tag{6.65}$$

For $p = 1$, the boundary value problem (6.56) reduces to 1-D diffusion equation discussed in §6.1.2, and Green's function (6.64) reduces to (6.16), that is,

$$G(x,t) = \frac{1}{2\sqrt{kt}} W\left(-\frac{x}{\sqrt{kt}}, -\frac{1}{2}, \frac{1}{2}\right) = \frac{1}{2\sqrt{\pi kt}} e^{-x^2/(4kt)}.$$

ANOTHER METHOD. (Kemppainen and Ruotsalainen [2008:141]) Green's function $G(x,t)$ for the boundary value problem (6.57) can be obtained by taking Laplace transform in the time variable and Fourier transform in the space variable of the fractional equation

$$\left(D_t^p - \frac{\partial^2}{\partial x^2}\right) G(x,t) = \delta(x,t), \tag{6.66}$$

which yields $\widetilde{\overline{G}}(\alpha, s) = \dfrac{1}{\alpha^2 + s^p}$. Using Laplace transform of the Mittag-Lefler function, defined by (6.60), we get Fourier transform of Green's function as

$$\widetilde{G}(\alpha, t) = t^p E_{p,p}\left(-\alpha^2 t^p\right),$$

which on inversion gives Green's function in terms of the Fox H-function (see Notations; Fox [1961], Prudnikov et al. [1990:626], Costabel, [2004]) as

$$G(x,t) = t^{p-1} |x|^{-2} H_{12}^{20}\left(|x|^2 t^{-p} \Big|_{(1,1),(1,1)}^{p,p}\right). \tag{6.67}$$

6.10.1. 1-D Fractional Diffusion Equation in Semi-Infinite Medium.
Consider the boundary value problem defined by the heat equation in the semi-infinite medium $x > 0$, where the boundary $x = 0$ is kept at a temperature $U_0 f(t)$ and the initial temperature in the whole medium is zero:

$$\frac{\partial^p u}{\partial t^p} = k \frac{\partial^2 u}{\partial x^2}, \quad 0 < x < \infty, \ t > 0,$$
$$u(x,0) = 0, \quad x > 0,$$
$$u(0,t) = U_0 f(t), \quad t > 0,$$
$$u(x,t) \to 0 \quad \text{as } x \to \infty.$$

After applying Laplace transform to this problem we get

$$\frac{d^2 \overline{u}}{dx^2} + \frac{s^p}{k} \overline{u}(x,s) = 0, \quad x > 0,$$
$$\overline{u}(0,s) = U_0 \overline{f}(s), \quad \text{and} \quad \overline{u}(x,s) \to 0 \quad \text{as } x \to \infty. \tag{6.68}$$

The solution of this problem in the transform domain is

$$\overline{u}(x,s) = U_0 \overline{f}(s) e^{-ax}, \quad \text{where } a = \sqrt{\frac{s^p}{k}}, \tag{6.69}$$

which on inversion gives the solution as

$$u(x,t) = U_0 \int_0^t f(t-\tau)\, g(x,\tau)\, d\tau = U_0 f(t) \star g(x,t), \qquad (6.70)$$

where $g(x,t) = \mathcal{L}^{-1}\{e^{-ax}\}$. For $\alpha = 1$, the solution (6.69) in the Laplace transform domain is

$$\overline{u}(x,s) = U_0 \overline{f}(s)\, e^{-x\sqrt{s/k}},$$

which on inversion gives

$$u(x,t) = U_0 \int_0^t f(t-\tau)g(x,\tau)\, d\tau = U_0\, f(t) \star g(x,t),$$

where

$$g(x,t) = \mathcal{L}^{-1}\{e^{-x\sqrt{s/k}}\} = \frac{x}{2\sqrt{\pi k t^3}}\, e^{-x^2/(4kt)}. \qquad (6.71)$$

In particular, for $p = 1$ and $f(t) = 1$, the solution (6.69) in the Laplace domain is

$$\overline{u}(x,s) = \frac{U_0}{s}\, e^{-x\sqrt{s/k}},$$

which on inversion gives the classical solution

$$u(x,t) = U_0\, \mathrm{erfc}\left(\frac{x}{2\sqrt{kt}}\right). \qquad (6.72)$$

6.11. 1-D Fractional Schrödinger's Diffusion Equation

Consider the boundary value problem involving 1-D fractional Schrödinger's equation for a fine particle of mass m defined by

$$
\begin{aligned}
i\hbar\, \frac{\partial^p u}{\partial t^p} &= -\frac{\hbar^2}{2m}\, \frac{\partial^2 u}{\partial x^2}, \quad x \in \mathbb{R},\ t > 0, \\
u(x,0) &= f(x), \quad x \in \mathbb{R}, \\
u(x,t) &\to 0 \quad \text{as } |x| \to \infty,
\end{aligned}
\qquad (6.73)
$$

where $u(x,t)$ is the wave function, $\hbar$ the Plank's constant (see §6.4), and $f(x)$ an arbitrary function. Applying Laplace transform followed by Fourier transform to the problem (6.73), we get

$$\widetilde{\overline{u}}(\alpha,s) = \frac{s^{p-1}\, \widetilde{f}(\alpha)}{s^p + k\alpha^2}, \quad \text{where } k = \frac{i\hbar}{2m}, \qquad (6.74)$$

and α and s are the variables of Fourier and Laplace transforms, respectively. Inverting (6.74) and using the convolution theorem of Fourier transform, we obtain

$$
\begin{aligned}
u(x,t) &= \frac{1}{\sqrt{2\pi}} \int_{-\infty}^{\infty} \tilde{f}(\alpha)\, E_{p,1}\left(-k\alpha^2 t^p\right)\, d\alpha \\
&= \mathcal{F}^{-1}\left\{\tilde{f}(\alpha)\, E_{p,1}\left(-k\alpha^2 t^p\right)\right\} = \int_{-\infty}^{\infty} G(x-x',t)\, f(x')\, dx',
\end{aligned}
\tag{6.75}
$$

where Green's function $G(x,t)$ is given by

$$
G(x,t) = \frac{1}{\sqrt{2\pi}}\, \mathcal{F}^{-1}\left\{ E_{p,1}\left(-k\alpha^2 t^p\right)\right\} = \frac{1}{2\pi} \int_{-\infty}^{\infty} E_{p,1}\left(-k\alpha^2 t^p\right)\, e^{i\alpha x}\, d\alpha,
\tag{6.76}
$$

where $E_{p,1}$ is the Mittag-Leffler function defined by (6.60). For $p = 1$, the solution (6.75) reduces to

$$
u(x,t) = \int_{-\infty}^{\infty} G(x-x',t)\, f(x')\, dx',
\tag{6.77}
$$

where Green's function $G(x,t)$ is

$$
\begin{aligned}
G(x,t) &= \frac{1}{2\pi} \int_{-\infty}^{\infty} E_{p,1}\left(-k\alpha^2 t^p\right)\, d\alpha = \frac{1}{2\pi} \int_{-\infty}^{\infty} e^{i\alpha x - k\alpha^2 t}\, d\alpha \\
&= \frac{1}{2\sqrt{\pi kt}}\, e^{-x^2/(4kt)}.
\end{aligned}
\tag{6.78}
$$

This solution matches with the classical solution defined by (6.36).

6.12. Eigenpairs and Dirac Delta Function

First, we will find eigenfunctions for boundary value problems, and then represent the Dirac δ-function as a sum of eigenfunctions. This not only leads to certain methods of explicitly finding fundamental solutions for some boundary value problems but also provides a different approach to define the Dirac δ-function.

Let a linear boundary value problem be defined in a domain D by

$$
L(u) = 0, \quad u = g(s), \quad \frac{\partial u}{\partial n} + k(s)u = h(s) \quad \text{on } \partial D,
$$

and let the eigenfunctions ϕ_n be given by $L(\phi_n) = \lambda_n \phi_n$, where λ_n are called the eigenvalues associated with the eigenfunctions ϕ_n. The form of the eigenpairs (λ_n, ϕ_n) depends on the coordinate system used for the boundary value problem. For example, if the coordinate system is rectangular cartesian, the eigenfunctions are of the form $\cos nx$, $\sin nx$, or e^{inx}, and the eigenvalues depend on n if the linear operator L contains constant coefficients. The eigenfunctions ϕ_n must be a complete set in the sense that for any function u there exists an integer N such that $\left\| u - \sum_{n=1}^{N} c_n \phi_n \right\| \le \varepsilon$, where $\varepsilon > 0$ is a preassigned arbitrarily small quantity and c_n are constants, and the norm is defined by $\|u\| = \sqrt{\int_V u\,\bar{u}\,dV}$, where $\bar{u}$ denotes the complex conjugate of u. A sufficiently continuous function u possesses an eigenfunction expansion to any degree of accuracy; thus

$$u = \lim_{N \to \infty} \sum_{n=0}^{N} c_n \phi_n.$$

Also, a complete set of eigenfunctions forms an orthonormal set with the property that $< \phi_m, \phi_n > = \delta_{mn}$, where δ_{mn} is the Kronecker delta. Now, we can define the Dirac δ-function on a region D in terms of the complete orthonormal set of eigenfunctions for the region D, as

$$\delta_D(\mathbf{x}, \mathbf{x}') = \sum_{n=0}^{\infty} \phi_n(\mathbf{x})\,\bar{\phi}_n(\mathbf{x}'), \tag{6.79}$$

where $\mathbf{x}$ and $\mathbf{x}'$ are two points in D, which are called the field and the source point, respectively.

Example 6.1. Consider the 1-D transient problem:

$$\frac{\partial^2 u}{\partial x^2} = \frac{1}{k}\frac{\partial u}{\partial t}, \quad -a < x < a, \tag{6.80}$$

with the initial and boundary conditions as $u(x, 0) = F(x)$ for $-a < x < a$, and $u(-a, t) = 0 = u(a, t)$ for $t > 0$. Using Bernoulli's separation method by assuming $u(x, t) = X(x)T(t)$, we get

$$\frac{1}{X}\frac{d^2 X}{dx^2} = \frac{1}{kT}\frac{dT}{dt} = -\lambda^2,$$

where λ is a real parameter. The boundary conditions and initial condition become: $X(-a) = 0 = X(a)$, and $T(0) = F(x)$. It will be found that only the choice of $\lambda > 0$ gives meaningful results, leading to the solution of the following two equations

$$\frac{d^2 X}{dx^2} + \lambda^2 X = 0, \quad \frac{dT}{dt} + k\lambda^2 T = 0,$$

which eventually leads to the solution

$$u(x, t) = \sum_{-\infty}^{\infty} A_n \sin\frac{n\pi x}{a} e^{-n^2\pi^2 kt/a^2}, \tag{6.81}$$

where

$$2a A_n = \int_{-\infty}^{\infty} F(x) \sin\frac{n\pi x}{a}\, dx. \tag{6.82}$$

The orthonormal set of spatial eigenpairs for this problem are $\left(\dfrac{n\pi}{a}, \dfrac{1}{2a}\sin\dfrac{n\pi x}{a}\right)$, and the orthonormal eigenfunctions are $\phi_n(x) = \frac{1}{2a}e^{in\pi x/a}$. Hence, the Dirac δ-function in the region $-a < x < a$ for the steady–state (as $t \to \infty$) 1-D Laplace's equation over the interval $-a < x < a$ is

$$\delta(x, x') = \frac{1}{4a^2}\sum_{n=-\infty}^{\infty} e^{in\pi x/a}e^{-in\pi x'/a} = \frac{1}{4a^2}\sum_{-\infty}^{\infty} e^{in\pi(x-x')/a}. \qquad (6.83)$$

Note that in (6.82) we have assumed that the eigenfunction expansion is periodic with period $2a$. Hence the Dirac δ-function (6.83) also has period $2a$. The solution (6.81) can also be obtained by using the Laplace transform method. We have graphed the real part of this function for the basic interval $-a < x < a$ in Fig. 6.5, using $n = 5, 10$, and 25 terms in (6.83), with $a = 3$, and $x' = 0$. The graphs, however, repeat outside this interval with a period $2a$. From the graphs in Fig. 6.5, it is obvious that the peak becomes infinitely higher and narrower at $x = 0$ as n increases. ∎

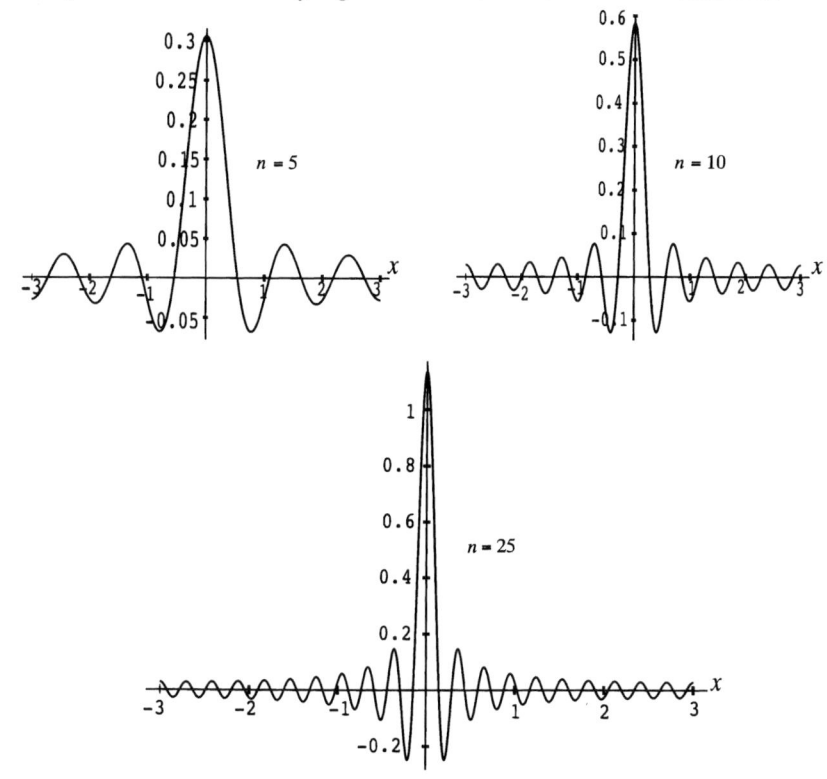

Fig. 6.5. Dirac δ-Function with 5, 10, and 25 Terms.

Example 6.2. Consider the 2-D Laplace's equation

$$\frac{\partial^2 u}{\partial x^2} + \frac{\partial^2 u}{\partial y^2} = 0$$

over the rectangle $R = \{-a < x < a, -b < y < b\}$. The eigenpairs for this region are given by

$$\lambda_n = -\frac{n^2\pi^2}{2\sqrt{ab}}\left(\frac{1}{a^2} + \frac{1}{b^2}\right), \quad \phi_n(x,y) = \frac{1}{2\sqrt{ab}}e^{in\pi x/a}e^{in\pi y/b}.$$

Note that

$$\frac{\partial^2\phi_n}{\partial x^2} + \frac{\partial^2\phi_n}{\partial y^2} = -\frac{n^2\pi^2}{2\sqrt{ab}}\left(\frac{1}{a^2} + \frac{1}{b^2}\right)\phi_n.$$

It can be easily verified that these eigenvalues are orthonormal by showing that $\langle\phi_m, \phi_n\rangle = \int_{-a}^{a}\int_{-b}^{b}\phi_m\bar{\phi}_n\,dy\,dx = \delta_{mn}$. The solution in the complex form of a double Fourier series over the rectangle is given by

$$u(x,y) = \frac{1}{2\sqrt{ab}}\sum_{n=-\infty}^{\infty}\sum_{m=-\infty}^{\infty}c_{nm}e^{in\pi x/a}e^{im\pi y/b}. \tag{6.84}$$

This representation of u is periodic both in x and y with periods $2a$ and $2b$, respectively. The Dirac δ-function for this region is

$$\delta(\mathbf{x}, \mathbf{x}') = \delta(x, x')\,\delta(y, y')$$
$$= \frac{1}{4ab}\sum_{n=-\infty}^{\infty}e^{in\pi(x-x')/a}e^{in\pi(y-y')/b}, \tag{6.85}$$

where $\mathbf{x} = (x, y)$ is the field point and $\mathbf{x}' = (x', y')$ the source point. The solution of the problem is given by

$$u(\mathbf{x}) = u(x,y) = \sum_{n_1=-\infty}^{\infty}\sum_{n_2=-\infty}^{\infty}c_{n_1n_2}\phi_{n_1n_2}. \tag{6.86}$$

6.13. Exercises

6.1. Find Green's function $G(x,t; x',t')$ for the operator $L \equiv \dfrac{\partial}{\partial t} - k\dfrac{\partial^2}{\partial x^2}$ in the region $\{0 < x < a; t > 0\}$ subject to the conditions $G(0,t; x',t') = G(a,t; x',t') = 0$ and $G(x,t; x',t') = 0$ for $t < t'$.

ANS. $G(x,t; x',t') = \dfrac{2}{a}H(t-t')\displaystyle\sum_{n=1}^{\infty}e^{-n^2\pi^2k(t-t')/a^2}\sin\dfrac{n\pi x}{a}\sin\dfrac{n\pi x'}{a}$.

6.2. Use Green's function to find the solution of the problem $\dfrac{\partial u}{\partial t} - k\dfrac{\partial^2 u}{\partial x^2} = 0$, subject to the boundary conditions $u(x,0) = 0$, $u(0,t) = T_0$, and $u(a,t) = 0$.

6.3. Use Green's function to find the solution of the problem $u_t - u_{xx} = 0$, $x \in \mathbb{R}$, $0 < t$, subject to the conditions $u(x,0) = H(x) - H(-x)$.

ANS. $u(x,t) = \mathrm{sgn}(x) \; \mathrm{erf}\left(\dfrac{x}{2\sqrt{t}}\right)$.

6.4. For the Fokker-Planck operator $\dfrac{\partial}{\partial t} - \dfrac{\partial}{\partial x}\left(\dfrac{\partial}{\partial x} + x\right)$ find Green's function in $\mathbb{R}$.

HINT. Use the transformation $u = v\,e^t$, $X = x\,e^t$ and $2T = e^{2t}$. Then the given operator reduces to the diffusion operator.

ANS. $G\left(x,t;x',t'\right) = \dfrac{H(t - t')}{\sqrt{2\pi\left(1 - e^{-2(t-t')}\right)}} \; \exp\left\{\dfrac{-\left(x - x'\,e^{-(t-t')}\right)^2}{2\left(1 - e^{-2(t-t')}\right)}\right\}$.

6.5. Solve the diffusion problem with a source function $f(x,t)$, defined by

$$u_t = k\,u_{xx} + f(x,t), \quad x \in \mathbb{R}, \ t > 0, \text{ such that } u(x,0) = 0 \text{ for } x \in \mathbb{R}.$$

ANS. $u(x,t) = \dfrac{1}{2\sqrt{\pi k}} \displaystyle\int_0^t (t - \tau)^{-1/2}\,d\tau \int_{-\infty}^{\infty} f(z,\tau)\,e^{-(x-z)^2/(4k(t-\tau))}\,dz.$

6.6. Solve the boundary value problem: $u_t = k\,u_{xx} - h\,u$, $0 < x < \infty$, $t > 0$, where h is a constant, such that $u(x,0) = u_0$ for $x > 0$, and $u(0,t) = 0$ and $u_x(0,t) \to 0$ uniformly in t as $x \to \infty$, $t > 0$.

ANS. $u(x,t) = u_0\,e^{-ht}\,\mathrm{erf}\sqrt{\dfrac{x}{4kt}}.$

6.7. Solve the boundary value problem: $u_t = k\,u_{xx}$, $0 < x < \infty$, $t > 0$, such that $u(x,0) = 0$ for $x > 0$, and $u(0,t) = v_0$ and $u_x(0,t) \to 0$ uniformly in t as $x \to \infty$, $t > 0$.

ANS. $u(x,t) = v_0\,\mathrm{erfc}\sqrt{\dfrac{x}{4kt}}.$

6.8. Solve $u_{xx} - u_t = A\,e^{-\alpha x}$, $A \geq 0$, $\alpha > 0$, where $u(0,t) = 0 = u(l,t)$ for $t > 0$, and $u(x,0) = f(x)$ for $0 < x < l$.

ANS. $u = v - \dfrac{A}{\alpha^2} - \dfrac{A}{\alpha^2}\left(e^{-\alpha l} - 1\right)\dfrac{x}{l} + \dfrac{A}{\alpha^2}\,e^{-\alpha x}$, where

$$v = \sum_{n=1}^{\infty} A_n \sin\dfrac{n\pi x}{l}\,e^{-n^2\pi^2 t/l^2}, \quad A_n = \dfrac{2}{l}\int_0^l g(x)\sin\dfrac{n\pi x}{l}\,dx, \text{ and}$$

$$g(x) = f(x) + \dfrac{A}{\alpha^2}\left[1 + \left(e^{-\alpha l} - 1\right)\dfrac{x}{l} - e^{-\alpha x}\right].$$

6.9. Solve $u_t = k^2\,u_{xx} + f(x,t)$, $0 < x < a$, subject to the boundary conditions $u(0,t) = 0 = u(a,t)$ for $0 \leq x \leq a$, and the initial condition $u(x,0) = 0$ for $t > 0$.

ANS. $u(x,t) = \displaystyle\sum_{n=1}^{\infty} u_n(t)\sin\dfrac{n\pi x}{a}$, where

$$u_n(t) = \int_0^t e^{n^2\pi^2 k^2 (\tau-t)/a^2} f_n(\tau)\, d\tau, \qquad f_n(t) = \frac{2}{a}\int_0^a f(y,t) \sin\frac{n\pi y}{a}\, dy.$$

6.10. Find the interior temperature of the cooling ball of Example 4.13, if

$$f(\rho, \phi) = \begin{cases} 1, & 0 \le \phi < \pi/2 \\ 0, & \pi/2 \le \phi < \pi. \end{cases}$$

Ans. $u(\rho, \phi, t) = \displaystyle\sum_{m,n=1}^{\infty} C_{mn} \frac{J_{n+1/2}(\alpha_{mn}\rho) P_n(\cos\phi)}{\sqrt{\alpha_{mn}\rho}} e^{-\alpha_{mn}^2 t},$

$$C_{mn} = \frac{(2n+1)P_n'(0)\sqrt{\alpha_{mn}} \int_0^1 \rho^{3/2} J_{n+1/2}(\alpha_{mn}\rho)\, d\rho}{2n(n+1)\int_0^1 \rho J_{n+1/2}^2(\alpha_{mn}\rho)\, d\rho}$$

$$= \frac{(2n+1)P_n'(0)\sqrt{\alpha_{mn}} \int_0^1 \rho^{3/2} J_{n+1/2}(\alpha_{mn}\rho)\, d\rho}{n(n+1)J_{n-1/2}^2(\alpha_{mn})},$$

where α_{mn} are the consecutive positive roots of $J_{n+1/2}(\alpha) = 0$.

6.11. Determine the steady-state temperature inside a solid hemisphere $0 \le \rho \le 1$, $0 \le \phi \le \pi/2$ (a) when the base $\phi = \pi/2$ is at $0°$ and the curved surface $\rho = 1$, $0 \le \phi < \pi/2$, is at $1°$; and (b) when the base $\phi = \pi/2$ is insulated, but the temperature on the curved surface is $f(\phi)$.

Hint: $\dfrac{\partial u}{\partial z} = \cos\phi \dfrac{\partial u}{\partial \rho} - \dfrac{\sin\phi}{\rho}\dfrac{\partial u}{\partial \phi} = 0$ on the base. Also use $\displaystyle\int_0^1 P_n(x)\, dx =$

$\dfrac{1}{n(n+1)} P_n'(0)$, and

$(x^2-1)P_n'(x) = \dfrac{n(n+1)}{2n+1}\big[P_{n+1}(x) - P_{n-1}(x)\big]$, where $P_n(x)$ are the Legendre polynomials.

Ans. (a) $u(\rho, \phi) = \displaystyle\sum_{n=0}^{\infty} \rho^{2n+1}[P_{2n}(0) - P_{2n+2}(0)]\, P_{2n+1}(\cos\phi).$

(b) $u(\rho, \phi) = \displaystyle\sum_{n=0}^{\infty} c_n \rho^{2n}\, P_{2n}(\cos\phi)$, where

$$c_n = (4n+1)\int_0^{\pi/2} f(\phi)\, P_{2n}(\cos\phi)\, \sin\phi\, d\phi.$$

6.12. Solve $u_t = u_{xx}$, $-\pi < x < \pi$, subject to the conditions $u(x,0) = f(x)$, $u(-\pi, t) = u(\pi, t)$, and $u_x(-\pi, t) = u_x(\pi, t)$, where $f(x)$ is a periodic function of period 2π. This problem describes the heat flow inside a rod of length 2π, which is shaped in the form of a closed circular ring.

Hint: Assume $X(x) = A\cos\omega x + B\sin\omega x.$

ANS. $\omega_n = n$; $u(x,t) = \displaystyle\sum_{n=0}^{\infty} e^{-n^2 t} \left(a_n \cos nx + b_n \sin nx \right)$, where

$$a_n = \frac{1}{\pi} \int_{-\pi}^{\pi} f(x) \cos nx \, dx, \quad b_n = \frac{1}{\pi} \int_{-\pi}^{\pi} f(x) \sin nx \, dx, \quad a_0 = \frac{1}{2\pi} \int_{-\pi}^{\pi} f(x) \, dx.$$

6.13. Solve the problem $u_t = \nabla^2 u$, $r < 1$, $0 < z < 1$, such that $u(r, z, 0) = 1$, $u(1, z, t) = 0$, and $u(r, 0, t) = 0 = u(r, 1, t)$. This problem describes the temperature distribution inside a homogeneous isotropic solid circular cylinder.

ANS. $u(r, z, t) = \displaystyle\sum_{m,n=1}^{\infty} C_{mn} e^{-(\lambda_m^2 + n^2 \pi^2) t} J_0(\lambda_m r) \sin n\pi z$, where λ_m are the zeros of J_0, and $C_{mn} = \dfrac{4\left(1 - (-1)^n\right)}{n\pi \lambda_m J_1(\lambda_m)}$.

6.14. Find the steady-state temperature in a solid circular cylinder of radius 1 and height 1 under the conditions that the flat faces are kept at $0°$ and the curved surface at $1°$. ANS. $u(r, z) = 4 \displaystyle\sum_{\substack{n=1 \\ n \text{ odd}}}^{\infty} \frac{I_0(n\pi r)}{I_0(n\pi)} \frac{\sin n\pi z}{mn}$.

6.15. Solve the steady-state problem of temperature distribution in a half-cylinder $0 \le r \le 1$, $0 \le \theta \le \pi$, $0 \le z \le 1$, where the flat faces are kept at $0°$ and the curved surface at $1°$. ANS. $u(r, \theta, z) = \dfrac{16}{\pi^2} \displaystyle\sum_{\substack{m,n=1 \\ m,n \text{ odd}}}^{\infty} \frac{I_m(n\pi r)}{I_m(n\pi)} \frac{\sin n\pi z}{mn} \sin m\theta$.

6.16. Solve $\dfrac{\partial u}{\partial t} = \dfrac{\partial}{\partial x}\left(x \dfrac{\partial u}{\partial x}\right)$, $0 < x < 1$, $t > 0$, subject to the conditions $u(x, 0) = f(x)$ and $u(1, t) = 0$.
HINT: Set $4x = r^2$.
ANS. $u(x, t) = \displaystyle\sum_{n=1}^{\infty} C_n e^{-\lambda_n^2 t / 4} J_0(\lambda_n \sqrt{x})$, where λ_n are the zeros of $J_0(\lambda)$, and
$$C_n = \frac{\int_0^1 f(x) J_0(\lambda_n \sqrt{x}) \, dx}{\int_0^1 [J_0(\lambda_n \sqrt{x})]^2 \, dx}.$$

6.17. Solve the dimensionless partial differential equation governing the plane wall transient heat conduction $u_t = u_{xx}$, $0 < x < 1$, subject to the boundary conditions $u(0, t) = 1$, $u(1, t) = 0$ for $t > 0$, and the initial condition $u(x, 0) = 0$ for $0 < x < 1$. Also find the half-space solution.

HINT. Since the nonhomogeneous boundary condition does not allow us to determine the eigenfunctions, as in Exercise 6.15 above, we proceed as follows: First, we find a particular solution of the problem, which satisfies only the boundary conditions. Although there is more than one way to determine the particular solution, we can, for example, take the steady-state case, where the equation becomes $\tilde{u}_{xx} = 0$, which, after integrating twice, has the general solution $\tilde{u}(x) = c_1 x + c_2$, with the boundary conditions $\tilde{u}(0) = 1$, $\tilde{u}(1) = 0$.

Thus, $c_1 = -1$, $c_2 = 1$, and the steady-state solution is $\widetilde{u}(x) = 1 - x$. Next, we formulate a homogeneous problem by writing $u(x,t)$ as a sum of the steady-state solution $\widetilde{u}(x)$ and a transient term $v(x,t)$, i.e., we take $u(x,t) = \widetilde{u}(x) + v(x,t)$, or $v(x,t) = u(x,t) - \widetilde{u}(x)$, thus reducing the problem to finding $v(x,t)$; that is, we solve the following problem: $v_t = v_{xx}$, with the boundary conditions $v(0,t) = u(0,t) - \widetilde{u}(0) = 0$, $v(1,t) = u(1,t) - \widetilde{u}(1) = 0$, and the initial condition $v(x,0) = u(x,0) - \widetilde{u}(x) = x - 1$, by the Bernoulli's separation method as in the above Exercise 6.15.

ANS. $v(x,t) = -\dfrac{2}{\pi} \displaystyle\sum_{n=1}^{\infty} \dfrac{1}{n} e^{-n^2\pi^2 t} \sin n\pi x$, and

$$u(x,t) = 1 - \frac{x}{l} - \frac{2}{\pi}\sum_{n=1}^{\infty}\frac{1}{n}e^{-n^2\pi^2 t/l^2}\sin\frac{n\pi x}{l},\ \text{where } l \text{ is the thickness}$$

of the plate. The solution for the half-space is derived by letting $l \to \infty$. Since

$$\lim_{l\to\infty} u(x,t) = 1 - \frac{2}{\pi}\sum_{n=1}^{\infty}\frac{l}{n\pi}e^{-n^2\pi^2 t/l^2}\sin\frac{n\pi x}{l}\cdot\frac{\pi}{l},\ \text{let } n\pi/l = \xi \text{ and } \pi/l = d\xi.$$

Then $\displaystyle\lim_{l\to\infty} u(x,t) = 1 - \frac{2}{\pi}\int_0^{\infty}\frac{1}{\xi}e^{-\xi^2 t}\sin\xi t\,d\xi = 1 - \mathrm{erf}\left(\frac{x}{2\sqrt{t}}\right) = \mathrm{erfc}\left(\frac{x}{2\sqrt{t}}\right).$

6.18. Derive Eq (6.21).

6.19. Solve the linearized Bergers equation $u_t = \nu\,u_{xx}$, $-\infty < x < \infty$, $t > 0$, where ν is the kinematic viscosity, such that

$$u(x,0) = \begin{cases} f_1(x), & x < 0, \\ f_2(x), & x > 0. \end{cases}$$

HINT. Use Fourier transform with respect to x.

ANS. $u(x,t) = \frac{1}{2}\left[f_1(x) + f_2(x)\right] - \frac{1}{2}\left[f_1(x) - f_2(x)\right]\mathrm{erf}\,\dfrac{x}{2\sqrt{\nu t}}.$

6.20. Show that the solution $u(x,t)$ for the equation $u_{xx} = u_t + xu$, $x \in \mathbb{R}$, $t > 0$, and satisfying the initial condition $u(x,0) = f(x)$ is

$$u(x,t) = \frac{1}{2\sqrt{\pi t}}\,e^{t^3/3 - xt}\int_{-\infty}^{\infty} f(s)\exp\left\{-\frac{(x - s - t^2)^2}{4}\right\}ds.$$

6.21. Show that the solution of the diffusion equation $u_t = u_{xx}$, $t \geq 0$ and satisfying the initial conditions $u(x,0) = 1$, $x > 0$; $u(x,0) = -1$, $x < 0$, with discontinuity at $t = 0$ is $u(x,t) = \mathrm{erf}\left(4x\sqrt{t}\right)$.

6.22. Solve the heat conduction problem $u_t = k\,u_{xx}$, $0 \leq x \leq a$, $t > 0$, subject to the Dirichlet boundary conditions $u(0,t) = 0 = u(a,t)$, and the initial condition $u(x,0) = f(x)$, $0 \leq x \leq a$.

ANS. $u(x,t) = \dfrac{2}{a}\displaystyle\sum_{n=1}^{\infty}e^{-kn^2\pi^2 t/a^2}\sin\frac{n\pi x}{a}\int f(s)\sin\frac{n\pi s}{a}\,ds.$

7

Hyperbolic Equations

The homogeneous equation for the wave operator (or the d'Alembertian) $\Box_c \equiv \dfrac{\partial^2}{\partial t^2} - c^2 \nabla^2$ describes, e.g., the dynamics of a vibrating string in $\mathbb{R} \times \mathbb{R}^+$, of a drum membrane or surface of a lake in $\mathbb{R}^2 \times \mathbb{R}^+$, or of a sound wave in air, or an electromagnetic wave in vacuum in $\mathbb{R}^3 \times \mathbb{R}^+$. Here c denotes the speed of sound or light. The wave equation propagates signals with velocities less than or equal to c. Unlike the diffusion equation, the wave equation is not affected by a change in the sign in the time variable, and so in this sense it is reversible. We will discuss Sturm-Liouville system for 1-D wave equation, which is followed by finding Green's functions for 1-D, 2-D and 3-D wave equations, and other applications.

7.1. 1-D Wave Equation

We will discuss some boundary value problems for 1-D wave equation by different methods.

7.1.1. Sturm-Liouville System for 1-D Wave Equation. Consider the problem

$$\frac{\partial^2 u}{\partial t^2} = \frac{\partial}{\partial x}\left(p(x)\frac{\partial u}{\partial x}\right) - q(x)u + F(x,t), \quad a \le x \le b, \ t > 0,$$

$$a_1 u(a,t) + a_2 u_x(a,t) = 0, \ b_1 u(b,t) + b_2 u_x(b,t) = 0, \tag{7.1}$$

$$u(x,0) = f(x), \ u_t(x,0) = g(x), \quad a < x < b,$$

where the function p, q and F, and the constants $a_{1,2}$ and $b_{1,2}$ are the same as in Eq (6.1). Let L denote the Sturm-Liouville operator. Then Eq (7.1) can be written as

$$u_{tt} = L[u] + F, \text{ where } L[u] \equiv \frac{\partial}{\partial x}\left(p(x)\,u_x\right) + q(x)u. \tag{7.2}$$

Using Bernoulli's separation method, let the solution of Eq (7.2) with $F = 0$ be of the form $u(x,t) = X(x)\,T(t) \ne 0$. Then Eq (7.2) reduces to

$$L[u] = \lambda X, \quad x \in I, \quad \frac{d^2 T}{dt^2} = \lambda T, \quad t > 0, \qquad (7.3)$$

where λ is the separation constant. The boundary conditions on X are

$$a_a X(a) + a_2 X'(a) = 0, \quad b_1 X(b) + b_2 X'(b) = 0. \qquad (7.4)$$

The system (7.3)-(7.4) is known as the associated Sturm-Liouville system for 1-D wave equation, which can be solved by finding the eigenpair $(\lambda_n, X_n(x))$, $n = 1, 2, \ldots$. Using the superposition principle, the solution of Eq (7.2) can be formally written as

$$u(x,t) = \sum_{n=1}^{\infty} X_n(x) T_n(t), \qquad (7.5)$$

where $T_n(t)$ are yet to be determined, as follows: Assume that the forcing term $F(x,t)$ can be expanded in terms of the eigenfunctions $X_n(x)$ as

$$F(x,t) = \sum_{n=1}^{\infty} X_n(x) f_n(t), \qquad (7.6)$$

where $f_n(t)$ are the generalized Fourier coefficients of $F(x,t)$, given by

$$f_n(t) = \int_a^b F(x,t) X_n(x) \, dx. \qquad (7.7)$$

Then substituting (7.5) and (7.6) into (7.2), we find that

$$\sum_{n=1}^{\infty} \ddot{T}_n(t) X_n(t) = L \left[\sum_{n=1}^{\infty} T_n(t) X_n(x) \right] + \sum_{n=1}^{\infty} X_n(x) f_n(t)$$

$$= \sum_{n=1}^{\infty} T_n(t) L [X_n(x)] + \sum_{n=1}^{\infty} X_n(x) f_n(t)$$

$$= [\lambda_n T_n(t) + f_n(t)] T_n(x),$$

which yields the second-order ordinary differential equation

$$\ddot{T}_n(t) + \mu^2 T_n(t) = f_n(t), \quad T_n(0) = f(x), \ T'(0) = g(x),$$

where $\mu_n^2 = -\lambda_n$. Applying the Laplace transform method to this equation we find that

$$T_n(t) = T_n(0) \cos(\mu_n t) + \frac{1}{\mu_n} \dot{X}(0) \sin(\mu_n t) + \frac{1}{\mu_n} \dot{X}(0) \sin(\mu_n t), \qquad (7.8)$$

where from the initial conditions in (7.1) we have

$$T_n(0) = \int_a^b X_n(s) f(s) \, ds, \quad \dot{T}_n(0) = \int_a^b X_n(s) g(s) \, ds.$$

Hence, from (7.5) and (7.7) the solution $u(x,t)$ of this problem is given by

$$u(x,t) = \sum_{n=1}^{\infty} X_n(x)\, T_n(t)$$

$$= \sum_{n=1}^{\infty} \left[\left\{ \int_a^b X_n(s)\, f(s)\, ds \right\} \cos{(\mu_n t)} + \frac{1}{\mu_n} \left\{ \int_a^b X_n(s)\, g(s)\, ds \right\} \sin{(\mu_n t)} \right]$$

$$+ \frac{1}{\mu_n} \int_a^b \sin{\mu_n (t - \tau)}\, f_n(\tau)\, d\tau$$

$$= \int_a^b \left[\sum_{n=1}^{\infty} X_n(x) X(s)\, \cos{(\mu_n t)} \right] f(s)\, ds$$

$$+ \int_a^b \left[\sum_{n=1}^{\infty} \frac{1}{\mu_n} X_n(x) X(s)\, \sin{(\mu_n t)} \right] g(s)\, ds$$

$$+ \int_0^t \int_a^b \left[\sum_{n=1}^{\infty} \frac{1}{\mu_n} X_n(x) X(s)\, \sin{(\mu_n t)} \right] F(s, \tau)\, ds\, d\tau. \tag{7.9}$$

Define two Green's functions G_1 and G_2 as

$$G_{1,2}(x, x'; t) = \sum_{n=1}^{\infty} X_n(x) X(x') \left\{ \begin{array}{c} \cos{(\mu_n t)} \\ \sin{(\mu_n t)} \end{array} \right\}. \tag{7.10}$$

Then the solution (7.9) can be written as

$$u(x,t) = \int_a^b \left[G_1(x, x'; t)\, f(x') + \frac{1}{\mu_n} G_2(x, x'; t)\, g(x') \right] dx'$$

$$+ \int_0^t \int_a^b \frac{1}{\mu_n} G_2(x, x'; t - \tau)\, F(x', \tau)\, dx'\, d\tau. \ \blacksquare \tag{7.11}$$

7.1.2. Vibrations of a Variable String. Consider the problem of vibrations of a string of density proportional to $(1 + x)^{-2}$ and defined by

$$\frac{1}{(1 + x)^2} \frac{\partial^2 u}{\partial t^2} - \frac{\partial^2 u}{\partial x^2} = 0, \quad 0 < x < 1, \ t > 0, \tag{7.12}$$

and subject to the initial conditions $u(x, 0) = f(x)$, $\dfrac{\partial u}{\partial t}(x, 0) = 0$, and the boundary conditions $u(0, t) = 0 = u(1, t)$. By Bernoulli's separation method $(u = X(x)\, T(t))$ we get

$$X'' + \frac{\lambda}{(1 + x)^2} X = 0, \qquad T'' + \lambda T = 0,$$

$$X(0) = 0 = X(1), \qquad T(0) = f(x), \ T'(0) = 0, \tag{7.13}$$

where λ is the separation constant. This is an eigenvalue problem. The equation in X has the solutions of the form $(1 + x)^a$, where $a(a - 1) + \lambda = 0$, or $a = \frac{1}{2}(1 \pm \sqrt{1 - 4\lambda})$. Then for $X(0) = 0$ we get

$$X(x) = (1 + x)^{1 + \sqrt{1-4\lambda}/2} - (1 + x)^{1 - \sqrt{1-4\lambda}/2}.$$

The condition $X(1) = 0$ gives $2^{1+\sqrt{1-4\lambda}/2} - 2^{1-\sqrt{1-4\lambda}/2} = 0$, or $2\sqrt{1-4\lambda} = 1$. If $\lambda < 1/4$, then $\sqrt{1 - 4\lambda}$ is real, and this equation has no solution in this case. If $\lambda = 1/4$, the two solutions are $(1+x)^{1/2}$ and $(1+x)^{1/2} \ln(1+x)$, which are no longer linearly independent; although the latter solution satisfies the condition $X(0) = 0$, it fails to satisfy $X(1) = 0$, thus showing that $\lambda = 1/4$ is not an eigenvalue. Finally, if $\lambda > 1/4$, the expression $\sqrt{1 - 4\lambda}$ is imaginary, but it has the solutions

$$(1 + x)^{(1+\sqrt{1-4\lambda})/2} = (1 + x)^{1/2} \exp\left\{ \frac{i}{2} \sqrt{4\lambda - 1}\, \ln(1 + x) \right\}$$

$$= (1 + x)^{1/2}\left[\cos\left(\sqrt{\lambda - \frac{1}{4}}\, \ln(1 + x) \right) + i\, \sin\left(\sqrt{\lambda - \frac{1}{4}}\, \ln(1 + x) \right) \right],$$

and the two linearly independent solutions are given by the real and imaginary parts of this expression. To satisfy the condition $X(0) = 0$ we take

$$X(x) = (1 + x)^{1/2} \sin\left(\sqrt{\lambda - \frac{1}{4}}\, \ln(1 + x) \right).$$

Then the condition $X(1) = 0$ gives $2^{1/2} \sin\left(\sqrt{\lambda - \frac{1}{4}}\, \ln 2 \right) = 0$, so $\sqrt{\lambda - \frac{1}{4}}\, \ln 2$ must be an integral multiple of π, i.e., $\sqrt{\lambda - \frac{1}{4}}\, \ln 2 = \pi$, which gives the eigenvalues

$$\lambda_n = \left(\frac{n\pi}{\ln 2} \right)^2 + \frac{1}{4}, \quad n = 1, 2, \dots,$$

and the corresponding eigenfunctions are

$$X_n(x) = (1 + x)^{1/2} \sin\frac{n\pi \ln(1 + x)}{\ln 2}.$$

We take the Fourier series $f(x) = \sum_{n=1}^{\infty} c_n X_n(x)$, where, since $p(x) = (1 + x)^{-2}$,

$$c_n = \frac{\int_0^1 f(x)(1 + x)^{-3/2} \sin\dfrac{n\pi \ln(1 + x)}{\ln 2}\, dx}{\int_0^1 (1 + x)^{-1} \sin^2\dfrac{n\pi \ln(1 + x)}{\ln 2}\, dx}$$

$$= \frac{2}{\ln 2} \int_0^1 f(x)(1 + x)^{-3/2} \sin\frac{n\pi \ln(1 + x)}{\ln 2}\, dx.$$

The Fourier series converges absolutely and uniformly if $f(0) = f(1) = 0$ and $\int_0^1 f'^2\, dx < +\infty$. Since Eq (7.13) with $T(0) = 1$, $T'(0) = 0$ has the solution $T(t) = \cos\sqrt{\lambda}\,t$, the problem (7.12) has the solution

$$u(x,t) = \sum_{n=1}^{\infty} c_n \cos\left(\sqrt{\frac{n^2\pi^2}{(\ln 2)^2} + \frac{1}{4}}\,t\right)(1+x)^{1/2}\sin\frac{n\pi\,\ln(1+x)}{\ln 2}. \tag{7.14}$$

This series converges uniformly, but to ensure continuous derivatives of u we must assume that $f \in C^2$, and $f(0) = f(1) = 0$, $f''(0) = f''(1) = 0$, and $\int_0^1 f'''^2\, dx < +\infty$. ∎

7.1.3. Green's Function for 1-D Wave Equation

We assume that the singularity is at the origin and at $t = 0$. In $\mathbb{R}$, Green's function $G_1(x;t)$ satisfies the equation

$$\frac{\partial^2 G_1}{\partial t^2} - c^2\frac{\partial^2 G_1}{\partial x^2} = \delta(x)\,\delta(t),$$

such that $G_1(x;t) \to 0$ as $|x| \to \infty$, $t > 0$. Applying the Laplace and Fourier transforms to this equation, we get

$$\left(s^2 + a^2c^2\right)\overline{\overline{G}}_1 = \frac{1}{\sqrt{2\pi}}, \quad \text{or} \quad \overline{\overline{G}}_1 = \frac{1}{\sqrt{2\pi}\,(s^2 + a^2c^2)},$$

which after the Laplace inversion gives

$$\widetilde{G}_1 = \frac{1}{ac\sqrt{2\pi}}\sin act.$$

After inverting the Fourier transform, we obtain

$$G_1(x;t) = \frac{1}{2c}\left[H(x + ct) - H(x - ct)\right].$$

However, if we reverse the process by carrying out the Fourier inversion prior to the Laplace inversion, we get

$$G_1(x,t) = \frac{1}{2c}H\left(t - |x|/c\right) = \frac{1}{2c}H(ct - |x|).$$

It is easy to check that the two solutions are equivalent. Thus, if a source is at the point $(x';t')$, the above solution becomes

$$G_1(x,t;x',t') = \frac{1}{2c}H\left[c(t - t') - (|x - x'|)\right]. \quad ∎ \tag{7.15}$$

7.2. 2-D Wave Equation

In $\mathbb{R}^2$, Green's function G_2 satisfies the equation

$$\frac{\partial^2 G_2}{\partial t^2} - c^2 \left(\frac{\partial^2 G_2}{\partial x^2} + \frac{\partial^2 G_2}{\partial y^2} \right) = \delta(x - x')\, \delta(y - y')\, \delta(t - t').$$

Applying the Laplace transform, we get

$$s^2 \overline{G}_2 - c^2 \left(\frac{\partial^2 \overline{G}_2}{\partial x^2} + \frac{\partial^2 \overline{G}_2}{\partial y^2} \right) = e^{-st'} \delta(x - x')\, \delta(y - y'),$$

or, using axial symmetry in polar cylindrical coordinates with $r^2 = (x - x')^2 + (y - y')^2$, we have

$$s^2 \overline{G}_2 - c^2 \left(\frac{d^2 \overline{G}_2}{dr^2} + \frac{1}{r} \frac{\partial \overline{G}_2}{\partial r} \right) = \frac{e^{-st'} \delta(r)}{2\pi r}. \tag{7.16}$$

The forcing term $\dfrac{e^{-st'} \delta(r)}{2\pi r}$ in Eq (7.16) can be considered as representing a source of strength $e^{-st'}/c^2$ at the point $r = 0$. The solution of the homogeneous part of Eq (7.16) is $A\, I_0(kr) + B\, K_0(kr)$, where I_0 and K_0 are the modified Bessel functions of the first and the second kind of order zero, respectively, and $k = s/c$. Since $I_0 \to \infty$ as $r \to \infty$, we take $A = 0$, and then B is obtained by equating the flux across a circle C_ε of radius ε to $e^{-st'}/c^2$. Thus,

$$-\lim_{\varepsilon \to 0} \int_{C_\varepsilon} B\, \frac{\partial K_0(sr/c)}{\partial r}\, ds = \lim_{\varepsilon \to 0} 2\pi\varepsilon\, B \left(\frac{s}{c} \right) K_1 \left(\frac{s\varepsilon}{c} \right) = 2\pi\, B = \frac{e^{-st'}}{c^2}.$$

Hence, $B = \dfrac{e^{-st'}}{2\pi c^2}$, and $\overline{G}_2(r, s; t') = \dfrac{e^{-st'}}{2\pi c^2} K_0 \left(\dfrac{sr}{c} \right)$. Since $\mathcal{L}^{-1}\{K_0(\alpha s)\} = \dfrac{H(t - \alpha)}{\sqrt{t^2 - \alpha^2}}$ (see Erdelyi et al. [1954], or Abramowitz and Stegun [1965]), on inversion we get

$$G_2(r, t; t') = \begin{cases} \dfrac{1}{2\pi c \sqrt{c^2(t - t')^2 - r^2}} & \text{for } r < c(t - t') \\[2ex] 0 & \text{for } r > c(t - t') \end{cases} \tag{7.17}$$

$$= \frac{H(c(t - t') - r)}{2\pi c \sqrt{c^2(t - t')^2 - r^2}}. \; \blacksquare$$

7.3. 3-D Wave Equation

In $\mathbb{R}^3$, Green's function G_3 satisfies the equation

$$\frac{\partial^2 G_3}{\partial t^2} - c^2 \left(\frac{\partial^2 G_3}{\partial x^2} + \frac{\partial^2 G_3}{\partial y^2} + \frac{\partial^2 G_3}{\partial z^2} \right) = \delta(x - x')\delta(y - y')\delta(z - z')\delta(t - t').$$

Applying the Laplace transform, we get

$$s^2\overline{G}_3 - c^2\left(\frac{\partial^2\overline{G}_3}{\partial x^2} + \frac{\partial^2\overline{G}_3}{\partial y^2} + \frac{\partial^2\overline{G}_3}{\partial z^2}\right) = e^{-st'}\delta(x-x')\delta(y-y')\delta(z-z'),$$

or, using the spherical symmetry with $r^2 = (x-x')^2 + (y-y')^2 + (z-z')^2$, we have

$$s^2\overline{G}_3 - c^2\left(\frac{d^2\overline{G}_3}{dr^2} + \frac{2}{r}\frac{\partial\overline{G}_3}{\partial r}\right) = \frac{e^{-st'}\delta(r)}{4\pi r^2}. \tag{7.18}$$

Set $r\overline{G}_3 = \overline{V}$ in Eq (7.18). Then it reduces to

$$s^2\overline{V} - c^2\frac{\partial^2\overline{V}}{\partial r^2} = \frac{e^{-st'}\delta(r)}{4\pi r}, \qquad r = |\mathbf{x}-\mathbf{x}'|.$$

The solution of this equation is $\overline{V} = A\,e^{sr/c} + B\,e^{-sr/c}$. Since $e^{sr/c}$ becomes unbounded as $r \to \infty$, we take $A = 0$. Thus, $\overline{G}_3 = B\,e^{-sr/c}/r$. To evaluate B, the forcing term, which is $\dfrac{e^{-st'}\delta(r)}{4\pi r^2}$ in Eq (7.18), can be considered as representing a source of strength $e^{-st'}/c^2$ at the point $r = 0$. Then by equating $e^{-st'}/c^2$ to the flux across a sphere S_ε of radius ε and letting $\varepsilon \to 0$, we have

$$-\lim_{\varepsilon\to 0}\iint_{\partial S_\varepsilon} B\frac{\partial}{\partial r}\left(\frac{e^{-sr/c}}{r}\right) dS = \frac{e^{-st'}}{c^2},$$

or
$$-4\pi B\lim_{\varepsilon\to 0}\varepsilon^2\left[-\frac{s}{c}\frac{e^{-sr/c}}{r} - \frac{e^{-sr/c}}{r^2}\right]_{r=\varepsilon} = \frac{e^{-st'}}{c^2},$$

which gives $4\pi B = e^{-st'}/c^2$, or $\overline{G}_3 = \dfrac{e^{-st'-sr/c}}{4\pi rc^2}$. On inversion, this yields

$$G_3(r;t,t') = \frac{1}{4\pi rc^2}\delta\left(t-t'-\frac{r}{c}\right),$$

or $G_3(x,y,z,t;x',y',z',t')$

$$= \frac{1}{4\pi rc^2}\delta\left(t-t'-\frac{1}{c}\sqrt{(x-x')^2+(y-y')^2+(z-z')^2}\right) \tag{7.19}$$

$$= \frac{1}{4\pi rc}\delta\left(c(t-t')-r\right), \qquad r = |\mathbf{x}-\mathbf{x}'|. \ \blacksquare$$

The graphs of Green's functions G_1, G_2 and G_3 are presented in Figs. 7.1, 7.2 and

7.3, respectively.

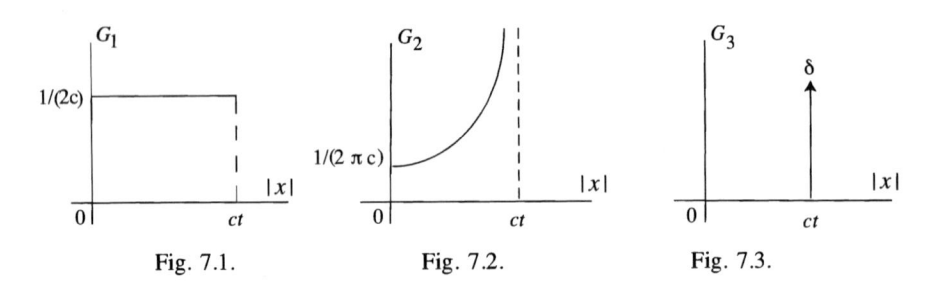

Fig. 7.1.   Fig. 7.2.   Fig. 7.3.

7.4. 2-D Axisymmetric Wave Equation

Consider the 2-D wave equation in the polar cylindrical coordinates:

$$u_{tt} - c^2 \nabla^2 u = f(r,t), \quad 0 < r < \infty, \quad \text{where } \nabla^2 \equiv \frac{1}{r} \frac{\partial}{\partial r} \left(r \frac{\partial}{\partial r} \right),$$

subject to the initial conditions $u(r,0) = 0 = u_t(r,0)$. The Green's function $G(r,t)$ satisfies the system

$$G_{tt} - c^2 \nabla^2 G = \frac{1}{2\pi r} \delta(r)\delta(t), \quad G(r,0) = 0 = G_t(r,0).$$

Applying the zero-order Hankel transform with respect to r, followed by the Laplace transform with respect to t, to this system, we get

$$\overline{\widehat{G}}(\sigma, s) = \frac{c^2}{2\pi \left(s^2 + c^2\sigma^2\right)},$$

where s and σ are the variables of the Laplace and Hankel transform, respectively. Inverting the Laplace transform we get

$$\widehat{G}(\sigma, t) = \frac{c}{2\pi\sigma} \sin(c\sigma t),$$

which after inverting the Hankel transform gives

$$G(r,t) = \frac{c\,H(t)}{2\pi} \mathcal{H}_0^{-1}\left\{\frac{1}{\sigma} \sin(c\sigma t)\right\} = \frac{c\,H(t)\,\mathcal{H}(ct - r)}{2\pi\sqrt{c^2 t^2 - r^2}} . \blacksquare \qquad (7.20)$$

7.5. Vibrations of a Circular Membrane

Consider the problem of the transverse vibrations of a stretched circular membrane of unit radius fixed on its boundary, defined by

$$\frac{\partial^2 u}{\partial t^2} - c^2\left[\frac{\partial^2 u}{\partial r^2} + \frac{1}{r}\frac{\partial u}{\partial r} + \frac{1}{r}\frac{\partial^2 u}{\partial \theta}\right] = 0, \quad r < 1, \ t > 0, \qquad (7.21)$$

$$u(r,\theta,0) = f(r,\theta), \quad \frac{\partial u}{\partial r}(r,\theta,0) = 0, \quad u(1,\theta,t) = 0.$$

Using Bernoulli's separation method ($u = R(r)\Theta(\theta)T(t)$), the solutions are of the form $R(r) \cos m\theta \, \cos c\sqrt{\lambda}\,t$ and $R(r) \sin m\theta \, \cos c\sqrt{\lambda}\,t$, where $m, n = 0, 1, 2, \ldots$, and R is a solution of the eigenvalue problem

$$(r\,R')' - \frac{m^2}{r}\,R + \lambda r R = 0,$$

$$R(1) = 0; \quad R \text{ and } R' \text{ bounded.}$$

(7.22)

The eigenfunctions of this problem are $J_m\left(\sqrt{\lambda_k^{(m)}}\right)$, while the eigenvalues $\lambda_k^{(m)}$ are the solutions of the equation $J_m\left(\sqrt{\lambda_k^{(m)}}\right) = 0$. Thus,

$$u(r, \theta, t) = \frac{1}{2} \sum_{k=1}^{\infty} C_{k0}\, J_0\left(\sqrt{\lambda_k^{(0)}}\,r\right) \cos\left(c\sqrt{\lambda_k^{(0)}}\,t\right) +$$

$$+ \sum_{n=1}^{\infty} \sum_{m=1}^{\infty} (C_{km} \cos m\theta + D_{km} \sin m\theta)\, J_m\left(\sqrt{\lambda_k^{(m)}}\,r\right) \cos\left(c\sqrt{\lambda_k^{(m)}}\,t\right),$$

(7.23)

where

$$C_{km} = \frac{1}{a_m} \int_0^1 \int_{-\pi}^{\pi} f(r, \theta) \cos m\theta\, J_m\left(\sqrt{\lambda_k^{(m)}}\,r\right) r\, dr\, d\theta,$$

$$D_{km} = \frac{1}{a_m} \int_0^1 \int_{-\pi}^{\pi} f(r, \theta) \sin m\theta\, J_m\left(\sqrt{\lambda_k^{(m)}}\,r\right) r\, dr\, d\theta,$$

$$a_m = \pi \int_0^1 \left[J_m\left(\sqrt{\lambda_k^{(m)}}\,r\right)\right]^2 r\, dr,$$

and the series in (7.23) converge uniformly to the solution $u(r, \theta, t)$. The special case for a circular drum is discussed in Examples 4.12 on a circular drum.

7.6. 3-D Wave Equation in a Cube

Consider the problem in a cube $D = \{0 < x, y, z < \pi\}$:

$$\frac{\partial^2 u}{\partial t^2} - c^2 \nabla^2 u \equiv \frac{\partial^2 u}{\partial t^2} - c^2\left(\frac{\partial^2 u}{\partial x^2} + \frac{\partial^2 u}{\partial y^2} + \frac{\partial^2 u}{\partial z^2}\right) = 0 \quad \text{in } D \text{ for } t > 0,$$

$$u(0, y, z, t) = u(\pi, y, z, t) = 0,$$

$$u(x, 0, z, t) = u(x, \pi, z, t) = 0,$$

$$u(x, y, 0, t) = u(x, y, \pi, t) = 0,$$

$$u(x, y, z, 0) = f(x, y, z),$$

$$\frac{\partial u}{\partial t}(x, y, z, 0) = g(x, y, z).$$

(7.24)

We begin with the identity

$$\frac{\partial u}{\partial t}\left[\frac{\partial^2 u}{\partial t^2} - c^2 \nabla^2 u\right] = \frac{\partial}{\partial t}\left[\frac{1}{2}\left(\frac{\partial u}{\partial t}\right)^2 + \frac{c^2}{2}|\nabla u|^2\right] - \nabla \cdot \left[c^2 \frac{\partial u}{\partial t}\nabla u\right],$$

and integrate it over a 4-D domain bounded by the sides of the cube D, the initial plane $t = 0$, and a 3-D surface S which passes through a point (x_0, y_0, z_0, t_0) (see Fig. 7.4).

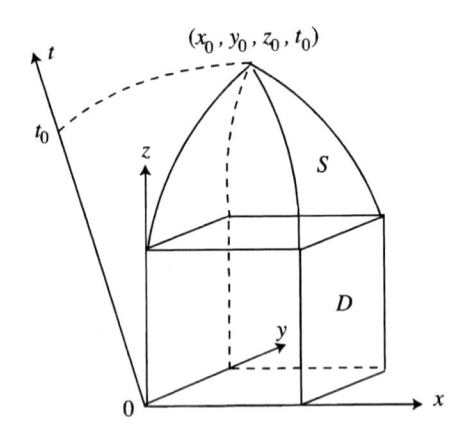

Fig. 7.4. 4-D Cube.

Applying the divergence theorem (see §1.7) and using the conditions in (7.24), we find that for $f = g = 0$,

$$\iiint_S \left\{\frac{1}{2}\left[\left(\frac{\partial u}{\partial t}\right)^2 + c^2 |\nabla u|^2\right] n_i - c^2 \frac{\partial u}{\partial t}\left[\frac{\partial u}{\partial x}n_x + \frac{\partial u}{\partial y}n_y + \frac{\partial u}{\partial z}n_z\right]\right\} dS = 0,$$
$$(7.25)$$

where n_x, n_y, n_z are the components of the upward normal $\mathbf{n}$ to the surface S such that $n_t > 0$ on S. Completing the squares, the integrand in (7.25) becomes

$$\frac{1}{2n_t}\left[\frac{\partial u}{\partial t}n_t - c^2\left(\frac{\partial u}{\partial x}\right)^2\right] + \frac{c^2}{2n_t}\left(n_t^2 - c^2 n^2\right)|\nabla u|^2 +$$
$$\frac{c^4 n^2}{2n_t}\left[\left(\frac{\partial u}{\partial x} - \frac{1}{n^2}\frac{\partial u}{\partial n}n_x\right)^2 + \left(\frac{\partial u}{\partial y} - \frac{1}{n^2}\frac{\partial u}{\partial n}n_y\right)^2 + \left(\frac{\partial u}{\partial z} - \frac{1}{n^2}\frac{\partial u}{\partial n}n_z\right)^2\right],$$

where $\dfrac{\partial u}{\partial n} = \dfrac{\partial u}{\partial x}n_x + \dfrac{\partial u}{\partial y}n_y + \dfrac{\partial u}{\partial z}n_z$ and $n^2 = n_x^2 + n_y^2 + n_z^2$. This integrand is nonnegative if

$$n_t^2 - c^2 n^2 \geq 0. \qquad (7.26)$$

If the surface S has the property (7.26), then (7.25) implies that the integrand is zero. Thus, for the steepest surface S with property (7.26) through the given point

(x_0, y_0, z_0, t_0), we must have

$$n_t^2 - c^2 \left(n_x^2 + n_y^2 + n_z^2 \right) = 0 \quad \text{on } S, \tag{7.27}$$

except at the point (x_0, y_0, z_0, t_0). Thus, the surface S is the right cone with slope $1/c$, and

$$t = t_0 - \frac{1}{c} \sqrt{(x - x_0)^2 + (y - y_0)^2 + (z - z_0)^2}. \tag{7.28}$$

This cone is known as the *characteristic cone* with (x_0, y_0, z_0, t_0) as its vertex. Its cross-section by a plane $t = \text{const}$ is the sphere of radius $c(t - t_0)$ and center (x_0, y_0, z_0). Now, on this cone S Eq (7.25) becomes

$$\iiint_S \frac{1}{2n_t} \left\{ \left[\frac{\partial u}{\partial t} n_t - c^2 \frac{\partial u}{\partial n} \right]^2 + c^4 n^2 \left[\left(\frac{\partial u}{\partial x} - \frac{n_x}{n^2} \frac{\partial u}{\partial n} \right)^2 + \right. \right.$$
$$\left. \left. \left(\frac{\partial u}{\partial y} - \frac{n_y}{n^2} \frac{\partial u}{\partial n} \right)^2 + \left(\frac{\partial u}{\partial z} - \frac{n_z}{n^2} \frac{\partial u}{\partial n} \right)^2 \right] \right\} dS = 0.$$

Since the integrand is the sum of continuous nonnegative functions, each term must vanish everywhere on S. Also, since $n_t - c^2 n^2 = 0$ and $n_t^2 + n^2 = 1$, we find from the first term (which vanishes) that

$$\frac{\partial u}{\partial t} = c \sqrt{1 + c^2} \left(\frac{\partial u}{\partial x} n_x + \frac{\partial u}{\partial y} n_y + \frac{\partial u}{\partial z} n_z \right) \quad \text{on } S.$$

Since at the vertex (x_0, y_0, z_0, t_0) of the cone S the spatial part (n_x, n_y, n_z) of the outward normal has all possible limiting directions, and therefore no fixed direction, we conclude that $\dfrac{\partial u}{\partial x} = \dfrac{\partial u}{\partial y} = \dfrac{\partial u}{\partial z} = 0$ at (x_0, y_0, z_0, t_0). Again, repeating this argument at each point (x_0, y_0, z_0, t), $t \leq t_0$, we conclude that $u(x_0, y_0, z_0, t_0) = 0$. Hence, the solution u at the vertex (x_0, y_0, z_0, t_0) is uniquely determined by the boundary conditions imposed below and on the characteristic cone S through (x_0, y_0, z_0, t_0). Any surface whose outward normal satisfies the condition (7.27) is known as the *characteristic surface*.

By using Bernoulli's separation method it can be shown that the formal solution of this problem is

$$u(x, y, z, t) = \sum_{l=1}^{\infty} \sum_{m=1}^{\infty} \sum_{n=1}^{\infty} \left[C_{lmn} \cos \sqrt{l^2 + m^2 + n^2} \, ct + \right.$$
$$\left. D_{lmn} \frac{\sin \sqrt{l^2 + m^2 + n^2} \, ct}{\sqrt{l^2 + m^2 + n^2}} \right] \sin lx \, \sin my \, \sin nz, \tag{7.29}$$

where

$$C_{lmn} = \frac{8}{\pi^3} \int_0^\pi \int_0^\pi \int_0^\pi f(x, y, z) \sin lx \, \sin my \, \sin nz \, dx \, dy \, dz,$$

$$D_{lmn} = \frac{8}{\pi^3} \int_0^\pi \int_0^\pi \int_0^\pi g(x, y, z) \sin lx \, \sin my \, \sin nz \, dx \, dy \, dz.$$

7.7. Schrödinger Wave Equation

According to de Broglie's postulation[1] 'every corpuscular motion can be treated as a wave motion such that the wavelength λ is inversely proportional to the product of the electronic mass m and velocity c of a particle', i.e., $\lambda = \dfrac{h}{mc}$, where h is the Plank's constant (see §6.4). The total energy E of the particle is related to the wave frequency ν by $E = h\nu$. Since the kinetic energy is given by $E = mc^2/2$, the potential energy V is related to E by $E = V + mc^2/2$, or

$$h\nu = V + \frac{h^2}{2m\lambda^2}. \tag{7.30}$$

Consider a wave function U defined by

$$U(x,t) = A\,\cos 2\pi\left(\frac{x}{\lambda} - \nu t - \tau\right),$$

where τ denote the phase shift such that it is a phase lag if $t \geq -2\pi + \tau$ and a phase advance if $t \leq 2\pi + \tau$. Then

$$U_t = 2\pi\nu A\,\sin 2\pi\left(\frac{x}{\lambda} - \nu t - \tau\right), \quad U_{tt} = -4\pi^2\nu^2 A\cos 2\pi\left(\frac{x}{\lambda} - \nu t - \tau\right),$$

$$U_x = -\frac{2\pi}{\lambda}A\sin 2\pi\left(\frac{x}{\lambda} - \nu t - \tau\right), \quad U_{xx} = -\frac{4\pi^2}{\lambda^2}A\cos 2\pi\left(\frac{x}{\lambda} - \nu t - \tau\right),$$

which gives

$$U_{tt} = -4\pi^2\nu^2 U, \quad U_{xx} = -\frac{4\pi^2}{\lambda^2}U.$$

Hence,

$$h\nu U = -\frac{h}{4\pi^2\nu}U_{tt}, \tag{7.31}$$

$$\frac{h^2}{4m\lambda^2}U = -\frac{h^2}{8\pi^2 m}U_{xx}. \tag{7.32}$$

Multiplying Eq (7.30) by U and substituting into Eqs (7.31) and (7.32), we get

$$\frac{\partial^2 U}{\partial x^2} - \frac{8\pi^2 m}{h^2}V U = \frac{2}{\lambda^2\nu^2}\frac{\partial^2 U}{\partial t^2},$$

which is 1-D Schrödinger's wave equation. In 3-D this equation becomes

$$\nabla^2 U - \frac{8\pi^2 m}{h^2}V U = \frac{2}{\lambda^2\nu^2}\frac{\partial^2 U}{\partial t^2}. \tag{7.33}$$

---

[1] As mentioned in Sagan [1989:292], this postulation was made by de Broglie in 1924 and provided a basis for quantum mechanics.

Now, let $U = A \cos 2\pi \left(\dfrac{x}{\lambda} - \nu t - \tau \right) u(x, y, z)$. Then Eq (7.33) gives

$$\nabla^2 u + 8\pi^2 \left(\frac{1}{\lambda^2} - \frac{m}{h^2} V \right) u = 0.$$

Since $E = \dfrac{h^2}{m\lambda^2}$, or $\dfrac{1}{\lambda^2} = \dfrac{mE}{h^2}$, the above equation becomes

$$\nabla^2 u + \frac{8\pi^2 m}{h^2} (E - V) u = 0, \tag{7.34}$$

where $u(x, y, z) \equiv u(\mathbf{x}) \to 0$ as $\cos |\mathbf{x}| \to \infty$ such that $\displaystyle\iiint_{-\infty}^{\infty} u^2 \, dx \, dy \, dz < +\infty$,
and the norm $\|u\|$ is defined by $\displaystyle\iiint_{-\infty}^{\infty} u^2(\mathbf{x}) \, dx \, dy \, dz = 1,$ (or $\displaystyle\iiint_{-\infty}^{\infty} |u|^2 \, dx \, dy \, dz$
$= 1$ in the case when u is complex). This condition can be expressed in spherical coordinates as an iterated integral

$$\int_0^r r^2 R^2 \, dr \int_0^{2\pi} \Phi^2 \, d\phi \int_0^{\pi} \Theta^2 \sin\theta \, d\theta = 1. \tag{7.35}$$

The function u is sometimes called a *probability wave*; that is, it is a function for which $\displaystyle\iiint_{-\infty}^{\infty} u^2 \, dx \, dy \, dz$ represents the probability that a particle is found in the volume element $dx \, dy \, dz$.

7.8. Hydrogen Atom

The hydrogen atom consists of one nucleus of charge e and one electron of charge $-e$ with the electronic mass m. Hence, the potential energy is $V = -e^2/r$, where r is the distance between nucleus and electron. Thus, Eq (7.34) becomes

$$\nabla^2 u + \frac{8\pi^2 m}{h^2} \left(h\nu + \frac{e^2}{r} \right) u = 0, \tag{7.36}$$

where we have chosen spherical coordinates such that the nucleus is at the origin, i.e., the source point is $(r', \theta', \phi') = \mathbf{0}$. The spherical coordinate system (see Fig. 7.5) is defined by $x = r \sin\theta \cos\phi$, $y = r \sin\theta \sin\phi$, $z = r \cos\theta$. Using Bernoulli's separation method, with $u = R(r)\Theta(\theta)\Phi(\phi)$, where $r > 0, 0 \le \theta \le \pi, 0 \le \phi < 2\pi$, Eq (7.36) becomes

$$\frac{R'' + 2R'/r}{R} + \frac{\cot\theta}{r^2} \frac{\Theta'}{\Theta} + \frac{1}{r^2} \frac{\Theta''}{\Theta} + \frac{1}{r^2 \sin^2\theta} \frac{\Phi''}{\Phi} + \frac{8\pi^2 m}{h^2} \left(h\nu + \frac{e^2}{r} \right) = 0,$$

or

$$\frac{r^2 R''}{R} + \frac{2r\, R'}{R} + r^2 \frac{8\pi^2 m}{h^2}\left(h\nu + \frac{e^2}{r}\right) = -\cot\theta\,\frac{\Theta'}{\Theta} - \frac{\Theta''}{\Theta} - \frac{1}{\sin^2\theta}\frac{\Phi''}{\Phi}. \quad (7.37)$$

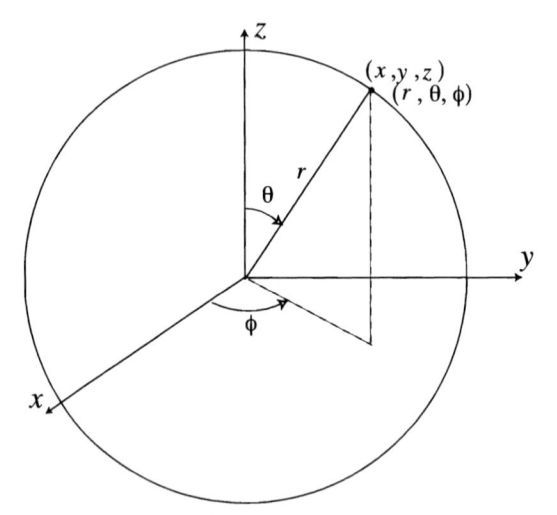

Fig. 7.5 Spherical Coordinates.

Since both sides of this equation must be equal to a constant, which we take as κ, we get

$$r^2 R'' + 2r\,R' - \kappa\,R + \frac{8\pi^2 m}{h^2}r^2\left(h\nu + \frac{e^2}{r}\right)R = 0, \quad (7.38)$$

To separate equations in Θ and Φ, we take the constant as $-\mu^2$. Then the right side of Eq (7.37) gives two equations:

$$\sin^2\theta\,\Theta'' + \sin\theta\cos\theta\,\Theta' + \kappa\sin^2\theta\,\Theta - \mu^2\Theta = 0, \quad (7.39)$$

$$\Phi'' + \mu^2\Phi = 0, \quad (7.40)$$

To solve Eq (7.39), let $t = \cos\theta$, $\Theta(\theta) = P(t)$, and $\kappa = n(n+1)$. Then this equation becomes

$$\frac{d}{dt}\left[(1-t^2)\frac{dP}{dt}\right] + \left[n(n+1) - \frac{\mu^2}{1-t^2}\right]P = 0. \quad (7.41)$$

This equation would be Legendre's equation if it were not for the term $\dfrac{\mu^2}{1-t^2}$. The solutions Eq (7.41) are associated Legendre polynomials $\Theta = P_n^\mu(\cos\theta)$, where $P_n^\mu(t)$, and $\mu = 0, \pm1, \pm2, \dots$. For associated Legendre's equation and its properties, see §C.2. The solution of Eq (7.40) are $\Phi = A\cos\mu\phi + B\sin\mu\phi$, where ϕ is a periodic function of period 2π, i.e., $\Phi(\phi) = \Phi(2\pi + \phi)$. Now, to solve Eq (7.38), first we divide it by r^2, and get

$$R'' + \frac{2}{r} R' + \left[a + \frac{b}{r} - \frac{n(n+1)}{r^2} \right] R = 0,$$

where $a = \dfrac{8\pi^2 m\nu}{h}$ and $b = \dfrac{ae^2}{h\nu}$. Introducing a new variable $\rho = 2r\sqrt{-a}$, and taking $K = \dfrac{b}{2\sqrt{-a}} = \dfrac{2\pi me^2}{h\sqrt{-2mh\nu}}$, the above equation becomes

$$\frac{d^2R}{d\rho^2} + + \frac{2}{\rho} \frac{dR}{d\rho} - \left(\frac{1}{4} - \frac{K}{\rho} + \frac{n(n+1)}{\rho^2} \right) R = 0,$$

or

$$\frac{d}{d\rho} \left(\rho^2 R' \right) - \frac{\rho^2 - 4K\rho + 4n(n+1)}{4} R = 0. \tag{7.42}$$

This equation is closely related to Laguerre's equation, and its solution is

$$R(\rho) = \rho^n\, e^{-\rho/2}\, L_{K+n}^{2n+1}(\rho), \tag{7.43}$$

where $n \in \mathbb{Z}$, K is any number, and $L_k^\sigma(\rho)$ are the generalized Laguerre polynomials defined by $L_k^\sigma(\rho) = \dfrac{\rho^{-\sigma}\, e^{\rho}}{k!} \dfrac{d^k}{d\rho^k} \left(\rho^{n+\sigma}\, e^{-\rho} \right)$ (for Laguerre polynomials and associated Laguerre polynomials, see §C.4 and §C.5). If we impose the condition that K be an integer N, then we find that $E = -\dfrac{2\pi^2 m\, e^4}{h^2 N^2}$, $N = 1, 2, \ldots$, are the possible values of the kinetic energy such that these values represent a discrete energy spectrum with a limit point at $E = 0$. For $N = 1$ we get a *normal quantum state* (or *ground state*, as it is commonly called) of the hydrogen atom, which is the most stable state occupied by the spectrum. For $N > 1$, we have a manifold of higher quantum states, described as follows: Assuming that $m \le k$, we have $2n + 1 \le N + n$, that is, $n \le N - 1$, which means that for every N there are N possible values of n, namely $n = 0, 1, 2, \ldots, N - 1$. Since $\mu \le n$, we will have $\mu = 0, 1, 2, \ldots, n$ for every n. Hence, for every N there are $\dfrac{N(N+1)}{2}$ independent quantum states. Hence, the solutions of Eq (7.37) are given by

$$u_{N,n\mu}(r, \theta, \phi) = \frac{h^2 N}{4\pi^2 m\, e^2}\, r^2\, P_n^\mu(\cos\theta)\, \exp\left[-\frac{2\pi^2 m\, e^2}{h^2 N} r \right] \times$$

$$\times\, L_{N+n}^{2n+1} \left(\frac{4\pi^2 m\, e^2}{h^2 N} r \right) (A\, \cos\mu\phi + B\, \sin\mu\phi), \tag{7.44}$$

where the constants A and B can be determined from the boundary conditions (to be provided), and N is called the *principal quantum number*, n the *azimuthal quantum number*, and μ the *magnetic quantum number*.

7.8.1. Harmonic Oscillator. For a vibrating object the governing equation involves the potential energy V which is defined by $v = \frac{1}{2}kt^2 = \frac{1}{2}m\omega^2 t^2$, where ω is the angular frequency of vibrations $(= \sqrt{k/m})$, k being the spring constant or force unit. The classical vibration frequency is $\nu = \dfrac{1}{2\pi}\sqrt{k/m}$, which gives $\omega = 2\pi\nu$. The total energy is the sum of the potential and kinetic energies. We will now find the form of the Hamiltonian operator. The kinetic energy T can be written in the operator form as $T = -\dfrac{\hbar^2}{2m}\dfrac{d^2}{dt^2}$, whereas the potential energy is $2\pi^2\nu^2 mt^2$. Thus, the Hamiltonian operator is

$$\mathbf{H} = -\frac{\hbar^2}{2m}\frac{d^2}{dt^2} + 2\pi^2\nu^2 mt^2 = -\frac{\hbar^2}{2m}\frac{d^2}{dt^2} + \frac{1}{2}m\omega^2 t^2. \tag{7.45}$$

If we set $b = \dfrac{2\pi\nu m}{\hbar}$, then Schrödinger's equation is $\mathbf{H}u = Eu$, or

$$\frac{d^2u}{dt^2} + \left(\frac{2mE}{\hbar^2} - \frac{m\omega}{\hbar}^2 t^2\right) u = 0. \tag{7.46}$$

This equation is nonlinear as the potential varies with t^2. We seek a solution of the form

$$u(t) = c\,e^{-bt^2}, \tag{7.47}$$

where b and c are constants. Then since

$$\frac{du}{dt} = -2bct\,e^{-bt^2}, \text{ and } \frac{d^2u}{dt^2} = \left(-2bc + 4b^2ct^2\right)e^{-bt^2},$$

the term

$$\left(\frac{2mE}{\hbar^2} - \frac{m\omega}{\hbar}^2 t^2\right) = \left(-\frac{2mcE}{\hbar^2} + \frac{m^2c\omega^2}{\hbar^2}t^2\right)e^{-bt^2}.$$

Substituting the value of this term and that of $\dfrac{d^2}{dt^2}$ in Eq (7.46) and equating the terms in t^2, we obtain $\dfrac{m^2c\omega^2}{\hbar^2}t^2\,e^{-bt^2} = 4b^2ct^2\,e^{-bt^2}$, which yields $b = \dfrac{m\omega}{2\hbar}$. Again, equating the constant terms in Eq (7.46) we find that $E = b\dfrac{\hbar^2}{m}$, which gives $E = \frac{1}{2}\omega\hbar^2$. Hence, $u(t) = ce^{-m\omega t^2/(2\hbar)}$, which is the solution of the Schrödinger equation (7.46) for the harmonic oscillator in its lowest energy state.

7.9. 1-D Fractional Nonhomogeneous Wave Equation

Consider the boundary value problem

$$\frac{\partial^p u}{\partial t^p} - c^2\frac{\partial^2 u}{\partial x^2} = F(x,t), \quad x \in \mathbb{R}, \ t > 0,$$
$$u(x,0) = f(x), \quad u_t(x,0) = g(x), \quad x \in \mathbb{R}, \tag{7.48}$$

where c is the wave constant and $1 < p \leq 2$ (for fractional derivatives, see Appendix E). Applying the Laplace transform followed by the Fourier transform to the problem (7.48), we get

$$\widetilde{\widetilde{u}}(\alpha, s) = \frac{\widetilde{f}(\alpha)\, s^{p-1}}{s^p + c^2\alpha^2} + \frac{\widetilde{g}(\alpha)\, s^{p-2}}{s^p + c^2\alpha^2} + \frac{\widetilde{\overline{F}}(\alpha, s)}{s^p + c^2\alpha^2}, \tag{7.49}$$

where s is the variable of the Laplace transform and α the variable of the Fourier transform. First, inverting the Laplace transform we obtain

$$\widetilde{u}(\alpha, t) = \widetilde{f}(\alpha)\, \mathcal{L}^{-1}\left\{\frac{s^{p-1}}{s^p + c^2\alpha^2}\right\} + \widetilde{g}(\alpha)\, \mathcal{L}^{-1}\left\{\frac{s^{p-2}}{s^p + c^2\alpha^2}\right\}$$

$$+ \mathcal{L}^{-1}\left\{\frac{\widetilde{\overline{F}}(\alpha, s)}{s^p + c^2\alpha^2}\right\},$$

which, in view of the formula (6.59), gives

$$\widetilde{u}(\alpha, t) = \widetilde{f}(\alpha)\, E_{p,1}\left(-c^2\alpha^2 t^p\right) + \widetilde{g}(\alpha)\, E_{p,2}\left(-c^2\alpha^2 t^p\right)$$

$$+ \int_0^t \widetilde{F}(\alpha, t-\tau)\, \tau^{p-1}\, E_{p,p}\left(-c^2\alpha^2 t^p\right)\, d\tau, \tag{7.50}$$

where $E_{p,1}$ is the Mittag-Leffler function defined by (6.60). Next, taking inverse Fourier transform, we get

$$u(x, t) = \frac{1}{\sqrt{2\pi}} \int_{-\infty}^{\infty} \widetilde{f}(\alpha)\, E_{p,1}\left(-c^2\alpha^2 t^p\right)\, e^{i\,\alpha x}\, d\alpha$$

$$+ \frac{1}{\sqrt{2\pi}} \int_{-\infty}^{\infty} \widetilde{g}(\alpha)\, t\, E_{p,2}\left(-c^2\alpha^2 t^p\right)\, e^{i\,\alpha x}\, d\alpha$$

$$+ \frac{1}{\sqrt{2\pi}} \int_0^t \tau^{p-1}\, d\tau \int_{-\infty}^{\infty} \widetilde{F}(\alpha, t-\tau)\, E_{p,p}\left(-c^2\alpha^2 t^p\right)\, e^{i\,\alpha x}\, d\alpha. \tag{7.51}$$

In the case when $p = 2$, the fractional wave equation in (7.48) reduces to the classical 1-D wave equation discussed in §7.1. In this case since

$$E_{2,1}\left(-c^2\alpha^2 t^p\right) = \cosh\left(i\,c\alpha t\right) = \cos(c\alpha t),$$

$$t E_{2,2}\left(-c^2\alpha^2 t^p\right) = \frac{t \sinh(i\,c\alpha t)}{i\,c\alpha t} = \frac{1}{c\alpha}\sin(c\alpha t),$$

the solution (7.51) reduces to

$$u(x, t) = \frac{1}{\sqrt{2\pi}} \int_{-\infty}^{\infty} \widetilde{f}(\alpha)\, \cos(c\alpha t)\, e^{i\,\alpha x}\, d\alpha$$

$$+ \frac{1}{\sqrt{2\pi}} \int_{-\infty}^{\infty} \widetilde{g}(\alpha)\, \frac{\sin(c\alpha t)}{c\alpha}\, e^{i\,\alpha x}\, d\alpha$$

$$+ \frac{1}{c\sqrt{2\pi}} \int_0^t d\tau \int_{-\infty}^{\infty} \tilde{F}(\alpha, \tau) \frac{\sin c\alpha(t-\tau)}{\alpha} e^{i\alpha x} d\alpha$$

$$= \frac{1}{2} |f(x-ct) + f(x+ct)| + \frac{1}{2c} \int_{x-ct}^{x+ct} g(y) \, dy$$

$$+ \frac{1}{2c} \int_0^t d\tau \int_{x-c(t-\tau)}^{x+c(t-\tau)} F(y, \tau) \, dy. \tag{7.52}$$

If we take $c^2 = k$ and $g(x) = 0$ in the problem (7.48), then it reduces to 1-D fractional diffusion equation discussed in §6.10, such that Eq (7.49) reduces to

$$\widetilde{\bar{u}}(\alpha, s) = \frac{\tilde{f}(\alpha) \, s^{p-1}}{s^p + k\alpha^2} + \frac{\widetilde{\bar{F}}(\alpha, s)}{s^p + k\alpha^2},$$

which, in view of formula (6.59), gives

$$\tilde{u}(\alpha, s) = \tilde{f}(\alpha) E_{\alpha,1} \left(-k\alpha^2 t^p\right)$$

$$+ \int_0^1 (t-\tau)^{p-1} E_{p,p} \left(-k\alpha^2 (t-\tau)^p\right) \tilde{F}(\alpha, t) \, d\tau.$$

Taking the inverse Fourier transform of this result and applying convolution theorem of the Fourier transform, we get

$$u(x,t) = \frac{1}{\sqrt{2\pi}} \int_{-\infty}^{\infty} \tilde{f}(\alpha) E_{p,1} \left(-k\alpha^2 t^p\right) e^{i\alpha x} d\alpha$$

$$+ \frac{1}{\sqrt{2\pi}} \int_0^t d\tau \int_{-\infty}^{\infty} \tilde{F}(\alpha, \tau) (t-\tau)^{p-1} E_{p,p} \left(-k\alpha^2 (t-\tau)^p\right) e^{i\alpha x} d\alpha$$

$$= \int_{-\infty}^{\infty} G_1 (x-x', t) \, f(x') \, dx'$$

$$+ \int_0^t (t-\tau)^{p-1} d\tau \int_{-\infty}^{\infty} G_2 (x-x', t-\tau) \, F(x', \tau) \, dx', \tag{7.53}$$

where

$$G_1(x,t) = \frac{1}{2\pi} \int_{-\infty}^{\infty} E_{p,1} \left(-k\alpha^2 t^p\right) e^{i\alpha x} d\alpha, \tag{7.54a}$$

$$G_2(x,t) = \frac{1}{2\pi} \int_{-\infty}^{\infty} E_{p,p} \left(-k\alpha^2 t^p\right) e^{i\alpha x} d\alpha. \tag{7.54b}$$

For $\alpha = 1$, the solution (7.53) reduces to the classical solution:

$$u(x,t) = \int_{-\infty}^{\infty} G_1 (x-x', t-\tau) f(x') \, dx'$$

$$+ \int_0^t d\tau \int_{-\infty}^{\infty} G_2 (x-x', t-\tau) F((x', \tau) \, dx', \tag{7.55}$$

where
$$G_1(x,t) = G_2(x,t) = \frac{1}{2\sqrt{\pi kt}}\, e^{-x^2/(4kt)}. \tag{7.56}$$

7.10. Applications of the Wave Operator

We will discuss some examples which exhibit the importance of Green's function in certain physical problems.

7.10.1. Cauchy Problem for the 2-D and 3-D Wave Equation.

The Cauchy problem for this wave equation is governed by

$$u_{tt} = c^2 \left(u_{xx} + u_{yy} \right), \quad -\infty < x, y < \infty,\ t > 0,$$

subject to the initial conditions $u(x,y,0) = 0$, $u_t(x,y,0) = f(x,y)$ for $-\infty < x, y < \infty$, where u and its first partial derivatives vanish at infinity, and c is a constant. We apply the 2-D Fourier transform to this system, and get

$$\frac{d^2\widetilde{u}}{dt^2} + c^2\sigma^2\widetilde{u} = 0, \quad \text{where } \sigma^2 = \alpha^2 + \beta^2,$$

$$\widetilde{u}(\alpha,\beta,0) = 0, \quad \left.\frac{d\widetilde{u}}{dt}\right|_{t=0} = \widetilde{f}(\alpha,\beta).$$

The solution of this system in the transform domain is

$$\widetilde{u}(\alpha,\beta,t) = \widetilde{f}(\alpha,\beta)\,\frac{\sin(c\,\sigma t)}{c\,\sigma},$$

which on inversion gives the solution

$$\begin{aligned}
u(x,y,t) &= \frac{1}{2\pi c}\int_{-\infty}^{\infty}\int_{-\infty}^{\infty} \widetilde{f}(\boldsymbol{\sigma})\,\frac{\sin(c\,\sigma t)}{c\,\sigma}\, e^{i\boldsymbol{\sigma}\cdot\mathbf{r}}\, d\boldsymbol{\sigma}\\
&= \frac{1}{4\,i\,\pi c}\int_{-\infty}^{\infty}\int_{-\infty}^{\infty}\frac{\widetilde{f}(\boldsymbol{\sigma})}{\sigma}\Bigg[\exp\Big\{i\,\sigma\,\Big(\frac{\boldsymbol{\sigma}\cdot\mathbf{r}}{\sigma}+ct\Big)\Big\}-\\
&\qquad\qquad - \exp\Big\{i\,\sigma\,\Big(\frac{\boldsymbol{\sigma}\cdot\mathbf{r}}{\sigma}-ct\Big)\Big\}\Bigg]\,d\boldsymbol{\sigma},
\end{aligned} \tag{7.57}$$

where $\boldsymbol{\sigma} = (\alpha,\beta)$ are the variables of the double Fourier transform and $\mathbf{r} = (x,y)$ such that $\boldsymbol{\sigma}\cdot\mathbf{r} = \alpha x + \beta y$. In this solution the terms $\exp\{i\,(\boldsymbol{\sigma}\cdot\mathbf{r}\pm\gamma ct)$ represent, for any given $\boldsymbol{\sigma}$, the *plane traveling wave solution* of the wave equation.

The solution of the Cauchy problem for the 3-D wave equation

$$u_{tt} = c^2 \left(u_{xx} + u_{yy} + u_{zz} \right), \quad -\infty < x, y, z < \infty,\ t > 0,$$

and subject to the initial conditions $u(x,y,z,0) = 0$, $u_t(x,y,z,0) = f(x,y,z)$,

$-\infty < x, y, z < \infty$, can be similarly found as

$$u(x, y, z, t) = \frac{1}{4\pi\, i\, c\sqrt{2\pi}} \int_{-\infty}^{\infty}\int_{-\infty}^{\infty}\int_{-\infty}^{\infty} \frac{\widetilde{f}(\zeta)}{\zeta}\left[\exp\left\{ i\zeta\left(\frac{\zeta\cdot\mathbf{r}}{\zeta} + ct \right) \right\} \right.$$
$$\left. - \exp\left\{ i\zeta\left(\frac{\zeta\cdot\mathbf{r}}{\zeta} - ct \right) \right\} \right] d\zeta,$$
$$(7.58)$$

where $\zeta = (\alpha, \beta, \gamma)$ are the variables of the triple Fourier transform and $\mathbf{r} = (x, y, z)$ such that $\zeta \cdot \mathbf{r} = \alpha x + \beta y + \gamma z$.

Let $\mathbf{k}$ denote σ or ζ. Then for any given $\mathbf{k}$ the terms $e^{i\,(\mathbf{k}\cdot\mathbf{r}\pm kct)}$ define a plane traveling wave structure, since these terms remain constant on the planes $\mathbf{k}\cdot\mathbf{r} = \pm kct$ which move normal to themselves in the direction of $\pm\mathbf{k}$ at constant speed c. This wave structure, represented by the solution (7.58), is the sum total of all such plane waves at different wave numbers $\mathbf{k}$.

As a particular case, when $f(x, y, z) = \delta(x)\,\delta(y)\,\delta(z)$, we find that $\widetilde{f}(\mathbf{k}) = (2\pi)^{-3/2}$, and the solution (7.58) becomes

$$u(x, y, z, t) = \frac{1}{8\pi^3} \int_{-\infty}^{\infty}\int_{-\infty}^{\infty}\int_{-\infty}^{\infty} \frac{\sin(k\, ct)}{kc}\, e^{i\,(\mathbf{k}\cdot\mathbf{r})}\, d\mathbf{k},$$

which, by transforming to spherical coordinates (r, θ, ϕ) where the z-axis is taken along the direction of r such that $\mathbf{k} \cdot \mathbf{r} = kr\,\cos\theta$, reduces to

$$\begin{aligned}
u(r, t) &= \frac{1}{8\pi^3} \int_0^{2\pi} d\phi \int_0^{\pi} d\theta \int_0^{\infty} e^{i\,kr\,\cos\theta}\, \frac{\sin ckt}{ck}\, k^2 \sin\theta\, dk \\
&= \frac{1}{2\pi^2 cr} \int_0^{\infty} \sin(ckt)\, \sin(kr)\, dk \\
&= \frac{1}{8\pi^2 cr} \Re\left\{ \int_{-\infty}^{\infty} \left[e^{i\,k(ct-\tau)} - e^{i\,k(ct+\tau)} \right] dk \right\} \\
&= \frac{1}{4\pi cr} \left[\delta(ct - \tau) - \delta(ct + \tau) \right] \\
&= \frac{1}{4\pi cr}\delta(t - \tau) = \frac{1}{4\pi c^2 r}\,\delta\left(t - \frac{\tau}{c} \right),
\end{aligned} \qquad (7.59)$$

where $\delta(ct + \tau) = 0$ since $ct + \tau > 0$ for $t > 0$.

7.10.2. d'Alembert Solution of the Cauchy Problem for the Wave Equation. METHOD 1: (By Bernoulli's separation method) First, we will consider the homogeneous solution of the problem of a vibrating string governed by the 1-D wave equation

$$\frac{\partial^2 u}{\partial t^2} = c^2 \frac{\partial^2 u}{\partial x^2}, \quad 0 < x < l, \qquad (7.60)$$

subject to the boundary and initial conditions

$$u(0,t) = 0 = u(l,t), \quad t > 0,$$
$$u(x,0) = f(x), \quad u_t(x,0) = g(x), \quad 0 < x < l,$$

where $f \in C^1(0,l)$ is a given function. For a complete solution see Example 4.1. ∎

METHOD 2: (By integral transform method) Let us now consider the nonhomogeneous Cauchy problem

$$u_{tt} - c^2 u_{xx} = f(x,t), \quad -\infty < x < \infty, \ t > 0,$$
$$u(x,0) = g(x), \quad u_t(x,0) = h(x), \quad -\infty < x < \infty, \tag{7.61}$$

where $f(x,t)$ represents the source term. Applying the Fourier transform, followed by the Laplace transform, to this system, we get

$$\overline{\widetilde{u}}(\alpha, s) = \frac{s\,\widetilde{g}(\alpha) + \widetilde{h}(\alpha) + \overline{\widetilde{f}}(\alpha, s)}{s^2 + c^2\alpha^2}.$$

The inverse Laplace transform gives

$$\widetilde{u}(\alpha, t) = \widetilde{g}(\alpha)\cos(c\alpha t) + \frac{\widetilde{h}(\alpha)}{c\alpha}\sin(c\alpha t) + \frac{1}{c\alpha}\mathcal{L}^{-1}\left\{\frac{c\alpha}{s^2 + c^2\alpha^2}\,\overline{\widetilde{f}}(\alpha, s)\right\}$$

$$= \widetilde{g}(\alpha)\cos(c\alpha t) + \frac{\widetilde{h}(\alpha)}{c\alpha}\sin(c\alpha t) + \frac{1}{c\alpha}\int_0^t \sin c\alpha(t-\tau)\,\widetilde{f}(\alpha, \tau)\,d\tau.$$

which, after inverting the Fourier transform, gives the solution as

$$u(x,t) = \frac{1}{2\sqrt{2\pi}}\int_{-\infty}^{\infty}\left(e^{i\,c\alpha t} + e^{-i\,c\alpha t}\right)e^{i\,\alpha x}\,\widetilde{g}(\alpha)\,d\alpha$$

$$+ \frac{1}{2\sqrt{2\pi}}\int_{-\infty}^{\infty}\left(e^{i\,c\alpha t} - e^{-i\,c\alpha t}\right)e^{i\,\alpha x}\,\frac{\widetilde{h}(\alpha)}{i\,c\alpha}\,d\alpha$$

$$+ \frac{1}{2c\sqrt{2\pi}}\int_0^t d\tau\int_{-\infty}^{\infty}\frac{e^{i\,c\alpha(t-\tau)} + e^{-i\,c\alpha(t-\tau)}}{i\,\alpha}\,e^{i\,\alpha x}\,\widetilde{f}(\alpha, \tau)\,d\alpha$$

$$= \frac{1}{2}\left[g(x+ct) + g(x-ct)\right] + \frac{1}{2c}\int_{x-ct}^{x+ct} h(y)\,dy$$

$$+ \frac{1}{2c\sqrt{2\pi}}\int_0^t d\tau\int_{-\infty}^{\infty}\widetilde{f}(\alpha, \tau)\,d\alpha\int_{x-c(t-\tau)}^{x+c(t-\tau)} e^{i\,\alpha y}\,dy$$

$$= \frac{1}{2}\left[g(x+ct) + g(x-ct)\right] + \frac{1}{2c} + \frac{1}{2c}\int_{x-ct}^{x+ct} h(y)\,dy$$

$$+ \frac{1}{2c}\int_0^t d\tau\int_{x-c(t-\tau)}^{x+c(t-\tau)} f(y, \tau)\,dy. \ ∎ \tag{7.62}$$

7.10.3. Free Vibration of a Large Circular Membrane. To find the displacement function for the membrane, we consider the axisymmetric wave equation

$$u_{rr} + \frac{1}{r} u_r = c^2 u_{tt}, \quad 0 \le r < \infty, \ t > 0,$$

$$u(r, 0) = f(r), \quad u_t(r, 0) = g(r), \quad 0 \le r < \infty,$$

where f and g are arbitrary functions, and $c^2 = T/\rho = \text{const}$, T being the tension in the membrane, and ρ the surface density of the membrane. Applying the zero-order Hankel transform to this system we get

$$\widehat{u}(\sigma, t) = \widehat{f}(\sigma) \cos(c\sigma t) + \frac{\widehat{g}(\sigma)}{c\sigma} \sin(c\sigma t),$$

which on inversion gives the solution as

$$u(r, t) = \int_0^\infty y \widehat{f}(y) \cos(cyt) J_0(yr) \, dy + \frac{1}{c} \int_0^\infty y \widehat{g}(y) \sin(cyt) J_0(yr) \, dy. \ \blacksquare$$

$$(7.63)$$

7.10.4. Hyperbolic or Parabolic Equations in Terms of Green's Functions. To solve the wave equation in terms of Green's function, consider

$$u_{tt} - c^2 \nabla^2 u = f(\mathbf{x}, t) \quad \text{in } D, \tag{7.64}$$

subject to the conditions $u(\mathbf{x}, 0) = 0 = u_t(\mathbf{x}, 0)$ and $u(\mathbf{x}, t) = 0$ on the boundary ∂D. The Green's function $G(\mathbf{x}, t; \mathbf{x}', t')$ for this problem satisfies the equation

$$G_{tt} - c^2 \nabla^2 G = \delta(\mathbf{x} - \mathbf{x}') \, \delta(t - t') \quad \text{in } D, \tag{7.65}$$

and the conditions $G(\mathbf{x}, \mathbf{x}'; 0) = 0 = G_t(\mathbf{x}, \mathbf{x}'; 0)$ and $G(\mathbf{x}, t; \mathbf{x}', t') = 0$ on ∂D. Multiplying Eq (7.64) by G and Eq (7.65) by u and subtracting, we get

$$G u_{tt} - c^2 G \nabla^2 u - u G_{tt} + c^2 u \nabla^2 G = G f(\mathbf{x}, t) - u \delta(\mathbf{x} - \mathbf{x}') \, \delta(t - t').$$

Integrating this equation with respect to t from 0 to t', we have

$$\int_0^{t'} \left[(G u_{tt} - u G_{tt}) + c^2 \left(u \nabla^2 G - G \nabla^2 u \right) \right] dt$$

$$= \int_0^{t'} \left[G f(\mathbf{x}, t) - u \delta(\mathbf{x} - \mathbf{x}') \, \delta(t - t') \right] dt,$$

which gives

$$(G u_t - u G_t) \Big|_0^{t'} - \int_0^{t'} \left[(G_t u_t - u_t G_t) - c^2 \left(u \nabla^2 G - G \nabla^2 u \right) \right] dt$$

$$= \int_0^{t'} \left[G f(\mathbf{x}, t) - u \delta(\mathbf{x} - \mathbf{x}') \, \delta(t - t') \right] dt.$$

Since, in view of the boundary and initial conditions, $u(\mathbf{x}, 0) = 0 = u_t(\mathbf{x}, 0)$, and $G(\mathbf{x}, \mathbf{x}'; 0) = G(\mathbf{x}, \mathbf{x}'; t', t') = 0 = G_t(\mathbf{x}, \mathbf{x}'; 0) = G_t(\mathbf{x}, \mathbf{x}'; t', t')$, we find that

$$c^2 \int_0^{t'} \left(u \nabla^2 G - G \nabla^2 u \right) dt = \int_0^{t'} G f(\mathbf{x}, t) \, dt - u(\mathbf{x}, t') \, \delta(\mathbf{x} - \mathbf{x}').$$

Integrating this equation over the region D, we obtain

$$c^2 \iiint_D \int_0^{t'} \left(u \nabla^2 G - G \nabla^2 u \right) dt \, d\mathbf{x}$$

$$= \iiint_D \int_0^{t'} G f(\mathbf{x}, t) \, dt - \iiint_D u(\mathbf{x}, t') \, \delta(\mathbf{x} - \mathbf{x}') \, d\mathbf{x}.$$

Applying Green's second identity (1.16) after interchanging the time integral with the space integral, we get

$$c^2 \int_0^{t'} \iint_{\partial D} \left(u \frac{\partial G}{\partial n} - G \frac{\partial u}{\partial n} \right) dS \, dt = \int_0^{t'} \iiint_D G f(\mathbf{x}, t) \, d\mathbf{x} \, dt - u(\mathbf{x}', t').$$

Since u and G both vanish on the boundary ∂D, we have

$$u(\mathbf{x}', t') = \int_0^{t'} \iiint_D G(\mathbf{x}, \mathbf{x}'; t, t') \, f(\mathbf{x}, t) \, d\mathbf{x} \, dt.$$

Interchanging $\mathbf{x}$ and t with $\mathbf{x}'$ and t', respectively, and noting the symmetry of $G(\mathbf{x}, \mathbf{x}'; t, t')$ with respect to $\mathbf{x}$ and $\mathbf{x}'$ and with respect to t and t', we obtain

$$u(\mathbf{x}, t) = \int_0^t \iiint_{D'} G(\mathbf{x}, \mathbf{x}'; t, t') \, f(\mathbf{x}', t') \, d\mathbf{x}' \, dt', \qquad (7.66)$$

where D' is the domain of integration in the $\mathbf{x}'$ coordinate system. For the general problem of Eq (7.64), subject to the conditions

$$u(\mathbf{x}, 0) = \phi(\mathbf{x}), \quad u_t(\mathbf{x}, 0) = \psi(x), \quad \alpha \frac{\partial u}{\partial n} + \beta u(\mathbf{x}, t) = g(\mathbf{x}, t) \quad \text{on } \partial D',$$

the solution is given by

$$u(\mathbf{x}, t) = \int_0^t \iiint_{D'} G(\mathbf{x}, \mathbf{x}'; t, t') \, f(\mathbf{x}', t') \, d\mathbf{x}' \, dt'$$

$$+ \iiint_{D'} \left[\psi(\mathbf{x}') G(\mathbf{x}, \mathbf{x}'; t, 0) - \phi(\mathbf{x}') \frac{\partial G(\mathbf{x}, \mathbf{x}'; t, 0)}{\partial t'} \right] d\mathbf{x}'$$

$$+ \frac{c^2}{\alpha} \int_0^t \iint_{\partial D'} G(\mathbf{x}, \mathbf{x}'; t, t') \, g(\mathbf{x}', t') \, dS \, dt', \qquad \alpha \neq 0,$$

or $\quad u\left(\mathbf{x}, t\right) = \int_0^t \iiint_{D'} G\left(\mathbf{x}, \mathbf{x}'; t, t'\right) f(\mathbf{x}', t')\, d\mathbf{x}'\, dt'$

$$+ \iiint_{D'} \left[\psi(\mathbf{x}')\, G\left(\mathbf{x}, \mathbf{x}'; t, 0\right) - \phi(\mathbf{x}') \frac{\partial G\left(\mathbf{x}, \mathbf{x}'; t, 0\right)}{\partial t'} \right] d\mathbf{x}'$$

$$- \frac{c^2}{\beta} \int_0^{t'} \iint_{\partial D'} \frac{\partial G\left(\mathbf{x}, \mathbf{x}'; t, t'\right)}{\partial n}\, g\left(\mathbf{x}', t'\right)\, dS\, dt', \quad \beta \neq 0$$

$$(7.67)$$

Similarly, for the diffusion problem $u_t - k \nabla^2 u = f(\mathbf{x}, t)$ in D, subject to the conditions $u(\mathbf{x}, 0) = \phi(\mathbf{x})$, $\alpha \dfrac{\partial u}{\partial n} + \beta\, u(\mathbf{x}, t) = g(\mathbf{x}, t)$ on ∂D, $\alpha \neq 0$, the solution in terms of its Green's function is given by

$$u\left(\mathbf{x}, t\right) = \int_0^t \iiint_{D'} G\left(\mathbf{x}, \mathbf{x}'; t, t'\right) f(\mathbf{x}', t')\, d\mathbf{x}'\, dt'$$

$$+ \iiint_{D'} \phi(\mathbf{x}')\, G\left(\mathbf{x}, \mathbf{x}'; t, 0\right)\, d\mathbf{x}'$$

$$+ \frac{k}{\alpha} \int_0^t \iint_{\partial D'} G\left(\mathbf{x}, \mathbf{x}'; t, t'\right) g\left(\mathbf{x}', t'\right)\, dS'\, dt', \quad \alpha \neq 0,$$

or

$$u\left(\mathbf{x}, t\right) = \int_0^t \iiint_{D'} G\left(\mathbf{x}, \mathbf{x}'; t, t'\right) f(\mathbf{x}', t')\, d\mathbf{x}'\, dt'$$

$$+ \iiint_{D'} \phi(\mathbf{x}')\, G\left(\mathbf{x}, \mathbf{x}'; t, 0\right)\, d\mathbf{x} \qquad\qquad (7.68)$$

$$- \frac{k}{\beta} \int_0^t \iint_{\partial D'} \frac{\partial G\left(\mathbf{x}, \mathbf{x}'; t, t'\right)}{\partial n}\, g\left(\mathbf{x}', t'\right)\, dS'\, dt', \quad \beta \neq 0. \ \blacksquare$$

We will establish (7.68) by using the Laplace transform technique in §7.11 given below.

We remark that the singularity solutions are synonymous to fundamental solutions that are the free space Green's functions, or Green's functions for the whole space. On the other hand, the solutions of the equation $L\left(\dfrac{\partial}{\partial t}, \dfrac{\partial}{\partial x} \right) = \delta\left(\mathbf{x} - \mathbf{x}'\right) \delta(t - t')$ are known as the causal Green's functions, or simply Green's functions. We will not discuss the fundamental solutions for partial differential operators; details about these solutions are available in Friedlander [1982], Kythe [1996] and Vladimirov [1984].

7.11. Laplace Transform Method

This method can be used effectively to solve parabolic or hyperbolic equations. For example, consider the equation $\left[L\left(\dfrac{\partial}{\partial t} \right) - \nabla^2 \right] u = f(\mathbf{x}, t)$, where $L\left(\dfrac{\partial}{\partial t} \right)$ is a linear differential operator for $t > 0$. Applying the Laplace transform with respect to t to this equation and assuming homogeneous initial condition, we get

$$\left[L(s) - \nabla^2\right]\bar{u} = F(\mathbf{x}, s). \tag{7.69}$$

The transformed operator is self-adjoint. The Green's function satisfies the equation

$$\left[L(s) - \nabla^2\right]\overline{G} = e^{-st'}\delta(\mathbf{x} - \mathbf{x}'). \tag{7.70}$$

Multiplying (7.69) by $\overline{G}$ and (7.70) by $\bar{u}$ we obtain

$$\bar{u}\,\nabla^2\overline{G} - \overline{G}\,\nabla^2\bar{u} = \overline{G}F(\mathbf{x}, s) - \bar{u}\,e^{-st'}\delta(\mathbf{x} - \mathbf{x}').$$

Integrating both sides of this equation over the region under consideration and using the divergence theorem (§1.7), we have

$$\iiint_D \left[\bar{u}\,\nabla^2\overline{G} - \overline{G}\,\nabla^2\bar{u}\right] d\mathbf{x} = \iint_{\partial D}\left(\bar{u}\,\frac{\partial \overline{G}}{\partial n} - \overline{G}\,\frac{\partial \bar{u}}{\partial n}\right) dS$$

$$= \iiint_D \left[\overline{G}F(\mathbf{x}, s) - \bar{u}\,e^{-st'}\delta(\mathbf{x} - \mathbf{x}')\right] d\mathbf{x},$$

which yields

$$\bar{u}(\mathbf{x}') = e^{st'}\left[\iiint_D \overline{G}F(\mathbf{x}, s)\,d\mathbf{x} - \iint_{\partial D}\left(\bar{u}(\mathbf{x}')\,\frac{\partial \overline{G}}{\partial n} - \overline{G}\,\frac{\partial \bar{u}}{\partial n}\right) dS\right].$$

The right side of this equation consists of either known functions or the expressions disappears because of the boundary condition. Thus,

For the Dirichlet problem $\overline{G} = 0$ and $\bar{u}$ is known on ∂D.

For the Neumann problem, $\dfrac{\partial \overline{G}}{\partial n} = 0$ and $\dfrac{\partial \bar{u}}{\partial n}$ is known on ∂D. Hence, the solution is known for the transform domain, and the solution to the problem can be found by carrying out the Laplace inversion.

In the case of the Robin problem, let $u + \lambda\dfrac{\partial u}{\partial n} = \phi(\mathbf{x}_B, t)$ be satisfied on the boundary, where $\mathbf{x}_B$ is any point on the boundary. The corresponding Green's function is the solution of (7.70) with $\phi(\mathbf{x}_B, t) = 0$. Hence

$$\iint_{\partial D}\left(\bar{u}\,\frac{\partial \overline{G}}{\partial n} - \overline{G}\,\frac{\partial \bar{u}}{\partial n}\right) dS$$

$$= \iint_{\partial D}\left[\frac{\partial \overline{G}}{\partial n}\left(\overline{\phi}(\mathbf{x}_B, s) - \lambda\frac{\partial \bar{u}}{\partial n}\right) - \overline{G}\frac{\partial \bar{u}}{\partial n}\right] dS$$

$$= \iint_{\partial D}\left[\frac{\partial \overline{G}}{\partial n}\left(\overline{\phi}(\mathbf{x}_B, s) - \lambda\frac{\partial \bar{u}}{\partial n}\right) + \lambda\frac{\partial \overline{G}}{\partial n}\frac{\partial \bar{u}}{\partial n}\right] dS \tag{7.71}$$

$$= \iint_{\partial D}\overline{\phi}(\mathbf{x}_B, s)\frac{\partial \overline{G}}{\partial n}\,dS.$$

Once again the right side of Eq (7.71) is defined in terms of known functions. The final solution can be found by inverting the Laplace transform.

Example 7.1. Solve the boundary value problem

$$u_{tt} - c^2\, u_{xx} = e^{-|x|}\, \sin t, \quad u(x,0) = 0, \quad u_t(x,0) = e^{-|x|}.$$

Applying the Laplace transform to this problem, we get

$$\bar{u}_{xx}(x,s) - \frac{s^2}{c^2}\, \bar{u}(x,s) = -\frac{2+s^2}{c^2\,(1+s^2)}\, e^{-|x|}.$$

Green's function for this problem in the transform domain is

$$G(x,x';s) = -\frac{c}{2s}\, e^{-s|x-x'|/c},$$

where s is the variable of the transform. This gives the solution in the transform domain as

$$\bar{u}(x,s) = \frac{2+s^2}{2cs\,(1+s^2)} \int_{-\infty}^{\infty} e^{-|x'|-s|x-x'|/c}\, dx'.$$

CASE 1. For $x < 0$, we have

$$\bar{u}(x,s) = \frac{2+s^2}{2cs\,(1+s^2)} \left\{ \int_{-\infty}^{x} e^{x'-s(x-x')/c}\, dx' + \int_{x}^{0} e^{x'+s(x-x')/c}\, dx' \right.$$
$$\left. + \int_{0}^{\infty} e^{-x'+s(x-x')/c}\, dx' \right\}$$

$$= \frac{2+s^2}{s\,(1+s^2)\,(s^2-c^2)} \left[s\, e^x - c\, e^{sx/c} \right]$$

$$= B\left(\frac{1}{s-c} - \frac{1}{s+c} \right) e^x - \frac{e^x}{(1+c^2)\,(1+s^2)}$$

$$+ \left[\frac{2}{cs} - B\left(\frac{1}{s-c} + \frac{1}{s+c} \right) - \frac{cs}{(1+c^2)\,(1+s^2)} \right] e^{sx/c},$$

where $B = \dfrac{2+c^2}{2c\,(1+c^2)}$. Hence, after inverting the solution for $x < 0$ is

$$u(x,t) = B\left(e^{ct+x} - e^{-ct+x} \right) - \frac{e^x\, \sin t}{1+c^2}$$
$$+ H(ct+x) \left[\frac{2}{c} - B\left(e^{ct+x} + e^{-ct-x} \right) - \frac{c}{1+c^2}\, \cos(t - x/c) \right].$$

CASE 2. For $x > 0$, the solution in the transform domain is

$$\bar{u}(x,s) = \frac{2+s^2}{2cs\,(1+s^2)} \left\{ \int_{-\infty}^{0} e^{x'-s(x-x')/c}\,dx' + \int_{0}^{x} e^{-x'-s(x-x')/c}\,dx' \right.$$
$$\left. + \int_{x}^{\infty} e^{-x'+s(x-x')/c}\,dx' \right\}$$

$$= \frac{2+s^2}{s\,(1+s^2)} \left[s\,e^{-x} - c\,e^{-sx/c} \right]$$

$$= B\left(\frac{1}{s-c} - \frac{1}{s+c} \right) e^{-x} - \frac{e^{-x}}{(1+c^2)\,(1+s^2)}$$
$$+ \left[\frac{2}{cs} - B\left(\frac{1}{s-c} + \frac{1}{s+c} \right) - \frac{cs}{(1+c^2)\,(1+s^2)} \right] e^{-sx/c},$$

which after inverting gives the solution for $x > 0$ as

$$u(x,t) = B\left(e^{ct-x} - e^{-ct-x} \right) - \frac{e^{-x}\,\sin t}{1+c^2}$$
$$+ H(ct-x) \left[\frac{2}{c} - B\left(e^{ct-x} + e^{-ct+x} \right) - \frac{c}{1+c^2}\,\cos\left(t - x/c \right) \right]. \blacksquare$$

7.12. Quasioptics and Diffraction

It is known that the electromagnetic radiation is represented by rays of light, and reflections at plane and curved mirrors follow the rule: $\theta_{in} = \theta_{out}$, while refraction follows Snell's law: $n_1 \sin\theta_1 = n_2 \sin\theta_2$, where n_1, n_2 are different refraction indices of the two media and θ_1, θ_2 the angles from the normal of the incident and refracted waves, respectively. In some problems diffraction is included, and one such problem is discussed below.

7.12.1. Diffraction of Monochromatic Waves.
The phenomenon of diffraction of waves plays an important role not only in the case of light waves but in all kinds of waves, since diffraction is the result of all waves of given wavelength along all obstructing paths. In the case of light waves it affects the visibility of objects around us. For example, we cannot see the motion of a single molecule even under a microscope, because it is much smaller than a polystyrene bead or a grain of pollen. The objects are seen through a microscope since they block some of the light that illuminates them from below as we look down upon them. If an object is smaller than one half the wavelength of light, the light around it eliminates most of the shadow it casts and we do not see it. However, some objects which are fluorescent and emit light are visible and refraction no longer makes them invisible. For example, individual DNA molecules are visible only when they are tainted with fluorescent dyes, but they are not visible in normal microscope light because the width of the DNA helix is much smaller than the wavelength of light.

If we consider the contribution of an infinitely small neighborhood around a path of waves, we obtain what is known as a wave front, and if we add all wave fronts (i.e., integrate over all paths from the source to the detector which is a point on the screen), we obtain a procedure to study the phenomenon of diffraction. A diffraction pattern can be determined by the phase and the amplitude of each wavefront, which enables us to find at each point the distance for each source on the incoming wavefront. If this distance for each source differs by an integer multiple of wavelength, then all wavefronts will be in phase, which will result in constructive interference. But if this distance is an integer plus one half of the wavelength, there will be a complete destructive interference. Usually such values of minima and maxima are determined to explain diffraction results.

The simplest type of diffraction occurs in a 2-D problem, like the water waves since they propagate only on water surface. In the case of light waves we can neglect one dimension if the diffracting object extends in the direction of light over a distance far greater than its wavelength. However, in the case of light emanating through small circular holes we must consider the 3-D problem. Some qualitative observations about diffraction are as follows:

(i) The angular spacing in diffraction patterns is inversely proportional to the dimension of the object causing diffraction, i.e., the smaller the diffracting object, the wider the resulting diffraction pattern, and conversely. This pattern depends on the sines of the angles.

(ii) The diffraction angles are invariant under scaling, i.e., they depend only on the ratio of the wavelength to the size of the diffracting object.

(iii) If the diffracting object has a periodic structure, like a diffracting grating, the diffraction becomes sharper.

The problem of computing the pattern of a diffracted wave depends on the phase of each of the simple sources on the incoming wave front. There are two types of diffraction: One is known as the *Fraunhofer diffraction* or far-field diffraction, in which the point of observation is far from the diffracting object. It is mathematically simpler that the other type, known as the *Fresnel diffraction* or near-field diffraction, in which the point of observation is closer to the diffracting object.

The best way to study diffraction of light waves is presented by multi-slit arrangements which can be regarded as multiple simple wave sources if the slits are very narrow. A light wave going through a very narrow slit extends to infinity in one direction and this situation results in a 2-D problem. The simplest case is that of a two-slit arrangement, where the slits are a distance a apart (Fig. 7.6.). Then the maxima and minima in the amplitude are determined by the path difference to the first slit and to the other slit.

(a) **Fraunhofer Approximation.** If the observer is far away from the slits, the difference in path lengths to the two slits as seen from the image is $\Delta s = a \sin \theta$. The maxima in the intensity occur if this path length difference Δs is an integer multiple of wavelength, i.e., if $a\, sin\theta = m\,\lambda$, where m is the number of the order of each maximum, a the distance between the slits and θ the angle at which constructive interference occurs.

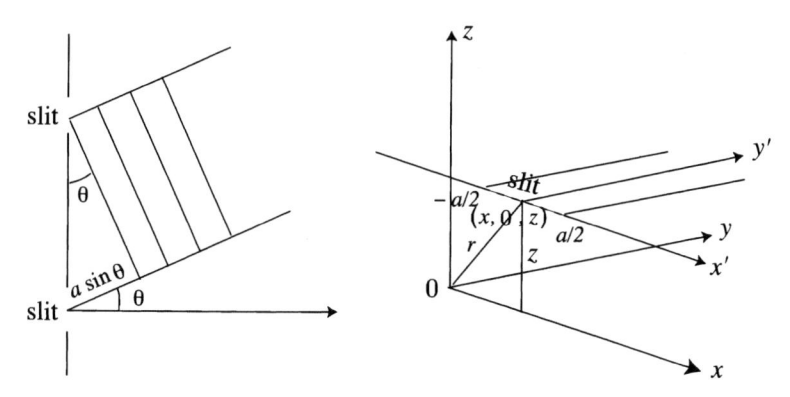

Fig. 7.6. Fraunhofer Diffraction. Fig. 7.7. Fresnel Approximation.

A radial wave is represented by $E(r) = \dfrac{A}{r}\cos(k\,r - \omega t + \phi)$, where $k = 2\pi/\lambda$, ω the frequency and ϕ the phase of the wave at the slits. The function $E(r)$ is the real part of the complex wave $\Psi(r)$ which is defined as $\Psi(r) = \dfrac{A}{r}\,e^{i\,(kr-\omega t+\phi)}$. The amplitude of this complex wave is given by $|\Psi(r)| = A/r$. For M slits the radial wave at a point x on the screen which is at a distance l away is given by

$$E_M(r) = A\,e^{i\,(-\omega t+\phi)}\sum_{m=0}^{M-1}\frac{e^{i\,k\sqrt{l^2+(x-ma)^2}}}{\sqrt{l^2+(x-ma)^2}}. \tag{7.72}$$

Since $\sqrt{l^2+(x-ma)^2} = l\left[1 + \dfrac{(x-ma)^2}{l^2}\right]^{1/2} \approx l + \dfrac{x^2}{2l} - \dfrac{max}{l} + \dfrac{m^2a^2}{2l}$, we can neglect the term of the order $a^2/(2l)$, and then the complex wave is defined by

$$\Psi(r) = \frac{A}{l}\exp\left\{i\left(k\left(l+\frac{x^2}{2l}\right) - \omega t + \phi\right)\right\}\sum_{m=0}^{M-1}e^{-i\,kxma/l}$$

$$= \frac{A}{l}\exp\left\{i\left(k\left(l+\frac{x^2}{2l}\right) - \omega t + \phi\right)\right\}\frac{\sin\dfrac{Mkax}{2l}}{\sin\dfrac{kax}{2l}}, \tag{7.73}$$

where the above sum is a finite geometric series. If we set $I_0 = A/l$, the intensity of this wave is given by

$$I(x) = |\Psi|^2 = I_0\left(\frac{\sin\dfrac{Mkax}{2l}}{\sin\dfrac{kax}{2l}}\right). \tag{7.74}$$

(b) Fresnel Approximation. Let $\Psi(r)$ define a monochromatic complex plane wave of wavelength α passing through a slit of width a centered at the point $(x, 0, z)$. The 3-D situation can be reduced to a 2-D problem if we consider a coordinate system (x', y') through the point $(x, 0, z)$ so that the slit is defined by $\{-a/2 < x' < a/2, -\infty < y' < \infty\}$ (see Fig. 7.7). Notice that this slit is located on the (x, z)-plane and the light beam is diffracted along the y'-axis. This complex wave is defined by

$$\Psi(r) = \int_{\text{slit}} \frac{i}{r\lambda} \Psi_0 \, e^{-ikr} \, d(\text{slit}), \qquad (7.75)$$

where Ψ_0 is a constant, and

$$r = \sqrt{(x-x')^2 + y'^2 + z^2} = z\sqrt{1 + \frac{(x-x')^2 + y'^2}{z^2}} \approx z + \frac{(x-x')^2 + y'^2}{2z}.$$

Since the factor $1/r$ in (7.75) is non-oscillatory, its contribution to the magnitude of the intensity is very small compared to the exponential factor. We will, therefore, approximate $1/r$ by $1/z$. Then (7.75) becomes

$$
\begin{aligned}
\Psi(x) &= \frac{i\,\Psi_0}{\lambda z} \int_{-a/2}^{a/2} \int_{-\infty}^{\infty} e^{-ik\left(z + \frac{(x-x')^2 + y'^2}{2z}\right)} \, dx' \, dy' \\
&= \frac{i\,\Psi_0}{\lambda z} e^{-ikz} \int_{-a/2}^{a/2} e^{-ik(x-x')^2/(2z)} \, dx' \int_{-\infty}^{\infty} e^{-iky'^2/(2z)} \, dy' \\
&= \frac{i\,\Psi_0}{\lambda z} e^{-ikz} e^{-ikx^2/(2z)} \left[\int_{-a/2}^{a/2} e^{ikxx'/z} e^{-ikx'^2/(2z)} \, dx' \right] \sqrt{\frac{2\pi z}{ik}} \\
&= \Psi_0 \sqrt{\frac{i}{z\lambda}} e^{-ikz} e^{-ikx^2/(2z)} \left[\int_{-a/2}^{a/2} e^{ikxx'/z} e^{-ikx'^2/(2z)} \, dx' \right],
\end{aligned}
\qquad (7.76)
$$

where we have used the formula $\displaystyle \int_{-\infty}^{\infty} e^{-iky'^2/(2z)} \, dy' = \sqrt{\frac{2\pi z}{ik}}$. Since the quantity kx'^2/z is small in the Fresnel approximation, we can approximate the factor $e^{-ikx'^2/(2z)} \approx 1$, and the terms e^{-ikz} can also be neglected if we take z as a very small positive quantity, say $z = \varepsilon$. Then (7.76) reduces to

$$
\begin{aligned}
\Psi(x) &= \Psi_0 \sqrt{\frac{i}{\varepsilon\lambda}} e^{-ikx^2/(2\varepsilon)} \int_{-a/2}^{a/2} e^{ikxx'/\varepsilon} \, dx' \\
&= \Psi_0 \sqrt{\frac{i}{\varepsilon\lambda}} e^{-ikx^2/(2\varepsilon)} \frac{e^{ikax/(2\varepsilon)} - e^{-ikax/(2\varepsilon)}}{ikx/\varepsilon} \\
&= \Psi_0 \sqrt{\frac{i}{\varepsilon\lambda}} e^{-ikx^2/(2\varepsilon)} \frac{\sin\left(\dfrac{kax}{2\varepsilon}\right)}{\dfrac{kx}{2\varepsilon}} = \Psi_0 \sqrt{\frac{i}{\varepsilon\lambda}} e^{-ikx^2/(2\varepsilon)} \operatorname{sinc}\left(\frac{kax}{2\varepsilon}\right)
\end{aligned}
$$

$$= \Psi_0 \sqrt{\frac{i}{\varepsilon\lambda}}\, e^{-i\,kx^2/(2\varepsilon)}\, \mathrm{sinc}\left(\frac{\pi a x}{\varepsilon\lambda}\right), \quad \text{where } k = 2\pi/\lambda. \tag{7.77}$$

Since $\lim_{x\to 0} \mathrm{sinc}(x) = 1$, the sinc term in the above expression is almost equal to 1 in this near-field approximation, and hence, can be neglected. This leads to

$$\Psi(x) = \frac{2\lambda\,\Psi_0}{\pi} \sqrt{\frac{i}{\varepsilon\lambda}}\, e^{-i\,kx^2/(2\varepsilon)},$$

which defines a diffracted wave similar to (2.26) in Example 2.8. The intensity of the diffracted wave (7.77) is given by $I = I_0 \left[\mathrm{sinc}\left(\frac{\pi a x}{\varepsilon\lambda}\right) \right]^2$, where $I_0 = \frac{|\Psi_0|}{\varepsilon\lambda}$.

7.13. Exercises

7.1. Use Green's function to solve the following problem (see §7.1):
$u_{tt} - c^2 u_{xx} = e^{-x} \sin t$, $u(x,0) = 0$, $u_t(x,0) = e^{-x}$ for $x > 0$, $t > 0$, $u(0,t) = 0$, and $\lim_{x\to\infty} u(x,t) = 0$.

ANS. $u(x,t) = B \left(e^{-(x-ct)} - e^{-(x+ct)} \right) - \dfrac{1}{1+c^2}\, e^{-x} \sin t - H(ct - x)$

$\times \left[B \left(e^{-(x-ct)} + e^{(x-ct)} \right) - \dfrac{1}{1+c^2} \cos\left(t - \dfrac{x}{c} \right) \right]$, where $B = \dfrac{c^2 + 2}{2c\,(1+c^2)}$.

7.2. For the wave operator show that

$$\int_{-\infty}^{\infty} G_3\left(x, y, z, t; x', y', z', t'\right)\, dz' = G_2\left(x, y, t; x', y', t'\right).$$

HINT. From (7.19), we have

$$F\left(x, y, t; x', y', t'\right) = \int_{-\infty}^{\infty} G_3\left(x, y, z, t; x', y', z', t'\right)\, dz'$$

$$= \frac{1}{4\pi c^2} \int_{-\infty}^{\infty} \frac{\delta\left(t - t' - \dfrac{1}{c}\sqrt{(x - x')^2 + (y - y')^2 + (z - z')^2}\right)}{\sqrt{(x - x')^2 + (y - y')^2 + (z - z')^2}}\, dz'$$

$$= \frac{1}{4\pi c^2} \int_{-\infty}^{\infty} \frac{\delta\left(t - t' - \dfrac{1}{c}\sqrt{(x - x')^2 + (y - y')^2 + u^2}\right)}{\sqrt{(x - x')^2 + (y - y')^2 + u^2}}\, du,$$

where $u = z - z'$. Letting $r^2 + u^2 = c^2 v^2$, $r^2 = (x - x')^2 + (y - y')^2$, and noting that the integrand is an even function of u, we get

$$F\left(x, y, t; x', y', t'\right) = \frac{1}{2\pi c} \int_0^{\infty} \frac{\delta\left(t - t' - v\right)}{\sqrt{c^2 v^2 - r^2}}\, dv = \frac{1}{2\pi c\,\sqrt{c^2\,(t - t')^2 - r^2}},$$

for $t - t' > r/c$. Hence, $F(x, y, t; x', y', t') = G_2(x, y, t; x', y', t')$.

7.3. Show that for the wave operator

$$\int_{-\infty}^{\infty} G_2(x, y, t; x', y', t')\, dz' = G_1(x, t; x', t').$$

HINT. Follow the method in Exercise 7.2.

7.4. Derive (7.26) by using the Bernoulli's separation method.

7.5. Find Green's function for the operator $L \equiv \dfrac{\partial^2}{\partial x^2} + \dfrac{\partial^2}{\partial y^2} + 2\alpha\, \dfrac{\partial}{\partial x} + 2\beta\, \dfrac{\partial}{\partial y}$ in the region $\{0 < x < a, 0 < y < b\}$ subject to Dirichlet boundary conditions.

ANS. $G(x, y; x', y')$

$$= -4\, e^{\alpha(x'-x) + \beta(y'-y)} \sum_{m=1}^{\infty} \sum_{n=1}^{\infty} \frac{ab \sin \dfrac{m\pi x}{a} \sin \dfrac{n\pi y}{b} \sin \dfrac{m\pi x'}{a} \sin \dfrac{n\pi y'}{b}}{\pi^2 \left(m^2 b^2 + n^2 a^2 \right) + a^2 b^2 \left(\alpha^2 + \beta^2 \right)}.$$

ANS. $u(x, t) = T_0 \left(1 - \dfrac{x}{a} \right) - 2T_0 \sum_{n=1}^{\infty} \dfrac{1}{n\pi} e^{-n^2 \pi^2 kt/a^2} \sin \dfrac{n\pi x}{a}.$

7.6. Solve Dirichlet problem for Laplace's equation in $\mathbb{R}^3$.

SOLUTION. Taking $\mathbf{x}_1 = \mathbf{0} = (0, 0, 0)$, Green's function becomes symmetrical about the origin. Hence, the differential equation becomes

$$\frac{\partial^2 G}{\partial r^2} + \frac{2}{r}\frac{\partial G}{\partial r} + k^2 G = \frac{\delta(r)}{4\pi r^2}, \qquad \lim_{r \to \infty} G = 0,$$

and the additional that the flux of G across a small sphere with center at the origin be equal to -1. Substituting $G = w/r$, the differential equation for G becomes $\dfrac{d^2 w}{dr^2} + k^2 w = \dfrac{\delta(r)}{4\pi r}$. Applying Fourier sine transform to this equation, we obtain $\sqrt{\dfrac{2}{\pi}}\, \alpha\, c_0 + \left(k^2 - \alpha^2 \right) w_s = \sqrt{2\pi}\, \dfrac{\alpha}{4\pi}$, where $c_0 = \lim_{r \to 0} w$. Thus, $w_s = \sqrt{\dfrac{2}{\pi}} \left(c_0 - \dfrac{1}{4\pi} \right) \dfrac{\alpha}{\alpha^2 - k^2}$. On inverting the sine transform, we have $w = \left(c_0 - \dfrac{1}{4\pi} \right) e^{-ikr}$, and therefore, $G(r) = \left(c_0 - \dfrac{1}{4\pi} \right) \dfrac{e^{-ikr}}{r}$. Using the flux condition, we get $-\lim_{\varepsilon \to 0} \displaystyle\int_S \left(\dfrac{dG}{dr} \right)_{r=\varepsilon} dS = -1$, where S is a sphere of radius ε with center at the origin, and $dS = \varepsilon^2 \sin \phi\, d\phi\, d\theta$ is the infinitesimal element of the surface. Then

$$\lim_{\varepsilon \to 0} \left(c_0 - \frac{1}{4\pi} \right) \int_0^{2\pi} \int_0^{\pi} e^{-ikr} \left(\frac{1}{\varepsilon} + \frac{1}{\varepsilon^2} \right) \varepsilon^2 \sin \phi\, d\phi\, d\theta = -1,$$

whence we have $\left(c_0 - \dfrac{1}{4\pi} \right) 4\pi = -1$, which gives $c_0 = 0$. Hence, $G(r) = -\dfrac{e^{-ikr}}{4\pi r}$.

7.7. Use double Fourier transform to solve the telegrapher's equation

$u_{tt} + a\, u_t + b\, u = c^2\, u_{xx}$, $-\infty < x, t < \infty$ subject to the conditions $u(0,t) = f(t)$, $u_x(0,t) = g(t)$, $\infty < t < \infty$, where a, b, c are constants and $f(t)$ and $g(t)$ are arbitrary functions of t.

ANS. $u(x,t) = \dfrac{1}{2\pi} \displaystyle\int_{-\infty}^{\infty} \left\{ \widetilde{f}(\alpha)\, \cos(\mu x) + \dfrac{\widetilde{g}(\alpha)}{\mu} \sin(\mu x)\, e^{i\,\alpha t}\, d\alpha \right\}$, where $\mu^2 = \left(\alpha^2 - i\,a\alpha - b \right)/c^2$.

7.8. Show that Green's function $G(x,t)$ for the Bernoulli-Euler problem for an elastic foundation, defined by

$b\, G_{xxxx} + k\, G + m\, G_{tt} = a\, \delta(x)\, \delta(t)$, $x \in \mathbb{R}$, $t > 0$, and subject to the initial conditions $G(x,0) = 0 = G_t(x,0)$, where $b = EI$, k and a are constants, is given by

$$G(x,t) = \dfrac{a}{2\pi m} \int_{-\infty}^{\infty} \dfrac{\sin pt}{p}\, e^{i\,\alpha x}\, d\alpha, \quad p = \sqrt{a^2\alpha^4 + \omega^2},\ a^2 = b/m \text{ and } \omega^2 = k/m.$$

HINT. Use double Laplace and Fourier transforms.

7.9. Find Green's function for 1-D Klein-Gordon's equation

$u_{tt} - c^2\, u_{xx} + a^2\, u = f(x,t)$, $x \in \mathbb{R}$, $t > 0$, where a and c are constants, and subject to the initial conditions $u(x,0) = 0 = u_t(x,0)$ for all $x \in \mathbb{R}$.

HINT. Apply Laplace transform followed by Fourier transform to $G_{tt} - c^2\, G_{xx} + a^2\, G = \delta(x)\, \delta(t)$, giving $\widetilde{\overline{G}}(\alpha, s) = \dfrac{1}{\sqrt{2\pi}} \dfrac{1}{s^2 + \sqrt{c^2\alpha^2 + a^2}}$.

ANS. $G(x,t) = \dfrac{1}{2\pi c} \displaystyle\int_{-\infty}^{\infty} \left(\alpha^2 + \dfrac{a^2}{c^2} \right)^{-1/2} \sin\left(ct\sqrt{\alpha^2 + \dfrac{a^2}{c^2}} \right) = \dfrac{1}{2c} H(ct - |x|)\, J_0\left(\dfrac{a}{c}\sqrt{c^2 t^2 - x^2} \right)$. Note that the above Green's function reduces to that for the wave equation in the limit as $a \to 0$.

7.10. Show that Green's function for the differential equation $L[u](t) = f(t)$, $t > 0$, where $L = \dfrac{d^2}{dt^2} + 2\lambda\dfrac{d}{dt} + \omega_0{}^2$, $\omega_0 > \lambda > 0$, is the differential operator for a (damped) harmonic oscillator subject to an external force and initially at rest after being subjected to a sudden impulse at $t = 0$, i.e., when $f(t) = \delta(t)$, is given by $G(t) = \dfrac{H(t)}{\alpha} e^{-\lambda t}\, \sin \alpha t$, $\alpha = \sqrt{\omega_0{}^2 - \lambda^2}$.

HINT. Use the Laplace transform method to solve $L[G](t) = \delta(t)$.

7.11. Solve $u_{tt} = c^2\, u_{xx}$, $0 < x < l$, subject to the boundary conditions $u(0,t) = u_1$, $u(l,t) = u_2$ for $t > 0$, where u_1, u_2 are prescribed quantities, and the initial conditions $u(x,0) = g(x)$, $u_t(x,0) = h(x)$ for $0 \leq x \leq l$.

ANS. $u(x,t) = U(x) + v(x,t)$, where $U(x) = u_1 + (u_2 - u_1)\frac{x}{l}$, describes the steady-state solution (static deflection), and $v(x,t)$ is the solution of the problem in Exercise 4.9, with $f(x)$ replaced by $g(x) - U(x)$.

7.12. Solve $u_{tt} = c^2 (u_{xx} + u_{yy})$ in the rectangle $R = \{(x,y) : 0 < x < a, \ 0 < y < b\}$, subject to the condition $u = 0$ on the boundary of R for $t > 0$, and the initial conditions $u(x,y,0) = f(x,y)$, $u_t(x,y,0) = g(x,y)$. This problem describes a vibrating rectangular membrane. ANS. $u(x,y,t) =$

$$\sum_{m,n=1}^{\infty} (A_{mn} \cos \lambda_{mn}t + B_{mn} \sin \lambda_{mn}) \sin \frac{m\pi x}{a} \sin \frac{n\pi y}{b},$$

where $A_{mn} = \dfrac{4}{ab} \displaystyle\int_0^b \int_0^a f(x,y) \sin \frac{m\pi x}{a} \sin \frac{n\pi y}{b} \, dx \, dy,$

$$B_{mn} = \dfrac{4}{ab\lambda_{mn}} \int_0^b \int_0^a g(x,y) \sin \frac{m\pi x}{a} \sin \frac{n\pi y}{b} \, dx \, dy,$$

for $m, n = 1, 2, \ldots$; and the eigenvalues are $\lambda_{mn} = c\pi \sqrt{\dfrac{m^2}{a^2} + \dfrac{n^2}{b^2}}$.

7.13. Solve the problem of transverse vibrations of a semi-infinite string initially at rest, defined by 1-D wave equation $u_{tt} = c^2 u_{xx}$, $0 < x < \infty$, $t > 0$, and subject to the boundary and initial conditions: $u(0,t) = U_0 f(t)$, $t \geq 0$; $u(x,t) \to 0$ as $x \to \infty$, $t \geq 0$; and $u(x,0) = 0 = u_t(x,0)$ for $0 \leq x < \infty$, where U_0 is a constant.

HINT. Use Laplace transform. ANS. $u(x,t) = H\left(t - \dfrac{x}{c}\right) f\left(t - \dfrac{x}{c}\right)$.

7.14. Solve $u_{xx} - u_{tt} = e^{-a^2\pi^2 t} \sin a\pi x$, subject to the conditions $u(x,0) = 0$, $u_t(x,0) = 0$, and $u(0,t) = u(1,t) = 0$, where a is an integer.

ANS. $\dfrac{1}{a^2\pi^2(1 + a^2\pi^2)} \left[\cos a\pi t - e^{-a^2\pi^2 t} - a\pi \sin a\pi t \right] \sin a\pi x$.

8

Elliptic Equations

We will derive Green's functions for Laplace's and Helmholtz's equations. In previous sections we have used a source (of strength $+1$) for parabolic and hyperbolic equations, but now we will use a sink (of strength -1) for elliptic equations. The method is to first develop Green's function $G(\mathbf{x})$ for the sink (source) at the origin, and then obtain the general Green's function $G(\mathbf{x} - \mathbf{x}') \equiv G(\mathbf{x}; \mathbf{x}')$ for the sink (source) at a point $\mathbf{x}' \neq \mathbf{0}$ by simply replacing $\mathbf{x}$ in $G(\mathbf{x})$ by $\mathbf{x} - \mathbf{x}'$. Note that in $\mathbb{R}$ the Laplacian operator is an ordinary differential operator, which has already been studied in Chapter 3. We will, therefore, determine Green's function only in $\mathbb{R}^2$ and $\mathbb{R}^3$ by different methods.

8.1. Green's Function for 2-D Laplace's Equation

In $\mathbb{R}^2$, in order to solve Laplace's equation $\nabla^2 u = 0$, we use polar cylindrical coordinates and have

$$\frac{1}{r} \frac{\partial}{\partial r} \left(r \frac{\partial u}{\partial r} \right) = \frac{\delta(r)}{2\pi r}, \quad r^2 = x^2 + y^2.$$

METHOD 1. (By direct method) Multiply both sides of this equation by r and integrate from 0 to r, to get $\dfrac{\partial u}{\partial r} = \dfrac{1}{2\pi r}$, which gives

$$u = \frac{1}{2\pi} \log r + B.$$

Since it is not possible to normalize u by requiring that it must vanish as $r \to \infty$, we normalize it such that $u = 0$ at $r = 1$, which results in $B = 0$. Hence, Green's function in this case with singularity at the origin is given by

$$G(\mathbf{x}) = \frac{1}{2\pi} \log r = \frac{1}{2\pi} \log |\mathbf{x}|, \tag{8.1}$$

and for an arbitrary singular point $\mathbf{x}'$ by

$$G(\mathbf{x} - \mathbf{x}') = \frac{1}{2\pi} \log |\mathbf{x} - \mathbf{x}'|. \quad \blacksquare \qquad (8.2)$$

METHOD 2. (Based on the physical concepts) Since the operator ∇^2 is invariant under a rotation of the coordinate axes, we seek a solution that depends only on $r = |\mathbf{x}|$. For $r > 0$, the function $G(r)$ satisfies the homogeneous equation $\nabla^2 G = 0$, which in $\mathbb{R}^2$, in polar cylindrical coordinates, is

$$\frac{1}{r}\frac{\partial}{\partial r}\left(r\frac{\partial G}{\partial r}\right) = 0 \quad \text{for } r \neq 0, \text{ where } r = |\mathbf{x}| = \sqrt{x^2 + y^2}.$$

This equation has a solution $G(r) = A \ln r + B$. To evaluate A, we consider the flux across a circle C_ε of radius ε with center at the origin. Then

$$-\int_{C_\varepsilon} \left[\frac{\partial G}{\partial r}\right]_{r=\varepsilon} ds = -2\pi A = -1,$$

which gives $A = 1/2\pi$. If we normalize as before, we have $B = 0$, and Green's function is the same as (8.1). If the source is at $\mathbf{x}'$, then Green's function is the same as given by (8.2). $\blacksquare$

METHOD 3. (Method of images) The problem of finding Green's function $G(\mathbf{x}; \mathbf{x}') = G(x, y; x', y')$ inside some region D bounded by a closed curve Γ with the homogeneous boundary condition $G = 0$ on Γ amounts to that of finding the electrostatic potential due to a point charge at the point (x', y') inside a grounded conductor in the shape of the boundary Γ. Consider, for example, the region D to be the half-plane $x > 0$. Then Green's function is given by the point charge of strength -1 at (x', y') together with an equal but opposite charge (of strength $+1$) at the point $(x', -y')$ which is the image of the point (x', y') in the x-axis (Fig. 8.1). Thus, Green's function for the right half-plane such that $G = 0$ on the x-axis and $G \to 0$ as $r \to \infty$, where $r = |\mathbf{x} - \mathbf{x}'| = \sqrt{(x - x')^2 + (y - y')^2}$, is given by

$$G(x, y; x', y') = \frac{1}{4\pi} \ln\left\{\frac{(x - x')^2 + (y - y')^2}{(x - x')^2 + (y + y')^2}\right\}. \qquad (8.3)$$

Since the algebraic sum of the charges over the entire (x, y)-plane is zero, the condition $G \to 0$ as $r \to \infty$ holds. However, the method of images cannot be used to find Green's function for the upper half-plane $y > 0$ subject to the condition $G_y = 0$ on the x-axis and $G \to 0$ as $r \to \infty$ (see the explanation at the end of §8.12.2: Neumann problem for the half-space). But we can use the method of images to determine Green's function for the quarter-plane $x > 0, y > 0$, subject to the conditions $G = 0$ on $y = 0$ and $G_y = 0$ on $x = 0$. As seen in Fig. 8.1, we have the charge of strength $+1$ at each $(x', -y')$ and $(-x', y')$, and the charge of strength -1 at each (x', y') and $(-x', -y')$. Thus, Green's function for the quarter-plane is

$$G(x, y; x', y') = \frac{1}{4\pi} \ln\left\{\frac{[(x - x')^2 + (y - y')^2][(x + x')^2 + (y + y')^2]}{[(x - x')^2 + (y + y')^2][(x + x')^2 + (y - y')^2]}\right\}. \qquad (8.4)$$

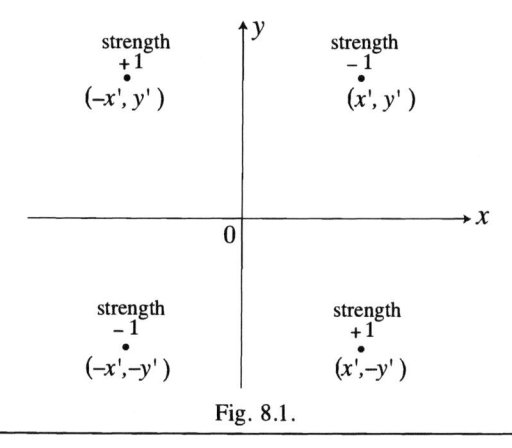

Fig. 8.1.

8.2. 2-D Laplace's Equation in a Rectangle

Consider $\nabla^2 u = 0$ in the rectangle $R = \{0 < x < a, 0 < y < b\}$. Then Green's function $G(x, y; x', y')$ for this rectangle satisfies the equation

$$\nabla^2 G(x, y; x', y') = \delta(x - x')\,\delta(y - y'), \tag{8.5}$$

subject to the boundary conditions

$$G(0, y; x', y') = G(a, y; x', y') = G(x, 0; x', y') = G(x, b; x', y') = 0.$$

The double Fourier series representations for $G(x, y; x', y')$ and $\delta(x - x')\,\delta(y - y')$ are of the form

$$G(x, y; x', y') = \sum_{m=0}^{\infty} \sum_{n=0}^{\infty} A_{mn} \sin \frac{m\pi x}{a} \sin \frac{n\pi y}{b},$$

$$\delta(x - x')\,\delta(y - y') = \sum_{m=0}^{\infty} \sum_{n=0}^{\infty} B_{mn} \sin \frac{m\pi x}{a} \sin \frac{n\pi y}{b}, \tag{8.6}$$

and satisfy the prescribed boundary conditions. To determine A_{mn}, we first find B_{mn}:

$$B_{mn} = \frac{4}{ab} \sin \frac{m\pi x'}{a} \sin \frac{n\pi y'}{b}.$$

Then substituting (8.6) into Eq (8.5), we get

$$-\sum_{m=0}^{\infty} \sum_{n=0}^{\infty} A_{mn} \left(\frac{m^2 \pi^2}{a^2} + \frac{n^2 \pi^2}{b^2} \right) \sin \frac{m\pi x}{a} \sin \frac{n\pi y}{b}$$

$$= \frac{4}{ab} \sum_{m=0}^{\infty} \sum_{n=0}^{\infty} \sin \frac{m\pi x'}{a} \sin \frac{n\pi y'}{b} \sin \frac{m\pi x}{a} \sin \frac{n\pi y}{b}.$$

Hence,

$$A_{mn} = -\frac{4ab}{m^2\pi^2b^2 + n^2\pi^2a^2} \sin\frac{m\pi x'}{a} \sin\frac{n\pi y'}{b},$$

and

$$G\left(x, y; x', y'\right)$$

$$= -\sum_{m=0}^{\infty}\sum_{n=0}^{\infty}\frac{4ab}{m^2\pi^2b^2 + n^2\pi^2a^2} \sin\frac{m\pi x'}{a} \sin\frac{n\pi y'}{b} \sin\frac{m\pi x}{a} \sin\frac{n\pi y}{b}. \blacksquare$$

$$(8.7)$$

8.3. Green's Function for 3-D Laplace's Equation

The functions u which satisfy Laplace's equation $\nabla^2 u\left(\mathbf{x}\right) = \delta\left(\mathbf{x}\right)$ in a domain D subject to homogeneous boundary conditions are known as Green's functions for Laplace's equation.[3] We will give some basic methods for finding Green's function for this operator.

METHOD 1. (By direct method) Assuming that the source point $\mathbf{x}'$ is at the origin, Green's function $G(\mathbf{x}) \equiv G(\mathbf{x}; 0)$ for the Laplacian ∇^2 in $\mathbb{R}^3$ satisfies the equation

$$\frac{\partial^2 u}{\partial x^2} + \frac{\partial^2 u}{\partial y^2} + \frac{\partial^2 u}{\partial z^2} = \delta(x)\,\delta(y)\,\delta(z).\qquad(8.8)$$

Since the problem has radial symmetry, we transform Eq (8.8) into spherical coordinates, which becomes

$$\frac{\partial^2 u}{\partial r^2} + \frac{2}{r}\frac{\partial u}{\partial r} = \frac{\delta(r)}{4\pi r^2}.$$

Multiplying both sides of this equation by r^2, we get

$$\frac{\partial}{\partial r}\left(r^2\frac{\partial u}{\partial r}\right) = \frac{\delta(r)}{4\pi}.$$

Integrating both sides of this equation from 0 to r, we find that

$$r^2\frac{\partial u}{\partial r} = \frac{1}{4\pi}, \quad \text{or} \quad u = -\frac{1}{4\pi r},$$

where the arbitrary constant due to integration is taken to be zero, in order to satisfy the requirement that u must vanish as $r \to \infty$. Hence, Green's function in this case with singularity (source) at the origin is given by

$$G(\mathbf{x}) = -\frac{1}{4\pi r} = -\frac{1}{4\pi|\mathbf{x}|},\qquad(8.9)$$

---

[3]In some books the equation is taken as $-\nabla^2 u(\mathbf{x}) = \delta(\mathbf{x})$, to signify the source of strength $+1$ at the origin. This practice is prevalent in physics, and to get the results in this case Green's functions derived here must be multiplied by -1.

and for an arbitrary singular point $\mathbf{x}'$ by

$$G(\mathbf{x} - \mathbf{x}') = -\frac{1}{4\pi\, r} = -\frac{1}{4\pi\, |\mathbf{x} - \mathbf{x}'|}. \qquad \blacksquare \qquad (8.10)$$

METHOD 2. (Based on the physical concepts) Again, since the operator ∇^2 is invariant under a rotation of the coordinate axes, we seek a solution that depends only on $r = |\mathbf{x}|$. For $r > 0$, the function $G(r)$ satisfies the homogeneous equation $\nabla^2 G = 0$, which in the spherical coordinates is

$$\frac{1}{r^2}\frac{\partial}{\partial r}\left(r^2 \frac{\partial G}{\partial r}\right) = 0.$$

This equation has a solution $G(r) = A/r + B$. If we require this solution to vanish at infinity, then $B = 0$, and thus, $G(r) = A/r$. To determine A, we consider a negative source of unit strength at $x = 0$, which is given by the flux across a sphere of radius ε as

$$-\iint_{\partial S_\varepsilon} \frac{\partial G}{\partial r}\, dS = \int_0^{2\pi}\int_0^{\pi} \frac{A}{\varepsilon^2}\, \varepsilon^2 \sin\phi\, d\phi\, d\theta = 4\pi A = -1, \qquad (8.11)$$

where ∂S_ε is the surface of the sphere S_ε. Physically, the above equation represents the conservation of the charge, i.e., the flux of the electric field through the closed surface ∂S_ε (of area $4\pi\varepsilon^2$) is equal to the charge in the interior of S_ε. Using Eq (8.11), we find that $A = -1/(4\pi)$, and thus, Green's function of the three-dimensional Laplacian is given by

$$G(\mathbf{x}) = -\frac{1}{4\pi r} = -\frac{1}{4\pi|\mathbf{x}|}.$$

If the source is at $\mathbf{x}'$, Green's function is the same as in (8.10).

8.3.1. Laplace's Equation in a Rectangular Parallelopiped.

Green's function for Laplace's equation in a rectangular parallelopiped D, defined by $x = \pm a/2, y = \pm b/2, z = \pm c/2$, subject to the condition $u = 0$ on the boundary ∂D is determined by constructing the lattices which correspond to the given domain D and have the vertices $\left((k + \frac{1}{2})a, (m + \frac{1}{2})b, (n + \frac{1}{2})c\right), k, m, n = 0, \pm 1, \pm 2, \ldots$, which reflect the point (x', y', z') respectively across the lattice-planes. This generates a system of points $\{ka + (-1)^k x', mb + (-1)^m y', nc + (-1)^n z'\}$. We assume that at each of these points a unit mass is connected, which is positive or negative according as $k + m + n$ is even or odd. Thus, the potential of such a mass distribution is zero in the lattice-planes, since the contributions of the individual mass units cancel each other there, and we get the following expression for Green's function:

$$G(x, y, z; x', y', z') = \frac{1}{4\pi}\sum_{k=-\infty}^{\infty}\sum_{m=-\infty}^{\infty}\sum_{n=-\infty}^{\infty} \frac{(-1)^{k+m+n}}{\sqrt{N(k, m, n; x, y, z; x', y', z')}},$$

$$(8.12)$$

where $\quad N(k,m,n;x,y,z;x',y',z')$

$$= \left[ka + (-1)^k x' - x\right]^2 + \left[mb + (-1)^m y' - y\right]^2 + \left[nc + (-1)^n z' - z\right]^2.$$

The convergence in (8.12) can be at least conditional, and so we denote an expression of the form $\phi(k+1) - \phi(k)$, where $\phi(k)$ is any function of k, by $\Delta_k \phi(k)$. Then for fixed k and m, and omitting the factor $(-1)^{k+m}$, we can write the summation over n in (8.12) as

$$N'(k,m) = \sum_{n=\pm 1, \pm 2, \ldots} \Delta_n \frac{1}{\sqrt{N(k,m,n)}} = -\sum_{n=0,\pm 2,\pm 4,\ldots} \Delta_n \frac{1}{\sqrt{N(k,m,n)}},$$

since $\lim_{|n| \to \infty} N(k,m,n) = \infty$. We apply the same transformation to the sums over k and m. Since $\lim_{|m| \to \infty} N(k,m) = 0$, we have

$$N''(k) = \sum_{m=\pm 1, \pm 2, \ldots} \Delta_m N'(k,m) = -\sum_{m=0,\pm 2,\pm 4,\ldots} \Delta_m N'(k,m),$$

and so

$$G(x,y,z;x',y',z') = \frac{1}{4\pi} \sum_{k=\pm 1,\pm 3,\ldots} \Delta_k N''(k) = -\frac{1}{4\pi} \sum_{k=0,\pm 2,\pm 4,\ldots} \Delta_k N''(k),$$

(8.13)

because $\lim_{|k| \to \infty} N''(k) = 0$. Combining these results we get

$$G(x,y,z;x',y',z') = \pm \frac{1}{4\pi} \sum_k \sum_m \sum_n \Delta_k \Delta_m \Delta_n \frac{1}{\sqrt{N(k,m,n)}},$$

where each summation index runs over either all the even or all the odd integers from $-\infty$ to ∞, and the plus or minus sign in the beginning on the left side is chosen according as the summation is taken over all even or all odd integers. The results (8.12) and (8.13) are valid only if no $N(k,m,n)$ vanishes.

REPRESENTATION IN TERMS OF RIEMANN ϑ-FUNCTIONS. The sum (8.12) can be represented as an integral over the θ-functions, as follows: Since the Laplace transform of e^{-t^2} is given by $\dfrac{2}{\sqrt{\pi}} \displaystyle\int_0^\infty e^{-st^2}\,dt = \dfrac{1}{\sqrt{s}}$, we substitute for s the expression $N(k,m,n;x,y,z,x',y',z')$ and get

$$G(x,y,z;x',y',z') = \frac{1}{2\pi^{3/2}} \sum_k \sum_m \sum_n \Delta_k \Delta_m \Delta_n \int_0^\infty e^{Nt^2}\,dt.$$

Interchanging the summations and integration, we find that

$$G(x,y,z;x',y',z') = \frac{1}{2\pi^{3/2}} \int_0^\infty \sum_k \sum_m \sum_n \Delta_k \Delta_m \Delta_n\, e^{Nt^2}\,dt$$

$$= \frac{1}{2\pi^{3/2}} \int_0^\infty f_1\, f_2\, f_3\,dt,$$

(8.14)

where
$$f_1 = \sum_{k=-\infty}^{\infty} (-1)^k \exp\left\{-t^2 \left[ka + (-1)^k x' - x\right]^2\right\},$$

$$f_2 = \sum_{m=-\infty}^{\infty} (-1)^m \exp\left\{-t^2 \left[mb + (-1)^m y' - y\right]^2\right\},$$

$$f_3 = \sum_{n=-\infty}^{\infty} (-1)^n \exp\left\{-t^2 \left[nc + (-1)^n z' - z\right]^2\right\}.$$

We will express the function G in terms of the Riemann ϑ-functions defined by

$$\vartheta_m(\zeta, \tau) = \vartheta(\zeta, \tau) = \sum_{\nu=-\infty}^{\infty} e^{i\pi r^2 \tau} e^{2i\pi\nu\zeta}.$$

But first we will show that the integral in (8.14) converges to zero on the interval $[0, \infty)$. The proof is as follows: Since

$$\int_0^\infty f_1 f_2 f_3 \, dt = \int_0^\infty \sum_{k=-\infty}^{\infty} \int_0^\infty f_2 f_3 \exp\left\{-t^2 \left[ka + (-1)^k x' - x\right]^2\right\} dt,$$

we have

$$\left| \sum_{|k|>p} (-1)^k \exp\left\{-t^2 \left[ka + (-1)^k x' - x\right]^2\right\} \right| < e^{a^2 t^2/4} \sum_{|k|>p} (-1)^k e^{-a^2 t^2 k^2/2}$$

$$< \frac{2e^{-a^2 t^2/4}}{a^2} \sum_{|k|>p} (-1)^k \frac{1}{k^2} e^{-a^2 t^2/4} \frac{1}{p^2} < \frac{4}{pa^2} e^{-a^2 t^2/4}.$$

Thus, the integral over the interval $[0, \infty)$ converges to zero with increasing p, and f_2, f_3 remain uniformly bounded in the interval $[0, \infty)$. This completes the proof. Now, to represent Green's function G in terms of ϑ_{00}-function, we find that

$$f_1 = e^{-t^2(x-x')^2} \vartheta_{00}\left(-\frac{2at^2 i (x - x')}{\pi}, \frac{4a^2 t^2 i}{\pi}\right)$$

$$- e^{-t^2(x+x')^2} \vartheta_{00}\left(-\frac{2at^2 i (x + x')}{\pi}, \frac{4a^2 t^2 i}{\pi}\right)$$

$$f_2 = e^{-t^2(y-y')^2} \vartheta_{00}\left(-\frac{2bt^2 i (y - y')}{\pi}, \frac{4b^2 t^2 i}{\pi}\right)$$

$$- e^{-t^2(y+y')^2} \vartheta_{00}\left(-\frac{2bt^2 i (y + y')}{\pi}, \frac{4b^2 t^2 i}{\pi}\right)$$

$$f_3 = e^{-t^2(z-z')^2} \vartheta_{00}\left(-\frac{2ct^2 i (z - z')}{\pi}, \frac{4c^2 t^2 i}{\pi}\right)$$

$$- e^{-t^2(z+z')^2} \vartheta_{00}\left(-\frac{2ct^2 i (z + z')}{\pi}, \frac{4c^2 t^2 i}{\pi}\right).$$

To each we apply the formula

$$\vartheta_{00}(\zeta, \tau) = \frac{e^{-i\pi\zeta^2/\tau}}{\sqrt{-i\tau}} \vartheta_{00}\left(\frac{\zeta}{\tau}, -\frac{1}{\tau}\right),$$

taking the principal value of the square root, and setting $q_x = e^{-\pi^2/(4a^2t^2)}$, $q_y = e^{-\pi^2/(4b^2t^2)}$, and $q_z = e^{-\pi^2/(4c^2t^2)}$. Then

$$
\begin{aligned}
f_1 &= \frac{\sqrt{\pi}}{2at}\left[\vartheta_{00}\left(-\frac{x-x'}{2a}, \frac{\pi i}{4a^2t^2}\right) - \vartheta_{00}\left(-\frac{x+x'}{2a}, \frac{\pi i}{4a^2t^2}\right)\right] \\
&= \frac{\sqrt{\pi}}{2at}\left[\sum_{k=-\infty}^{\infty} q_x^{k^2} e^{-k(x-x')\pi i/a} - \sum_{k=-\infty}^{\infty} q_x^{k^2} e^{-k(x+x')\pi i/a}\right] \\
&= \frac{\sqrt{\pi}}{at}\sum_{k=1}^{\infty} q_x^{k^2}\left(\cos\frac{k(x-x')\pi}{a} - \cos\frac{k(x+x')\pi}{a}\right) \\
&= \frac{\sqrt{2\pi}}{at}\sum_{k=1}^{\infty} q_x^{k^2}\sin\frac{k\pi x}{a}\sin\frac{k\pi x'}{a};
\end{aligned}
$$

and similar expressions for f_2 and f_3. Thus, (8.14) becomes

$$
\begin{aligned}
G(x, y, z; x', y', z') &= \frac{4}{abc}\int_0^{\infty}\frac{1}{t^2}\sum_{k=1}^{\infty}\sum_{m=1}^{\infty}\sum_{n=1}^{\infty}\sin\frac{k\pi x}{a}\sin\frac{k\pi x'}{a}\times \\
&\quad \times \sin\frac{m\pi y}{b}\sin\frac{m\pi y'}{b}\sin\frac{n\pi z}{c}\sin\frac{n\pi z'}{c}\exp\left\{-\frac{\pi}{4t^2}\left[\frac{k^2}{a^2}+\frac{m^2}{b^2}+\frac{n^2}{c^2}\right]\right\},
\end{aligned}
$$

which, after setting $1/t^2 = \tau$, gives

$$
\begin{aligned}
G(x, y, z; x', y', z') &= \frac{2}{abc}\int_0^{\infty}\sum_{k=1}^{\infty}\sum_{m=1}^{\infty}\sum_{n=1}^{\infty}\exp\left\{-\frac{\pi^2\tau}{4}\left[\frac{k^2}{a^2}+\frac{m^2}{b^2}+\frac{n^2}{c^2}\right]\right\}\times \\
&\quad \times \sin\frac{k\pi x'}{a}\sin\frac{k\pi x}{a}\sin\frac{k\pi x'}{a}\sin\frac{m\pi y}{b}\sin\frac{m\pi y'}{b}\sin\frac{n\pi z}{c}\sin\frac{n\pi z'}{c} \\
&= \frac{8}{abc\pi^2}\sum_{k=1}^{\infty}\sum_{m=1}^{\infty}\sum_{n=1}^{\infty} \\
&\quad \frac{\sin\dfrac{k\pi x'}{a}\sin\dfrac{k\pi x}{a}\sin\dfrac{k\pi x'}{a}\sin\dfrac{m\pi y}{b}\sin\dfrac{m\pi y'}{b}\sin\dfrac{n\pi z}{c}\sin\dfrac{n\pi z'}{c}}{\dfrac{k^2}{a^2}+\dfrac{m^2}{b^2}+\dfrac{n^2}{c^2}}.
\end{aligned}
$$

If $\tau = 1/t^2$, then

$$G(x, y, z; x', y', z')$$
$$= \frac{1}{32abc} \int_0^\infty \left\{ \left[\vartheta_{00} \left(-\frac{x - x'}{2a}, \frac{\pi i \tau}{4a^2} \right) - \vartheta_{00} \left(-\frac{x + x'}{2a}, \frac{\pi i \tau}{4a^2} \right) \right] \right.$$
$$\left[\vartheta_{00} \left(-\frac{y - y'}{2b}, \frac{\pi i \tau}{4b^2} \right) - \vartheta_{00} \left(-\frac{y + y'}{2b}, \frac{\pi i \tau}{4b^2} \right) \right] \qquad (8.15)$$
$$\left. \left[\vartheta_{00} \left(-\frac{z - z'}{2c}, \frac{\pi i \tau}{4c^2} \right) - \vartheta_{00} \left(-\frac{z + z'}{2c}, \frac{\pi i \tau}{4c^2} \right) \right] \right\}. \blacksquare$$

8.4. Harmonic Functions

A real-valued function $u(x, y) \in C^2(D)$ is said to be *harmonic* in a region D iff $\nabla^2 u = 0$ in D. Some properties of harmonic functions in $\mathbb{R}^2$ are as follows:

(i) The function $\dfrac{1}{r} = \dfrac{1}{\sqrt{(x - x_0)^2 + (y - y_0)^2}}$ is harmonic in a region that does not contain the point (x_0, y_0).

(ii) If $u(x, y)$ is a harmonic function in a simply connected region D, then there exists a conjugate harmonic function $v(x, y)$ in D such that $u(x, y) + i\, v(x, y)$ is an analytic function of $z = x + iy = (x, y)$ in D. In view of the Cauchy-Riemann equations[4],

$$v(x, y) - v(x_0, y_0) = \int_{(x_0, y_0)}^{(x, y)} (-u_y \, dx + u_x \, dy),$$

where $(x_0, y_0) = z_0$ is a given point in D. This property is also true if D is multiply connected. However, in that case the conjugate function $v(x, y)$ can be multiple-valued, as we see by considering $u(x, y) = \log r = \log \sqrt{x^2 + y^2}$ defined on a region D containing the origin which has been indented by a small circle centered at the origin. Then, in view of the Cauchy-Riemann equations, we get

$$v(x, y) - v(x_0, y_0) = \tan^{-1} \frac{y}{x} \pm 2n\pi + \text{const}, \quad n = 1, 2, \dots,$$

which is multiple-valued.

(iii) Since derivatives of all orders of an analytic function exist and are themselves analytic, any harmonic function will have continuous partial derivatives of all orders, i.e., a harmonic function belongs to the class $C^\infty(D)$, and a partial derivative of any order is again harmonic.

(iv) A harmonic function must satisfy the mean-value theorem, where the mean value at a point is evaluated for the circumference or the area of the circle around

---

[4] The Cauchy-Riemann equations for a function $w = f(z) = u(x, y) + iv(x, y)$, $z = x + iy$, which is an analytic on a region D, are $u_x = v_y$, $u_y = -v_x$.

that point. If u is harmonic on a region containing the closed disk $\bar{B}(r, z_0)$, where $z_0 = x_0 + iy_0$, then

$$u(x_0, y_0) = \frac{1}{2\pi} \int_0^{2\pi} u\left(z_0 + r\, e^{i\theta}\right) d\theta.$$

(v) In view of the maximum modulus theorem, the maximum (and also the minimum) of a harmonic function u in a region D occurs only on the boundary ∂D.

Theorem 8.1. (Maximum Principle) *A nonconstant function that is harmonic inside a bounded region D with boundary Γ and continuous in the closed region $\bar{D} = D \cup \Gamma$ attains its maximum and minimum values only on the boundary Γ of the region.*

Thus, u has a maximum (or minimum) at $z_0 \in D$, i.e., if $u(z) \leq u(z_0)$ (or $u(z) \geq u(z_0)$) for z in a neighborhood $B(\varepsilon, z_0)$ of z_0, then $u = $ const in $B(\varepsilon, z_0)$.

(v) The value of a harmonic function u at an interior point in terms of the boundary values u and $\dfrac{\partial u}{\partial n}$ is given by Green's third identity (1.19).

(vi) A harmonic function satisfies the mean-value theorem, according to which the value of a harmonic function at a point is equal to the average value of the function on a circle in D with its center at the point.

Proofs of these results are available in textbooks on complex analysis, e.g., Carrier, Krook, and Pearson [1966]. The functions u that satisfy Laplace's equation $\nabla^2 u = 0$ are known as harmonic functions. We have the following result which is a corollary of Theorem 2.2.

Theorem 8.2. Green's function for the Laplacian is symmetric, i.e.,

$$G(\mathbf{x}, \mathbf{y}) = G(\mathbf{y}, \mathbf{x}).$$

PROOF. Since the Laplacian is self-adjoint, its Green's function is symmetric, by Corollary 2.2. ∎

8.5. 2-D Helmholtz's Equation

The Helmholtz operator is defined as $(\nabla^2 + k^2)$, where $k > 0$ is real. Helmholtz's equation $(\nabla^2 + k^2)\, u = 0$ is also called the *reduced wave scattering equation* and k is known as the real (positive) acoustic wave number. Green's function for this operator satisfies the equation

$$(\nabla^2 + k^2)G(\mathbf{x} - \mathbf{x}') = \delta(\mathbf{x} - \mathbf{x}'). \tag{8.16}$$

In some books this operator is taken as $-\left(\nabla^2 + k^2\right)$, which signifies a source of strength $+1$.

For the operator $\nabla^2 + k^2$ in $\mathbb{R}^2$ we solve the equation

$$\left(\nabla^2 + k^2\right) G(x, y) = \delta(x)\,\delta(y).$$

Since the right side is zero everywhere except at the origin, where a negative source (sink) of unit strength exists, we find a solution with singularity at the origin such that

$$-\lim_{\varepsilon \to 0} \int_{r=\varepsilon} \frac{\partial G}{\partial n}\, ds = -1, \tag{8.17}$$

where r is the distance from the origin. The solution is symmetric with respect to the origin, and therefore, dependent only on r. Thus, the equation $\left(\nabla^2 + k^2\right) u = 0$ reduces to

$$\frac{\partial^2 G}{\partial r^2} + \frac{1}{r}\frac{\partial G}{\partial r} + k^2 G = 0 \quad \text{for } r \neq 0.$$

Its solution is

$$G(r) = A H_0^{(1)}(kr) + B H_0^{(2)}(kr),$$

where $H_0^{(1)}(kr) = J_0(kr) + i Y_0(kr)$ and $H_0^{(2)}(kr) = J_0(kr) - i Y_0(kr)$ are the Hankel functions of the first and second kind, respectively. Applying the condition (8.17), we find that

$$
\begin{aligned}
-\lim_{\varepsilon \to 0} \int_{r=\varepsilon} \frac{\partial G}{\partial n}\, ds &= -\lim_{\varepsilon \to 0} 2\pi\varepsilon \left[\frac{\partial}{\partial r} \left\{ A H_0^{(1)}(kr) + B H_0^{(2)}(kr) \right\} \right]_{r=\varepsilon} \\
&= \lim_{\varepsilon \to 0} 2\pi\varepsilon \left[k \left\{ (A + B) J_1(kr) + (A - B) i Y_1(kr) \right\} \right]_{r=\varepsilon} \\
&= 2\pi \left(-\frac{2i}{\pi} \right)(A - B) = -1,
\end{aligned}
$$

which gives $A - B = -i/4$. Since the contribution from the coefficient $(A + B)$ vanishes, we can assign it any value. Traditionally, B is taken zero. Hence, Green's function is

$$G(r) = -\frac{i\, H_0^{(1)}(kr)}{4}. \tag{8.18}$$

8.5.1. Closed-Form Green's Function for Helmholtz's Equation. Consider the equation $\nabla^2 u(\mathbf{r}) + k^2 u(\mathbf{r}) = -f(\mathbf{r})$. The eigenfunctions $\phi_n(\mathbf{r})$ are the solutions of the homogeneous equation, i.e.,

$$\nabla^2 \phi_n(\mathbf{r}) + k_n^2 \phi_n(\mathbf{r}) = -f(\mathbf{r}). \tag{8.19}$$

Green's function $G(\mathbf{r}_1, \mathbf{r}_2)$ satisfies the (point source) equation

$$\nabla^2 G(\mathbf{r}_1, \mathbf{r}_2) + k^2 G(\mathbf{r}_1, \mathbf{r}_2) = -\delta(\mathbf{r}_1 - \mathbf{r}_2).$$

Expanding Green's function in a series of eigenfunctions of the homogeneous equation (8.19), we get

$$G(\mathbf{r}_1, \mathbf{r}_2) = \sum_{n+0}^{\infty} a_n(\mathbf{r}_2)\phi_n(\mathbf{r}_1), \tag{8.20}$$

which, after substituting into (8.20), gives

$$\sum_{n=0}^{\infty} (k_n^2 - k^2) \, a_n(\mathbf{r}_2) \, \phi_n(\mathbf{r}_1) = \sum_{n=0}^{\infty} \phi_n(\mathbf{r}_1)\phi_n(\mathbf{r}_2), \qquad (8.21)$$

where we have replaced $\delta(\mathbf{r}_1 - \mathbf{r}_2)$ by its eigenfunction expansion (see §3.4). Then, by using the orthogonality of the eigenfunction $\phi_n(\mathbf{r}_1)$ we determine a_n which, when substituted into (8.21), yield the required Green's function (in closed form):

$$G(\mathbf{r}_1, \mathbf{r}_2) = \sum_{n=0}^{\infty} \frac{\phi_n(\mathbf{r}_1) \, \phi_n(\mathbf{r}_2)}{k_n^2 - k^2}, \qquad (8.22)$$

which is symmetric in $\mathbf{r}_1$ and $\mathbf{r}_2$. Hence, the solution of the nonhomogeneous Helmholtz's equation is given by

$$u(\mathbf{r}_1) = \int G(\mathbf{r}_1, \mathbf{r}_2) \, f(\mathbf{r}_2) \, d\mathbf{r}_1. \qquad (8.23)$$

If the given equation is $L[u](\mathbf{r}) + \lambda \, u(\mathbf{r}) = -f(\mathbf{r})$, where L is a self-adjoint operator, then the above Green's function becomes

$$G(\mathbf{r}_1, \mathbf{r}_2) = \sum_{n=0}^{\infty} \frac{\phi_n(\mathbf{r}_1) \, \phi_n(\mathbf{r}_2)}{\lambda_n^2 - \lambda^2}, \qquad (8.24)$$

where (λ_n, ϕ_n) are orthonormalized eigenpairs of the associated homogeneous equation. ∎

8.6. Green's Function for 3-D Helmholtz's Equation

In $\mathbb{R}^3$, it is obvious from physical considerations that Green's function for the Helmholtz operator must be spherically symmetric. Then for $\mathbf{x}' = 0$ Green's function $G(\mathbf{x})$ in the spherical coordinates satisfies the equation

$$\frac{1}{r^2} \frac{d}{dr} \left(r^2 \frac{dG}{dr} \right) + k^2 G = \frac{\delta(r)}{4\pi r^2}. \qquad (8.25)$$

If we substitute $G = w/r$, then this equation reduces to $\dfrac{d^2 w}{dr^2} + k^2 w = \dfrac{\delta(r)}{4\pi r}$, which, after applying the Fourier sine transform, gives

$$\sqrt{\frac{2}{\pi}} \, \alpha \, c_0 + (k^2 - \alpha^2) \, \tilde{w}_s = \sqrt{\frac{2}{\pi}} \int_0^{\infty} \frac{\delta(r)}{4\pi r} \sin(\alpha r) \, dr$$

$$= \sqrt{\frac{2}{\pi}} \frac{1}{4\pi} \lim_{r \to 0} \frac{\sin(\alpha r)}{r} = \frac{\alpha}{4\pi} \sqrt{\frac{2}{\pi}},$$

where $c_0 = \lim_{r \to 0} w$. Then $\widetilde{w}_s = \sqrt{\dfrac{2}{\pi}} \left(c_0 - \dfrac{1}{4\pi}\right) \dfrac{\alpha}{\alpha^2 - k^2}$, which, on inverting yields $w = \left(c_0 - \dfrac{1}{4\pi}\right) e^{-ikr}$. Thus, $G(r) = \left(c_0 - \dfrac{1}{4\pi}\right) \dfrac{e^{-ikr}}{r}$. Using the flux condition, we get $-\lim_{\varepsilon \to 0} \int_S \left[\dfrac{dG}{dr}\right]_{r \to \varepsilon} dS = -1$, where S is a sphere of radius ε with center at the origin and $dS = \varepsilon^2 \sin\phi \, d\phi \, d\theta$ is the infinitesimal element of this surface. Thus, since

$$\lim_{\varepsilon \to 0} \left(c_0 - \dfrac{1}{4\pi}\right) \int_0^{2\pi} \int_0^{\pi} \left(\dfrac{ik\,e^{-ik\varepsilon}}{\varepsilon} + \dfrac{e^{-ik\varepsilon}}{\varepsilon^2}\right) \varepsilon^2 \sin\phi \, d\phi \, d\theta = -1,$$

we get $\left(c_0 - \dfrac{1}{4\pi}\right) 4\pi = -1$, which gives $c_0 = 0$. Hence,

$$G(r) = -\dfrac{e^{-ikr}}{4\pi r}. \tag{8.26}$$

The physical interpretation of this Green's function is as follows: If we consider the flux of the quantity $G(r)$ across a sphere of radius r and take the limit as $\varepsilon \to 0$, we obtain the limiting value -1. This implies the presence of a sink of unit strength at the origin. If instead of Eq (8.25) we were to find the solution of the equation

$$\dfrac{1}{r^2} \dfrac{d}{dr}\left(r^2 \dfrac{dG}{dr}\right) + k^2 G = -\dfrac{\delta(r)}{4\pi r^2}, \tag{8.27}$$

the solution $G(r) = \dfrac{e^{-ikr}}{4\pi r}$ would then represent Green's function due to a source of unit strength at the origin. We point out that the flux of a quantity u away from the source is always given by $-\dfrac{\partial u}{\partial n}$, where n is the outward unit normal to the surface. ∎

Theorem 8.3. *Green's function for the Helmholtz Operator is symmetric, that is,*

$$G(\mathbf{x}, \mathbf{y}) = G(\mathbf{y}, \mathbf{x}).$$

The proof is similar to that of Theorem 8.2 and is left as an exercise.

8.7. 2-D Poisson's Equation in a Circle

In the polar cylindrical coordinates the problem in a circle of radius a is defined by a Poisson's equation as

$$\nabla^2 u \equiv \dfrac{\partial^2 u}{\partial r^2} + \dfrac{1}{r} \dfrac{\partial u}{\partial r} + \dfrac{1}{r^2} \dfrac{\partial^2 u}{\partial \theta^2} = F(r, \theta), \quad r < a, \tag{8.28}$$

$$u(a, 0) = 0.$$

This problem can always be solved by Bernoulli's separation method by taking $u = R(r)\Theta(\theta)$, and is left as an exercise. However, we will use Fourier transform method to solve this problem. Since the basis is $\{r^n \cos n\theta, r^n \sin n\theta\}$, the solution, if any, can be formally written as a trigonometric Fourier series

$$u(r,\theta) = \frac{1}{2} a_0(r) + \sum_{n=1}^{\infty} \left[a_n(r) \cos n\theta + b_n(r) \sin n\theta \right],$$

where

$$a_n(r) = \frac{1}{\pi} \int_{-\pi}^{\pi} u(r,\theta) \cos n\theta \, d\theta, \quad b_n(r) = \frac{1}{\pi} \int_{-\pi}^{\pi} u(r,\theta) \sin n\theta \, d\theta.$$

Since u is a periodic function of period 2π, we take the finite Fourier transform of (8.28) and get $a_n'' + \frac{1}{r} a_n' - \frac{n^2}{r^2} a_n = A_n(r), b_n'' + \frac{1}{r} b_n' - \frac{n^2}{r^2} b_n = B_n(r)$, where

$$A_n(r) = \frac{1}{\pi} \int_{-\pi}^{\pi} F(r,\theta) \cos n\theta \, d\theta, \quad B_n(r) = \frac{1}{\pi} \int_{-\pi}^{\pi} F(r,\theta) \sin n\theta \, d\theta. \quad (8.29)$$

The boundary conditions imply that $a_n(a) = b_n(a) = 0$. At the singular point $r = 0$, let us require that both $a_n(r)$ and $b_n(r)$ be bounded. Then from (8.29) we have

$$(r\,a_n)' - \frac{n^2}{r} a_n = r\,A_n, \quad (r\,b_n)' - \frac{n^2}{r} b_n = r\,B_n, \quad a_n(a) = b_n(a) = 0,$$

which is a Sturm-Liouville system. Using above conditions on a_n and b_n at $r = a$ and $r = 0$, we get

$$a_0(r) = \int_0^r \ln\left(\frac{a}{r}\right) A_0(r')\, r'\, dr' + \int_r^a \ln\left(\frac{a}{r}\right) A_0(r')\, r'\, dr',$$

$$a_n(r) = \frac{1}{2\pi}\left\{ \int_0^r \left[\left(\frac{a}{r}\right)^n - \left(\frac{r}{a}\right)^n \right] \left(\frac{r'}{a}\right)^n A_n(r')\, r'\, dr' \right.$$
$$\left. + \int_r^a \left(\frac{r}{a}\right)^n \left[\left(\frac{a'}{r}\right)^n - \left(\frac{r'}{a}\right)^n \right] A_n(r')\, r'\, dr' \right\}, \quad \text{for } n \geq 1, \quad (8.30)$$

$$b_n(r) = \frac{1}{2\pi}\left\{ \int_0^r \left[\left(\frac{a}{r}\right)^n - \left(\frac{r}{a}\right)^n \right] \left(\frac{r'}{a}\right)^n B_n(r')\, r'\, dr' \right.$$
$$\left. + \int_r^a \left(\frac{r}{a}\right)^n \left[\left(\frac{a'}{r}\right)^n - \left(\frac{r'}{a}\right)^n \right] B_n(r')\, r'\, dr' \right\},$$

Now, by Schwarz's inequality (§1.9.3), we have for $n \geq 2$

$$\left| a_n(r)^2 \right| \leq \frac{a^2}{4n\,(n^2 - 1)} \int_0^a [A_n(r)]^2 \, r'\, dr',$$

$$\left| b_n(r)^2 \right| \leq \frac{a^2}{4n\,(n^2 - 1)} \int_0^a [B_n(r)]^2 \, r'\, dr'.$$

If $\int_0^a \int_{-\pi}^{\pi} F(r,\theta)\, r\, dr\, d\theta < +\infty$, then the Fourier series solution of the problem (8.29) is

$$u(r,\theta) = \frac{1}{2} a_0(r) + \sum_{n=1}^{\infty} \left[a_n(r) \cos n\theta + b_n \sin n\theta \right], \tag{8.31}$$

which converges uniformly, i.e., $u \in C^2$ and satisfies the boundary condition.

NOTE. If the boundary condition is changed to $u(a,\theta) = f(\theta)$, and if $F = 0$, then the solution by Bernoulli's separation method is

$$u(r,\theta) = \frac{1}{2} a_0(r) + \sum_{n=1}^{\infty} \left(\frac{r}{a}\right)^n \left[a_n(r) \cos n\theta + b_n \sin n\theta \right], \tag{8.32}$$

where a_n and b_n are the Fourier coefficients of $f(\theta)$.

Green's function for this problem is obtained as follows: Substituting (8.30) into the solution (8.31), we get

$$u(r,\theta) = \frac{1}{2} \int_0^r A_0(r') \ln\left(\frac{a}{r}\right) r'\, dr' + \sum_{n=1}^{\infty} \int_0^r \frac{1}{2n} \left[\left(\frac{a}{r}\right)^n - \left(\frac{r}{a}\right)^n \right] \left(\frac{r'}{a}\right)^n .$$
$$\cdot \left[A_n(r') \cos n\theta + B_n(r') \sin n\theta \right] r'\, dr' +$$
$$+ \frac{1}{2} \int_r^a A_0(r') \ln\left(\frac{a'}{r}\right) r'\, dr' + \sum_{n=1}^{\infty} \int_r^a \frac{1}{2n} \left[\left(\frac{a'}{r}\right)^n - \left(\frac{r'}{a}\right)^n \right] \left(\frac{r}{a}\right)^n .$$
$$\cdot \left[A_n(r') \cos n\theta + B_n(r') \sin n\theta \right] r'\, dr', \tag{8.33}$$

which after interchanging summation and integration gives

$$u(r,\theta) = \int_{-\pi}^{\pi} \int_0^a G(r,\theta;r',\theta')\, F(r',\theta')\, r'\, dr'\, d\theta',$$

where for $r' < r$

$$G(r,\theta;r',\theta') = \frac{1}{2\pi} \left\{ \ln\left(\frac{a}{r}\right) + \sum_{n=1}^{\infty} \frac{1}{n} \left[\left(\frac{a}{r}\right)^n - \left(\frac{r}{a}\right)^n \right] \left(\frac{r'}{a}\right)^n \cos n(\theta - \theta') \right\}, \tag{8.34}$$

while for $r' > r$ we find that $G(r,\theta;r',\theta') = G(r',\theta';r,\theta)$. The series in the above solution are uniformly convergent in θ' for $r' < r$. If we use the identity

$$\frac{1}{2} + \sum_{n=1}^{\infty} r^n \cos n\theta = \frac{1 - r^2}{2\left[1 - 2r \cos\theta + r^2\right]},$$

then

$$\sum_{n=1}^{\infty} \frac{1}{n} r^n \cos n\theta = \int_0^r \frac{\cos\theta - z'}{1 - 2z' \cos\theta + z'^2}\, dz' = -\frac{1}{2} \ln\left[1 - 2r \cos\theta + r^2\right].$$

Thus, from (8.34) we find that

$$G(r,\theta;r',\theta') = \frac{1}{2\pi}\left\{\ln\left(\frac{a}{r}\right) + \frac{1}{4\pi}\ln\left[1 - \frac{2rr'}{a^2}\cos(\theta - \theta') + \left(\frac{rr'}{a^2}\right)^2\right] - \right.$$
$$\left. - \frac{1}{4\pi}\ln\left[1 - \frac{2r'}{r}\cos(\theta - \theta') + \frac{r'^2}{r^2}\right]\right.$$
$$= \frac{1}{4\pi}\left[\ln\left\{a^2 - 2rr'\cos(\theta - \theta') + \frac{r^2 r'^2}{a^2}\right\} - \right.$$
$$\left. - \ln\left\{r^2 - 2rr'\cos(\theta - \theta') + r'^2\right\}\right]$$
$$= \frac{1}{4\pi}\ln\left\{\frac{a^2 - 2rr'\cos(\theta - \theta') + \dfrac{r^2 r'^2}{a^2}}{r^2 - 2rr'\cos(\theta - \theta') + r'^2}\right\}, \quad \text{for } r' < r. \quad (8.35)$$

This Green's function is continuous at $r' = r$; in fact, it belongs to the class C^2 and satisfies Laplace's equation $\nabla^2 G = 0$ at all points except the point (r', θ'). Moreover, $\left|\dfrac{\partial G}{\partial r}\right|$ and $\left|\dfrac{\partial G}{\partial\theta}\right|$ have finite (improper) integrals, in spite of the singularities, and so we can differentiate u under the integral sign and get

$$\frac{\partial u}{\partial r} = \iint \frac{\partial G}{\partial r} F(r',\theta')\, r'\, dr'\, d\theta', \quad \frac{\partial u}{\partial\theta} = \iint \frac{\partial G}{\partial\theta} F(r',\theta')\, r'\, dr'\, d\theta', \quad (8.36)$$

provided $F(r',\theta')$ is bounded near (r,θ). For $r' > r$, the Green's function is $G(r,\theta;r',\theta') = G(r',\theta';r,\theta)$.

NOTES. 1. The expression $\ln\left[r^2 - 2rr'\cos(\theta - \theta') + r'^2\right] r'\, dr'\, d\theta'$ becomes $\ln\left\{(x - x')^2 + (y - y')^2\right\} dx'\, dy'$ in the rectangular cartesian coordinates. Thus, the above Green's function (8.35) in rectangular cartesian coordinates is

$$G(x,y;x',y') = \frac{1}{4\pi}\left[\ln\left\{(x - x')^2 + (y - y')^2\right\} - \right.$$
$$\left. - \ln\left\{a^2 - 2(xx' + yy') + \frac{(x^2 + y^2)(x'^2 + y'^2)}{a^2}\right\}\right], \quad (8.37)$$

2. Since $\nabla^2\left[\ln\left\{a^2 - 2rr'\cos(\theta - \theta') + \dfrac{r^2 r'^2}{a^2}\right\}\right] = 0$ throughout the circle, we conclude that for any bounded domain D the function

$$v(r,\theta) = \frac{1}{4\pi}\iint_D \ln\left\{r^2 - 2rr'\cos(\theta - \theta') + r'^2\right\} F(r',\theta')\, r'\, dr'\, d\theta'$$

is a particular solution of Poisson's equation $\nabla^2 v = F(r,\theta)$ in D, provided $F \in C^2$. In rectangular cartesian coordinates, this function becomes

$$v^*(x,y) = \frac{1}{4\pi}\iint_D \ln\left[(x - x')^2 + (y - y')^2\right] F^*(x',y')\, dx'\, dy',$$

where $v^*(x,y) = v^*(r\cos\theta, r\sin\theta) = v(r,\theta)$ and $F^*(x,y) = F^*(r\cos\theta, r\sin\theta) = F(r,\theta)$. Hence, we have the following result:

Theorem 8.4. *For any bounded domain D with boundary ∂D, the solution of the nonhomogeneous problem*

$$\nabla^2 u = F \quad in\ D,\ subject\ to\ the\ condition \quad u = 0 \quad on\ \partial D, \tag{8.38}$$

is of the form

$$u(x,y) = \iint_D G(x,y;x',y')\,F(x',y')\,dx'\,dy', \tag{8.39}$$

where $G(x,y;x',y')$ is Green's function for Poisson's equation in D.

Physically, this Green's function represents the potential at a point (x,y) due to a point source at (x',y'), or the displacement at (x,y) of a circular membrane due to a point force at (x',y').

From (8.37) it is clear that Green's function in rectangular cartesian coordinates is of the form

$$G(x,y;x',y') = \frac{1}{4\pi}\left[\ln\left\{(x-x')^2 + (y-y')^2\right\} + \gamma(x,y;x',y')\right], \tag{8.40}$$

where, for each $(x',y') \in D$, the term γ is the solution of the problem

$$\nabla^2\gamma = 0 \quad in\ D.$$
$$\gamma = \frac{1}{4\pi}\ln\left\{(x-x')^2 + (y-y')^2\right\} \quad for\ (x,y) \in \partial D. \tag{8.41}$$

Example 8.1. Let D be the circle of radius a. Then the problem (8.40) in polar cylindrical coordinates becomes

$$\nabla^2\gamma = 0 \quad for\ r < a,$$
$$\gamma(a,\theta;r',\theta') = \frac{1}{4\pi}\ln\left[r'^2 - 2r'a\cos(\theta - \theta') + a^2\right].$$

Since the Fourier series for the problem gives

$$\frac{1}{4\pi}\ln\left[r'^2 - 2r'a\cos(\theta - \theta') + a^2\right] = \frac{1}{2\pi}\ln a - \frac{1}{2\pi}\sum_{n=1}^{\infty}\frac{1}{n}\left(\frac{r'}{a}\right)^n \cos n(\theta - \theta'),$$

we find that

$$\gamma(r,\theta;r',\theta') = \frac{1}{2\pi}\ln a - \frac{1}{2\pi}\sum_{n=1}^{\infty}\frac{1}{n}\left(\frac{r}{a}\right)^n\left(\frac{r'}{a}\right)^n \cos n(\theta - \theta')$$
$$= \frac{1}{2\pi}\ln a + \frac{1}{4\pi}\ln\left[1 - 2\frac{rr'}{a^2}\cos(\theta - \theta') + \frac{r^2 r'^2}{a^4}\right]$$
$$= \frac{1}{4\pi}\ln\left[a^2 - 2rr'\cos(\theta - \theta') + \frac{r^2 r'^2}{a^2}\right],$$

which when substituted in (8.40) gives (8.35). ∎

NOTES. 1. It is important to observe that the introduction of Green's function reduces the nonhomogeneous problem (8.28) to a 2-parameter family (8.41) of a homogeneous problem.

2. The problem (8.28) can be reduced to a homogeneous problem by introducing a particular solution v of the problem $\nabla^2 v = F$; thus, for example, by introducing the function v or v^* given above, and setting $w = u - v$ or $w = u - v^*$. The function w satisfies the boundary value problem

$$\nabla^2 w = 0 \quad \text{on } D, \text{ such that } w = u - v \text{ on } \partial D,$$

so that after finding w we obtain $u = w + v$ or $u = w + v^*$.

Example 8.2. Let $\nabla^2 u = 1$ in the square $S : \{0 < x < \pi, \, 0 < y < \pi\}$, such that $u = 0$ on the sides $x = 0, x = \pi, y = 0, y = \pi$. Let $w = u - v$. Then

$$\nabla^2 w = 0 \quad \text{on } S,$$
$$w(x,0)w(x,\pi) = \tfrac{1}{2}x(x - \pi), \quad w(0,y) = w(\pi,y) = 0.$$

Using Bernoulli's separation method, the solution of this system is

$$w(x,y) = \frac{4}{\pi} \sum_{k=1}^{\infty} \frac{\sin(2k-1)x \, \cosh(2k-1)\left(y - \frac{\pi}{2}\right)}{(2k-1)^2 \, \cosh(2k-1)\frac{\pi}{2}},$$

and thus,

$$u(x,y) = \tfrac{1}{2}x(x - \pi) + \frac{4}{\pi} \sum_{k=1}^{\infty} \frac{\sin(2k-1)x \, \cosh(2k-1)\left(y - \frac{\pi}{2}\right)}{(2k-1)^2 \, \cosh(2k-1)\frac{\pi}{2}}. \; \blacksquare$$

8.8. Method for Green's Function in a Rectangle

The above analysis leads to the following methods to determine Green's function for the parabolic or elliptic operators in compact domains.

8.8.1. METHOD 1: To solve the problem

$$\nabla^2 u = F \quad \text{in } R : \{0 < x < a, \, 0 < y < b\},$$
$$u = 0 \quad \text{on the sides } x = 0, x = a, y = 0, y = b,$$

follow these steps:

STEP 1. Use the Fourier sine transform:

$$b_n(y) = \frac{2}{a} \int_0^a u(x,y) \sin nx \, dx, \quad B_n(y) = \frac{2}{a} \int_0^a F(x,y) \sin nx \, dx.$$

STEP 2. This reduces the problem to

$$b_n'' - n^2 b_n = B_n \quad \text{for } 0 < x < a,$$
$$b_n(0) = b_n(a) = 0,$$

which has the solution

$$b_n(y) = \frac{1}{n \sinh nb} \left[\int_0^y \sinh n(b-y) \sinh ny' B_n(y')) \, dy' \right.$$
$$\left. + \int_y^b \sinh ny \sinh n(b-y') B_n(y') \, dy' \right].$$

STEP 3. Write the Fourier series for u by substituting the definition of B_n, and interchanging integration and differentiation. This yields the formal solution

$$u(x,y) = \int_0^a \int_0^b G(x,y;x',y') \, F(x',y') \, dx' \, dy',$$

where

$$G(x,y;x',y') = \begin{cases} \displaystyle\sum_{n=1}^{\infty} \frac{2 \sinh n(b-y) \sinh by' \sin nx \sin nx'}{na \sinh nb}, & y' \leq y, \\[2ex] \displaystyle\sum_{n=1}^{\infty} \frac{2 \sinh ny \sinh n(b-y') \sin nx \sin nx'}{na \sinh nb}, & y \leq y', \end{cases}$$

or in terms of the elliptic functions

$$G(x,y;x',y') = \frac{1}{2\pi} \sum_{n=1}^{\infty} \frac{1}{n} \left\{ e^{-n|y-y'|} + \right.$$

$$+ \frac{e^{-n(2b+|y-y'|)} + e^{-n(2b-|y-y'|)} - e^{-2(2b-y-y')} - e^{-n(y+y')}}{1 - e^{-2nb}} \times$$

$$\left. \times \left[\cos n(x-x') - \cos n(x+x') \right] \right\}$$

$$= \frac{1}{4\pi} \ln \left[1 + e^{-2|y-y'|} - 2 e^{-|y-y'|} \cos(x+x') \right] -$$

$$- \frac{1}{4\pi} \ln \left[1 + e^{-2|y-y'|} - 2 e^{-|y-y'|} \cos(x-x') \right] +$$

$$+ \sum_{n=1}^{\infty} \frac{e^{-n(2b+|y-y'|)} + e^{2b-|y-y'|)} - e^{-n(2b-y-y')} - e^{-n(y-y')}}{na \left(1 - e^{-2nb}\right)} \sin nx \sin nx',$$

where the infinite series and all its derivatives are uniformly convergent in x and y for any fixed $(x',y') \in D$. ∎

8.8.2. METHOD 2. To find the solution of the homogeneous boundary value problem $\nabla^2 u = 0$ in D, such that $u = g$ on ∂D, in terms of the known Green's function, we follow these steps:

STEP 1. Use the above Method 1 to reduce the nonhomogeneous problem to a homogeneous problem.

STEP 2. Choose any point $(x_0, y_0) \in D$. Let $h(x, y) \in C^2(D)$ such that $h = g$ on ∂D and $h = 0$ in the neighborhood of (x_0, y_0).

STEP 3. Set $w = u - h$. Then w satisfies $\nabla^2 w = -\nabla^2 h$ in D, and $w = 0$ on ∂D. Hence,

$$w(x_0, y_0) = \iint_D G(x, y; x', y') \nabla^2 h(x', y') \, dx' \, dy'.$$

STEP 4. Since $h = 0$ near the point (x_0, y_0), apply the divergence theorem (§1.7) and the above relation for $w(x_0, y_0)$, which gives

$$w(x_0, y_0) = - \oint_{\partial D} \frac{\partial G}{\partial n}(x_0, y_0; x', y') \, g(x', y') \, ds,$$

where $\dfrac{\partial G}{\partial n}$ is the directional derivative of G in the direction of the unit vector n perpendicular to the boundary ∂D.

STEP 5. Since $h(x_0, y_0) = 0$, so $w(x_0, y_0) = u(x_0, y_0)$. As (x_0, y_0) is an arbitrary point in D, the solution $u(x, y)$ of this problem is given by

$$u(x, y) = - \oint_{\partial D} \frac{\partial G}{\partial n}(x_0, y_0; x', y') \, g(x', y') \, ds. \ \blacksquare \qquad (8.42)$$

Example 8.3. For the circle $r < a$, the Green's function is given by (8.35). Then

$$\frac{\partial G}{\partial n} = \frac{\partial G}{\partial r'}(r, \theta; a, \theta')$$

$$= \frac{1}{4\pi} \left[\frac{2r^2/a^2 - 2r\cos(\theta - \theta')}{r^2 - 2ra\cos(\theta - \theta') + a^2} - \frac{2a - 2r\cos(\theta - \theta')}{r^2 - 2ra\cos(\theta - \theta') + a^2} \right]$$

$$= \frac{1}{2a\pi} \frac{a^2 - r^2}{r^2 - 2ra\cos(\theta - \theta') + a^2}.$$

Thus, (8.42) gives

$$u(r, \theta) = \frac{a^2 - r^2}{2\pi} \int_{-\pi}^{\pi} \frac{g(\theta') \, d\theta'}{r^2 - 2ra\cos(\theta - \theta') + a^2}, \quad r < a, \qquad (8.43)$$

which the *Poisson's integral formula* for the circle. $\blacksquare$

8.9. Poisson's Equation in a Cube

Consider the problem

$$\nabla^2 u \equiv \frac{\partial^2 u}{\partial x^2} + \frac{\partial^2 u}{\partial y^2} + \frac{\partial^2 u}{\partial z^2} = F(x, y, z), \quad \text{in the cube } D : \{0 < x, y, z < \pi\},$$

$$u(0, y, z) = u(\pi, y, z) = 0,$$
$$u(x, 0, z) = u(x, \pi, z) = 0,$$
$$u(x, y, 0) = u(x, y, \pi) = 0, \tag{8.44}$$

where $u \in C^2$. Expand u and F in triple Fourier sine series:

$$u(x, y, z) = \sum_{l=1}^{\infty} \sum_{m=1}^{\infty} \sum_{n=1}^{\infty} a_{lmn} \sin lx \sin my \sin nz,$$

$$F(x, y, z) = \sum_{l=1}^{\infty} \sum_{m=1}^{\infty} \sum_{n=1}^{\infty} A_{lmn} \sin lx \sin my \sin nz,$$

substitute into Poisson's equation in (8.44), which gives $a_{lmn} = \dfrac{A_{lmn}}{l^2 + m^2 + n^2}$.
Thus,

$$u(x, y, z) = \sum_{l=1}^{\infty} \sum_{m=1}^{\infty} \sum_{n=1}^{\infty} \frac{A_{lmn}}{l^2 + m^2 + n^2} \sin lx \sin my \sin nz. \tag{8.45}$$

This series for $u(x, y, z)$ and its first and second derivatives converge absolutely and uniformly, provided the series for F does so. By Parseval's equality (§1.9.2), the solution (8.45) can be written formally as

$$u(x, y, z) = \int_0^\pi \int_0^\pi \int_0^\pi G(x, y, z; x', y', z') F(x', y', z') \, dx' \, dy' \, dz'$$

or
$$= \int_0^\pi \int_0^\pi \int_0^\pi G(\mathbf{x}, \mathbf{x}') F(\mathbf{x}') \, d\mathbf{x}', \tag{8.46}$$

where Green's function $G(\mathbf{x}, \mathbf{x}')$ is of the form

$$G(\mathbf{x}, \mathbf{x}') = \frac{8}{\pi^3} \sum_{l=1}^{\infty} \sum_{m=1}^{\infty} \sum_{n=1}^{\infty} \frac{\sin lx \sin my \sin nz \sin lx' \sin my' \sin nz'}{l^2 + m^2 + n^2}. \tag{8.47}$$

Let $r = |\mathbf{x} - \mathbf{x}'| = \sqrt{(x - x')^2 + (y - y')^2 + (z - z')^2}$, and choose a constant a depending on the fixed point $\mathbf{x}$ such that the sphere $r \le a$ is inside the cube D. Define the function

$$\psi(r) = \begin{cases} \dfrac{1}{2r} \left[5 \left(1 - \dfrac{r}{a} \right)^4 - 3 \left(1 - \dfrac{r}{a} \right)^5 \right] & \text{for } r < a, \\ 0 & \text{for } r \ge a. \end{cases}$$

It is easy to show that the function $\psi(r) - 1/r$ has continuous second derivatives, bounded third derivatives, and square integrable fourth derivatives. Let us change the variables $\xi = x' - x$, $\eta = y' - y$, $\zeta = z' - z$, and set

$$\frac{\pi^3}{8} B_{lmn} = \int_0^\pi \int_0^\pi \int_0^\pi \psi(r) \sin lx' \sin my' \sin nz' \, dx' \, dy' \, dz'$$

$$= \iiint_{\xi^2+\eta^2+\zeta^2<a^2} \psi\left(\sqrt{\xi^2 + \eta^2 + \zeta^2}\right) \times$$

$$\times \sin l(x + \xi) \sin m(y + \eta) \sin n(z + \zeta) \, d\xi \, d\eta \, d\zeta$$

$$= \iiint_{\xi^2+\eta^2+\zeta^2<a^2} \psi \, (\sin l\xi \cos lx + \cos l\xi \sin lx) \times$$

$$\times (\sin m\eta \cos my + \cos m\eta \sin my) \times$$

$$\times (\sin n\zeta \cos nz + \cos n\zeta \sin nz) \, d\xi \, d\eta \, d\zeta$$

$$= \sin lx \sin my \sin nz \iiint_{\xi^2+\eta^2+\zeta^2<a^2} \psi \cos l\xi \cos m\eta \cos n\zeta \, d\xi \, d\eta \, d\zeta.$$

Since ψ is even in ξ, η, ζ, the integrals of $\psi \sin l\xi$, $\psi \sin m\eta$ and $\psi \sin n\zeta$ are zero. Further, since

$$\cos l\xi \cos m\eta \cos n\zeta = \frac{1}{4} \left[\cos(l\xi + m\eta + n\zeta) + \cos(l\xi + m\eta - n\zeta) + \right.$$

$$\left. + \cos(l\xi - m\eta + n\zeta) + \cos(l\xi - m\eta - n\zeta) \right],$$

and since ψ is even in ξ, η, ζ, the integral of ψ times each of these cosines gives the same result. Then

$$\frac{\pi^3}{8} B_{lmn} = \iiint_{\xi^2+\eta^2+\zeta^2<a^2} \psi\left(\sqrt{\xi^2 + \eta^2 + \zeta^2}\right) \cos(l\xi + m\eta + n\zeta) \, d\xi \, d\eta \, d\zeta.$$

In terms of the spherical coordinates (r, θ, ϕ) with origin at (x, y, z) and polar axis in the direction of the vector with components (l, m, n), we have $\sqrt{\xi^2 + \eta^2 + \zeta^2} = r$, and $l\xi + m\eta + n\zeta = r\mu \cos\theta$, where $\mu = \sqrt{l^2 + m^2 + n^2}$, and thus,

$$\frac{\pi^3}{8} B_{lmn} = \int_0^a \int_0^\pi \int_0^{2\pi} \psi(r) \cos(r\mu \cos\theta) \, r^2 \sin\theta \, dr \, d\theta \, d\phi$$

$$= 2\pi \int_0^a \psi(r) \left[-\frac{\sin(r\mu \cos\theta)}{r\mu} \right]_0^\pi r^2 \, d\theta$$

$$= \frac{4\pi}{\mu} \int_0^a \psi(r) r \sin(r\mu) \, dr.$$

Then, using the definition of $\psi(r)$, we get

$$B_{lmn} = \frac{32}{\pi^2 \mu^2} \left[1 - \frac{60(2 + \cos a\mu)}{a^4 \mu^4} + \frac{180 \sin a\mu}{a^5 \mu^5} \right] \sin lx \, \sin my \, \sin nz.$$

Thus,

$$\frac{1}{4\pi} \psi(r) = \frac{8}{\pi^3} \sum_{l=1}^{\infty} \sum_{m=1}^{\infty} \sum_{n=1}^{\infty} \left[\frac{1}{r^2} - \frac{60(2 + \cos a\mu)}{a^4 \mu^6} + \frac{180 \sin a\mu}{a^5 \mu^7} \right] \times$$

$$\times \sin lx \, \sin my \, \sin nz \, \sin lx' \, \sin my' \, \sin nz'.$$

Comparing this with (8.47) we find that

$$G(x, y, z; x', y', z') = \frac{1}{4\pi} \psi(r) - \frac{480}{\pi^3} \sum_{l=1}^{\infty} \sum_{m=1}^{\infty} \sum_{n=1}^{\infty} \left[\frac{2 + \cos a\mu}{a^4 \mu^6} - \frac{3 \sin a\mu}{a^5 \mu^7} \right] \times$$

$$\times \sin lx \, \sin my \, \sin nz \, \sin lx' \, \sin my' \, \sin nz'. \quad (8.48)$$

The function $G - \dfrac{1}{4\pi r} \in C^2$, and Green's function G vanishes on the faces of the cube D. Also, $\nabla^2(1/r) = 0$ for $\neq 0$, and $\nabla^2 [\psi(r)] = \dfrac{1}{r} \dfrac{d^2}{dr^2} (r\psi)$ for $r \neq 0$, and since the Laplacian of the series in (8.48) is the series of $\dfrac{1}{4\pi} \nabla^3 \psi$, which implies that $\nabla^2 G = 0$. Hence, Green's function G which satisfies $\nabla^2 G = 0$ in the cube D, and $G = 0$ on ∂D, is given by

$$G(x, y, z; x', y', z') = \frac{1}{4\pi \sqrt{(x - x')^2 + (y - y')^2 + (z - z')^2}}$$

$$- \gamma(x, y, z; x', y', z') = \frac{1}{4\pi r} - \gamma(\mathbf{x}; \mathbf{x}'), \quad (8.49)$$

where $r = |\mathbf{x} - \mathbf{x}'|$, and γ is a regular solution of Laplace's equation, i.e., $\nabla^2 \gamma = 0$ in the cube D and $\gamma = \dfrac{1}{4\pi r}$ on ∂D.

8.10. Laplace's Equation in a Sphere

Consider the boundary value problem in a sphere of radius a:

$$\frac{\partial^2 u}{\partial r^2} + \frac{2}{r} \frac{\partial u}{\partial r} + \frac{1}{r^2 \sin \theta} \frac{\partial}{\partial \theta} \left(\sin \theta \frac{\partial u}{\partial \theta} \right) + \frac{1}{r^2 \sin^2 \theta} \frac{\partial^2 u}{\partial \phi^2} = 0, \quad r < a,$$

$$u(a, \theta, \phi) = f(\theta, \phi), \quad (8.50)$$

which is the same as (4.11). This problem has been solved by Bernoulli's separation method in §4.3.3, which leads to Eqs (4.12), (4.13), and (4.14). The values of λ for which there is a solution P of Eq (4.14) such that P and $\sqrt{1-t^2}\,P'$ remain bounded at both $t = 1, -1$, must be found in the following two cases:

CASE 1. $m \neq 0$: Eq (4.14) with $\lambda = 0$ has two solutions: $(1+t)^{m/2}(1-t)^{-m/2}$ and $(1-t)^{m/2}(1+t)^{-m/2}$, and Green's function is given by

$$
G(t,t') = \begin{cases}
\dfrac{1}{2m}\left[\dfrac{(1+t')(1-t)}{(1-t')(1+t)}\right]^{m/2}, & t' \le t, \\[4mm]
\dfrac{1}{2m}\left[\dfrac{(1+t)(1-t')}{(1-t)(1+t')}\right]^{m/2}, & t' \ge t.
\end{cases}
\tag{8.51}
$$

Obviously, $G(t,t') \le \dfrac{1}{2m}$, which implies that $\displaystyle\int_{-1}^{1} [G(t,t')]^2 \, dt\,dt' < +\infty$, and the set of eigenfunctions is complete in this cases.

CASE 2. $m = 0$: Eq (4.14) with $\lambda = 0$ has only one solution: $P = 1$, which is continuously differentiable in $[-1,1]$. Hence, $\lambda = 0$ is an eigenvalue. Green's function $G_0(t,t')$ which is the solution of the problem:

$$
\left[\left(1-t^2\right) G'\right]' - G = F \quad \text{for } -1 < t < 1,
$$
$$
G \text{ and } G' \text{ bounded,}
$$

is given by

$$
G_0(t,t') = \sum_{k=1}^{\infty} \frac{u_k(t)\,u_k(t')}{(1+\lambda_k)\displaystyle\int_{-1}^{1} [u_k(t)]^2 \, dt},
\tag{8.52}
$$

where λ_k are the eigenvalues. The eigenfunctions for this case satisfy the equation

$$
\frac{d}{dt}\left[\left(1-t^2\right)\frac{dP}{dt}\right] + \lambda\,P = 0,
\tag{8.53}
$$

a solution of which in the neighborhood of $t = 1$ in powers of $(t-1)$ is

$$
P = (t-1)^{\alpha} \sum_{k=0}^{\infty} c_k\,(t-1)^k.
$$

Substituting this expression for P into Eq (8.53) we get a power series which must be identically zero:

$$
-2c_0\,\alpha^2 (t-1)^{\alpha-1} - \sum_{k=0}^{\infty}\Big\{2(k+\alpha+1)^2 c_{k+1} +
$$
$$
+ \Big[(k+\alpha+1)(k+\alpha)\,c_k - \lambda\Big] c_k \Big\}(t-1)^{k+\alpha} = 0.
$$

If $c_0 \neq 0$, we must have $\alpha = 0$. Since the above equation has a double root $\alpha = 0$, there is a second solution of the form $\ln(1 - t)$ at $t = 1$. But this solution, being unbounded, must be discarded. Now, we set the coefficient of $(t - 1)^k$ equal to zero, and get $c_{k+1} = -\dfrac{[k(k+1) - \lambda]}{2(k+1)^2} c_k$, $k = 0, 1, 2, \ldots$, which, by recursion, gives

$$c_k = \frac{(-1)^k \left[k(k-1) - \lambda\right] \left[(k-1)(k-2) - \lambda\right] \cdots \left[2 - \lambda\right] \left[-\lambda\right]}{2^k (k!)^2} c_0, \quad k \geq 1.$$

If we apply the ratio test to the series $\sum\limits_{k=0}^{\infty} c_k (t - 1)^k$, we find that it converges for $|t - 1| < 2$. This also holds for the derivatives of this series. Bur for $k + 1 > \lambda$, we have $-\dfrac{c_{k+1}}{c_k} > \dfrac{1}{2}\dfrac{k-1}{k+1}$, which shows that this series will, in general, approach $\pm\infty$ as $t \to 1$, except in the case when the series terminates. Thus, if for some integer n we have $n\, c_n \neq 0$ but $c_{n+1} = c_{n+2} = \cdots = 0$, then the above recursion formula is true iff $\lambda = n(n + 1)$, $n = 0, 1, 2, \ldots$. Hence, we obtain a solution bounded in $[-1, 1]$ iff $\lambda = n(n + 1)$. Setting $c_0 = 1$, the eigenfunctions corresponding the eigenvalues $\lambda_n = n(n + 1)$ are given by

$$P_n(t) = \sum_{n=0}^{\infty} \frac{(n + k)!}{2^k (n - k)! (k!)^2} (1 - t)^k,$$

where $P_n(t)$ are the Legendre polynomials of degree n in t; a few of these polynomials are: $P_0(t) = 1$, $P_1(t) = t$, $P_2(t) = \frac{3}{2} t^2 - \frac{1}{2}$, $P_3(t) = \frac{5}{2} t^3 - \frac{3}{2} t$. These polynomials satisfy the orthogonality condition: $\int_{-1}^{1} t^k P_k(t)\, dt = 0$, for $k = 0, 1, 2, \ldots, n - 1$; they are normalized by taking $P_n(t) = 1$, and defined by Rodrigue's formula:

$$P_n(t) = \frac{1}{2^n\, n!} \frac{d^n}{dt^n} \left(t^2 - 1\right)^n.$$

Hence, the formal solution of the problem (8.50) is given by

$$u(r, \theta, \phi) = \int_0^{2\pi} \int_0^{\pi} G(r, \theta, \phi; a, \theta', \phi')\, f(\theta', \phi')\, \sin\theta\, d\theta\, d\phi, \tag{8.54}$$

where

$$G(r, \theta, \phi; a, \theta', \phi') = \sum_{n=0}^{\infty} \left(\frac{r}{a}\right)^n G_n(\theta, \phi; \theta', \phi'), \tag{8.55}$$

$$G_n(\theta, \phi; \theta', \phi') = \frac{2n + 1}{4\pi} P_n(\cos\theta)\, P_n(\cos\theta')$$

$$+ \sum_{m=1}^{\infty} \frac{(2n + 1)(n - m)!}{2\pi\, (n + m)!} P_n^m(\cos\theta)\, P_n^m(\cos\theta')\, \cos m(\theta - \theta'), \tag{8.56}$$

$$f(\theta, \phi) = \sum_{n=0}^{\infty} \left[\frac{1}{2} a_{n0} P_0(\cos \theta) + \sum_{m=1}^{\infty} (a_{nm} \cos m\phi + n_{nm} \sin m\phi) P_n^m(\cos \theta) \right],$$

$$a_{nm} = \frac{(2n+1)(n-m)!}{2\pi\,(n+m)!} \int_0^{2\pi} \int_0^{\pi} f(\theta, \phi) P_n^m(\cos \theta) \cos m\phi \sin \theta \, d\theta \, d\phi,$$

$$b_{nm} = \frac{(2n+1)(n-m)!}{2\pi\,(n+m)!} \int_0^{2\pi} \int_0^{\pi} f(\theta, \phi) P_n^m(\cos \theta) \sin m\phi \sin \theta \, d\theta \, d\phi.$$

For a fixed n, the function $G_n(\theta, \phi; \theta', \phi')$ has the property:

$$\int_0^{2\pi} \int_0^{\pi} G_n(\theta, \phi; \theta', \phi') r^k P_k^m(\cos \theta') \, (a \cos m\phi' + b \sin m\phi') \sin \theta' \, d\theta' \, d\phi'$$

$$= \begin{cases} r^n P_n^m(\cos \theta) \, (a \cos m\phi + b \sin m\phi), & \text{for } k = n, \\ 0 & \text{for } k \neq n. \end{cases}$$

Since $P_n(1) = 1$, and $P_n^m(t)$ has a factor $\left(1 - t^2\right)^{m/2}$ for $m \geq 1$, so $P_n^M(1) = 0$ for $m \geq 1$. Thus, $G_n(0, \phi'; \theta'', \phi'') = \dfrac{2n+1}{4\pi} P_n(\cos \theta'')$, and

$$G_n(\theta, \phi; \theta', \phi') = \frac{2n+1}{4\pi} P_n(\cos \theta''), \tag{8.57}$$

where θ'' is such that $\cos \theta'' = \cos \theta \cos \theta' + \sin \theta \sin \theta' \cos(\phi - \phi')$. Substituting the above value of G_n into (8.55) we get

$$G(r, \theta, \phi; a, \theta', \phi') = \frac{1}{4\pi} \sum_{n=0}^{\infty} (2n+1) \left(\frac{r}{a}\right)^n P_n(\cos \theta'')$$

$$= \frac{1}{4\pi} \left(2r \frac{\partial}{\partial r} + 1\right) \sum_{n=0}^{\infty} \left(\frac{r}{a}\right)^n P_n(\cos \theta''). \tag{8.58}$$

Example 8.4. Consider the Laplace equation in spherical coordinates, independent of ϕ:

$$r \frac{\partial^2 (ru)}{\partial r^2} + \frac{1}{\sin \theta} \frac{\partial}{\partial \theta} \left(\sin \theta \frac{\partial u}{\partial \theta}\right) = 0. \tag{8.59}$$

Using Bernoulli's separation method, let $u = R(r)\Theta(\theta)$, which after substitution in Eq (8.59) gives

$$\frac{r}{R} \frac{d^2(rR)}{dr^2} = -\frac{1}{\sin \theta} \frac{d}{d\theta} \left(\sin \theta \frac{d\Theta}{d\theta}\right). \tag{8.60}$$

Since each side of Eq (8.60) must be equal to a constant, say α^2, this equation is equivalent to the following two equations:

$$r \frac{d^2(rR)}{dr^2} - \alpha^2 R = 0, \tag{8.61}$$

$$\frac{1}{\sin \theta} \frac{d}{d\theta} \left(\sin \theta \frac{d\Theta}{d\theta}\right) + \alpha^2 \Theta = 0. \tag{8.62}$$

Eq (8.61) can be written as

$$r^2 \frac{d^2 R}{dr^2} + 2r \frac{dR}{dr} - \alpha^2 R = 0,$$

and has the general solution as $R(r) = Ar^n + Br^m$, where $n = -\frac{1}{2} + \sqrt{\alpha^2 + \frac{1}{4}}$ and $m = -\frac{1}{2} - \sqrt{\alpha^2 + \frac{1}{4}}$. Thus, $m = -n - 1$, and we can write $\alpha^2 = n(n+1)$, where n is arbitrary. Then $R(r) = Ar^n + Br^{-n-1}$. Thus, $R = r^n$ and $R = r^{-n-1}$ are particular solutions of Eq (8.61) with $\alpha^2 = n(n+1)$, i.e., of the equation

$$r^2 \frac{d^2 R}{dr^2} + 2r \frac{dR}{dr} - n(n+1)R = 0.$$

Again, with this value of α^2 Eq (8.62) becomes

$$\frac{1}{\sin\theta} \frac{d}{d\theta}\left(\sin\theta \frac{d\Theta}{d\theta}\right) + n(n+1)\Theta = 0,$$

where $n > 0$ is an integer (see Example 4.2), so that its particular solution is $\Theta(\theta) = P_n(\cos\theta)$. Hence, $u(r,\theta) = r^n P_n(\cos\theta)$ and $u(r,\theta) = \frac{1}{r^{n+1}} P_n(\cos\theta)$, where n is a positive integer, are particular solutions of Eq (8.59), where the first of these solutions, obtained in Example 4.2, is known as a *solid zonal harmonic*. ∎

8.11. Poisson's Equation and Green's Function in a Sphere

Consider the problem in a sphere $r < a$, which is governed by Eq (8.50). Then expanding u and F as Fourier series in $P_n^m(\cos\theta)\cos m\phi$ and in $P_n^m(\cos\theta)\sin m\phi$ we find that

$$a_{nm}(r) = \frac{(2n+1)(n-m)!}{2\pi(n+m)!} \int_{-\pi}^{\pi}\int_0^{\pi} u(r,\theta,\phi)P_n^m(\cos\theta)\cos m\phi \, \sin\theta \, d\theta \, d\phi,$$

$$b_{nm}(r) = \frac{(2n+1)(n-m)!}{2\pi(n+m)!} \int_{-\pi}^{\pi}\int_0^{\pi} u(r,\theta,\phi)P_n^m(\cos\theta)\sin m\phi \, \sin\theta \, d\theta \, d\phi,$$

$$A_{nm}(r) = \frac{(2n+1)(n-m)!}{2\pi(n+m)!} \int_{-\pi}^{\pi}\int_0^{\pi} F(r,\theta,\phi)P_n^m(\cos\theta)\cos m\phi \, \sin\theta \, d\theta \, d\phi,$$

$$B_{nm}(r) = \frac{(2n+1)(n-m)!}{2\pi(n+m)!} \int_{-\pi}^{\pi}\int_0^{\pi} F(r,\theta,\phi)P_n^m(\cos\theta)\sin m\phi \, \sin\theta \, d\theta \, d\phi.$$

$$(8.63)$$

Then taking the finite Fourier transform of Eq (8.63) and integrating by parts we get

$$a_{nm}'' + \frac{2}{r}a_{nm}' - \frac{n(n+1)}{r^2}a_{nm} = A_{nm}; \quad a_{nn}(a) = 0, \quad a_{nm} \text{ bounded},$$

$$b_{nm}'' + \frac{2}{r}b_{nm}' - \frac{n(n+1)}{r^2}b_{nm} = B_{nm}; \quad b_{nn}(a) = 0, \quad a_{nm} \text{ bounded}.$$

These two boundary value problems have the solutions

$$a_{nm}(r) = \int_0^a G_n(r,r')\, A_{nm}(r)\, r'^2\, dr',$$

$$b_{nm}(r) = \int_0^a G_n(r,r')\, B_{nm}(r)\, r'^2\, dr',$$

(8.64)

where

$$G_n(r,r') = \begin{cases} \dfrac{1}{(2n+1)\,a}\left(\dfrac{r'}{a}\right)^n\left[\left(\dfrac{r}{a}\right)^{-n-1} - \left(\dfrac{r}{a}\right)^n\right] & \text{for } r' \le r, \\[3mm] \dfrac{1}{(2n+1)\,a}\left(\dfrac{r}{a}\right)^n\left[\left(\dfrac{r'}{a}\right)^{-n-1} - \left(\dfrac{r'}{a}\right)^n\right] & \text{for } r' \ge r. \end{cases}$$

(8.65)

Then the solution of the problem (8.63) is

$$u(r,\theta,\phi) = \sum_{n=0}^{\infty}\left[\frac{1}{2}\, a_{n0}\, P_n(\cos\theta) + \right.$$

$$\left. + \sum_{m=1}^{n}\left(a_{nm}\,\cos m\phi + b_{nm}\,\sin m\phi\right) P_n^m(\cos\theta)\right], \quad (8.66)$$

where the series converge uniformly iff $\iiint F^2 r^2 \sin\theta\, dr\, d\theta\, d\phi < +\infty$. Now, if we substitute the values of $A_{nm}(r)$ and $B_{nm}(r)$ from (8.63) into (8.64) and interchange integration and summation in (8.66), we obtain

$$u(r,\theta,\phi) = \int_{-\pi}^{\pi}\int_0^{\pi}\int_0^a G(r,\theta,\phi;r',\theta',\phi)\, F(r',\theta',\phi')\, r'^2 \sin\theta'\, dr'\, d\theta'\, d\phi',$$

(8.67)

where for $r' < r$

$$G(r,\theta,\phi;r',\theta',\phi.) =$$

$$= \sum_{n=0}^{\infty}\frac{1}{(2n+1)\,a}\left(\frac{r'}{a}\right)^n\left[\left(\frac{r}{a}\right)^{-n-1} - \left(\frac{r}{a}\right)^n\right] G_n(\theta,\phi;\theta',\phi'), \quad (8.68)$$

where $G_n(\theta,\phi;\theta',\phi)$ is defined in (8.57). Thus, by (8.58)

$$G(r,\theta,\phi;r',\theta',\phi') = \frac{1}{4\pi a}\sum_{n=0}^{\infty}\left(\frac{r'}{a}\right)^n\left[\left(\frac{r}{a}\right)^{-n-1} - \left(\frac{r}{a}\right)^n\right] P_n(\cos\theta''). \quad (8.69)$$

For $r' > r$, just interchange r and r' in the above formulas.

8.12. Applications of Elliptic Equations

Some examples of these applications are as follows:

8.12.1. Dirichlet Problem for Laplace's Equation. The solution of the boundary value problem $\nabla^2 u = f(\mathbf{x})$ in D such that $u = g(\mathbf{x}_s)$ on ∂D, where $\mathbf{x}_s$ is an arbitrary point on the boundary, is given by

$$u(\mathbf{x}') = \iiint_D f(\mathbf{x})G(\mathbf{x}, \mathbf{x}')\,d\mathbf{x} + \iint_{\partial D} g(\mathbf{x}_s)\frac{\partial G(\mathbf{x}, \mathbf{x}')}{\partial n}\,dS. \tag{8.70}$$

PROOF. Let $\mathbf{x}'$ denote an arbitrary point in D; then by (1.16) we have

$$\iiint_D (u\nabla^2 v - v\nabla^2 u)\,d\mathbf{x} = \iint_{\partial D} \left(u\frac{\partial v}{\partial n} - v\frac{\partial u}{\partial n}\right)dS.$$

If we replace v by $G(\mathbf{x}, \mathbf{x}')$ we get

$$\iiint_D \left(u\nabla^2 G(\mathbf{x}, \mathbf{x}') - G(\mathbf{x}, \mathbf{x}')\nabla^2 u\right)d\mathbf{x}$$
$$= \iint_{\partial D} \left(u\frac{\partial G(\mathbf{x}, \mathbf{x}')}{\partial n} - G(\mathbf{x}, \mathbf{x}')\frac{\partial u}{\partial n}\right)dS.$$

which yields

$$\iiint_D (u\,\delta(\mathbf{x}, \mathbf{x}') - f(\mathbf{x})G(\mathbf{x}, \mathbf{x}'))\,d\mathbf{x} = \iint_{\partial D} g(\mathbf{x}_s)\frac{\partial G(\mathbf{x}, \mathbf{x}')}{\partial n}\,dS,$$

and the result follows. ∎

8.12.2. Neumann Problem for Laplace's Equation. Find the solution of $\nabla^2 u = F(\mathbf{x})$ in D such that $\dfrac{\partial u}{\partial n} = P(\mathbf{x}_s)$ on ∂D, where $\mathbf{x}_s$ is an arbitrary point on the boundary. First, we note that the solution to the problem stated here is not always possible, for if we apply the divergence theorem (§1.7) to ∇u, we have

$$\iiint_D \nabla \cdot \nabla u\,d\mathbf{x} = \iiint_D \nabla^2 u\,d\mathbf{x} = \iint_{\partial D} \frac{\partial u}{\partial n}\,dS,$$

which implies that

$$\iiint_D F(\mathbf{x})\,d\mathbf{x} = \iint_{\partial D} P(\mathbf{x}_s)\,dS. \tag{8.71}$$

This consistency condition is necessary for the existence of the solution to the Neumann problem. We note that the homogeneous problem, i.e., when $F(x)$ and $P(x)$ are both zero, always has a nontrivial solution $u = C$ =(const). Hence, an arbitrary constant can always be added to any solution of the Neumann problem. Green's function in this case needs to be modified such that it satisfies the consistency condition. If $F(\mathbf{x}) = \delta(\mathbf{x} - \mathbf{x}')$ and $P(\mathbf{x}_s) = 0$, the consistency condition is not satisfied. However, the consistency condition is satisfied if take $F(\mathbf{x}) = \delta(\mathbf{x} - \mathbf{x}') - 1/V$, where V is the volume of D. This modified Green's function, also known as the *Neumann's function* (see §10.3), satisfies

$$\nabla^2 G(\mathbf{x}, \mathbf{x}') = \delta(\mathbf{x} - \mathbf{x}') - \frac{1}{V}, \quad \frac{\partial G(\mathbf{x}, \mathbf{x}')}{\partial n} = 0.$$

Proceeding as in §8.12.1, Green's second identity (1.16) in the present case yields

$$\iiint_D \left[u \left\{ \delta\left(\mathbf{x} - \mathbf{x}'\right) - \frac{1}{V} \right\} - F(\mathbf{x}) G(\mathbf{x}, \mathbf{x}') \right] d\mathbf{x}$$
$$= \iint_{\partial D} \left(u(\mathbf{x}_s) \frac{\partial G(\mathbf{x}_s, \mathbf{x})}{\partial \mathbf{n}} - P(\mathbf{x}_s) G(\mathbf{x}_s, \mathbf{x}') \right) dS,$$

or

$$u(\mathbf{x}') = \iiint_D \left[\frac{u}{V} + F(\mathbf{x}) G(\mathbf{x}, \mathbf{x}') \right] d\mathbf{x} - \iint_{\partial D} P(\mathbf{x}_s) G(\mathbf{x}, \mathbf{x}') \, dS. \qquad (8.72)$$

In the solution (8.72) the only unknown integral is $\iiint_D \frac{u}{V} \, d\mathbf{x}$ which is equal to the average value of u in D and is a constant. Since we can add an arbitrary constant to any solution of the Neumann problem, we can express the solution (8.72) as

$$u(\mathbf{x}') = \iiint_D F(\mathbf{x}) G(\mathbf{x}, \mathbf{x}') \, d\mathbf{x} - \iint_{\partial D} P(\mathbf{x}_s) G(\mathbf{x}_s, \mathbf{x}') \, dS + C.$$

This solution is valid only when the consistency condition (8.71) is satisfied.

In the one-dimensional case the corresponding problem is

$$y'' = f(x), \quad y'(0) = \alpha, \ y'(l) = \beta.$$

If we integrate this equation from 0 to l, we get

$$y' \Big|_0^l = \int_0^l f(x) \, dx, \quad \text{or} \quad \beta - \alpha = \int_0^l f(x) \, dx. \qquad (8.73)$$

Thus, the solution to the one-dimensional Neumann problem does not always exist. The condition (8.73) is a consistency condition required for the solution to exist.

The solution obtained by the method of images satisfying the differential equation $\nabla^2 u(x, y; x', y') = \delta(x - x') \delta(y - y')$, subject to the Neumann condition $u_y(x, 0; x'y') = 0$, is given by

$$u(x, y; x', y') = \frac{1}{4\pi} \ln \left\{ \left[(x - x')^2 + (y - y')^2 \right] \left[(x - x')^2 + (y + y')^2 \right] \right\}.$$

It is clear that the algebraic sum of charges is not zero and $u(x, y; x', y')$ does not approach zero as $r \to \infty$ and does not satisfy the consistency condition (8.71) for the Neumann problem in the half-space. Thus, we cannot use Green's second identity (1.16), as we have done above, for the half-space problem because in this case the area is unbounded.

8.12.3. Robin Problem for Laplace's Equation. Find the solution of $\nabla^2 u = f(\mathbf{x})$ in D such that $u + \alpha \dfrac{\partial u}{\partial n} = g(\mathbf{x}_s)$ on ∂D, where $\mathbf{x}_s$ is an arbitrary point on the boundary. The solution for the Robin problem is given by

$$u(\mathbf{x}') = \iiint_D f(\mathbf{x}) G(\mathbf{x}, \mathbf{x}') dx - \frac{1}{\alpha} \iint_{\partial D} g(\mathbf{x}_s) G(\mathbf{x}, \mathbf{x}') \, dS$$

$$= \iiint_D f(\mathbf{x}) G(\mathbf{x}, \mathbf{x}') \, dx + \iint_{\partial D} g(\mathbf{x}_s) \frac{\partial G(\mathbf{x}, \mathbf{x}')}{\partial n} \, dS.$$

Green's function in this case is the solution of the following problem:

$$\nabla^2 G(\mathbf{x}, \mathbf{x}') = \delta(\mathbf{x} - \mathbf{x}') \text{ in } D \text{ such that } G(\mathbf{x}, \mathbf{x}') + \alpha \frac{\partial G(\mathbf{x}, \mathbf{x}')}{\partial n} = 0 \text{ on } \partial D.$$

The proof is similar to the previous two cases and is left as an exercise.

8.12.4. Dirichlet Problem for Helmholtz's Equation in Terms of Green's Function. We will solve the Dirichlet problem, which in this case is

$$(\nabla^2 + k^2) u(\mathbf{x}) = F(\mathbf{x}) \quad \text{in } D, \quad u(\mathbf{x}_s) = K(\mathbf{x}_s) \quad \text{on } \partial D. \tag{8.74}$$

Green's function satisfies the following conditions:

$$(\nabla^2 + k^2) G(\mathbf{x}, \mathbf{y}) = \delta(\mathbf{x} - \mathbf{y}) \quad \text{in } D, \quad G(\mathbf{x}_s, \mathbf{y}) = 0 \quad \text{on } \partial D, \tag{8.75}$$

where $\mathbf{x}_s$ is any point on the boundary ∂D. Multiplying the first equation in (8.74) by $G(\mathbf{x}, \mathbf{y})$ and the first equation in (8.75) by $u(\mathbf{x})$ and subtracting, we get

$$G(\mathbf{x}, \mathbf{y}) \nabla^2 u(\mathbf{x}) - u(\mathbf{x}) \nabla^2 G(\mathbf{x}, \mathbf{y}) = F(\mathbf{x}) G(\mathbf{x}, \mathbf{y}) - u(\mathbf{x}) \delta(\mathbf{x} - \mathbf{y}). \tag{8.76}$$

Integrating both sides of Eq (8.76) over D and using the divergence theorem (§1.7), we have

$$\iiint_D \left[G(\mathbf{x}, \mathbf{y}) \nabla^2 u(\mathbf{x}) - u(\mathbf{x}) \nabla^2 (G(\mathbf{x}, \mathbf{y})) \right] dx$$

$$= \iint_{\partial D} \left(G(\mathbf{x}, \mathbf{y}) \frac{\partial u(\mathbf{x})}{\partial n} - u(\mathbf{x}) \frac{\partial G(\mathbf{x}, \mathbf{y})}{\partial n} \right) dS$$

$$= \iiint_D [F(\mathbf{x}) G(\mathbf{x}, \mathbf{y}) - u(\mathbf{x}) \delta(\mathbf{x} - \mathbf{y})] dx$$

$$= \iiint_D F(\mathbf{x}) G(\mathbf{x}, \mathbf{y}) \, dx - u(\mathbf{y}).$$

Thus,

$$u(\mathbf{y}) = \iiint_D F(\mathbf{x})G(\mathbf{x},\mathbf{y})\,d\mathbf{x} + \iint_{\partial D} K(\mathbf{x}_s)\frac{\partial G(\mathbf{x}_s,\mathbf{y})}{\partial n}\,dS, \qquad (8.77)$$

since $G(\mathbf{x}_s,\mathbf{y}) = 0$, and $u(\mathbf{x}_s) = K(\mathbf{x}_s)$. The solution (8.77) implies that if both $F(\mathbf{x}) = K(\mathbf{x}_s) = 0$, then only a trivial solution exists. But we note that if D is the square $R : \{0 < x, y < a\}$, then $u = \phi(x,y) = \dfrac{1}{2}\sin(n\pi x/a)\sin(n\pi y/a)$ is a nontrivial solution of $(\nabla^2 + 2n^2\pi^2/a^2)u(x,y) = 0$, with $u = 0$ on the boundary. A consequence of this result is:

Theorem 8.5. *If $u(x,y)$ is a solution of*

$$(\nabla^2 + 2n^2\pi^2/a^2)\,u(x,y) = F(x,y) \quad \text{in } R, \text{ and } u = 0 \text{ on } \partial R, \qquad (8.78)$$

where R is the square $\{0 < x, y < a\}$, then $\phi(x,y) = \dfrac{1}{2}\sin(n\pi x/a)\sin(n\pi y/a)$, and $F(x,y)$ satisfies the condition

$$\iint_R F(x,y)\phi(x,y)\,dx\,dy = 0.$$

PROOF. Let $u(x,y)$ be a solution of Eq (8.78). Multiplying both sides of Eq (8.78) by $\phi(x,y)$ and noting that $(\nabla^2 + 2n^2\pi^2/a^2)\,\phi(x,y) = 0$, we get

$$\phi(x,y)(\nabla^2 + 2n^2\pi^2/a^2)u(x,y) - u(x,y)(\nabla^2 + 2n^2\pi^2/a^2)\phi(x,y) = \phi(x,y)F(x,y),$$

or $\phi(x,y)\nabla^2 u(x,y) - u(x,y)\nabla^2\phi(x,y) = \phi(x,y)F(x,y)$. Integrating both sides over R and applying the divergence theorem (§1.7) to the left side, we find that

$$\int_{\partial R}\left[(\phi(x,y)\frac{\partial u(x,y)}{\partial n} - u(x,y)\frac{\partial(\phi(x,y))}{\partial n}\right]ds = \iint_R \phi(x,y)F(x,y)\,dx\,dy.$$

However, the left side is zero because both $u(x,y)$ and $\phi(x,y)$ vanish on ∂R. Obviously, the solution is not unique, since $u(x,y) + C\phi(x,y)$ is also a solution for any arbitrary constant C.

The solution of the Neumann or the Robin boundary value problem can be derived in the same manner as that for Laplace's equation.

8.12.5. Dirichlet Problem for Laplace's Equation in the Half-Plane.

Let D be the half-plane $y > 0$. Then Green's function associated with the Dirichlet boundary condition $u = 0$ on the boundary $y = 0$ is given by (8.3). Hence, for $u(x,0) = f(x)$ we find from (8.70) that

$$\begin{aligned}
u(x',y') &= -\frac{1}{4\pi}\int_{-\infty}^{\infty} f(x)\frac{\partial}{\partial y}\left[\ln\frac{(x-x')^2+(y-y')^2}{(x-x')^2+(y+y')^2}\right]_{y=0} dx\\
&= -\frac{1}{4\pi}\int_{-\infty}^{\infty} f(x)\frac{-4y'}{(x-x')^2+y'^2}\,dx\\
&= \frac{y'}{\pi}\int_{-\infty}^{\infty}\frac{f(x)}{(x-x')^2+y'^2}\,dx.
\end{aligned}$$

After interchanging (x', y') and (x, y), we get

$$u(x, y) = \frac{y}{\pi} \int_{-\infty}^{\infty} \frac{f(x)}{(x - x')^2 + y^2} \, dx.$$

Clearly the boundary condition at $y = 0$ is not satisfied. For an explanation, see Exercise 8.9. ∎

8.12.6. Dirichlet Problem for Laplace's Equation in a Circle. To solve $\nabla^2 u = 0$, $r = \sqrt{x^2 + y^2} < 0$, $u(a, 0) = f(\theta)$, let Γ be the circle $r = a$, and let $u(a, \theta) = f(\theta)$. In this case Green's function is given in Exercise 8.2. Since $\dfrac{\partial}{\partial n} = \dfrac{\partial}{\partial r}$ on the circle Γ, we find from (8.70) that

$$
\begin{aligned}
u(r', \theta') &= \frac{1}{4\pi} \int_0^{2\pi} f(\theta) \frac{\partial}{\partial r} \left[\ln \frac{a^2[r^2 - 2rr'\cos(\theta - \theta') + r'^2]}{r^2 r'^2 - 2rr'a^2\cos(\theta - \theta') + a^4} \right]_{r=a} a \, d\theta \\
&= \frac{1}{4\pi} \int_0^{2\pi} f(\theta) \left[\frac{2r - 2r'\cos(\theta - \theta')}{r^2 - 2rr'\cos(\theta - \theta') + r'^2} \right. \\
&\qquad\qquad \left. - \frac{2rr'^2 - 2a^2r'\cos(\theta - \theta')}{r^2 r'^2 - 2rr'a^2\cos(\theta - \theta') + a^4} \right]_{r=a} a \, d\theta \\
&= \frac{a^2 - r'^2}{2\pi} \int_0^{2\pi} \frac{f(\theta)}{a^2 - 2ar'\cos(\theta - \theta') + r'^2} \, d\theta.
\end{aligned}
$$

After interchanging (r', θ') and (r, θ), we get

$$u(r, \theta) = \frac{a^2 - r^2}{2\pi} \int_0^{2\pi} \frac{f(\theta)}{a^2 - 2ar'\cos(\theta - \theta') + r'^2} \, d\theta.$$

Clearly the boundary condition at $y = 0$ is not satisfied. For an explanation, see Exercise 8.9. ∎

8.12.7. Dirichlet Problem for Laplace's Equation in the Quarter Plane. To find Green's function for the problem

$$\frac{\partial^2 \phi}{\partial x^2} + \frac{\partial^2 \phi}{\partial y^2} = 0, \ x, y > 0, \quad \phi(x, 0) = f(x), \quad \frac{\partial \phi}{\partial x}(0, y) = g(y),$$

we use Green's function to solve this problem. In this case Green's function $G = G(x, y; x', y')$ satisfies the system

$$\frac{\partial^2 G}{\partial x^2} + \frac{\partial^2 G}{\partial y^2} = \delta(x - x')\delta(y - y'), \ x, y > 0, \quad G(x, 0) = 0, \quad \frac{\partial G}{\partial x}(0, y) = 0.$$

METHOD OF IMAGES. The boundary conditions imply that $G = 0$ on the x-axis and that there is no flux across the y-axis. The distribution of sources and sinks pertaining to these conditions is as follows: a source of strength 1 at the points $(x', -y')$ and $(-x', -y')$ each, and a sink of strength -1 at the points (x', y') and $(-x', y')$ each (see Fig. 8.2).

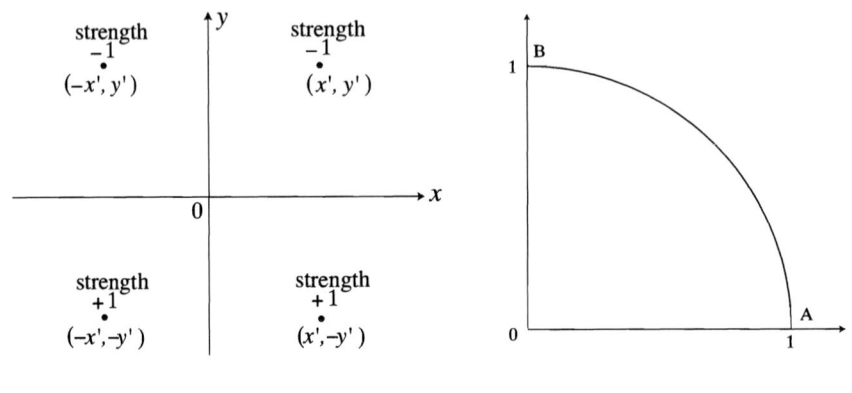

Fig. 8.2.   Fig. 8.3.

Green's function corresponding to the sources and sinks so marked in Fig. 8.2 is given by

$$G(x, y; x', y') = \frac{1}{4\pi} \ln\left\{ \frac{[(x - x')^2 + (y - y')^2][(x + x')^2 + (y - y')^2]}{[(x + x')^2 + (y + y')^2][(x - x')^2 + (y + y')^2]} \right\}, \quad (8.79)$$

where

$$G(x, 0; x', y') = \frac{1}{4\pi} \ln\left\{ \frac{[(x - x')^2 + y'^2][(x + x')^2 + y'^2]}{[(x + x')^2 + y'^2][(x - x')^2 + y'^2]} \right\} = \frac{1}{4\pi} \ln(1) = 0,$$

$$G(0, y; x', y') = \frac{1}{4\pi} \ln\left\{ \frac{[x'^2 + (y - y')^2][x'^2 + (y - y')^2]}{[x'^2 + (y + y')^2][x'^2 + (y + y')^2]} \right\}$$

$$= \frac{1}{2\pi} \ln\left\{ \frac{[x'^2 + (y - y')^2]}{[x'^2 + (y + y')^2]} \right\},$$

$$G_x(x, y; x', y') = \frac{1}{\pi} \left\{ \frac{x - x'}{(x - x')^2 + (y - y')^2} + \frac{x + x'}{(x + x')^2 + (y - y')^2} \right.$$
$$\left. - \frac{x + x'}{(x + x')^2 + (y + y')^2} - \frac{x - x'}{(x - x')^2 + (y + y')^2} \right\},$$

$$G_x(0, y; x', y') = \frac{1}{\pi} \left\{ \frac{-x'}{x'^2 + (y - y')^2} + \frac{x'}{x'^2 + (y - y')^2} - \frac{x'}{x'^2 + (y + y')^2} \right.$$
$$\left. - \frac{-x'}{x'^2 + (y + y')^2} \right\} = 0,$$

$$G_y(x, y; x', y') = \frac{1}{2\pi} \left\{ \frac{y - y'}{(x - x')^2 + (y - y')^2} + \frac{y - y'}{(x + x')^2 + (y - y')^2} \right.$$

$$-\frac{y+y'}{(x+x')^2+(y+y')^2}-\frac{y+y'}{(x-x')^2+(y+y')^2}\Big\},$$

$$G_y(x,0;x',y')=-\frac{y'}{\pi}\left\{\frac{1}{(x-x')^2+y'^2}+\frac{1}{(x+x')^2+y'^2}\right\}.$$

From §8.12.1, the solution of the problem is given by

$$\phi(x',y')=\int_\Gamma\left(\phi\frac{\partial G}{\partial n}-G\frac{\partial\phi}{\partial n}\right),$$

where Γ is the boundary of the quarter plane traversed in the counterclockwise direction and G is defined above. Note that $\dfrac{\partial}{\partial n}=-\dfrac{\partial}{\partial x}$ on the y-axis, $\dfrac{\partial}{\partial n}=-\dfrac{\partial}{\partial y}$ on the x-axis, $\dfrac{\partial\phi}{\partial x}=g(y)$ on the y-axis, and $\phi(x,0)=f(x)$. Also, since $G(x,0;x',y')=0$ on OA, $G_x(x,0;x',y')=0$ on BO (see Fig. 8.3), $ds=-dy$, and y varies from ∞ to 0, the solution is given by

$$\phi(x',y')=\lim_{\text{OA=OB}\to\infty}\int_{\text{OABO}}[-f(x)]\left[\frac{\partial G}{\partial y}\right]_{y=0}dx+g(y)\,[G]_{x=0}\,(-dy)$$

$$=\frac{y'}{\pi}\int_0^\infty f(x)\left[\frac{1}{(x-x')^2+y'^2}+\frac{1}{(x+x')^2+y'^2}\right]dx$$

$$+\frac{1}{2\pi}\int_0^\infty g(y)\ln\left\{\frac{x'^2+(y-y')^2}{x'^2+(y+y')^2}\right\}dy.$$

The solution $\phi(x,y)$ is obtained by interchanging (x',y') and (x,y). Clearly the boundary condition at $y=0$ is not satisfied. For an explanation, see Exercise 8.9. ∎

8.12.8. Vibration Equation for the Unit Sphere.
Consider the unit sphere $x^2+y^2+z^2\le 1$. Using spherical coordinates the eigenvalue problem is defined by

$$\nabla^2 u+\lambda u=\frac{1}{r^2\sin\theta}\left[\frac{\partial}{\partial r}\left(r^2 u_r\sin\theta\right)+\frac{\partial}{\partial\phi}\left(\frac{u_\phi}{\sin\theta}\right)+\frac{\partial}{\partial\theta}\left(u_\theta\sin\theta\right)\right]+\lambda u=0.$$

Using Bernoulli's separation method by taking $u=Y(\theta,\phi)R(r)$, we get

$$\frac{\left(r^2 R'\right)'+\lambda^2 r^2 R}{R}=-\frac{1}{Y\sin\theta}\left[\frac{\partial}{\partial\phi}\left(\frac{Y_\phi}{\sin\theta}\right)+\frac{\partial}{\partial\theta}\left(Y_\theta\sin\theta\right)\right]=k,$$

where k is a constant, whose value must be determined in such a way that the differential equation

$$\nabla^2 Y+kY\equiv\frac{1}{\sin\theta}\left[\frac{\partial}{\partial\phi}\left(\frac{Y_\phi}{\sin\theta}\right)+\frac{\partial}{\partial\theta}\left(Y_\theta\sin\theta\right)\right]+kY=0$$

has a solution which is continuous on the entire surface of the sphere. Thus, this solution must be periodic in ϕ, of period 2π, and regular at $\theta = 0$ and $\theta = \pi$, i.e., at both these points it must approach a limit which is independent of ϕ. This requirement is satisfied for the values of $k = n(n+1), n = 0, 1, 2, \ldots$, which are the eigenvalues for this problem. Hence, the solutions are the spherical harmonics $Y_n(\theta, \phi)$ (see Chapter 9).

The equation for $R(r)$, $(r^2 R')' - n(n+1)R + \lambda r^2 R = 0$, has the solutions $R_n(\sqrt{\lambda} r) = \dfrac{J_{n+1}(\sqrt{\lambda} r)}{\sqrt{r}}$, which are regular at $r = 0$.

8.13. Exercises

8.1. Find Green's function for the operator $\nabla^2 - k^2$ in $\mathbb{R}, \mathbb{R}^2$, and $\mathbb{R}^3$.

ANS. $-\dfrac{1}{2k} e^{-k|x-x'|}$ in $\mathbb{R}$; $-\dfrac{1}{2\pi} K_0(kr)$ in $\mathbb{R}^2$, where $r = |\mathbf{x} - \mathbf{x}'|$; and $-\dfrac{1}{4\pi r} e^{-kr}$ in $\mathbb{R}^3$, where $r = |\mathbf{x} - \mathbf{x}'|$.

8.2. Find Green's function for the Laplace operator in the circle $r \leq a$ subject to the Dirichlet boundary conditions, where $r = \sqrt{x^2 + y^2}$.

ANS. In rectangular cartesian coordinates: $G(x, y; x', y') = \dfrac{1}{2\pi} \ln \dfrac{a\, r_1}{r'\, r_2}$, where $r'^2 = x'^2 + y'^2, r_1^2 = (x-x')^2 + (y-y')^2$, and $r_2^2 = \left(x - \dfrac{a^2 x'}{r'^2}\right)^2 + \left(y - \dfrac{a^2 y'}{r'^2}\right)^2$.

In series form: $G(r, \theta; r', \theta') = \begin{cases} \ln \dfrac{r}{a} + a_n(r; r') \cos n(\theta - \theta') & \text{for } r' < r, \\[2mm] \ln \dfrac{r'}{a} + a_n(r'; r) \cos n(\theta - \theta') & \text{for } r' > r, \end{cases}$

where $a_n(r; r') = \dfrac{1}{2n\pi} \left[\left(\dfrac{r}{a}\right)^n - \left(\dfrac{a}{r}\right)^n\right]\left(\dfrac{r'}{a}\right)^n$ for $n \geq 1$. Note that this result can be obtained from the first answer by using the formula

$$1 + 2\sum_{n=0}^{\infty} \left(\dfrac{r}{a}\right)^n \cos n\theta = \dfrac{a^2 - r^2}{a^2 - 2ar\cos n\theta + r^2}.$$

8.3. Find Green's function for the Laplace operator in the sphere $r \leq a$, where $r = \sqrt{x^2 + y^2 + z^2}$.

ANS. $G(\mathbf{x}; \mathbf{x}') = -\dfrac{1}{4\pi} \left\{ \dfrac{1}{|\mathbf{x} - \mathbf{x}'|} - \dfrac{a\,|\mathbf{x}'|}{\big| |\mathbf{x}'|^2 \mathbf{x} - a^2 \mathbf{x}' \big|} \right\}$.

8.4. Find Green's function for the Laplace operator in the semicircle $x^2 + y^2 \leq a^2$, $y \geq 0$.

ANS. $G(x, y; x', y') = \dfrac{1}{2\pi} \ln \dfrac{r_1 R_2}{r_2 R_1}$, where $x = r\cos\theta, y = r\sin\theta, x' =$

$r' \cos \theta', y' = r' \sin \theta',$

$r'^2 = x'^2 + y'^2, r_1^2 = (x - x')^2 + (y - y')^2 = r^2 - 2rr' \cos(\theta - \theta') + r'^2,$

$r_2^2 = \left(x - \dfrac{a^2 x'}{r'^2}\right)^2 + \left(y - \dfrac{a^2 y'}{r'^2}\right)^2 = r^2 - 2\dfrac{a^2 r}{r'} \cos(\theta - \theta') + \dfrac{a^4}{r'^2},$

$R_1^2 = (x - x')^2 + (y - y')^2 = r^2 - 2rr' \cos(\theta + \theta') + r'^2,$

$R_2^2 = \left(x - \dfrac{a^2 x'}{r'^2}\right)^2 + \left(y + \dfrac{a^2 y'}{r'^2}\right)^2 = r^2 - 2\dfrac{a^2 r}{r'} \cos(\theta + \theta') + \dfrac{a^4}{r'^2}.$

8.5. Use Green's function to find the solution of the problem $\nabla^2 u = 0$, in the region $\{x^2 + y^2 \leq a^2, y > 0\}$, subject to the boundary conditions $u(x, y) = k$ on $x^2 + y^2 = a^2$ and on $y = 0$ for $x < 0$, and $u(x, y) = 0$ on $y = 0$ for $x > 0$.

ANS. $u(x, y) = k - \dfrac{k}{\pi} \left[\tan^{-1} \left(\dfrac{a - x}{y} \right) - \tan^{-1} \left(\dfrac{x^2 + y^2 - ax}{ay} \right) \right].$

8.6. Find Green's function for the Laplace operator in the upper hemisphere $r = \sqrt{x^2 + y^2 + z^2} \leq a, z \geq 0.$

ANS. $G\left(\mathbf{x}; \mathbf{x}'\right) = -\dfrac{1}{4\pi \left|\mathbf{x} - \mathbf{x}'\right|} + \dfrac{a \left|\mathbf{x}'\right|}{4\pi \left| \left|\mathbf{x}'\right|^2 \mathbf{x} - a^2 \mathbf{x}'\right|} + \dfrac{1}{4\pi \left|\mathbf{x} - \mathbf{x}_1'\right|}$

$\qquad -\dfrac{a \left|\mathbf{x}'\right|}{4\pi \left| \left|\mathbf{x}'\right|^2 \mathbf{x} - a^2 \mathbf{x}_1'\right|}$, where $\mathbf{x}_1' = (x', y', -z').$

8.7. Use Green's function for the half-plane (see §8.1) to solve the following problems: (a) $\nabla^2 u = 0, x \in \mathbb{R}, 0 < y$, such that $u(x, 0) = g(x).$

(b) $\nabla^2 u = 0, x \in \mathbb{R}, 0 < y$, such that $u(x, 0) = k.$

(c) $\nabla^2 u = 0, x \in \mathbb{R}, 0 < y$, such that $u(x, 0) = |x|/x.$

ANS. (a) $u(\mathbf{x}') = \displaystyle\int_{-\infty}^{\infty} g\left(\mathbf{x}_s\right) \left(-\dfrac{\partial G(\mathbf{x}, \mathbf{x}')}{\partial y} \right) dx;$ (b) $u(x, y) = k;$ and

(c) $u(x, y) = \dfrac{2}{\pi} \tan^{-1}(x/y) = 1 - \dfrac{2\theta}{\pi}.$

8.8. Solve Laplace's equation $\dfrac{\partial^2 u}{\partial r^2} + \dfrac{1}{r}\dfrac{\partial u}{\partial r} + \dfrac{1}{r^2}\dfrac{\partial^2 u}{\partial \theta^2} = 0$, subject to the conditions $u(r, 0) = f(r)$ and $u(r, \pi) = 0$ for $r > 0.$

HINT. Reduce the partial differential equation to an ordinary differential equation by applying the finite Fourier sine transform. Find Green's function for the transformed problem, i.e., solve the problem

$r^2 \dfrac{d^2 G\left(r, r'\right)}{dr^2} + r\dfrac{G\left(r, r'\right)}{dr} - n^2 G\left(r, r'\right) = \delta\left(r, r'\right), \quad \lim_{r \to 0} G\left(r, r'\right) = 0,$

$\lim_{r \to 0} G\left(r, r'\right) = \text{finite}, \quad \dfrac{dG\left(r, r'\right)}{r}\bigg|_{r'+} - \dfrac{dG\left(r, r'\right)}{r}\bigg|_{r'-} = \dfrac{1}{r^2}.$

Green's function is given by

$$G(r, r') = -\frac{1}{2nr'} \begin{cases} \left(\dfrac{r}{r'}\right)^n, & r \leq r', \\[2mm] \left(\dfrac{r'}{r}\right)^n, & r \geq r'. \end{cases}$$

Obtain the solution to the transformed problem by using Green's function and then invert the transform.

ANS. $u(r, \theta) = \dfrac{1}{\pi} \displaystyle\int_0^\infty \dfrac{r \sin\theta \, f(r')}{r^2 - 2rr' \cos\theta + r'^2} \, dr' = \dfrac{y}{\pi} \displaystyle\int_0^\infty \dfrac{f(x')}{(x - x')^2 + y^2} \, dx'.$

Note that if we remove the restriction on u being zero on half of the x-axis, and take $u = f(x)$ on the entire x-axis, then we must add the solution for $x < 0$, to the above solution. Hence, the solution under this condition is

$$u(x, y) = \frac{y}{\pi} \int_{-\infty}^\infty \frac{f(x')}{(x - x')^2 + y^2} \, dx'.$$

8.9. Solve $\dfrac{\partial^2 u}{\partial x^2} + \dfrac{\partial^2 u}{\partial y^2} = 0, \; y > 0, \; u(x, 0) = f(x),$ where $f(x)$ is integrable for all x and approaches zero as $x \to \pm\infty$.

HINT. Apply Fourier transform with respect to x and solve $\dfrac{d^2 \tilde{u}}{dy^2} - \alpha^2 \tilde{u} = 0.$

ANS. $u(x, y) = \dfrac{y}{\pi} \displaystyle\int_{-\infty}^\infty \dfrac{f(\eta)}{(x - \eta)^2 + y^2} \, d\eta.$

Note that the solution does not directly satisfy the boundary condition at $y = 0$. However, it can be shown that $\lim_{y \to 0} u(x, y) = f(x)$. Since

$$u(x, y) = \frac{y}{\pi} \int_{-\infty}^\infty \frac{f(\eta)}{(x - \eta)^2 + y^2} \, d\eta$$

$$= \frac{1}{\pi} \left[f(\eta) \tan^{-1}\left(\frac{\eta - x}{y}\right) \right]_{-\infty}^\infty - \frac{1}{\pi} \int_{-\infty}^\infty f'(\eta) \tan^{-1}\left(\frac{\eta - x}{y}\right) d\eta,$$

the problem reduces to showing that

$$-\lim_{y \to 0} \frac{1}{\pi} \int_{-\infty}^\infty f'(\eta) \tan^{-1}\left(\frac{\eta - x}{y}\right) d\eta = f(x).$$

The left side can be written as

$$-\lim_{y \to 0} \frac{1}{\pi} \left[\int_{-\infty}^x f'(\eta) \tan^{-1}\left(\frac{\eta - x}{y}\right) d\eta + \int_x^\infty f'(\eta) \tan^{-1}\left(\frac{\eta - x}{y}\right) d\eta \right].$$

If we interchange the order of integration and the limit process and let $y \to 0$, then in the first integral $\eta - x < 0$ and therefore, $\lim_{y \to 0} \tan^{-1}\left(\dfrac{\eta - x}{y}\right) = -\dfrac{\pi}{2},$

and in the second integral $\lim_{y \to 0} \tan^{-1}\left(\dfrac{\eta - x}{y}\right) = \dfrac{\pi}{2}.$ Hence,

$$-\lim_{y\to 0}\frac{1}{\pi}\int_{-\infty}^{\infty}f(\eta)\tan^{-1}\left(\frac{\eta-x}{y}\right)d\eta = \frac{1}{2}\left[\int_{-\infty}^{x}f'(\eta)\,d\eta - \int_{x}^{\infty}f'(\eta)\,d\eta\right]$$

$$= \frac{1}{2}\left[f(x)+f(x)\right] = f(x).$$

8.10. Find Green's function for the boundary value problem in the rectangle $\{0 < x < a, 0 < y < b\}$:

$G_{xx} + G_{yy} = \delta(x-x')\,\delta(y-y')$, $0 < x, x' < a$, $0 < y, y' < b$, such that
$G(0,y) = 0 = G(a,y)$, $G(x,0) = 0 = G(x,b)$.

HINT. Set $G(x,y;x',y') = \displaystyle\sum_{m=1}^{\infty}\sum_{n=1}^{\infty} A_{mn}\sin\frac{m\pi x}{a}\sin\frac{n\pi y}{b}$, in the given equation and find A_{mn} from

$$\left[\left(\frac{m\pi}{a}\right)^2 + \left(\frac{n\pi}{b}\right)^2\right]\left(\frac{ab}{4}\right)a_{mn} = \sin\frac{m\pi x'}{a}\sin\frac{n\pi y'}{b}.$$

8.11. Use Rodrigue's formula and show that

(a) $\displaystyle\int_0^{\pi} P_n(\cos\theta)^2\sin\theta\,d\theta = \frac{(2n)!}{2^{2n}(n!)^2}\int_{-1}^{1}(1-t^2)^n\,dt = \frac{2}{2n+1}$.

(b) $\displaystyle\int_0^{\pi} P_n^m(\cos\theta)^2\sin\theta\,d\theta = \frac{2}{2n+1}\frac{(n+m)!}{(n-m)!}$.

8.12. Find Green's function for the Laplacian in $\mathbb{R}$.

ANS. Green's function for 1-D Laplace's equation satisfies the equation $\dfrac{d^2G}{dx^2} = \delta(x,x')$, whose general solution for a fixed x' is $G(x) = \dfrac{1}{2}|x-x'| + Ax + B$. If we require the symmetry about x', i.e., $G = G(|x-x'|)$, then $A = 0$, and set $B = 0$, then Green's function is given by $G(x) = \dfrac{1}{2}|x-x'|$.

8.13. Derive Green's function $G(x,y) = G(x,y;x',y')$ for the half-plane by Fourier transform for Laplace's equation.

HINT. Solve the problem

$$\frac{\partial^2 G}{\partial x^2} + \frac{\partial^2 G}{\partial y^2} = \delta(x-x')\delta(y-y'), \quad G(x,0) = 0,$$

$$\lim_{x\to\pm\infty}G(x,y),\ G_x(x,y) = 0, \quad \lim_{y\to\infty}G(x,y),\ G_y(x,y) = 0.$$

ANS. $G(x,y) = \dfrac{1}{4\pi}\log\left\{\dfrac{(x-x')^2+(y-y')^2}{(x-x')^2+(y+y')^2}\right\}$, same as in §8.1, Method 2.

8.14. Verify the solution given in §8.12.3 for the Robin problem $\nabla^2 u = f(x)$ in D such that $u + \alpha\dfrac{\partial u}{\partial n} = g(x_s)$ on ∂D, where x_s is an arbitrary point on the boundary ∂D.

8.15. Find Green's function for the problem

$$\nabla^2 u(r,\theta) = F(r,\theta) \quad \text{for } r < a,$$

$$u(a,\theta) + \left.\frac{\partial u(r,\theta)}{\partial r}\right|_{r=a} = 0, \quad \lim_{r\to 0} u(r,\theta) < +\infty.$$

ANS. $G\left(r,\theta;r',\theta'\right) = \sum_{n=0}^{\infty} G_n\left(r,\theta;r',\theta'\right)$, where

$$G_0\left(r,\theta;r',\theta'\right) = \begin{cases} \dfrac{1}{2\pi}\left[\ln\left(\dfrac{r}{a}\right) - \dfrac{1}{a}\right], & r' < r, \\[2ex] \dfrac{1}{2\pi}\left[\ln\left(\dfrac{r'}{a}\right) - \dfrac{1}{a}\right], & r' > r; \end{cases}$$

and

$$G_n\left(r,\theta;r',\theta'\right)$$
$$= \begin{cases} -\dfrac{1}{2n\pi(n+a)}\left[(n-a)\left(\dfrac{r}{a}\right)^n + (n+a)\left(\dfrac{a}{r}\right)^n\right]\left(\dfrac{r'}{a}\right)^n \cos n\psi, & r' < r, \\[2ex] -\dfrac{1}{2n\pi(n+a)}\left[(n-a)\left(\dfrac{r'}{a}\right)^n + (n+a)\left(\dfrac{a}{r'}\right)^n\right]\left(\dfrac{r}{a}\right)^n \cos n\psi, & r' > r, \end{cases}$$
and $\psi = \theta - \theta'$.

8.16. Find Green's function for the Laplacian in the sector $0 < r < a, 0 < \theta < \pi/3$.

HINT. In complex coordinate system the sector is defined by $0 < |z| < a$, $0 < \arg\{z\} < \pi/3$, $z = x + i\,y$. The conformal mapping $w = z^3$, $w = u + i\,v$, maps the sector onto a semi-circle. Use the result of Exercise 8.4.

ANS. $G\left(r,\theta;r',\theta'\right) = \dfrac{1}{4\pi}\left[\ln\left\{\dfrac{A\,B}{C\,D}\right\}\right]$, where $A = r^6 - 2r^3 r'^3 \cos 3\left(\theta - \theta'\right) + r'^6$, $B = r^6 r'^6 - 2a^6 r^3 r'^3 \cos 3\left(\theta + \theta'\right) + a'^{12}$, $C = r^6 - 2r^3 r'^3 \cos 3\left(\theta + \theta'\right) + r'^6$, $D = r^6 r'^6 - 2a^6 r^3 r'^3 \cos 3\left(\theta - \theta'\right)$.

8.17. Solve the following problem:

$$\frac{\partial^2 \phi}{\partial x^2} + \frac{\partial^2 \phi}{\partial y^2} = 0, \quad x,y > 0, \quad \phi(x,0) = f(x), \quad \frac{\partial \phi}{\partial x}(0,y) = g(y),$$

by using the Fourier transform.

HINT. Break up the problem int two parts: Part (1) $\phi_{xx} + \phi_{yy} = 0$, $\phi(x,0) = 0$, $\phi_x(0,y) = g(y)$. Apply the cosine transform with respect to x and note that
$$\mathcal{F}_c^{-1} \frac{e^{\alpha y}}{\alpha} = \sqrt{\frac{1}{2\pi}}\left[\gamma + \ln\left(x^2 + y^2\right) - 2\ln x\right], \text{ where } \gamma \text{ is the Euler gamma.}$$
Part (2) $\phi_{xx} + \phi_{yy} = 0$, $\phi(x,0) = f(x)$, $\phi_x(0,y) = 0$. Apply the sine transform with respect to y. Add the two solutions.

8.18. Show that Green's function $G(x,y)$ for the operator $L = \dfrac{\partial^2}{\partial x\,\partial y} + \gamma^2$ in $\mathbb{R}^2$ is given by $G(x,y) = H(x)H(y)\,J_0\left(2\gamma\sqrt{xy}\right)$, which has a pole at the origin.

HINT. Use the double Fourier transform.

Note that the classical solution for the equation $L[u](x, y) = 0$ is $u = J_0\left(2\gamma\sqrt{xy}\right)$ for $x, y > 0$.

8.19. Solve $u_{xx} + u_{yy} = 0$, under the conditions $u(0, y) = 0 = u(\pi, y)$, $u(x, 0) = \sin x$, $\lim\limits_{y\to\infty} u(x, y) < +\infty$.

ANS. $u = e^{-y} \sin x$.

8.20. Solve $r^2 u_{rr} + r u_r + u_{\theta\theta} = 0$, such that $u(b, \theta) = f(\theta)$, $u(r, \theta + 2\pi) = u(r, \theta)$, and $\lim\limits_{r\to 0} u(r, \theta) < +\infty$ (circular disc problem).

HINT: Separate the variables and show that the only relevant part of the solution reduces to $u(r, \theta) = c_0 + \sum_\alpha r^\alpha (A(\alpha) \cos\alpha\theta + B(\alpha) \sin\alpha\theta)$. Note that under the given conditions $u(r, \theta)$ must have a Fourier series representation in θ and, therefore, $\alpha = n$ is a positive integer.

8.21. Solve $u_{xx} + u_{yy} = 0$, under the conditions $u(x, 0) = 0 = u(x, \pi)$, $u(0, y) = 0$, and $u(\pi, y) = \cos^2 y$.

ANS. $u = \sum\limits_{n=1}^{\infty} C_n \sinh nx \sin ny$, where $C_n = \dfrac{2}{\pi \sinh n\pi} \int_0^\pi \cos^2 y \sin ny \, dy =$

$\dfrac{1}{\pi \sinh n\pi} \left[1 - (-1)^n\right] \left[\dfrac{1}{n} - \dfrac{n}{n^2 - 4}\right]$, $n \neq 2$, and $C_2 = 0$.

8.22. Solve $u_{rr} + \dfrac{1}{r} u_r + \dfrac{1}{r^2} u_{\theta\theta} = 0$, subject to the conditions $u = 0$ for $\theta = 0$ or $\pi/2$, and $u_r = \sin\theta$ at $r = a$.

ANS. $u = \sum\limits_{n=1}^{\infty} C_n r^{2n} \sin 2n\theta$, where $C_n = \dfrac{4(-1)^{n+1}}{\pi(4n^2 - 1)a^{2n}}$.

8.23. Solve $r u_{rr} + u_r + r u_{zz} = 0$, $u(a, z) = u_0$, under the conditions $u(a, 0) = u_0$, $u(r, 0) = 0 = u(r, h)$, and $\lim\limits_{r\to 0} u(r, z) < +\infty$ (steady-state temperature in a finite cylinder).

ANS. $u = \dfrac{4u_0}{\pi} \sum\limits_{n=0}^{\infty} \dfrac{I_0\left((2n+1)\pi r/h\right)}{(2n+1) I_0\left((2n+1)\pi a/h\right)} \sin\dfrac{(2n+1)\pi z}{h}$.

8.24. Solve Poisson's equation $u_{xx} + u_{yy} = -1$, $0 < x, y < 1$, subject to the Dirichlet boundary conditions $u(0, y) = 0 = u(1, y) = u(x, 0) = u(x, 1)$.

ANS. $u(x, y) = \dfrac{16}{\pi^4} \sum\limits_{\substack{j,k=1 \\ j,k \, \text{odd}}}^{\infty} \dfrac{\sin j\pi x \sin k\pi y}{jk \left(j^2 + k^2\right)}$.

8.25. Show that $G(r_1, r_2) = \dfrac{e^{r_1 - r_2}}{4\pi \, |r_1 - r_2|}$ is Green's function for the differential equation $\left(\nabla^2 + k^2\right) G(r_1, r_2) = -\delta(r_1 - r_2)$.

HINT. Show that (i) $G(r_1, r_2)$ satisfies the homogeneous differential equation for $r_1 \neq r_2$, and (ii) $\int \left(\nabla^2 + k^2\right) G(r_1, r_2) \, dr_1 = \begin{cases} 0, & r_1 \neq r_2, \\ -1, & r_1 = r_2, \end{cases}$ for small $|r_1 - r_2|$.

8.26. Show that Green's functions for the radial part of 3-D Helmholtz's equation, with singularity or source at the origin, are: (a) $G(r) = -\dfrac{1}{4\pi} y_0(kr)$ for the standing wave; and (b) $G(r) = \dfrac{i}{4\pi} h_0^{(1)}(kr)$ for the outgoing wave.

HINT. Assume the time dependence $e^{i\omega t}$.

Note that $y_n(x) = \sqrt{\dfrac{\pi}{2x}} Y_{n+1/2}(x) = (-1)^n \sqrt{\dfrac{\pi}{2x}} J_{n-1/2}(x)$, and

$h_n^{(1)}(x) = \sqrt{\dfrac{\pi}{2x}} H_{n+1/2}^{(1)}(x)$ are the spherical Bessel functions.

9

Spherical Harmonics

We will start with a short historical account of the development of the spherical harmonics. Laplace's equation in a sphere was solved by Bernoulli's separation method in §4.3.3 and §8.10. The spherical harmonics are the angular portion (θ and ϕ portions of the spherical coordinates) of a set of solutions of Laplace's equation. These harmonics are useful in many theoretical and physical applications, namely, in physics, seismology, geodesy, spectral analysis, magnetic fields, quantum mechanics and others. A detailed account of various approaches to spherical harmonics can be found in Courant and Hilbert [1968] and MacRobert [1967].

9.1. Historical Sketch

The development of this subject started with Laplace's own account of a function $Y_n^m(\theta, \phi)$ as a set of spherical harmonics that form an orthogonal system. This special research was developed by Laplace in 1782 in connection with the Newtonian potential for the law of universal gravitation in $\mathbb{R}^3$, when he determined that the gravitational potential $P(\mathbf{x})$ associated with a set of point-masses m_i located at points $\mathbf{x}_i \in \mathbb{R}^3$ is defined by $P(\mathbf{x}) = \sum_i \dfrac{m_i}{|\mathbf{x} - \mathbf{x}_i|}$, where each term in this summation is a Newtonian potential at the respective point mass. About the same time Legendre had determined that the expansion for the Newtonian potential in powers of $r = |\mathbf{x}|$ and $r_1 = |\mathbf{x}_1|$ is given by

$$|\mathbf{x} - \mathbf{x}_1|^{-1} = P_0(\cos \gamma)\, \frac{1}{r_1} + P_1(\cos \gamma)\, \frac{r}{r_1^2} + P_2(\cos \gamma)\, \frac{r^2}{r_1^3} + \cdots,$$

where γ is the angle between $\mathbf{x}$ and $\mathbf{x}_1$, and P_n are the well known Legendre polynomials, which are also a special case of spherical harmonics. The name 'solid spherical harmonics' was introduced in 1867 by William Thomson (Lord Kelvin) and Peter G. Tait in their book *Treatise on Natural Philosophy* to describe these functions which are the solution of homogeneous Laplace's equation $\nabla^2 u = 0$ in the sphere. The term 'Laplace's coefficients' is generally used to describe the zonal spherical harmonics as introduced by Laplace and Legendre.

9.2. Laplace's Solid Spherical Harmonics

A function $F(x, y, z)$ is said to be *homogeneous of degree n* if for any constant factor λ we have $F(\lambda x, \lambda y, \lambda z) = \lambda^n F(x, y, z)$. The function $F(x, y, z)$ is called a *solid spherical harmonic of degree n* if $F(x, y, z)$ is homogeneous of degree n and if $\nabla^2 F(x, y, z) = 0$. By this definition the spherical harmonics are the angular portion of the solution of Laplace's equation in a sphere.

Recall that Laplace's equation in spherical coordinates $(r, \theta, \phi), r > 0, 0 \le \theta \le \pi, 0 \le \phi < 2\pi$, is

$$\nabla^2 u \equiv \frac{1}{r^2} \frac{\partial}{\partial r} \left(r^2 \frac{\partial u}{\partial r} \right) + \frac{1}{r^2 \sin \theta} \frac{\partial}{\partial \theta} \left(\sin \theta \frac{\partial u}{\partial \theta} \right) + \frac{1}{r^2 \sin^2 \theta} \frac{\partial^2 u}{\partial \phi^2} = 0, \quad (9.1)$$

which is the same as Eq (8.50). In §4.3.3 and §8.10 we have used Bernoulli's separation method to determine Green's function for the sphere. Let the solution of Eq (9.1) be the form $u = R(r)\Theta(\theta)\Phi(\phi)$. Substituting it into this equation we obtain

$$\frac{r \sin^2 \theta}{R} \frac{d^2(rR)}{dr^2} + \frac{\sin \theta}{\Theta} \frac{d}{d\theta} \left(\sin \theta \frac{d\Theta}{d\theta} \right) = -\frac{1}{\Phi} \frac{d^2 \Phi}{d\phi^2}.$$

Since the left side of this equation does not contain ϕ and the right side contains ϕ only, it must be a constant, say m^2. Then the above equation is equivalent to the two equations

$$\frac{d^2 \Phi(\phi)}{d\phi^2} + m^2 \Phi = 0, \quad (9.2)$$

$$\frac{r}{R} \frac{d^2(rR)}{dr^2} + \frac{1}{\Theta \sin \theta} \frac{d}{d\theta} \left[\sin \theta \frac{d\Theta}{d\theta} \right] - \frac{m^2}{\sin^2 \theta} = 0. \quad (9.3)$$

The general solution of Eq (9.2) is $\Phi = A \cos m\phi + B \sin m\phi$. Since the first term of (9.3) does not involve θ and the other two terms do not involve r, then each part must be constant, which we take as $n(n + 1)$ (see Example 4.11), and Eq (9.3) breaks up into

$$r \frac{d^2(rR)}{dr^2} - n(n + 1)R = 0, \quad (9.4)$$

$$\frac{1}{\sin \theta} \frac{d}{d\theta} \left(\sin \theta \frac{d\Theta}{d\theta} \right) + \left[n(n + 1) - \frac{m^2}{\sin^2 \theta} \right] \Theta = 0. \quad (9.5)$$

The solution of Eq (9.4) is is $R = A_1 r^n + B_1 r^{-n-1}$. If in Eq (9.5) we set $\cos \theta = t$ and get

$$\frac{d}{dt} \left[(1 - t^2) \frac{d\Theta}{dt} \right] + \left[n(n + 1) - \frac{m^2}{1 - t^2} \right] \Theta = 0, \quad (9.6)$$

which has a particular solution $\Theta = (1 - t^2)^{m/2}\dfrac{d^m P_n(t)}{dt^m} = \sin^m\theta\dfrac{d^m P_n(t)}{dt^m}$, if m and n are positive integers and $m < n + 1$. A second but less useful particular solution of Eq (9.6) is $\Theta = (1 - t^2)^{m/2}\dfrac{d^m Q_n(t)}{dt^m}$. Combining these solutions the two particular solutions of Eq (9.1) are

$$u = r^n \left(A \cos m\phi + B \sin m\phi \right) \sin^m\theta \, \frac{d^m}{dt^m} P_n(t), \qquad (9.7)$$

$$u = r^{-n-1} \left(A \cos m\phi + B \sin m\phi \right) \sin^m\theta \, \frac{d^m}{dt^m} P_n(t). \qquad (9.8)$$

The function $\sin^m\theta\dfrac{d^m P_n(t)}{dt^m}$ or $(1 - t^2)^{m/2}\dfrac{d^m P_n(t)}{dt^m}$ is represented by $P_n^m(t)$ and is called the associated Legendre function of the first kind of order m and degree n. By differentiating the value of $P_n^m(t)$ we get the formula

$$P_n^m(t) = \frac{(2n)! \sin^m\theta}{2^n n! \, (n-m)!} \left[t^{n-m} - \frac{(n-m)(n-m-1)}{2 \cdot (2n-1)} t^{n-m-2} \right.$$
$$\left. + \frac{(n-m)(n-m-1)(n-m-2)(n-m-3)}{2 \cdot 4 \cdot (2n-1)(2n-3)} t^{n-m-4} - \cdots \right]$$

where the expression within the square brackets ends with the terms involving t^0 if $n - m$ is even and with the term involving t if $n - m$ is odd. Tables of some values of $P_n^m(t)$ are given in Appendix C.

The functions $\cos m\phi P_n^m(t) = \cos m\phi \sin^m\theta\dfrac{d^m P_n(t)}{dt^m}$ and $\sin m\phi P_n^m(t) = \sin m\phi \sin^m\theta\dfrac{d^m P_n(t)}{dt^m}$ are called *tesseral harmonics* of degree n and order m. There are obviously $2n + 1$ tesseral harmonics of order zero and degree n, viz.,

$$P_n(t), \quad \cos\phi \sin\theta\frac{dP_n(t)}{dt}, \qquad \sin\phi \sin\theta\frac{dP_n(t)}{dt}$$
$$\cos 2\phi \sin^2\theta\frac{d^2 P_n(t)}{dt^2}, \qquad \sin 2\phi \sin^2\theta\frac{d^2 P_n(t)}{dt^2}$$
$$\cos 3\phi \sin^3\theta\frac{d^3 P_n(t)}{dt^3}, \qquad \sin 3\phi \sin^3\theta\frac{d^3 P_n(t)}{dt^3}$$
$$\cdots \qquad\qquad\qquad \cdots$$
$$\cos m\phi \sin^m\theta\frac{d^m P_n(t)}{dt^m}, \qquad \sin m\phi \sin^m\theta\frac{d^m P_n(t)}{dt^m}.$$

Note that $P_n(t)$ is a tesseral harmonic of order zero and degree n and is sometimes written as $P_n^0(t)$. If each of these is multiplied by a constant and their sum taken, this sum is called a *surface spherical harmonic* of degree n, and is a solution of Eqs (9.4)-(9.5). This sum is represented by the functions $Y_n(t, \phi)$ or $Y_n(\theta, \phi)$. The functions $r^n Y_n^m(t, \phi)$ and $r^{-n-1}Y_n^m(t, \phi)$, which are called *solid spherical harmonics* of degree n, are solutions of Laplace's equation (9.1). Note that a tesseral harmonic is a special case of a surface spherical harmonic, and a zonal harmonic a special case of a tesseral harmonic.

The solutions of the Θ and Φ equations (9.2)-(9.3) are combined by multiplication to give the complete angular dependence of the wave functions. These angular functions are a product of the trigonometric functions, represented as a complex exponential $e^{i\,m\phi}$ and the associated Legendre functions $P_n^m(\cos\theta)$ of the first kind and order m, which can be represented as

$$Y_n^m(\theta,\phi) = N\,e^{i\,m\phi}\,P_n^m(\cos\theta),\tag{9.9}$$

where N is a normalization factor which is defined by almost similar expressions, but not uniquely, in different branches of science and technology (see §9.2.1), and θ and ϕ represent *colatitude* (or polar angle) and *latitude* (or azimuth), respectively. The colatitude ranges from 0 at the north pole to π at the south pole, and $\pi/2$ at the equator, whereas the latitude assumes all values $0 \le \phi < 2\pi$. The function $Y_n^m(\theta,\phi)$ are called *spherical harmonic functions* of degree n and order m. It can be verified that

$$\nabla^2 Y_n^m(\theta,\phi) = -\frac{n(n+1)}{r^2}\,Y_n^m(\theta,\phi).\tag{9.10}$$

The solutions of the R-equation (4.12) involving r are called the *radial wave functions* $R_n^m(r)$, and thus, the overall solutions are $u(r,\theta,\phi) = R_n^m(r)Y_N^m(\theta,\phi)$. If Laplace's equation (9.1) is solved on the surface of the sphere, the periodic boundary conditions in ϕ and the regularity conditions at both north and south poles are satisfied. This in turn ensures that the degree n and the order m are integers such that $n \ge 0$ and $|m| \le n$.

If the function u in Eq (9.1) were defined for $\theta \ge 0$, the resulting spherical harmonics would have been defined for integer order and non-integer degree. The general solution of Laplace's equation (9.1) in a sphere with center at the origin and radius a is a linear combination of the spherical harmonic functions $Y_n^m(\theta,\phi)$ multiplied by a scaling factor r^n:

$$u(r,\theta,\phi) = \sum_{n=0}^{\infty}\sum_{m=-n}^{n} c_n^m r^n\, Y_n^m(\theta,\phi),\tag{9.11}$$

where c_n^m are constant and the factors $r^n Y_n^m(\theta,\phi)$ are known as *solid spherical harmonics*. The representation (9.11) is valid in the sphere

$$r < a = \frac{1}{\limsup\limits_{n\to\infty} |c_n^m|^{1/n}}.$$

9.2.1. Orthonormalization.
Different normalizations for Laplace's spherical functions are used in various fields of science and technology. Some of the common conventions are as follows:

(i) In physics and seismology these functions are generally defined as

$$Y_n^m(\theta, \phi) = \sqrt{\frac{(2n+1)((n-m)!}{4\pi\,(n+m)!}}\; e^{i\,m\phi}\, P_n^m(\cos\theta). \tag{9.12}$$

These functions are orthonormal, i.e.,

$$\int_{\theta=0}^{\pi} \int_{\phi=0}^{2\pi} Y_n^m\, \overline{Y_{n'}^{m'}}\, dS = \delta_{nn'}\, \delta_{mm'}, \tag{9.13}$$

where $\overline{Y_{n'}^{m'}}$ denotes the complex conjugate of $Y_{n'}^{m'}$, $dS = d\theta\, d\phi$ is an element of the surface of the sphere, and $\delta_{nn'}$, $\delta_{mm'}$ is the Kronecker delta.

(ii) In geodesy and spectral analysis these functions are defined as

$$Y_n^m(\theta, \phi) = \sqrt{(2n+1)\,\frac{(n-m)!}{(n+m)!}}\; e^{i\,m\phi}\, P_n^m(\cos\theta), \tag{9.14}$$

such that

$$\frac{1}{4\pi} \int_{\theta=0}^{\pi} \int_{\phi=0}^{2\pi} Y_n^m\, \overline{Y_{n'}^{m'}}\, dS = \delta_{nn'}\, \delta_{mm'} \tag{9.15}$$

has unit value for $n = n'$ and $m = m'$.

(iii) In magnetics and quantum mechanics, these functions are defined as

$$Y_n^m(\theta, \phi) = \sqrt{\frac{(n-m)!}{(n+m)!}}\; e^{i\,m\phi}\, P_n^m(\cos\theta), \tag{9.16}$$

with the normalization

$$\frac{2n+1}{4\pi} \int_{\theta=0}^{\pi} \int_{\phi=0}^{2\pi} Y_n^m\, \overline{Y_{n'}^{m'}}\, dS = \delta_{nn'}\, \delta_{mm'}. \tag{9.17}$$

The functions $Y_n^m(\theta, \phi)$ defined by (9.16) are known as *Schmidt semi-normalized harmonics* in magnetics, and the normalization (9.17) is known as *Racah's normalization* in quantum mechanics.

Theorem 9.1. *The integral over the surface of the unit sphere of the product of two surface spherical harmonics of different degrees is zero, i.e.,*

$$\int_0^{2\pi} d\phi \int_{-1}^{1} Y_n(t, \phi)\, Y_m(t, \phi)\, d\theta = 0. \tag{9.18}$$

PROOF. As we have seen, $u = r^n Y_n(t, \phi)$ and $v = r^m Y_m(t, \phi)$ are solutions of Laplace's equation Hence, by Green's second identity (1.16) and formula (1.17), where $\nabla^2 u = 0 = \nabla^2 v$, we have on the surface Γ of the unit sphere

$$\int_\Gamma \left(u \frac{\partial v}{\partial n} - v \frac{\partial u}{\partial n} \right) dS = 0,$$

$$\frac{\partial v}{\partial n} = \frac{\partial v}{\partial r} = m r^{m-1} Y_m(t, \phi), \quad \frac{\partial u}{\partial n} = \frac{\partial u}{\partial r} = n r^{n-1} Y_n(t, \phi),$$

$$\frac{\partial v}{\partial n} - v \frac{\partial u}{\partial n} = (m - n) r^{n+m-1} Y_n(t, \phi) Y_m(t, \phi) = (m - n) Y_n(t, \phi) Y_m(t, \phi).$$

and thus,

$$(m - n) \int_\Gamma Y_n(t, \phi) Y_m(t, \phi) = (m - n) \int_0^{2\pi} d\phi \int_{-1}^1 Y_n(t, \phi) Y_m(t, \phi) \, dt = 0,$$

which for $n \neq m$ yields (9.18). ∎

All the above normalized spherical harmonics satisfy the following relation:

$$\overline{Y_{n'}^{m'}}(\theta, \phi) = (-1)^m Y_n^{-m}(\theta, \phi). \tag{9.19}$$

The proof is left as an exercise.

9.2.2. Condon-Shortley Phase Factor.

The associated Legendre functions $P_n^m(\cos\theta)$, which occur in the spherical harmonic functions $Y_n^m(\theta, \phi)$ as defined in (9.9) and subsequent conventions in different branches of science and technology, are related to the Legendre polynomials of the first kind $P_n(x)$, $x = \cos\theta$, by the relation:

$$P_n^m(x) = \left(1 - x^2\right)^{m/2} \frac{d^m}{dx^m} P_n(x) = \frac{\left(1 - x^2\right)^{m/2}}{2^n \, n!} \frac{d^{m+n}}{dx^{m+n}} \left(x^2 - 1\right)^n. \tag{9.20}$$

Since

$$P_n^{-m}(\cos\theta) = (-1)^m \frac{(n - m)!}{n + m)!} P_n^m(\cos\theta), \quad -n \le m \le n, \tag{9.21}$$

(see Exercise 9.6), the orthonormalization (9.12), and others, hold for both positive and negative values of m, within a possible factor $(-1)^m$. In view of (9.21) Y_n^{-m} and Y_n^m differ in phase by $(-1)^m$. This factor is known as the *Condon-Shortley phase factor* that occurs in some definitions of spherical harmonics, while there is a lack of inclusion of this factor in some other definitions of the associated Legendre polynomials. In fact, in certain quantum mechanical calculations, particularly in the quantum theory of angular momentum, this phase factor is usually associated with *positive* m spherical harmonic, while in many other applications of spherical harmonics it is a general practice to ignore these phase factor by using (9.12) for $m \geq 0$ and $Y_n^{-m} = Y_n^{m*}$ for negative subscript.

Omitting the Condon-Shortley convention in the definition of spherical harmonic function after using it in the definition of $P_n^m(\cos\theta)$ gives the representation (9.12), whereas using it in this definition will yield

$$Y_n^m(\theta, \phi) = (-1)^m \sqrt{\frac{(2n+1)((n-m)!}{4\pi \, (n+m)!}} \, e^{i\,m\phi} \, P_n^m(\cos\theta). \qquad (9.22)$$

The Condon-Shortley phase is not necessary in the definition of spherical harmonics but its inclusion simplifies the treatment of angular moment in quantum mechanics [Arfken, 1985:448]. The Condon-Shortley phase factor is never used in geodesy and magnetics in their definition of spherical harmonics. For more results on associated Legendre functions of the first kind with and without the inclusion of this factor, and a list of spherical harmonics, see §C.3.1.

Instead of deciding on whether to include or ignore the Condon-Shortley phase factor, a real basis of spherical harmonics is provided by setting

$$Y_n^m(\theta, \phi) = \begin{cases} Y_n^0(\theta, \phi) & \text{if } m = 0, \\[2mm] \dfrac{1}{\sqrt{2}} \left[Y_n^m(\theta, \phi) + (-1)^m Y_n^{-m}(\theta, \phi) \right] = \sqrt{2}\, N(n, m) P_n^m(\mu) \cos m\phi \\ \qquad \text{if } m > 0, \\[2mm] \dfrac{1}{i\sqrt{2}} \left[Y_n^{-m}(\theta, \phi) - (-1)^m Y_n^{-m}(\theta, \phi) \right] = \sqrt{2} N(n, m) P_n^{-m}(\mu) \sin m\phi \\ \qquad \text{if } m < 0, \end{cases}$$

where $\mu = \cos\theta$, $0 \leq \theta \leq \pi$, and $N(n, m)$ denote the normalization constants as a function of n and m. The real form has the associated Legendre function P_n^m for $m \geq 0$, while the imaginary form occurs only for $m < 0$. The spherical harmonics for $m > 0$ are said to be of *cosine type*, and those with $m < 0$ of the *sine type*. These functions have the same orthonormalization properties as described in §9.2.1.

9.2.3. Spherical Harmonics Expansion. Since Laplace's special harmonics form a complete set of orthonormal functions, they form an orthonormal basis of the Hilbert space of square-integrable functions. On the unit sphere any square-integrable function can be expanded as a linear combination of these special harmonics, i.e.,

$$u(\theta, \phi) = \sum_{n=0}^{\infty} \sum_{m=-n}^{n} c_n^m \, Y_n^m(\theta, \phi), \qquad (9.23)$$

where c_n^m are constants. This expansion is exact as long as $n \to \infty$ and converges uniformly to $u(\theta, \phi)$. Thus, a square-integrable function can be expanded in terms of the real harmonics $Y_n^m(\theta, \phi)$ in the form of the series (9.23). However, truncation errors arise when limiting the summation over n to a finite bandwidth of length N. The expansion (9.23) is valid in the sense of mean-square convergence, i.e., convergence in L_2, which implies that

$$\lim_{n\to\infty} \int_0^{2\pi} \int_0^\pi \left| u(\theta,\phi) - \sum_{n=0}^N c_n^m Y_n(\theta,\phi) \right|^2 \sin\theta \, d\theta \, d\phi = 0.$$

In the above expansion the coefficients c_n^m are analogous to the Fourier coefficients, and can be obtained by multiplying the above equation by the complex conjugate of $Y_n^m(\theta,\phi)$, then integrating over the solid angle Ω, and using the orthogonality relations of $Y_n^m(\theta,\phi)$. Thus, in the case of orthonormalized harmonics we have

$$c_n^m = \int_\Omega u(\theta,\phi) \overline{Y_n^m(\theta,\phi)} \, d\Omega = \int_0^{2\pi} \sin\theta \, u(\theta,\phi) Y_n^m(\theta,\phi) \, d\theta. \qquad (9.24)$$

If the coefficients c_n^m decay in n sufficiently rapidly, say exponentially, then the series (9.23) also converges uniformly to $u(\theta,\phi)$.

9.2.4. Addition Theorem.

The addition theorem for spherical harmonics is a generalization of the 2-D trigonometric identity $\cos(\theta - \theta') = \cos\theta\cos\theta' + \sin\theta\sin\theta'$. In this theorem the function $\cos(\theta - \theta')$ on the left side is replaced by the Legendre polynomials whereas the trigonometric functions on the right side are replaced by a product of the spherical harmonic functions and their complex conjugates. Thus,

Theorem 9.4. Let $\mathbf{x}$ and $\mathbf{y}$ be two unit vectors with spherical angular coordinates (θ, ϕ) and (θ', ϕ'), respectively. Then

$$P_n(\mathbf{x} \cdot \mathbf{y}) = \frac{4\pi}{2n+1} \sum_{m=-n}^n Y_{nm}(\theta,\phi) \overline{Y_{nm}(\theta',\phi')}, \qquad (9.25)$$

where P_n are the Legendre polynomials of first kind and order n. This result holds for both real and complex harmonics. In fact, it is valid for any orthogonal basis of special harmonics of degree n. For unit power harmonics the factor 4π appearing in (9.25) must be removed. A proof of this theorem can be found in Whittaker and Watson [1962:395], where the properties of the Poisson kernel in the unit sphere are used, and then the right-hand side is calculated.

As a particular case when $\mathbf{x} = \mathbf{y}$, the above theorem reduces to Unsöld's theorem [Unsöld, 1927], which states:

$$\sum_{m=-n}^n Y_{nm}(\theta,\phi) \overline{Y_{nm}(\theta',\phi')} = \frac{2n+1}{4\pi}. \qquad (9.26)$$

This result is a generalization of the 2-D trigonometric identity $\cos^2\theta + \sin^2\theta = 1$ for 3-D spherical harmonics.

9.2.5. Laplace's Coefficients. We have seen that for a point (x_1, y_1, z_1) the function

$$u = \left[(x - x_1)^2 + (y - y_1)^2 + (z - z_1)^2\right]^{-1/2}$$

is a solution of Laplace's equation $\nabla^2 u = 0$. This function u in spherical coordinates becomes

$$u = \left[r^2 - 2rr_1\left[\cos\theta\cos\theta_1 + \sin\theta\sin\theta_1 cos(\phi - \phi_1) + r_1^2\right]^{-1/2},\right.$$

which is a solution of Eq (9.1). If γ is the angle between the radius vectors r and r_1 for the points (x, y, z) and $(x_1, y_1, z1)$, then we can write $u = \left[r^2 - 2rr_1\cos\gamma + r_1^2\right]^{-1/2}$. Notice that $\cos\gamma = \cos\theta\cos\theta_1 + \sin\theta\sin\theta_1 cos(\phi - \phi_1)$. Also, $u = P_n(\cos\gamma)$ is a solution of Eq (9.6), and $u = r^n P_n(\cos\gamma)$ and $r^{-n-1} P_n(\cos\gamma)$ are solutions of Eq (9.1). If we transform our coordinates keeping the origin fixed and taking as our new polar axis the radius vector of (x_1, y_1, z_1), then γ becomes our new θ and $P_n(\cos\gamma)$ reduces to $P_n(\cos\theta)$, which is our new surface zonal harmonic, or a *Legendrian*, of degree n. It is a Legendrian with its axis not the original polar axis but the radius vector of (x_1, y_1, z_1). Since a Legendrian is a surface spherical harmonic,

$$P_n(\cos\gamma) = P_n\left[\cos\theta\cos\theta_1 + \sin\theta\sin\theta_1 cos(\phi - \phi_1)\right]$$

is a surface spherical harmonic of degree n. This harmonic is of a very special form since, being a determinate function of t, ϕ, t_1 and ϕ_1, it contains but two arbitrary constants if we regard it as a function of t and ϕ, instead of containing $2n+1$. This function is known as *Laplace's coefficient*[1] of degree n and denoted by $L_n(t, \phi, t_1, \phi_1)$. The radius vector (x_1, y_1, z_1) is called the *axis* of the Laplace's coefficient and the point where the axis cuts the surface of the unit sphere is called the *pole* of the Laplace's coefficient.

Theorem 9.2. *If the product of a surface spherical harmonic of degree n by a Laplace's coefficient of the same degree is integrated over the surface of the unit sphere, the result is equal to $\dfrac{4\pi}{2n+1}$ times the value of the spherical harmonic at the pole of the Laplace's coefficient, that is,*

$$\int_0^{2\pi} d\phi \int_0^\pi Y_n(t, \phi) L_n(t, \phi, t_1, \phi_1)\, d\theta = \frac{4\pi}{2n+1} Y_n(t_1, \phi_1). \tag{9.27}$$

PROOF. Transform to the axis of the Laplace's coefficient as a new axis, and let $Z_n(t, \phi)$ be the transformed spherical harmonic. Then $L_n(t, \phi, t_1, \phi_1)$ will become $P_n(t)$ and (9.27) will be proved if we can show that

$$\int_0^{2\pi} d\phi \int_{-1}^1 Z_n(t, \phi) P_n(t)\, dt = \frac{4\pi}{2n+1} Z_n(1, 0). \tag{9.28}$$

---

[1] During the nineteenth century this function was also known as the *Laplacian*, but this nomenclature was abandoned when Laplace's operator ∇^2 was referred to as 'Laplacian'.

Now,

$$Z_n(t,\phi)P_n(t) = A_0\left[P_n(t)\right]^2 + \sum_{m=1}^{n}\left(A_m\cos m\phi + B_m\sin m\phi\right)P_n^m(t)P_n(t),$$

$$\int_0^{2\pi} Z_n(t,\phi)P_n(t)\,d\phi = 2\pi A_0\left[P_n(t)\right]^2,$$

$$\int_{-1}^{1} dt\int_0^{2\pi} Z_n(t,\phi)P_n(t)\,d\phi = \frac{4\pi}{2n+1}A_0.$$

But $Z_n(1,0) = A_0$, because $P_n(1) = 1$ and $P_n^m(1)$ contains $(1-1)^{m/2}$ as a factor and is equal to zero. Hence (9.28) is proved. ∎

Theorem 9.3. *The Laplace's coefficient $L_n(t,\phi,t_1,\phi_1)$ has the following series development in terms of spherical harmonics:*

$$L_n(t,\phi,t_1,\phi_1) = P_n(t)P_n(t_1) + 2\sum_{m=1}^{n}\left[\frac{(n-m)!}{(n+m)!}P_n^m(t)P_n^m(t_1)\cos m(\phi-\phi_1)\right].$$
$$(9.29)$$

PROOF. Since

$$L_n(t,\phi,t_1,\phi_1) = P_n(\cos\gamma) = P_n(\cos\theta\cos\theta_1 + \sin\theta\sin\theta_1\cos(\phi-\phi_1)$$

$$= \sum_{n=0}^{\infty}\left[A_{0,n}P_n(t) + \sum_{m=1}^{n}\left(A_{m,n}\cos m\phi + B_{m,n}\sin m\phi\right)P_n^m(t)\right],$$

where

$$A_{0,n} = \frac{2n+1}{4\pi}\int_0^{2\pi}d\phi\int_{-1}^{1}L_n(t,\phi,t_1,\phi_1)P_n(t)\,dt$$

$$= \frac{2n+1}{4\pi}\frac{4\pi}{2n+1}P_n(t_1) = P_n(t_1),$$

$$A_{m,n} = \frac{2n+1}{2\pi}\frac{(n-m)!}{(n+m)!}\int_0^{2\pi}d\phi\int_{-1}^{1}L_n(t,\phi,t_1,\phi_1)\cos m\phi\,P_n^m(t)\,dt$$

$$= \frac{2(n-m)!}{(n+m)!}\cos m\phi_1\,P_n^m(t_1),$$

$$B_{m,n} = \frac{2n+1}{2\pi}\frac{(n-m)!}{(n+m)!}\int_0^{2\pi}d\phi\int_{-1}^{1}L_n(t,\phi,t_1,\phi_1)\sin m\phi\,P_n^m(t)\,dt$$

$$= \frac{2(n-m)!}{(n+m)!}\sin m\phi_1\,P_n^m(t_1),$$

and $A_{0,n} = A_{m,n} = B_{m,n} = 0$ for $n\neq m$. The result follows from these steps. ∎

Each term of the Laplace's coefficient L_n involves a numerical coefficient, plus a factor which is a function of t, a second factor which is the same function of t_1, and a third factor which is of the form $\cos m(\phi - \phi_1)$. Table 9.1 gives a first few Laplace's coefficients, where $C(m)$ denotes the coefficient of $\cos m(\phi - \phi_1)$ for $m = 0, 1, 2, 3, 4$.

Table 9.1.

	L_0	L_1	L_2	L_3	L_4
$C(0)$	1	t	$\frac{1}{4}(3^2 - 1)$	$\frac{1}{4}(5t^3 - t)$	$\frac{1}{64}(35t^4 - 30t^2 + 3)$
$C(1)$		$(1 - t^2)^{\frac{1}{2}}$	$3t(1 - t^2)^{\frac{1}{2}}$	$\frac{3}{8}(1 - t^2)^{\frac{1}{2}}(5t^2 - 1)$	$\frac{5}{8}(1 - t^2)^{\frac{1}{2}}(7t^3 - 3t)$
$C(2)$			$\frac{3}{4}(1 - t^2)$	$\frac{15}{4}t(1 - t^2)$	$\frac{5}{16}(1 - t^2)(7t^2 - 1)$
$C(3)$				$\frac{5}{8}(1 - t^2)^{\frac{3}{2}}$	$\frac{35}{8}(1 - t^2)^{\frac{3}{2}}$
$C(4)$					$\frac{35}{64}(1 - t^2)^2$

9.3. Surface Spherical Harmonics

Consider Laplace's equation (9.1) which we multiply by r^2 and substitute $\Theta(\theta)$ for u. Then we obtain

$$r\frac{\partial^2(r\Theta)}{\partial r^2} + \frac{1}{\sin\theta}\frac{\partial}{\partial\theta}\left(\sin\theta\frac{\partial\Theta}{\partial\theta}\right) + \frac{1}{\sin^2\theta}\frac{\partial^2\Theta}{\partial\phi^2} = 0. \tag{9.30}$$

In order to get surface spherical harmonics, we will solve this equation by setting

$$\Theta = r^n\,\Psi_n. \tag{9.31}$$

Note that if Θ is a solid spherical harmonic of degree n, then the functions Ψ_n which are related to Θ by the relation (9.31) are called *surface spherical harmonics*. The term 'surface' signifies the fact that for a constant value of r, the function Θ reduces to a function defined on the surface of a sphere of radius r, and therefore, Ψ_n defines Θ on the surface of the unit sphere since $\Psi = \Theta/r^n$. Now, substituting (9.31) into (9.30) and dividing by r^n, we obtain the equation for the surface spherical harmonics for a fixed r as:

$$\frac{1}{\sin\theta}\frac{\partial}{\partial\theta}\left(\sin\theta\frac{\partial\Psi_n}{\partial\theta}\right) + n(n + 1)\Psi_n + \frac{1}{\sin^2\theta}\frac{\partial^2\Psi_n}{\partial\phi^2} = 0. \tag{9.32}$$

As in solving Eq (9.3), we set $x = \cos\theta$ in (9.32) where $|x| \leq 1$, and get

$$\frac{\partial}{\partial x}\left[(1 - x^2)\frac{\partial\Psi_n}{\partial x}\right] + n(n + 1)\Psi_n + \frac{1}{1 - x^2}\frac{\partial^2\Psi_n}{\partial\phi^2} = 0.$$

An explicit expression for surface spherical harmonics is obtained if in this equation we put $\Psi_n(x, \phi) = X(x)\Phi(\phi)$ and after separating the variables equate both sides to $-m^2$. This leads to the equations

$$\Phi'' + m^2\Phi = 0, \tag{9.33}$$

$$\frac{d}{dx}\left[\left(1 - x^2\right)X'\right] + \left[n(n+1) - \frac{m^2}{1 - x^2}\right]X = 0. \tag{9.34}$$

We will show that Eq (9.34) is the associated Legendre's equation. If we make the substitution $X = \left(1 - x^2\right)^{m/2} y$, we obtain

$$\left(1 - x^2\right)y'' - 2(m + 1)xy' + (n - m)(n + m + 1)y = 0, \tag{9.35}$$

which when differentiated m times with respect to x by using Leibniz's rule[1] gives

$$\left(1 - x^2\right)u'' - 2(m + 1)xu' + (n - m)(n + m + 1)u = 0, \tag{9.36}$$

where $u = \dfrac{d^m}{dx^m}P_n(x)$. Now let $X(x) = (1 - x^2)^{m/2}u(x) = (1 - x^2)\dfrac{d^m P_n(x)}{dx^m}$. Then solving for u and differentiating, we get

$$u' = \left[X' + \frac{mxX}{1 - x^2}\right](1 - x^2)^{-m/2},$$

$$u'' = \left[X'' + \frac{2mxX'}{1 - x^2} + \frac{mX}{1 - x^2} + \frac{m(m + 2)x^2 X}{(1 - x^2)^2}\right](1 - x^2)^{-m/2}.$$

Substituting into Eq (9.36), we find that the new function X satisfies the differential equation (9.35), which is known as the associated Legendre's equation. Eq (9.35) reduces to Legendre's equation for $m = 0$. If n is an integer, and $y = P_n(x)$ are the Legendre polynomials, then the solution of this equation, and therefore of Eq (9.35), are given by $y = \dfrac{d^m}{dx^m}P_n(x)$, and finally the solution of Eq (9.34) are given by the associated Legendre functions of the first kind, i.e.,

$$X = P_n^m(x) = \left(1 - x^2\right)^{m/2}\frac{d^m}{dx^m}P_n(x). \tag{9.37}$$

Note that the solutions (9.37) are not always real-valued for $|x| > 1$. From the above form for $P_n^m(x)$ we will expect m to be non-negative. However, differentiating a negative number of times being not defined, we use Rodrigue's formula (C.2) to express $P_n(x)$; this limitation on m is relaxed and we may have $-n \le m \le n$; so m is both positive and negative. A few values of $P_n^m(x)$ are given in §C.2.2. A generating function for the associated Legendre functions is

$$\frac{(2m)! \, (1 - x^2)^{m/2}}{2^m \, m! \, (1 - 2xt + t^2)^{m+1/2}} = \sum_{k=0}^{\infty} P_{m+k}^m \, t^k,$$

---

[1] Leibniz's rule for the nth derivative of a product is defined in §1.8.1(b). Note that in the present case $D^m(1 - x^2)y'' = (1 - x^2)(D^{m+2}y) - 2mx(D^{m+1}y) - m(m - 1)(D^m y)$; and $D^m xy' = x(D^{m+1}y) + m(D^m y)$.

but, unlike the generating function for the Legendre polynomials, it is not used much. Recurrence formulas and orthogonality relation for $P_n^m(x)$ are given in §C.2. Using the definition in (9.37) and Rodrigue's formula (f.2) for $P_n(x)$, we have, with $\xi = x^2 - 1$,

$$\int_{-1}^{1} P_k^m(x) P_n^m(x)\, dx = \frac{(-1)^m}{2^{k+n}\, k!\, n!} \int_{-1}^{1} \xi^m \frac{d^{k+m}}{dx^{k+m}} \xi^k \frac{d^{n+m}}{dx^{n+m}} \xi^n\, dx.$$

If $k \neq n$, let us assume $k < n$. Notice that since the superscript is m for both polynomials, while integrating repeatedly all integrated parts will vanish so long as there is a factor $\xi = x^2 - 1$. If we integrate $(n+m)$-times, we obtain

$$\int_{-1}^{1} P_k^m(x) P_n^m(x)\, dx = \frac{(-1)^m (-1)^{n+m}}{2^{k+n}\, k!\, n!} \int_{-1}^{1} \frac{d^{n+m}}{dx^{n+m}} \left(\xi^m \frac{d^{k+m}}{dx^{k+m}} \xi^k \right) \xi^n\, dx.$$
$$(9.38)$$

If we expand the integrand on the right side by Leibniz's formula, we get

$$\xi^n \frac{d^{n+m}}{dx^{n+m}} \left(\xi^m \frac{d^{k+m}}{dx^{k+m}} \xi^k \right) = \xi^n \sum_{j=0}^{n+m} \frac{(n+m)!}{j!\,(n+m-j)!} \frac{d^{n+m-j}}{dx^{n+m-j}} \xi^m \frac{d^{k+m+j}}{dx^{k+m+j}} \xi^k.$$
$$(9.39)$$

Since the term ξ^m contain no power of x greater than x^{2m}, we must have $n + m - j \leq 2m$, or the derivative will vanish. Similarly, $k + m + j \leq 2k$. Thus, the conditions for a nonzero solution are: $j \geq n - m$, or $j \leq k - m$. If $k < n$, as we have assumed, there is no solution and the integral vanishes. The same result is obtained for $k > n$. Now, if $k = n$, we have only one term that corresponds to $j = n - m$. Substituting Eq (9.39) into Eq (9.38), we obtain

$$\int_{-1}^{1} [P_n^m(x)]^2\, dx = \frac{(-1)^{n+2m}(n+m)!}{2^{2n}\,(n!)^2 (2m)! (n-m)!} \int_{-1}^{1} \xi^n \left(\frac{d^{2m}}{dx^{2m}} \xi^m \right) \left(\frac{d^{2n}}{dx^{2n}} \xi^n \right) dx.$$
$$(9.40)$$

Since $\xi^m = (x^2 - 1)^m = x^{2m} - m x^{2m-1} + \cdots$, we have $\dfrac{d^{2m}}{dx^{2m}} \xi^m = (2m)!$, and eq (9.40) becomes

$$\int_{-1}^{1} [P_n^m(x)]^2\, dx = \frac{(-1)^{n+2m}(2n)!\,(n+m)!}{2^{2n}\,(n!)^2 (n-m)!} \int_{-1}^{1} \xi^n\, dx,$$

where the integral on the right equals $(-1)^n \displaystyle\int_0^{\pi} \sin^{2n+1} \theta\, d\theta = \dfrac{(-1)^n 2^{2n+1}(n!)^2}{(2n+1)!}$.

Combining these results we obtain the orthogonality relation

$$\int_{-1}^{1} P_k^m(x) P_n^m(x)\, dx = \frac{2}{2n+1} \frac{(n+m)!}{(n-m)!} \delta_{k,n},$$

which is the same as given in §C.2.4, or in spherical coordinates

$$\int_0^\pi P_k^m(\cos\theta) P_n^m(\cos\theta)\, d\theta = \frac{2}{2n+1}\frac{(n+m)!}{(n-m)!}\, \delta_{k,n}. \qquad (9.41)$$

A similar orthogonality relation for the associated Legendre functions with the same lower index but different upper indices is

$$\int_{-1}^1 P_n^m(x) P_n^k(x)(1-x^2)^{-1}\, dx = \frac{(n+m)!}{m(n-m)!}\, \delta_{m,k}, \qquad (9.42)$$

where a new wight factor $(x^2-1)^{-1}$ appears. The proof of this result is left as an exercise.

Example 9.1. We will solve the boundary value problem of steady temperature distribution in a spherical medium of unit radius. Let $u(r,\theta,\phi)$ represent the temperature, subject to the boundary condition $u(1,\theta,\phi) = f(\theta,\phi)$. The solutions of Eqs (9.33)-(9.34), written as

$$\Psi_n^m = (A\cos m\phi + B\sin m\phi)\, P_n^m(\cos\theta),$$

represents a surface spherical harmonic. Then the potential function $u = r^n\, \Psi_n^m$ can be represented as

$$u(r,\theta,\phi) = \sum_{n=0}^\infty \sum_{m=0}^n r^n\,(a_{nm}\cos m\phi + b_{nm}\sin m\phi)\, P_n^m(\cos\theta). \qquad (9.43)$$

We will solve Eq (9.43) subject to the prescribed boundary condition at $r=1$. The coefficients a_{nm} and b_{nm} in (9.43) will be determined by a Fourier-type expansion of the function $f(\theta,\phi)$:

$$f(\theta,\phi) = \sum_{n=0}^\infty \sum_{m=0}^n (a_{nm}\cos m\phi + b_{nm}\sin m\phi)\, P_n^m(\cos\theta). \qquad (9.44)$$

We will assume that this expansion is possible and the series on the right side of (9.44) converges uniformly to $f(\theta,\phi)$. If we multiply (9.44) by $\cos k\phi$, $k=0,1,2,\dots$, and integrate from $-\pi$ to π, we obtain

$$\int_{-\pi}^\pi f(\theta,\phi)\cos k\phi\, d\phi = \pi\sum_{n=0}^\infty a_{nk} P_n^k(\cos\theta). \qquad (9.45)$$

Similarly, multiplying (9.44) by $\sin k\phi$, $k=0,1,2,\dots$, we get

$$\int_{-\pi}^\pi f(\theta,\phi)\sin k\phi\, d\phi = \pi\sum_{n=0}^\infty b_{nk} P_n^k(\cos\theta). \qquad (9.46)$$

Since P_n^k are solutions of the associated Legendre's equation (9.34), we have

$$\frac{d}{dx}\left[(1-x^2)\frac{dP_n^k}{dx}\right] + \left[n(n+1) - \frac{k^2}{1-x^2}\right]P_n^k = 0,$$

$$\frac{d}{dx}\left[(1-x^2)\frac{dP_m^k}{dx}\right] + \left[m(m+1) - \frac{k^2}{1-x^2}\right]X = 0.$$

Now, we multiply the first equation by P_m^k and the second by P_n^k and subtract one from another, we get

$$P_m^k\frac{d}{dx}\left[(1-x^2)\frac{dP_n^k}{dx}\right] - P_n^k\frac{d}{dx}\left[(1-x^2)\frac{dP_m^k}{dx}\right] = [m(m+1) - n(n+1)]\,P_n^k P_m^k.$$

If we integrate both sides of this equation from -1 to 1 is zero, we find that the integral of the left side is zero, and we obtain

$$\int_{-1}^{1} P_n^k P_m^k\,dx = 0 \quad \text{if } n \neq m. \tag{9.47}$$

Substituting $x = \cos\theta$ in (9.47) we get

$$\int_0^\pi P_n^k(\cos\theta)P_m^k(\cos\theta)\,\sin\theta\,d\theta = 0 \quad \text{for } n \neq m. \tag{9.48}$$

Next, multiplying (9.45) and (9.46) by $\sin\theta P_j^k(\cos\theta)$ and integrating from 0 to π, we find that

$$\frac{1}{\pi}\int_0^\pi\int_{-\pi}^\pi f(\theta,\phi)\cos k\phi P_j^k(\cos\theta)\sin\theta\,dphi\,d\theta = a_{jk}\int_0^\pi \left[P_j^k(\cos\theta)\right]^2\sin\theta\,d\theta,$$

$$\frac{1}{\pi}\int_0^\pi\int_{-\pi}^\pi f(\theta,\phi)\sin k\phi P_j^k(\cos\theta)\sin\theta\,dphi\,d\theta = b_{jk}\int_0^\pi \left[P_j^k(\cos\theta)\right]^2\sin\theta\,d\theta.$$

Since (see Appendix C)

$$\int_0^\pi \left[P_j^k(\cos\theta)\right]^2\sin\theta\,d\theta = \int_{-1}^1 \left[P_j^k(x)\right]^2\,dx = \frac{2}{2j+1}\frac{(j+k)!}{(j-k)!}, \quad k \leq j,$$

we find that

$$a_{jk} = \frac{(2j+1)(j-k)!}{2\pi(j+k)!}\int_0^\pi\int_{-\pi}^\pi f(\theta,\phi)\cos k\phi P_j^k(\cos\theta)\sin\theta\,d\phi\,d\theta, \tag{9.49}$$

$$b_{jk} = \frac{(2j+1)(j-k)!}{2\pi(j+k)!}\int_0^\pi\int_{-\pi}^\pi f(\theta,\phi)\sin k\phi P_j^k(\cos\theta)\sin\theta\,d\phi\,d\theta, \tag{9.50}$$

and the coefficients for $k = 0$ is

$$a_{j0} = \frac{2j+1}{4\pi}\int_0^\pi\int_{-\pi}^\pi f(\theta,\phi)\cos k\phi P_j^k(\cos\theta)\sin\theta\,d\phi\,d\theta. \tag{9.51}$$

Having determined the coefficients we will use the trigonometric identity

$$\cos m\alpha \cos m\beta + \sin m\alpha \sin m\beta = \cos m(\alpha - \beta),$$

and the above values of the coefficients a_{jk} and b_{jk}, we finally get

$$
\begin{aligned}
u(r, \theta, \phi) &= \sum_{n=0}^{\infty} r^n \frac{2n+1}{4\pi} \int_0^{\pi} \int_{-\pi}^{\pi} f(\alpha, \beta) P_n(\cos \beta) \sin \beta \, d\alpha \, d\beta \, P_n(\cos \theta) \\
&+ \sum_{n=1}^{\infty} \sum_{m=1}^{n} r^n \frac{2n+1}{2\pi} \frac{(n-m)!}{(n+m)!} \int_0^{\pi} \int_{-\pi}^{\pi} f(\alpha, \beta) P_n^m(\cos \beta) \times \\
&\qquad\qquad \times \sin \beta \cos m(\phi - \alpha) \, d\alpha \, d\beta \, P_n^m(\cos \theta) \\
&= \int_0^{\pi} \int_{-\pi}^{\pi} \left\{ \sum_{n=0}^{\infty} r^n \frac{2n+1}{4\pi} f(\alpha, \beta) P_n(\cos \beta) \sin \beta \, P_n(\cos \theta) \times \right. \\
&\qquad\qquad \left. \times \sin \beta \cos m(\phi - \alpha) \, P_n^m(\cos \theta) \right\} d\alpha \, d\beta,
\end{aligned}
$$
$$(9.52)$$

where we have interchanged the order of summation and integration. ∎

9.3.1. Poisson Integral Representation. We start with Eq (9.52) and calculate the summation under the integral sign for $\theta = 0 = \phi$. Then we have

$$
\begin{aligned}
u(r, 0, 0) &= \frac{1}{4\pi} \int_0^{\pi} \int_{-\pi}^{\pi} \left\{ \sum_{n=0}^{\infty} r^n \, (2n+1) f(\alpha, \beta) P_n(\cos \alpha) \sin \alpha \, P_n(1) \right. \\
&+ \frac{1}{2\pi} \sum_{n=1}^{\infty} \sum_{m=1}^{n} r^n \frac{(2n+1)(n-m)!}{(n+m)!} f(\alpha, \beta) P_n^m(\cos \alpha) \times \\
&\qquad\qquad \left. \times \sin \alpha \cos m\beta \, P_n^m(1) \right\} d\beta \, d\alpha.
\end{aligned}
$$

Since $P_n^m(1) = 0$ for all $n \geq 1$, the second part of the above sum vanishes. Also, since $P_n(1) = 1$, we get

$$u(r, 0, 0) = \frac{1}{4\pi} \int_0^{\pi} \int_{-\pi}^{\pi} \left[\sum_{n=0}^{\infty} (2n+1) \, r^n f(\alpha, \beta) P_n(\cos \alpha) \sin \alpha \right] d\beta \, d\alpha. \quad (9.53)$$

Since $\displaystyle\sum_{n=0}^{\infty} 2n \, r^n P_n(x) = \frac{2rx - 2r^2}{(1 - 2rx + r^2)^{3/2}}$, $\displaystyle\sum_{n=0}^{\infty} r^n P_n(x) = \frac{1 - 2rx + r^2}{(1 - 2rx + r^2)^{3/2}}$, adding these results we obtain

$$\sum_{n=0}^{\infty} (2n+1) \, r^n P_n(x) = \frac{1 - r^2}{(1 - 2rx + r^2)^{3/2}}. \quad (9.54)$$

Substituting (9.54) into (9.53) we get

$$u(r,0,0) = \frac{1-r^2}{4\pi} \int_0^\pi \int_{-\pi}^\pi \frac{f(\alpha,\beta)\,\sin\alpha\,d\beta\,d\alpha}{(1-2r\cos\alpha+r^2)^{3/2}}. \tag{9.55}$$

This result holds for any point (r,θ,ϕ) on the sphere. Since the expression $\rho = \left(1-2r\cos\alpha+r^2\right)^{1/2}$ is equal to the distance between the reference point $(1,0,0)$ and the integration point (r,α,β), we choose the point $(r,0,0)$ to represent the north pole of the sphere. Hence, we have

$$u(r,\theta,\phi) = \frac{1-r^2}{4\pi} \int_0^\pi \int_{-\pi}^\pi \frac{f(\alpha,\beta)\,\sin\alpha\,d\beta\,d\alpha}{\rho^3}, \tag{9.56}$$

where ρ is the distance between the points $(1,0,0)$ and (r,θ,ϕ). Let γ be the angle at the origin between the lines L_1 and L_2 (see Fig. 9.1), where the direction cosines of the line L_1 are $(\sin\theta\cos\phi, \sin\theta\sin\phi, \cos\theta)$ and those of the line L_2 are $(\sin\alpha\cos\beta, \sin\alpha\sin\beta, \cos\alpha)$. Then $\cos\gamma = \cos\theta\cos\alpha + \sin\theta\sin\alpha\cos(\phi-\beta)$.

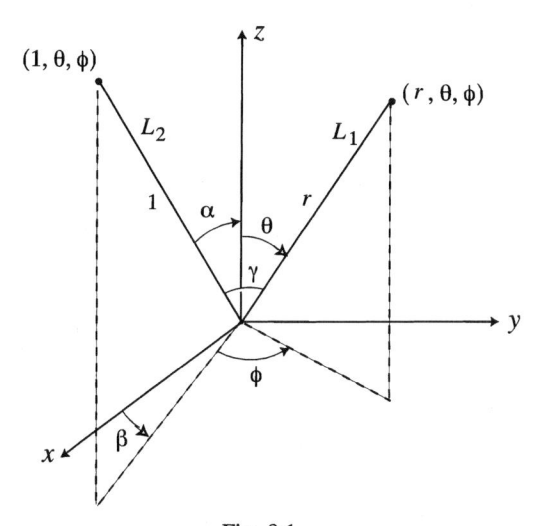

Fig. 9.1.

Thus, the Poisson integral representation of $u(r,\theta,\phi)$ for the interior of the unit sphere is given by

$$u(r,\theta,\phi) = \frac{1-r^2}{4\pi} \int_0^\pi \int_{-\pi}^\pi \frac{f(\alpha,\beta)\,\sin\alpha\,d\beta\,d\alpha}{[1-2\left(\cos\theta\cos\alpha+\sin\theta\sin\alpha\cos(\phi-\beta)\right)r+r^2]^{3/2}}, \tag{9.57}$$

and for the exterior of the unit sphere by

$$u(r,\theta,\phi) = \frac{r^2-1}{4\pi} \int_0^\pi \int_{-\pi}^\pi \frac{f(\alpha,\beta)\,\sin\alpha\,d\beta\,d\alpha}{[1-2\left(\cos\theta\cos\alpha+\sin\theta\sin\alpha\cos(\phi-\beta)\right)r+r^2]^{3/2}}. \tag{9.58}$$

9.3.2. Representation of a Function $f(\theta, \phi)$. We will consider the problem of representing a function of θ and ϕ for all points on the unit sphere, i.e., $-1 \leq \theta \leq 1$ and $0 \leq \phi \leq 2\pi$, in terms of surface spherical harmonics. Assume that

$$f(t, \phi) = \sum_{n=0}^{\infty} \left[A_{0,n} P_n(t) + \sum_{m=1}^{n} (A_{m,n} \cos m\phi P_n^m(t) + B_{m,n} \sin m\phi P_n^m(t)) \right].$$
(9.59)

Since $\int_0^{2\pi} \cos^2 m\phi \, d\phi = \int_0^{2\pi} \sin^2 m\phi \, d\phi = \pi$ anf $\int_0^{2\pi} d\phi = 2\pi$, we get

$$A_{0,n} = \frac{2n+1}{4\pi} \int_0^{2\pi} d\phi \int_{-1}^{1} f(t, \phi) P_n(t) \, dt,$$

$$A_{m,n} = \frac{2n+1}{4\pi} \frac{(n-m)!}{(n+m)!} \int_0^{2\pi} d\phi \int_{-1}^{1} f(t, \phi) \cos m\phi P_n^m(t) \, dt,$$

$$B_{m,n} = \frac{2n+1}{4\pi} \frac{(n-m)!}{(n+m)!} \int_0^{2\pi} d\phi \int_{-1}^{1} f(t, \phi) \sin m\phi P_n^m(t) \, dt,$$

Which gives the expansion (9.59). This development holds for all values of t and ϕ for all points on the unit sphere, provided only that the given function satisfies the conditions that would have been satisfied if it were to be developed into a Fourier series.

Example 9.2. To express $\sin^2 \theta \cos^2 \theta \sin \phi \cos \phi$ in terms of surface spherical harmonics, note that $F(t, \phi) = \frac{1}{2} t^2 (1 - t^2) \sin 2\phi$, and

$$A_{0,n} = \frac{2n+1}{8\pi} \int_{-1}^{1} t^2 (1 - t^2) P_n(t) \, dt \int_0^{2\pi} \sin 2\phi \, d\phi = 0,$$

$$A_{m,n} = \frac{2n+1}{4\pi} \frac{(n-m)!}{(n+m)!} \int_{-1}^{1} t^2 (1 - t^2) P_n^m(t) \, dt \int_0^{2\pi} \sin 2\phi \cos m\phi \, d\phi = 0,$$

$$B_{m,n} = \frac{2n+1}{4\pi} \frac{(n-m)!}{(n+m)!} \int_{-1}^{1} t^2 (1 - t^2) P_n^m(t) \, dt \int_0^{2\pi} \sin 2\phi \sin m\phi \, d\phi = 0, \ m \neq 2.$$

For $m = 2$ we have

$$B_{2,n} = \frac{2n+1}{4} \frac{(n-2)!}{(n+2)!} \int_{-1}^{1} t^2 (1 - t^2) \frac{d^2 P_n(t)}{dt^2} \, dt$$

$$= \frac{1}{2^n n!} \frac{2n+1}{4} \frac{(n-2)!}{(n+2)!} \int_{-1}^{1} t^2 (1 - t^2) \frac{d^{n+2} (t^2 - 1)^n}{dt^{n+2}} \, dt$$

$$= \frac{720}{2^n n!} \frac{2n+1}{4} \frac{(n-2)!}{(n+2)!} \int_{-1}^{1} \frac{d^{n-4} (t^2 - 1)^n}{dt^{n-4}} \, dt,$$

where, by repeated integration by parts,

$$\int_{-1}^{1} \frac{d^{n-4} (t^2 - 1)^n}{dt^{n-4}} \, dt = \begin{cases} 0 & \text{if } n > 4, \\ \int_{-1}^{1} (t^2 - 1)^4 \, dt = \dfrac{256}{315} & \text{if } n = 4, \end{cases}$$

Thus, $B_{2,n} = \dfrac{1}{2^4\,4!} \cdot \dfrac{9}{4} \dfrac{2!}{6!} \dfrac{4096}{7} = \dfrac{1}{105}$. In the same manner, we have $B_{2,3} = 0$ and $B_{2,2} = \dfrac{1}{42}$. Hence,

$$\sin^2\theta\cos^2\theta\sin\phi\cos\phi = \frac{1}{42}\sin 2\phi\, P_2^2(t) + \frac{1}{105}\sin 2\phi\, P_4^2(t),$$

which is the required answer. However, this can be verified by using the following steps:

$$\frac{1}{42}\sin 2\phi\, P_2^2(t) + \frac{1}{105}\sin 2\phi\, P_4^2(t)$$

$$= \frac{1}{42}\sin 2\phi\sin^2\theta\frac{d^2 P_2(t)}{dt^2} + \frac{1}{105}\sin 2\phi\sin^2\theta\frac{d^2 P_4(t)}{dt^2}$$

$$= \frac{1}{14}\sin^2\theta\sin 2\phi + \frac{1}{14}\sin^2\theta\sin 2\phi(7t^2 - 1)$$

$$= \frac{1}{2}t^2\sin^2\theta\sin 2\phi = \frac{1}{2}\cos^2\theta\sin^2\theta\sin 2\phi. \;\blacksquare$$

9.3.3. Addition Theorem for Spherical Harmonics.

The addition theorem between two different directions in the spherical coordinate system is given in §9.2.3. Another useful result is as follows.

Theorem 9.5. (Addition Theorem for spherical harmonics) *The following relations hold:*

$$P_n(\cos\gamma) = \frac{4\pi}{2n+1}\sum_{m=-n}^{n}(-1)^m Y_n^m(\theta_1,\phi_1)Y_n^{-m}(\theta_2,\phi_2), \qquad (9.60)$$

or equivalently,

$$P_n(\cos\gamma) = \frac{4\pi}{2n+1}\sum_{m=-n}^{n}(-1)^m Y_n^m(\theta_1,\phi_1)Y_n^{m*}(\theta_2,\phi_2); \qquad (9.61)$$

or in terms of the associated Legendre functions of the first kind:

$$P_n(\cos\gamma) = P_n(\cos\theta_1)P_n(\cos\theta_2)$$

$$+ 2\sum_{m=1}^{n}\frac{(n-m)!}{(n+m)!}P_n^m(\cos\theta_1)P_n^m(\cos\theta_2)\,\cos m(\phi_1 - \phi_2). \qquad (9.62)$$

PROOF. We will derive (9.61), and other result follow. Let a function $f(\theta,\phi)$ be expanded in a Laplace's series

$$f(\theta_1,\phi_1) = \begin{cases} Y_n^m(\theta_1,\phi_1) & \text{relative to } x_1, y_1, z_1, \\[1mm] \displaystyle\sum_{m=-n}^{n} a_{nm}\, Y_n^m(\gamma,\psi) & \text{relative to } x_2, y_2, z_2, \end{cases} \qquad (9.63)$$

where ψ is the azimuth angle, with any choice of the 0 of this angle (see Fig. 9.2). At $\gamma = 0$ we have

$$f(\theta_1, \phi_1)\Big|_{\gamma=0} = a_{n0}\sqrt{\frac{2n+1}{4\pi}}, \tag{9.64}$$

since $P_n(1) = 1$, and $P_n^m(1) = 0$ for $m \neq 0$. Multiplying (9.63) by $Y_n^{0*}(\gamma, \psi)$ and integrating over the surface S of the sphere, we get

$$\int_S f(\theta_1, \phi_1) Y_n^{0*}(\gamma, \psi)\, dS_{\gamma,\psi} = a_{n0},$$

which in view of (9.64) can be written as

$$\int_S Y_n^m(\theta_1, \phi_1) Y_n^{0*}(\gamma, \psi)\, dS = a_{n0}. \tag{9.65}$$

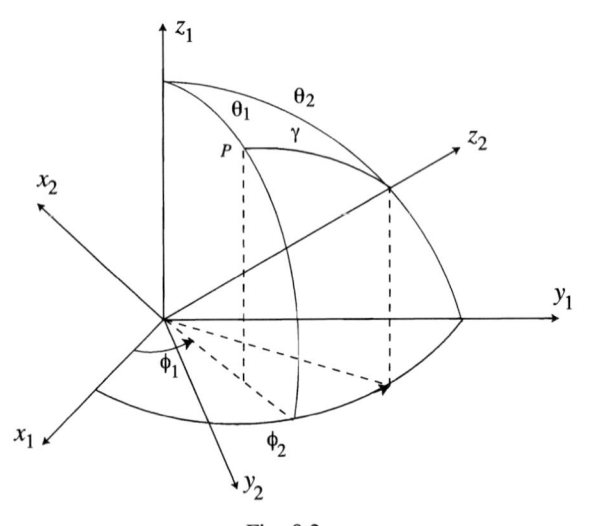

Fig. 9.2.

Let us assume that the polynomial $P_n(\cos\theta)$ has an expansion of the form

$$P_n(\cos\theta) = \sum_{m=-n}^{n} b_{nm} Y_n^m(\theta_1, \phi_1), \tag{9.66}$$

where b_{nm} depend on θ_2, ϕ_2, i.e., on the orientation of the z_2-axis. Multiplying the above integral by $Y_n^{m*}(\theta_1, \phi_1)$ and integrating with respect to θ_1 and ϕ_1 over S, we obtain $\int_S P_n(\cos\gamma) Y_n^{m*}(\theta_1, \phi_1)\, dS_{\theta_1,\phi_1} = b_{nm}$, which in terms of the spherical harmonics becomes

$$\sqrt{\frac{4\pi}{2n+1}} \int_S Y_n^0(\gamma, \psi) Y_n^{m*}(\theta_1, \phi_1)\, dS = b_{nm}, \tag{9.67}$$

where the subscripts have been dropped from the solid angle element dS. Since the range of integration is over all solid angles, the choice of the polar axis is relevant. Then comparing (9.85) and (9.67) we find that

$$b_{nm}^* = a_{n0}\sqrt{\frac{4\pi}{2n+1}} = \begin{cases} \dfrac{4\pi}{2n+1}f(\theta_1,\phi_1)\big|_{\gamma=0} & \text{by Eq (9.64)}, \\[2ex] \dfrac{4\pi}{2n+1}Y_n^m(\theta_2,\phi_2) & \text{by Eq (9.63)}. \end{cases}$$

Here the change of indices occurs because $\theta_1 \to \theta_2$ and $\phi_1 \to \phi_2$ for $\gamma \to 0$. Substituting into Eq (9.66) we obtain (9.61). ∎

Example 9.3. (An application of Theorem 9.5.) To determine Green's function for the 3-D Laplace's equation in spherical coordinates, let us assume that the source is on the polar axis at the point $r = 0, \theta = 0, \phi = 0$. Then using the expansion of the generating function for Legendre polynomials (see §C.1.5), i.e., $(1-2xt+t^2)^{-1/2} = \sum_{n=0}^{\infty} P_n(x)\,t^n$, $|t| < 1$, we obtain

$$\frac{1}{R} \equiv \frac{1}{|\mathbf{r}-\mathbf{k}a|} = \begin{cases} \displaystyle\sum_{n=0}^{\infty} P_n(\cos\gamma)\frac{a^n}{r^{n+1}}, & r > a, \\[3ex] \displaystyle\sum_{n=0}^{\infty} P_n(\cos\gamma)\frac{r^n}{a^{n+1}}, & r < a, \end{cases} \tag{9.68}$$

where $\mathbf{k}$ denotes the unit vector in the z-direction. Now, if we rotate the coordinate system such that the source is at (a,θ_2,ϕ_2) and the observation point at (r,θ_1,ϕ_1), then, in view of (8.10) the required Green's function is given by

$$G(r,\theta_1,\phi_1;a,\theta_2,\phi_2) = \frac{1}{4\pi|\mathbf{r}-\mathbf{k}a|}$$

$$= \begin{cases} \displaystyle\sum_{n=0}^{\infty}\sum_{m=-n}^{n}\frac{1}{2n+1}Y_n^{m*}(\theta_1,\phi_1)Y_n^m(\theta_2,\phi_2)\frac{a^n}{r^{n+1}}, & r > a, \\[3ex] \displaystyle\sum_{n=0}^{\infty}\sum_{m=-n}^{n}\frac{1}{2n+1}Y_n^{m*}(\theta_1,\phi_1)Y_n^m(\theta_2,\phi_2)\frac{r^n}{a^{n+1}}, & r < a. \end{cases} \tag{9.69}$$

9.3.4. Discrete Energy Spectrum. We have studied the Schrödinger wave equation for the hydrogen atom in §7.8. The geometry of the hydrogen atom is presented in Fig. 9.3. We will revisit that section and recall that we assumed a solution of Eq (7.36) in the form $u = R(r)\Theta(\theta)\Phi(\phi)$. Now we will find the solution of Eq (7.36) in a closed form, which will provide us with some information about the discrete energy spectrum and other quantum states of the hydrogen atom.

Let $\Phi(\phi) = A\cos m\phi + B\sin m\phi$, where $m = 1,2,\ldots$, and let $\Theta(\theta)$ satisfy Eq (4.88), or equivalently Eq (7.39) with the substitutions $t = \cos\theta$. Eq (7.41) has the solutions in terms of the associate Legendre functions of the first kind, i.e., $\Theta(\theta) = P_n^m(\theta)$, where $m \le n$, n being an integer (see Appendix C).

Eq (7.41) would be Legendre's equation if it were not for the term $\dfrac{\mu^2}{1 - t^2}$. For Legendre's equation and its properties, see §C.1. The solutions Eq (7.41) are associated Legendre polynomials $\Theta = P_n^\mu(\cos\theta)$, where $P_n^\mu(t)$, and $\mu = 0, \pm 1, \pm 2, \ldots$. For associated Legendre's equation and its properties, see §C.2. The solution of Eq (7.40) are $\Phi = A\cos\mu\phi + B\sin\mu\phi$, where ϕ is a periodic function of period 2π, i.e., $\Phi(\phi) = \Phi(2\pi + \phi)$. Eq (7.38) has been solved in §7.8, and the solution for the problem (7.37) is given by (7.44), namely,

$$
u_{N,n\mu}(r,\theta,\phi) = \frac{h^2 N}{4\pi^2 m\,e^2}\, r^2\, P_n^\mu(\cos\theta)\,\exp\left[-\frac{2\pi^2 m\,e^2}{h^2 N}\,r\right] \times
$$
$$
\times\, L_{N+n}^{2n+1}\left(\frac{4\pi^2 m\,e^2}{h^2 N}\,r\right)(A\cos\mu\phi + B\sin\mu\phi), \quad (9.70)
$$

where the constants A and B can be determined from the boundary conditions (to be provided), and N is called the *principal quantum number*, n the *azimuthal quantum number*, and μ the *magnetic quantum number*. In fact, if we look at the formal solution $u(r,\theta,\phi) = R(r)\Theta(\theta)\Phi(\phi)$, which represents the wave function, the term $R(r)$ is related to the principal quantum number, the term $\Theta(\theta)$ to the azimuthal (orbital) quantum number, and the term $\Phi(\phi)$ to the magnetic quantum number. The spin quantum number s is a property of the electron since it has an intrinsic spin.

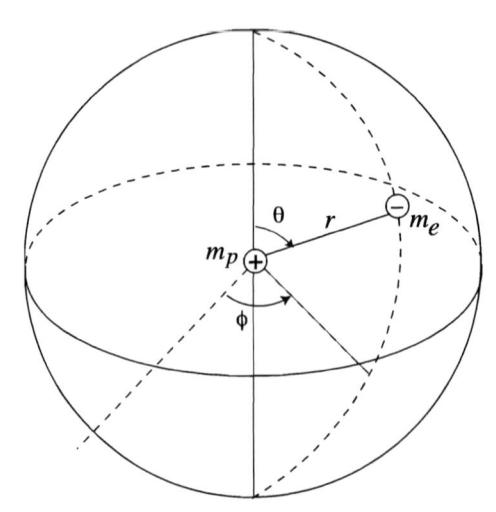

Fig. 9.3. Hydrogen Atom.

Thus, Bernoulli's separation method leads to three equations for the three spatial variables, and their solutions produce the above three quantum numbers associated with the hydrogen energy levels. The wave functions for certain spins of the hydrogen atom are listed below, where $a_0 = \dfrac{h^2}{me^2} = 0.529\,\text{Å}$ denotes the first Bohr radius which is the nuclear charge, and the subscript $1s, 2s, 3s$ indicate the first, second, and third state of its radial density:

$$u_{1s} = \frac{1}{\sqrt{\pi}} a_0^{-3/2} r\, e^{-r/a_0}, \quad u_{2s} = \frac{1}{4\sqrt{2\pi}} a_0^{-3/2} \left(2 - a_0^{-1}\right) e^{-r/(2a_0)},$$

$$u_{3s} = \frac{1}{81\sqrt{3\pi}} a_0^{-3/2} \left(27 - 18 r a_0^{-1} + 2 r^2 a_0^{-2}\right) e^{-r/(3a_0)}.$$

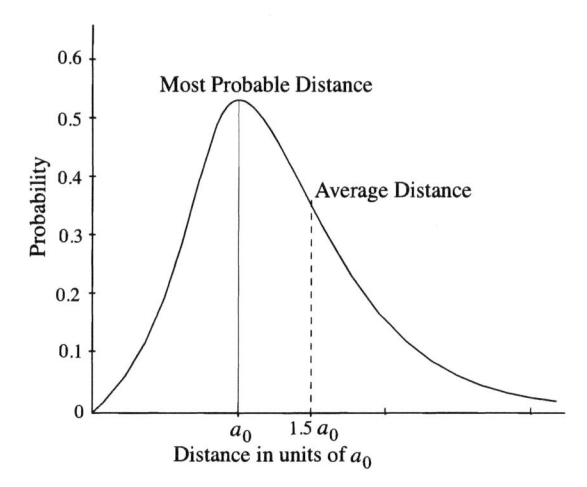

Fig. 9.4.

The probability of finding the electrons in terms of the three dimensions, i.e., its radial density, is proportional to the square of the respective wave function. Fig. 9.4 shows a plot of the radial density which indicates that the distance of the highest probability (most probable distance) is a_0.

To prove this statement, we find that squaring the $1s$ radial wave we obtain the probability function $P(r)$ as

$$P(r) = \frac{1}{\pi} a_0^{-3} r^2\, e^{-2r/a_0}. \tag{9.71}$$

Differentiating with respect to r and setting $P'(r) = 0$ gives

$$P'(r) = 0 = \frac{2}{\pi} r(r - a_0) a_0^{-2}\, e^{-2r/a_0},$$

which yields $r = 0$, or $r = a_0$. Since $r = 0$ is the absolute minimum (see Fig. 9.4), we find that $P''(a_0) = -8\pi a_0^{-3}\, e^{-2}$, which is negative. Hence, $P(r)$ has the relative maximum at the distance $r = a_0$.

For the $2s$ wave function, we use the value of u_{2s} given above. Then the probability of finding the point where the electron has a node follows from setting the probability $P(r) = 0$, and it occurs is at $r = 2a_0$, which can be seen from the $2 - r/a_0$ part of the wave function. Similarly, for the $3s$ function, the only part of the wave function

u_{3s}, and hence the square of this function, that can make $P(r) = 0$ is the term $\left(27 - 18ra_0^{-1} + 2r^2a_0^{-2}\right)$, or the polynomial $27 - 18x + 2x^2 = 0$ where $x = r/a_0$, i.e., at $x = \dfrac{18 \pm \sqrt{18^2 - 4(2)(27)}}{4} \approx 7.1$ or 1.9. Thus, the $3s$ wave has 2 nodes, at $r = 1.9$ and $r = 7.1$. In general the probability has $N - 1$ nodes where N is the principal quantum number. ∎

Example 9.4. To calculate the average probability distance for the hydrogen atom, we start with the normalized wave function u_{1s} and let r denote the spatial position coordinate of this distance. Then this distance, denoted by $|\mathbf{r}|$ is

$$
\begin{aligned}
|\mathbf{r}| &= \int_0^\infty \int_0^\pi \int_0^{2\pi} \frac{1}{\sqrt{\pi}} a^{-3/2} e^{-r/a_0} \, m \, \frac{q}{\sqrt{\pi}} a_0^{-3/2} \, e^{-r/a_0} \, r^2 \sin\theta \, dr \, d\theta \, d\phi \\
&= \frac{a_0^{-3}}{\pi} \int_0^\infty \int_0^\pi \int_0^{2\pi} r^3 e^{-2r/a_0} \, \sin\theta \, dr \, d\theta \, d\phi \\
&= \frac{a_0^{-3}}{\pi} \int_0^\infty r^3 e^{-2r/a_0} \, dr \int_0^\pi \int_0^{2\pi} \sin\theta \, dr \, d\theta \, d\phi \\
&= \frac{a_0^{-3}}{\pi} \frac{3! \, a_0^{-4}}{16} 4\pi = \tfrac{3}{2} a_0,
\end{aligned}
$$

where we have used the formula $\displaystyle\int_0^\infty x^n e^{-bx} \, dx = \frac{n!}{b^{n+1}}$. ∎

9.3.5. Further Developments.
Schrödinger used the model of a standing wave to represent the electron within the hydrogen atom and solved the resulting wave equation. The reasoning behind using the spherical coordinate system is based on the fact that the hydrogen atom has a spherically symmetric potential. The potential energy is due to a point charge.

The hydrogen atom, being an atom of the chemical element H, is an electronically neutral atom which contains a single positively charged proton and a single negatively-charged electron bound to the nucleus by the Coulomb force. In fact, the $1/r$ Coulomb factor leads to the Laguerre functions of the first kind. The hydrogen atom is very significant in the study of quantum mechanics and quantum field theory of a simple two-body problem. In 1914 Niels Bohr performed a number of simplifying assumptions and obtained the spectral frequencies of the hydrogen atom (see [Griffiths 1995]). The Bohr model explained only the spectral properties of the hydrogen atom, and his results were confirmed by an analytical solutions of the Schrödinger wave equation (7.36) by Bernoulli's separation method. These solutions use the fact that the Coulomb potential produced by the nucleus is isotropic, i.e., it is radially symmetric in space and depends only on the distance to the nucleus. They were useful in calculating hydrogen energy levels and frequencies of the hydrogen spectral lines. In this sense, the solutions of the Schrödinger equation are far more useful than the Bohr model, because these solutions also contain the shape of the electron's wave function (orbital part) for various quantum mechanical states, which eventually explains the anisotropic character of atomic bonds. The hydrogen spectrum showing the Bohr model is presented in Fig. 9.5.

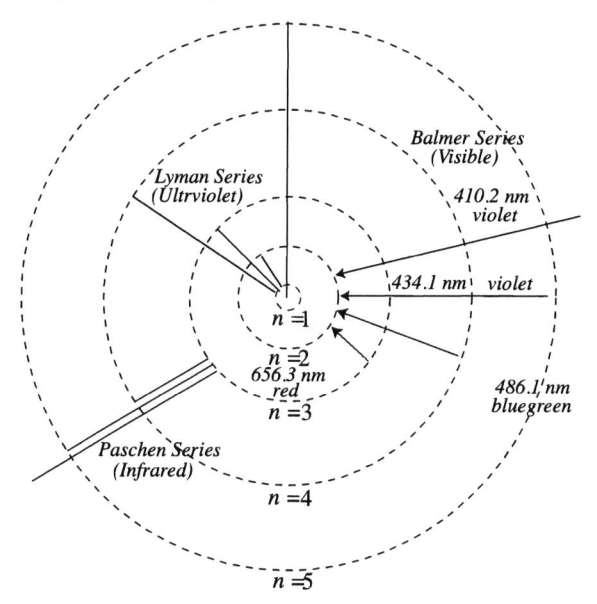

Fig. 9.5. Hydrogen Spectrum.

Pauli [1926] solved the problem of the hydrogen atom by using a rotational symmetry in four dimensions (the so called $O(4)$-symmetry) generated by the angular momentum and the Laplace-Runge-Lenz vector. The non-relativistic problem of the hydrogen atom was later solved by Duru and Kleinert [1979] using Feynman's path integral formulation of quantum mechanics. The energy eigenstates of the hydrogen atom are summarized as follows:

ENERGY LEVELS: $E_{jn} = \dfrac{-13.6\,e\,V}{n^2}\left[1 + \dfrac{\alpha^2}{n^2}\left(\dfrac{n}{j+\frac{1}{2}} - \dfrac{3}{4}\right)\right]$, where α is the fine-structure constant, and j denotes the total angular momentum eigenvalue, i.e., $l = \pm\frac{1}{2}$ depending on the direction of the electron spin; the value $-13.6eV$ is called the *Rydberg constant* that is determined in the Bohr model and it related to mass m and the charge q of the electron by the relation $-13.6eV = -\dfrac{m_e q_e^4}{8h^2\epsilon_0^2}$, or to the fine-structure constant by the relation $-13.6eV = -\dfrac{m_e c^2 \alpha^2}{2} = -\dfrac{0.51\,MeV}{2(137)^2}$. The normalized wave function is given in the spherical coordinates by

$u_{nlm}(r,\theta,\phi)$

$$= \sqrt{\left(\frac{2}{na_0}\right)^3 \frac{(n-l-1)!}{2n(n+l)!}}\; e^{r/(na_0)} \left(\frac{2r}{na_0}\right)^l L_{n-l-1}^{2l+1}\left(\frac{2r}{na_0}\right) Y_l^m(\theta,\phi), \quad (9.72)$$

where a_0 is the Bohr radius, L_{n-l-1}^{2l+1} are the associated Laguerre functions of the first kind and degree $n-l-1$, and Y_l^m are the spherical harmonic functions of degree l and order m, such that the quantum numbers take the following values: $n = 1, 2, \ldots$, $l = n - 1, n - 2, \ldots, 1, 0$, and $m = -l, \ldots, l$. The ground state is the state of the lowest energy in which the electron is first found; it corresponds to the principal quantum level $n = 1, l = 0$. The nodes of the wave function (9.72) are the spherical harmonics $Y_l^m(\theta, \phi)$. The quantum number determine the location of these nodes. There are $n - 1$ total nodes, of which l are angular nodes, and $n - l - 1$ remaining nodes are radial nodes. Out of the l angular nodes, m angular nodes go around the ϕ axis in the plane $z = 0$, and the remaining $l - m$ occur on the vertical θ axis.

The solutions of the Schrödinger equation fail to predict certain small but measurable deviations of the real spectrum of the hydrogen atom. For example, the mean speed of the hydrogen electron is only $1/137$th of the speed of light (c), and since the electron's wavelength is determined by its momentum, it is found that the orbitals with higher speed electrons exhibit contraction due to smaller wavelength. This requires a complete theoretical relativistic treatment of the problem which is not possible with the Schrödinger equation. Another significant feature is the internal magnetic field which exists due to moving electron and creates an electromagnetic field of the nucleus even when there is no external magnetic field. The spin of the electron creates an associated magnetic moment which interacts with this magnetic field. This effect, in the language of special relativity, is known as the *spin-orbit coupling*. These shortcomings in the Schrödinger equation treatment are overcome by the Dirac's relativistic equation which has predicted results that come closer to the experimental data. Since this subject is beyond the scope of this book, the interested reader is directed to related literature such as the books by Griffiths [1995], and Hecht [2000].

9.4. Exercises

9.1. Show that $u(x, t) = A \cos \dfrac{2\pi}{\lambda}(x - at - \varphi)$ is a solution of the equation $\dfrac{\partial^2 u}{\partial t^2} = a^2 \dfrac{\partial^2 u}{\partial x^2}$, where A is a constant, λ the wavelength and φ the phase of the wave represented by the solution.

9.2. Prove Theorem 9.1. If the time required to go through a complete vibration, called the period, is T, then set $a = \lambda/T = \lambda\omega$, where $\omega = 1/\lambda$ is called the angular frequency of the wave. Then show that the solution of the problem in Exercise 9.1 can be written as $u(x, t) = A \cos 2\pi \left(\dfrac{x}{\lambda} - \omega t - \phi \right)$.

9.3. Prove that
$$\int_0^\pi [P_n^m(\cos \theta)]^2 \sin \theta \, d\theta = \int_{-1}^1 [P_n^m(x)]^2 \, dx = \frac{2}{2n + 1} \frac{(n + m)!}{(n - m)!}, \quad m \leq n,$$
where $P_n^m(x)$ are the associated Legendre polynomials.

9.4. Write down all explicit representations of the associated Legendre polynomials $P_n^m(x)$ for $m \leq n \leq 3$.

ANS. $P_1^1(x) = \sqrt{1-x^2}$, $P_2^1(x) = 3\sqrt{1-x^2}$,

$$P_3^1(x) = \tfrac{1}{2}\sqrt{1-x^2}\left(15x^2 - 3\right), \quad P_2^2(x) = 3\left(1-x^2\right),$$

$$P_3^2(x) = 15x\left(1-x^2\right), \quad P_3^3(x) = 15\left(1-x^2\right)^{3/2}.$$

9.5. Prove the orthogonality relation (9.42) for associated Legendre polynomials, i.e., show that $\displaystyle\int_{-1}^{1} P_n^m(x)P_n^k(x)(1-x^2)^{-1}\,dx = \frac{(n+m)!}{m(n-m)!}\,\delta_{m,k}$.

9.6. Prove that $P_n^{-m}(x) = (-1)^m \dfrac{(n-m)!}{(n+m)!} P_n^m(x)$, where $P_n^m(x)$ is defined by

$$P_n^m(x) = \frac{1}{2^n n!}(1-x^2)^{m/2}\frac{d^{n+m}}{dx^{n+m}}(x^2-1)^n.$$

HINT. Apply Leibniz's formula to $(x+1)^n(x-1)^n$.

9.7. Show that $P_n'(0) = \begin{cases} 0, & n \text{ even,} \\[2mm] \dfrac{(-1)^{(n-1)/2}n!}{\{[(n-1)/2]!\,2^{(n-1)/2}\}^2}, & n \text{ odd.} \end{cases}$

HINT. Use either recurrence relation or expansion of the generating function.

9.8. Prove that $Y_n^m(0,\phi) = \sqrt{\dfrac{2n+1}{4\pi}}\,\delta_{m,0}$.

HINT. Use (9.1).

9.9. Introduce a new function $w(x)$ in Eq (9.75) by setting $u = x^{(1-m))/2}\,e^{x/2}\,w$, and

(a) Show that this results in the following equation:

$$x^2 w'' + rxw' - \frac{x^2 + [2(m-1)-4n]\,x + m^2 - 1}{4}\,w = 0;$$

(b) Show that the equation in (a) has a regular singular point at $x = 0$; and

(c) Find the series solution of this equation of the form $w = x^\lambda \sum_{k=0}^{\infty} c_k x^k$, and determine its radius of convergence.

ANS. (c) Recursion formula is $c_k = \dfrac{[2(\mu-1)-4m]\,c_{k-1} + c_{k-2}}{4k(k+\mu)}$, $\mu = 2\lambda-1$, c_0 arbitrary, $c_{-1} = 0$. The series converges everywhere on $\mathbb{R}$.

9.10. Prove Unsöld's theorem (9.29).

9.11. The Schrödinger equation for the harmonic oscillator in its lowest energy state is $\dfrac{d^2 u}{dt^2} + \left(\dfrac{2mE}{\hbar^2} - \dfrac{m\omega}{\hbar}^2 t^2\right) u = 0$, where u denotes the potential, and ω is the angular frequency of vibrations. Show that the solution of this equation is $u(t) = ce^{-m\omega t^2/(2\hbar)}$, where c is a constant.

HINT. The equation is nonlinear. Assume the solution of the form $u = ce^{-bt^2}$, and substituting it in the equation and equating the coefficients of t^2 and the constant terms to zero, determine b.

9.12. Prove the relation (9.19).

HINT. Use Exercise 9.6.

9.13. Prove that $P_n(\cos\theta) = (-1)^n \dfrac{r^{n+1}}{n!} \dfrac{\partial^n}{\partial z^n}\left(\dfrac{1}{r}\right)$.

HINT. Compare the Legendre polynomial expansion of the generating function with a Taylor series expansion.

9.14. Show that $P_n(\cos\theta) \geq -1$, and this result is sharp for $P_n(-1)$, n odd.

HINT. Use $P_n(-x) = (-1)^n P_n(x)$.

9.15. Use $P_n(\cos\theta) = \dfrac{1}{n!}\dfrac{d^n}{dt^n}(1 - 2t\cos\theta + t^2)^{-1/2}\Big|_{t=0}$ to show that $P_n(1) = 1$ and $P_n(-1) = (-1)^n$.

9.16. Prove that $P_n'(1) = \dfrac{d}{dx}P_n(x)\Big|_{x=1} = \dfrac{n(n+1)}{2}$,

9.17. Show that (i) $\int_{-1}^{1} x^m P_n(x)\,dx = 0$ for $m < n$, and (ii) $\int_{-1}^{1} x^n P_n(x)\,dx = \dfrac{2^{n+1}(n!)^2}{(2n+1)!}$.

HINT. Use Rodrigue's formula.

9.18. The associated Lengendre function $P_n^m(x)$ satisfies the self-adjoint equation

$$(1 - x^2)\frac{d^2 P_n^m(x)}{dx^2} - 2x\frac{dP_n^m(x)}{dx} + \left[n(n+1) - \frac{m^2}{1 - x^2}\right] = 0.$$

Use the equations for $P_n^m(x)$ and $P_n^k(x)$ to show that

$$\int_{-1}^{1} P - n^m(x) P_n^k(x)\frac{dx}{1 - x^2} = 0, \quad \text{for } m \neq k.$$

9.19. Show that $\sin\theta\dfrac{d}{d\theta}P_n(\cos\theta) = P_n^1(\cos\theta)$.

9.20. Show that

$$\int_0^\pi \left(\frac{dP_n^m}{d\theta}\frac{dP_{n'}^m}{d\theta} + \frac{m^2 P_n^m P_{n'}^m}{\sin^2\theta}\right)\sin\theta\,d\theta = \frac{2n(n+1)}{2n+1}\frac{(n+m)!}{(n-m)!}\delta_{nn'},$$

$$\int_0^\pi \left(\frac{P_n^1}{\sin\theta}\frac{dP_{n'}^1}{d\theta} + \frac{P_{n'}^1}{\sin\theta}\frac{dP_n^1}{d\theta}\right)\sin\theta\,d\theta = 0.$$

These integrals occur in problems of scattering of electromagnetic waves by spheres.

9.21. Show that $Y_n^m(0, \phi) = \sqrt{\dfrac{2n+1}{4\pi}}\, \delta_{m0}$.

9.22. Prove that $P_{n+1}(x)Q_{n-1}(x) - P_{n-1}(x)Q_{n+1}(x) = \dfrac{2n+1}{n(n+1)}\, x$.

9.23. Prove that $Q_n(\cos\theta) = (-1)^n \dfrac{r^{n+1}}{n!} \dfrac{\partial^n}{\partial z^n}\left[\dfrac{1}{2r}\ln\left(\dfrac{r-z}{r+z}\right)\right]$.

9.24. Show that the integral over the surface of the unit sphere of the product of two tesseral harmonics of the same degree but of different orders is zero.

HINT. $\displaystyle\int_0^{2\pi} \sin n\phi \cos m\phi \, d\phi = \int_0^{2\pi} \sin n\phi \sin m\phi \, d\phi$

$$= \int_0^{2\pi} \cos n\phi \cos m\phi \, d\phi = 0.$$

9.25. Show that $\displaystyle\int_{-1}^1 P_n^m(t)P_k^m(t)\, dt = \begin{cases} 0 & \text{if } n \neq k, \\[2mm] \dfrac{2}{2k+1}\dfrac{(k+m)!}{(k-m)!} & \text{if } n = k. \end{cases}$

9.26. Express $\cos^2\theta \sin^2\theta \sin\phi \cos^2\phi$ in terms of surface spherical harmonics.

HINT. Use the method of Example 9.2.

ANS. $\left[\frac{1}{6930}P_6^3(t) + \frac{1}{1540}P_4^3(t)\right]\sin 3\phi - \left[\frac{2}{693}P_6^1(t) - \frac{1}{770}P_4^1(t) - \frac{1}{63}P_2^1(t)\right]\sin\phi$.

9.27. Show that $\cos 2\phi = 2\cos 2\phi\left[\dfrac{5}{4!}P_2^2(t) + \dfrac{9\cdot 2!}{6!}P_4^2(t) + \dfrac{13\cdot 4!}{8!}P_6^2(t) + \cdots\right]$.

HINT. Follow the method of Example 9.2.

9.28. Show that if $S_n(x, y, z)$ is a solid spherical harmonic of degree n, then $\nabla^2\left[r^m S_n(x, y, z)\right] = m(2n + m + 1)r^{m-2}S_n(x, y, z)$.

HINT. $\nabla^2 S_n = 0, \nabla^2 r = \dfrac{2}{r}, \dfrac{\partial S_n}{\partial r} = \dfrac{nS_n}{r}$, and $\left(\dfrac{\partial r}{\partial x}\right)^2 + \left(\dfrac{\partial r}{\partial y}\right)^2 + \left(\dfrac{\partial r}{\partial z}\right)^2 = 1$.

9.29. Show that $P_n^m(0) = \begin{cases} (-1)^{(n-m)/2}\dfrac{(n+m)!}{2^n\left(\dfrac{n-n}{2}\right)!\left(\dfrac{n+n}{2}\right)!}, & n+m \text{ even}, \\[4mm] 0, & n+m \text{ odd}. \end{cases}$

9.30. Show that $P_n^n(\cos\theta) = (2n-1)!!\,\sin\theta, n = 0, 1, 2, \ldots$.

9.31. Show that $\displaystyle\sum_{n=0}^{\infty}\sum_{m=-n}^{n} Y_n^m(\theta_1,\phi_1)Y_n^{m*}(\theta 2,\phi_2) = \frac{1}{\sin\theta_1}\delta(\theta_1-\theta_2)\delta(\phi_1-\phi_2)$
$= \delta(\cos\theta_1 - \cos\theta_2)\delta(\phi_1-\phi_2)$. HINT. This is known as the spherical harmonic closure relation.

9.32. Solve the integral equation $\displaystyle f(x) = \int_{-1}^{1} \frac{\phi(s)}{(1-2xs+x^2)^{1/2}}\,ds,\ -1\leq x\leq 1,$
for the unknown function $\phi(x)$ if (a) $f(x) = x^{2k}$; and (b) $f(x) = x^{2k+1}$.

ANS. (a) $\phi(x) = \dfrac{4k+1}{2}P_{2k}(x)$; (b) $\phi(x) = \dfrac{4k+3}{2}P_{2k+1}(x)$.

10

Conformal Mapping Method

Green's functions are useful in solving Dirichlet problems of potential theory in itself and, in the case of conformal mapping, of a region onto a circular disk. In the latter case a relationship is needed between conformal mapping and Green's function for the circle. Although no unique expression is available for Green's function for a circle, yet an integral representation of Green's function for the disk leads to the Poisson integral representation. Besides certain Green's functions determined for Laplace's equation in a circle, semi-circle, and a sector which have been discussed in Chapter 8, there are other results for Green's function for a circle, which are presented in this chapter. Green's functions for the ellipse, certain half-plane regions and the parallel strip are determined, and an interpolation method for computation of Green's functions for convex regions is presented.

10.1. Definitions and Theorems

Let $\mathbb{C}$ denote the complex plane. If $a \in \mathbb{C}$ and $r > 0$, then we will denote by $B(r,a) = \{z \in \mathbb{C} : |z - a| < r\}$, $\bar{B}(r,a) = \{z \in \mathbb{C} : |z - a| \leq r\}$, and $\partial B(r,a) = \{z \in \mathbb{C} : |z - a| = r\}$ an open disk, a closed disk, and a circle, respectively, each of radius r and centered at a point a. Thus, $B(1,0)$ represents the open unit disk, sometimes denoted by U. A connected open set $D \subseteq \mathbb{C}$ is called a region (or domain), and ∂D or Γ denotes its boundary.

10.1.1. Cauchy-Riemann Equations.
Let $z = x + iy$ be a complex number. Then $\bar{z} = x - iy$, $x = \dfrac{z + \bar{z}}{2}$, and $y = \dfrac{z - \bar{z}}{2i}$. Also $\partial f = \dfrac{\partial f}{\partial z} = \dfrac{1}{2}(f_x - i f_y)$, $\bar{\partial} f = \dfrac{\partial f}{\partial \bar{z}} = \dfrac{1}{2}(f_x + i f_y)$. The Cauchy-Riemann equations for the function $f(z) = u(x,y) + i v(x,y)$ are: $u_x = v_y$, $u_y = -v_x$, or in polar form ($z = re^{i\theta}$), $u_r = \dfrac{1}{r} v_\theta$, $v_r = -\dfrac{1}{r} u_\theta$. and they satisfy the partial differential equations: $u_x v_x +$

$u_y \, v_y = 0$, and $\nabla^2 u = 0$, $\nabla^2 v = 0$.

The function $\log z = \ln |z| + i \arg\{z\}$ is a multiple-valued function, unless the principal values of $\arg\{z\}$ are taken, which range from $-\pi$ to π. If $z = re^{i\theta}$, then $\log z = \ln r + i\,\theta$ is a single-valued function for $r > 0$ and $-\pi < \theta < \pi$.

A function $f : D \mapsto \mathbb{C}$ is analytic on D iff $\bar{\partial} f = 0$, which is equivalent to the *Cauchy-Riemann equations* for the function $f(z) = u(x, y) + iv(x, y)$. Thus, $f'(z) = u_x + i\,v_x = v_y - i\,u_y$. The Cauchy-Riemann equations are the necessary conditions for $f(z)$ to be analytic on D. However, merely satisfying the Cauchy-Riemann equations alone is not sufficient to ensure the differentiability of $f(z)$ at a point in D. In view of the Cauchy-Riemann equations, property (ii) of §8.4 holds, that is,

$$v(x, y) - v(x_0, y_0) = \int_{(x_0, y_0)}^{(x, y)} (-u_y \, dx + u_x \, dy), \tag{10.1}$$

where $(x_0, y_0) = z_0$ is a given point in D. This property is also true if D is multiply connected, although in that case the conjugate function $v(x, y)$ can be multiple-valued, as we see by considering $u(x, y) = \log r = \log \sqrt{x^2 + y^2}$ defined on a region D containing the origin which has been indented by a small circle centered at the origin. Then, in view of (10.1),

$$v(x, y) - v(x_0, y_0) = \tan^{-1} \frac{y}{x} \pm 2n\pi + \text{const}, \quad n = 1, 2, \dots,$$

which is multiple-valued.

A simple closed curve, or the *Jordan contour*, Γ in $\mathbb{C}$ is a path $\gamma : [a, b] \mapsto \mathbb{C}$ such that $\gamma(t) = \gamma(s)$ iff $t = s$ or $|t - s| = b - a$. The Jordan curve theorem states that if Γ is a simple contour, then $\mathbb{C}\backslash\Gamma$ has two components, one called the interior of Γ, denoted by $\text{Int}\,(\Gamma)$, and the other called the exterior of Γ, denoted by $\text{Ext}\,(\Gamma)$, each of which has Γ as its boundary. Thus, if Γ is a Jordan contour, then $\text{Int}\,(\Gamma)$ and $\text{Ext}\,(\Gamma) \cup \{\infty\}$ are *simply connected* regions.

Let L_2 denotes the Hilbert space of all square-integrable analytic functions f in a simply connected region D with boundary Γ. A function $f(z)$ regular in D is said to belong to the class $L_2(D)$, denoted by $f \in L_2(D)$, if the integral $\iint_D |f(z)|^2 \, dS_z < +\infty$, where $dS_z = dx \, dy$ denotes an area element in D. If two functions $f, g \in L_2(D)$, then their inner product is defined by

$$\langle f, g \rangle = \iint_D f(z) \overline{g(z)} \, dS_z. \tag{10.2}$$

10.1.2. Conformal Mapping. A mapping f of a region D onto a region A is called *analytic* iff it is differentiable. The mapping f is called *conformal* if it is bijective and analytic. The *conformal mapping theorem* states that if a mapping $f : D \mapsto A$ is analytic and $f'(z_0) \neq 0$ for each $z \in D$, then f is conformal. The mapping f is conformal if it is analytic with a nonzero derivative. Two important properties are the following:

(i) If $f : D \mapsto A$ is conformal and bijective (i.e., one-to-one and onto), then $f^{-1} : A \mapsto D$ is also conformal and bijective.

(ii) If $f : D \mapsto A$ and $g : A \mapsto B$ are conformal, then the composition $f \circ g : D \mapsto B$ is conformal and bijective.

Property (i) is useful in solving boundary value problems (e.g., the Dirichlet problem) for a region D. The method involves finding a map $f : D \mapsto A$ such that A is a simply connected region on which the problem can be first solved, and then the result for the original problem is provided by f^{-1}. Since the Dirichlet problem involves harmonic functions, the following result on the composition of a harmonic function with a conformal map is useful: If u is harmonic on a region A and if $f : D \mapsto A$ is conformal, then $u \circ f$ is harmonic on D.

Theorem 10.1. (Riemann mapping theorem) *Let $D \subset \mathbb{C}$ be a simply connected region. Then there exists a bijective conformal map $f : D \mapsto U$, where U is the open unit disk. Moreover, the map f is unique provided that $f(z_0) = 0$ and $f'(z_0) > 0$ for $z_0 \in D$.*

This theorem implies that if D and A, both contained in $\mathbb{C}$, are any two simply connected regions, then there exists a bijective conformal map $g : D \mapsto A$. If $f : D \mapsto U$ and $h : A \mapsto U$, then $g = h^{-1} \circ f$ is bijective conformal. Thus, the regions D and A are said to be *conformal* if there exists a bijective conformal map between them. A bijective conformal map is also called a *univalent* map. A function $w = f(z)$ defining a univalent mapping is called a univalent function. Its inverse image is also a univalent function defined on the image region.

10.1.3. Symmetric Points. The points z and z^* are said to be *symmetric* with respect to a circle C through three distinct points z_1, z_2, z_3 iff

$$(z^*, z_1, z_2, z_3) = \overline{(z, z_1, z_2, z_3)}. \tag{10.3}$$

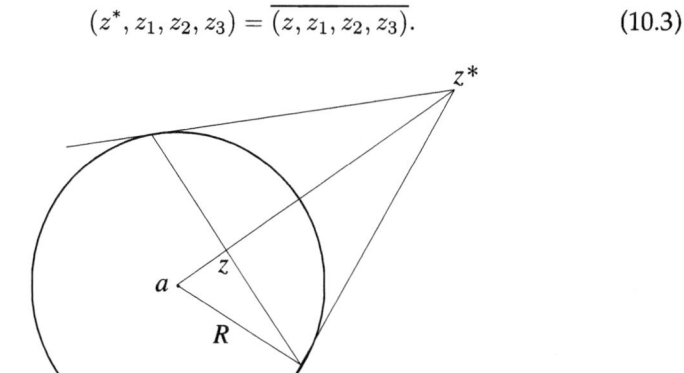

Fig. 10.1. Symmetry with Respect to a Circle.

The mapping that carries z into z^* is called a *reflection* with respect to C. The points z and z^* are also called the *inverse points* or *symmetric points* with respect to the circle C. If C is a straight line, then we choose $z_3 = \infty$, and the condition for symmetry (10.3) gives $\dfrac{z^* - z_1}{z_1 - z_2} = \dfrac{\bar{z} - \bar{z}_1}{\bar{z}_1 - \bar{z}_2}$. Let z_2 be any finite point on the line C. Then, since $|z^* - z_1| = |z - z_1|$, the points z and z^* are equidistant from the line C. Moreover, since $\Im\left\{\dfrac{z^* - z_1}{z_1 - z_2}\right\} = -\Im\left\{\dfrac{z - z_1}{z_1 - z_2}\right\}$, the line C is the perpendicular bisector of the line segment joining z and z^*. If C is the circle $|z - a| = R$, then, recalling that the conjugates of sums and products are sums and products of the conjugates, we find that

$$\overline{(z, z_1, z_2, z_3)} = \overline{(z - a, z_1 - a, z_2 - a, z_3 - a)}$$
$$= \left(\bar{z} - \bar{a}, \frac{R^2}{z_1 - a}, \frac{R^2}{z_2 - a}, \frac{R^2}{z_3 - a}\right)$$
$$= \left(\frac{R^2}{\bar{z} - \bar{a}}, z_1 - a, z_2 - a, z_3 - a\right) = \left(\frac{R^2}{\bar{z} - \bar{a}} + a, z_1, z_2, z_3\right).$$

Hence, in view of (10.3), the points z and $z^* = \dfrac{R^2}{\bar{z} - \bar{a}} + a$ are symmetric with respect to the circle C, i.e., $(z^* - a)(\bar{z} - \bar{a}) = R^2$. Also, $|z^* - a||z - a| = R^2$, and since $\dfrac{z^* - a}{z - a} > 0$, the points z and z^* are on the same ray from the center a (Fig 10.1). Note that the point symmetric to a is $z = \infty$.

10.1.4. Cauchy's Integral Formula. Let f be analytic on a region D, and let Γ be a simple closed contour in D that is homotopic to a point in D. Let $z_0 \in D$ be a point not on Γ. Then

$$f(z_0) \cdot I(\Gamma, z_0) = \frac{1}{2i\pi} \int_\Gamma \frac{f(z)}{z - z_0}\, dz, \tag{10.4}$$

where $I(\Gamma, z_0)$ denotes the *index* (or the winding number) of the curve Γ with respect to a point $z_0 \in \mathbb{C}$; i.e., it is the integer n that expresses n-times Γ winds around z_0, and is defined by

$$I(\Gamma, z_0) = \frac{1}{2i\pi} \int_\Gamma \frac{dz}{z - z_0},$$

or, simply by $I(\Gamma, z_0) = \begin{cases} \pm n & \text{if } z_0 \in \text{Int}(\Gamma) \\ 0 & \text{if } z_0 \in \text{Ext}(\Gamma) \end{cases}$, where the plus or minus sign is chosen according as the path Γ is traversed in the counterclockwise or clockwise direction. For $n = 1$, Cauchy's integral formula becomes

$$f(z_0) = \frac{1}{2i\pi} \int_\Gamma \frac{f(z)}{z - z_0}\, dz, \tag{10.5}$$

10.1.5. Mean-Value Theorem. The mean-value theorem for harmonic functions is discussed in §8.4. Thus, if we integrate in (10.1) on the circle $|z - z_0| = r_0$

and if the function $u(z)$ is harmonic on $|z - z_0| \leq r_0$, then this integral must vanish, which in polar coordinates is expressed as

$$\frac{1}{2\pi} \int_0^{2\pi} \frac{\partial u}{\partial r} r_0 \, d\theta = 0. \tag{10.6}$$

After multiplying the integral (10.6) by $\dfrac{dr}{r_0}$ and integrating from r_1 and r_2, we get

$$\frac{1}{2\pi} \int_0^{2\pi} u(r_1, \theta) d\theta = \frac{1}{2\pi} \int_0^{2\pi} u(r_2, \theta) d\theta. \tag{10.7}$$

These are the mean values of $u(z)$ taken on both circles $|z - z_0| = r_1$ and $|z| = r_2$, where $r_1 < r_2$. So long as $u(z)$ is harmonic on the larger circle, these mean values are equal. For $r_1 \to 0$ the left side of (10.7) takes the value $u(z_0)$ at the center, so that we finally obtain

$$u(z_0) = \frac{1}{2\pi} \int_0^{2\pi} u(r, \theta) d\theta. \tag{10.8}$$

This is the mean-value theorem of potential theory which states that for every function that is harmonic on a circle the value at the center is equal to the mean value of the function on the circumference (see §8.4.(iv)). An important consequence of the mean-value theorem is that a nonconstant function $u(z)$ harmonic in a region D takes neither a maximum nor a minimum value in the interior of D. Since the real and imaginary part of a regular analytic function $w = f(z) = u(z) + iv(z)$ are harmonic functions, the mean-value theorem also holds for analytic functions, and Cauchy's integral formula remains valid, i.e.,

$$f(z_0) = \frac{1}{2i\pi} \oint_{|z|=r} \frac{f(z)}{z - z_0} dz. \tag{10.9}$$

10.2. Dirichlet Problem

Let $z' \in D$ be a fixed point (known as the source point). Green's function for the Dirichlet problem in the region D with a logarithmic singularity at z' is the function $G(z, z')$ with the following properties:

(i) As a function of z, $G(z, z')$ is harmonic everywhere in D except at the point z'.

(ii) At the point z' the function $G(z, z')$ is defined by[1]

$$G(z, z') = \frac{1}{2\pi} \log \left(\frac{1}{r}\right) + g(z, z'), \qquad r = |z - z'|, \tag{10.10}$$

where $g(z, z')$ is harmonic everywhere in D.

---

[1] If the point z' is a sink of strength -1, then the definition (10.10) becomes $G(z, z') = \frac{1}{2\pi} \log r + g(z, z')$, $r = |z - z'|$; see (8.1).

(iii) $G(z, z') = 0$ if the point z' lies on the boundary $\partial D \equiv \Gamma$.

A direct consequence of conformal mapping is that the Dirichlet conditions are directly transferable from the physical model in the z-plane to the mathematical model in the w-plane. The Dirichlet problem for the region D can be solved in an explicit form by using Green's function. Since, in view of the property (ii), the function G becomes unbounded at $z = z'$, we can indent the point z' by a circle Γ_ε of small radius ε. Then the functions u and G become continuous in the region D_ε bounded by Γ and Γ_ε. An application of Green's second identity (1.16) (with $g = G$) yields

$$\int_{\Gamma + \Gamma_\varepsilon} \left(u \frac{\partial G}{\partial n} - G \frac{\partial u}{\partial n} \right) ds = 0, \tag{10.11}$$

where n denotes the outward normal to the boundary which is normal to Γ, exterior to D_ε, and interior to Γ_ε. Separating the integral (10.11) over two contours Γ and Γ_ε and using (10.10), we find that

$$\int_\Gamma \left(u \frac{\partial G}{\partial n} - G \frac{\partial u}{\partial n} \right) ds = \frac{1}{2\pi} \int_{\Gamma_\varepsilon} u \frac{\partial \log \left(\frac{1}{r} \right)}{\partial n} ds + \int_{\Gamma_\varepsilon} u \frac{\partial g}{\partial n} ds$$
$$- \frac{1}{2\pi} \int_{\Gamma_\varepsilon} \log \left(\frac{1}{r} \right) \frac{\partial u}{\partial n} ds - \int_{\Gamma_\varepsilon} g \frac{\partial u}{\partial n} ds \equiv I_1 + I_1 + I_2 + I_3 + I_4. \tag{10.12}$$

The interior normal to Γ_ε is along the radius r with its direction opposite to that of increasing r, and hence, $\dfrac{\partial}{\partial n} = -\dfrac{\partial}{\partial r}$, which yields

$$\frac{\partial \log \left(\frac{1}{r} \right)}{\partial n} = -\frac{\partial \log \left(\frac{1}{r} \right)}{\partial r} = \frac{1}{r}.$$

Also, since $r = \varepsilon$ on the circle Γ_ε, we get, in view of (10.8),

$$I_1 = \frac{1}{2\pi\varepsilon} \int_{\Gamma_\varepsilon} u \, ds = u(z')$$

for any ε by the mean value theorem. In I_3, since $r = \varepsilon$ and u is harmonic, we have

$$I_3 = \frac{1}{2\pi} \log \left(\frac{1}{\varepsilon} \right) \int_{\Gamma_\varepsilon} \frac{\partial u}{\partial n} ds = 0 \quad \text{for any } \varepsilon.$$

The remaining two integrals I_2 and I_4 tend to zero as $\varepsilon \to 0$. In fact, since $u, g, \dfrac{\partial u}{\partial n}$ and $\dfrac{\partial g}{\partial n}$ are bounded in the neighborhood of the point z', we find that

$$|I_2| \leq \max_{\Gamma_\varepsilon} \left| u \frac{\partial g}{\partial n} \right| \cdot 2\pi\varepsilon, \quad \text{and} \quad |I_4| \leq \max_{\Gamma_\varepsilon} \left| g \frac{\partial u}{\partial n} \right| \cdot 2\pi\varepsilon.$$

Hence, as $\varepsilon \to 0$, the relation (10.12) yields

$$\frac{1}{2\pi} \int_\Gamma \left(u \frac{\partial G}{\partial n} - G \frac{\partial u}{\partial n} \right) ds = u(z'). \tag{10.13}$$

Moreover, by condition (iii) the function G vanishes on Γ, and thus, from (10.13) we obtain

$$u(z') = \frac{1}{2\pi} \int_\Gamma u \frac{\partial G}{\partial n} ds. \tag{10.14}$$

Note that this equation is also a consequence of Green's third identity (1.19). If Green's function is known for the region D, then formula (10.14) can be used to solve the Dirichlet problem for any continuous or piecewise continuous, boundary values of the harmonic function $u(z)$. An alternate form of formula (10.14) is

$$u(z') = \frac{1}{2\pi} \int_\Gamma u \, dG(z, z'). \tag{10.15}$$

The relationship between Green's function and conformal mapping is established as follows: Let $w = f(z) = r\, e^{i\theta}$ map a simply connected region D conformally onto the open unit disk $U : |w| < 1$ such that the point $z' \in D$ goes into the point $w = 0$, i.e., $f(z') = 0$. If $f(z)$ has a simple zero at z', the function $F(z) = \dfrac{f(z)}{z - z'} \neq 0$ for all $z \in D$ and is regular everywhere in D. Thus, $\log F(z)$ is also regular analytic in D. Let us denote $\log F(z) = p + i\,q$. Then

$$f(z) = (z - z')\, e^{p + i q},$$

which yields, with $z - z' = r\, e^{i\phi}$,

$$\frac{1}{2\pi} \log \frac{1}{|f(z)|} = \frac{1}{2\pi} \left(\log \frac{1}{r} - p \right). \tag{10.16}$$

It can easily be verified that the function $\dfrac{1}{2\pi} \log \dfrac{1}{|f(z)|}$ satisfies all three properties for Green's function $G(z, z')$, namely, this function with a simple pole at z' is harmonic in D except at z', where it has a logarithmic singularity (in fact, it is a source of strength $+1$), and it is equal to $\dfrac{1}{2\pi} \log \dfrac{1}{r} + g(z)$, where $g(z) = -\dfrac{1}{2\pi}\, p$. Moreover, $|f(z)| = 1$ on the boundary Γ, and, therefore, $\dfrac{1}{2\pi} \log \dfrac{1}{|f(z)|} = 0$ there. Thus, we have shown that Green's function $G(z, z')$ and the mapping function $f(z)$ which produces the conformal map of the region D onto the unit disk U are related to each other by (see Courant and Hilbert [1968:378])

$$G(z, z') = \frac{1}{2\pi} \log \frac{1}{|f(z)|} = -\frac{1}{2\pi} \log |f(z)|, \tag{10.17a}$$

such that the point $z' \in D$ goes into the origin. If the singularity at $z = z'$ is a sink of strength -1, then this formula becomes

$$G(z, z') = \frac{1}{2\pi} \log |f(z)|. \tag{10.17b}$$

Since we have used a source of strength -1 in Chapter 8, we will employ formula (10.17b) whenever a relation between Green's function $G(z, z')$ and the conformal mapping function $f(z)$ is required. If Green's function for a region D is known, we can use (10.17a or 10.17b) and construct the function $f(z)$ which maps the region D conformally onto the unit disk. The method of accomplishing this using (10.17b), for example, is as follows: For each term in (10.16), we determine the respective conjugate harmonic function. The conjugate harmonic function for $\frac{1}{2\pi} \log \frac{1}{r}$ is $\frac{\phi}{2\pi}$, where $\phi = \arg\{z - z'\}$. Let $h(z)$ be conjugate to the function $g(z)$. Then, in view of (10.1),

$$h(z) = \int_{z'}^{z} \left(\frac{\partial p}{\partial x} \, dy - \frac{\partial p}{\partial y} \, dx \right) + C, \tag{10.18}$$

where C is an arbitrary constant, which corresponds to the rotation of the unit disk about $w = 0$. Hence, the required mapping function is given by

$$w = f(z) = r \, e^{i\phi} e^{-2\pi(g+ih)} = (z - z') e^{-2\pi(g+ih)}. \tag{10.19}$$

Note that the construction of Green's function $G(z, z')$ involves determining the harmonic function $g(z)$, whose boundary values are determined from the third property, namely, that $G(z, z') = 0$ on the boundary Γ. This means that $g(z)$ must take the values $\frac{1}{2\pi} \log \frac{1}{r}$ on Γ. Hence, the conformal mapping problem of transforming the region D onto the unit disk reduces to the solution of the Dirichlet problem with the boundary condition

$$g(x, y)\big|_{\Gamma} = \frac{1}{2\pi} \log r. \tag{10.20}$$

A similar result is obtained, with a minus sign, if we use (10.17a) instead. If D can be mapped conformally onto the unit disk, then the dependence of Green's function on z_0 can be given explicitly. Thus, if $w = f(z)$ is any function that maps D onto the unit disk, then we can use the mapping

$$z \mapsto e^{i\gamma} \frac{w - w_0}{1 - w \overline{w_0}},$$

where γ is an arbitrary real constant, which maps the unit disk onto itself such that the point $w(z_0)$ goes into the origin and $w_0 = f(z_0)$. Hence,

$$G(z, z') = \log \frac{w - w_0}{1 - w \overline{w_0}}. \tag{10.21}$$

10.2.1. Dirichlet Problem for a Circle in the (x, y)-Plane. Let the circle C be given by $|z| = R$, where $z = x + i\, y$. The problem is to find a solution for the boundary value problem

$$\nabla^2 u \equiv \frac{\partial^2 u}{\partial x^2} + \frac{\partial^2 u}{\partial y^2} = 0, \quad u = f(\theta) \quad \text{on } C,$$

where θ is the angular coordinate on C, i.e., $z = R\, e^{i\theta}$ on C. The constructive proof for the existence of the solution consists in deriving an expression for the solution. Let $F(\zeta)$ be an analytic function in the region enclosed by C such that $\Re\{F(\zeta)\}$ on $|\zeta| = R$ is equal to $f(\theta)$. Let z be a complex number in this region (at A in Fig 10.2). The point symmetric to z with respect to the circle C is z^*, which lies outside C (at B in Fig. 10.2), such that $z^*\, \bar z = R^2$. According to Cauchy's integral formula (10.5)

$$F(z_0) = \frac{1}{2\pi i} \int_C \frac{F(\zeta)}{\zeta - z}\, d\zeta, \quad \text{and} \quad 0 = \frac{1}{2\pi i} \int_C \frac{F(\zeta)}{\zeta - R^2/\bar z}\, d\zeta.$$

Subtracting we get

$$F(z) = \frac{1}{2\pi i} \int_C \frac{F(\zeta)\, (R^2 - z\, \bar z)}{\zeta\, (R^2 + z\, \bar z) - z\zeta^2 - zR^2}\, d\zeta.$$

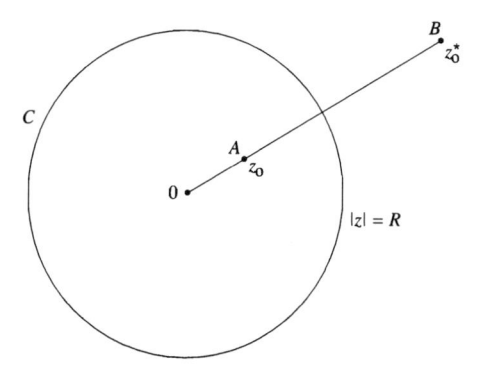

Fig. 10.2. Inverse (Symmetric) Points with Respect to the Circle $|z| = R$.

Since ζ lies on C and z_0 inside C, set $\zeta = R\, e^{i\theta}$, $z = r\, e^{i\phi}$, $r < \rho$. Then

$$F\left(r\, e^{i\theta}\right) = \frac{1}{2\pi} \int_0^{2\pi} F\left(R\, e^{i\theta}\right) \frac{R^2 - r^2}{R^2 - 2rR\cos(\theta - \phi) + r^2}\, d\theta.$$

Equating the real parts on both sides, we get

$$u(x, y) = \frac{R^2 - r^2}{2\pi} \int_0^\pi \frac{f(\theta)\, d\theta}{R^2 - 2rR\cos(\theta - \phi) + r^2}, \tag{10.22}$$

where $r^2 = x^2 + y^2$ and $\phi = \arctan(y/x)$. This formula is called *Poisson's integral representation* in the (x, y)-plane, which is the same as (8.43). This completes the proof of the existence of the solution for the Dirichlet problem.

A frequently encountered Green's function for a circle is Poisson's integral formula for the interior of a circle C of radius R. Let $z = re^{i\theta}$ and $z' = \rho e^{i\phi}$ be two distinct interior points of C. Then

$$G(z, z') = \frac{1}{4\pi} \log \left\{ \frac{R^2 + r^2\rho^2/R^2 - 2\rho\cos(\theta - \phi)}{r^2 + \rho^2 - 2r\rho\cos(\theta - \phi)} \right\},$$

$$\left. \frac{\partial G}{\partial \rho} \right|_{\rho=R} = -\frac{1}{2\pi R} \frac{R^2 - r^2}{R^2 + r^2 - 2Rr\cos(\theta - \phi)},$$

(10.23)

where G applies to the Dirichlet condition and $\dfrac{\partial G}{\partial \rho}$ to the Neumann condition (Korn [1961:530]).

10.3. Neumann Problem

We will consider the case when D is a circle with center at the origin and radius R. In this case we carry out the conformal mapping onto the unit disk by $w = \dfrac{z}{R}$, and Green's function (10.21) becomes

$$G(z, z') = \log \frac{R(z - z')}{R^2 - z\bar{z}'}.$$

Then $dG(z, z') = \left(\dfrac{1}{z - z'} + \dfrac{\bar{z}'}{R^2 - z\bar{z}'} \right) dz$. Since $|z|^2 = z\bar{z} = R^2$ on the boundary of the circle and $dz = iz\,d\theta$, we have $dG(z, z') = i\left(\dfrac{z}{z - z'} + \dfrac{\bar{z}}{\bar{z} - \bar{z}'} - 1 \right) d\theta$. Setting $z = Re^{i\theta}$ and $z' = \rho e^{i\phi}$, we find that

$$dG(z, z') = i\frac{R^2 - \rho^2}{R^2 + \rho^2 - 2R\rho\cos(\theta - \phi)}\,d\theta.$$

(10.24)

If we substitute (10.24) in (10.9), we obtain the Poisson integral

$$u(\rho e^{i\phi}) = \frac{1}{2\pi} \int_0^{2\pi} u(Re^{i\phi})\frac{R^2 - \rho^2}{R^2 + \rho^2 - 2R\rho\cos(\theta - \phi)}\,d\theta.$$

(10.25)

By a similar integral representation we can determine the harmonic function $v(z)$ which is conjugate to $u(z)$. In view of (10.1)

$$v(z) - v(0) = \int_0^z \frac{\partial u}{\partial n}\,ds.$$

When we apply this operation on (10.25) and follow through the corresponding integrations and differentiations, we get

$$v(\rho e^{i\phi}) - v(0)$$

$$= \frac{1}{2\pi} \int_0^{2\pi} u(R e^{i\phi}) \int_0^{\rho e^{i\phi}} \frac{\partial}{\partial u} \left[\frac{R^2 + \rho^2}{R^2 + \rho^2 - 2R\rho \cos(\theta - \phi)} \right] ds \, d\theta,$$

where the inner integral is taken on an arbitrary path that lies entirely in the interior of the circle. Note that

$$\frac{R^2 - \rho^2}{R^2 + \rho^2 - 2R\rho \cos(\theta - \phi)} = \frac{z}{z - z'} + \frac{\bar{z}}{\bar{z} - \bar{z}'} - 1$$

$$= \Re \left\{ \frac{2z}{z - z'} - 1 \right\} = \Re \left\{ \frac{z + z'}{z - z'} \right\}.$$

Thus,

$$\Im \left\{ \frac{z + z'}{z - z'} \right\} = \frac{-2R\rho \sin(\theta - \phi)}{R^2 + \rho^2 - 2R\rho \cos(\theta - \phi)}, \tag{10.26}$$

and hence

$$v(\rho e^{i\phi}) = v(0) - \frac{1}{2\pi} \int_0^{2\pi} u(R e^{i\theta}) \frac{2R \sin(\theta - \phi)}{R^2 + \rho^2 - 2R\rho \cos(\theta - \phi)} d\theta. \tag{10.27}$$

If we combine (10.25) and (10.27), we obtain Schwarz's formula:

$$f(\rho e^{i\phi}) = i v(0) + \frac{1}{2\pi} \int_0^{2\pi} u(R e^{i\phi}) \frac{R e^{i\theta} + \rho e^{i\phi}}{R e^{i\theta} - \rho e^{i\phi}} d\theta, \tag{10.28}$$

which allows us to determine the value of a complex potential function $f(z) = u(z) + i v(z)$ in a circle with prescribed boundary values $u(z)$ and $v(0)$.

Note that the above formulas allows us to determine the value of a complex potential function $f(z) = u(z) + i v(z)$ in a circle with prescribed boundary values $u(z)$ and $v(0)$, and find an explicit solution for the Neumann problem, for a disk, and thereby for all regions with known conformal mapping onto the unit disk. In this case the value of the normal derivative $\dfrac{\partial u}{\partial n}$ on the boundary of the region is prescribed, and we seek a function harmonic on D whose normal derivative takes this boundary value. For this problem to have a solution we require that

$$\oint_\Gamma \frac{\partial u}{\partial n} ds = 0. \tag{10.29}$$

Since, in view of (10.1),

$$v(z_2) = v(z_1) - \int_{z_1}^{z_2} \frac{\partial u}{\partial n} ds,$$

the value of $v(z)$ is determined on the boundary up to an additive constant. With this boundary value $v(z)$ the value of $u(z)$, which is a harmonic function conjugate to $v(z)$, can be determined in the interior of D by the formula (10.27). Thus,

$$u(\rho e^{i\phi}) = u(0) + \frac{1}{2\pi} \int_0^{2\pi} v(R e^{i\theta}) \frac{2R\rho \sin(\theta - \phi)}{R^2 + \rho^2 - 2R\rho \cos(\theta - \phi)} d\theta$$

$$= u(0) + \frac{1}{2\pi} \left[v(R e^{i\theta}) \log\left(R^2 + \rho^2 - 2R\rho \cos(\theta - \phi) \right) \right]_0^{2\pi} \quad (10.30)$$

$$- \frac{1}{2\pi} \int_0^{2\pi} \frac{\partial v(R e^{i\theta})}{\partial \theta} \log\left(R^2 + \rho^2 - 2R\rho \cos(\theta - \phi) \right) d\theta.$$

Since $v(R e^{i\theta})$ and $\log\left(R^2 + \rho^2 - 2R\rho \cos(\theta - \phi) \right)$ are periodic, the second term on the right side of (10.30) vanishes. Also,

$$\frac{\partial v}{\partial \theta} d\theta = dv = \frac{\partial u}{\partial n} ds = \frac{\partial u}{\partial n} R \, d\theta,$$

so that we finally obtain the following integral formula:

$$u(R e^{i\theta}) = u(0) - \frac{1}{2\pi} \int_0^{2\pi} \frac{\partial u}{\partial n} R \log\left(R^2 + \rho^2 - 2R\rho \cos(\theta - \phi) \right) d\theta, \quad (10.31)$$

which establishes a relationship between the boundary values of $\dfrac{\partial u}{\partial n}$ and the values of $u(z)$. We shall denote the expression

$$\log\left(R^2 + \rho^2 - 2R\rho \cos(\theta - \phi) \right) = \log\left\{ (z - z')(\bar{z} - \bar{z}') \right\} \equiv N(z, z') \quad (10.32)$$

and call $N(z, z')$ the *Neumann's function*. It plays the same role for the Neumann boundary value problem as Green's function does for the Dirichlet problem. This function represents a regular analytic function of z in D except for logarithmic singularities. Since $\bar{z} = \dfrac{R^2}{z}$, we have

$$N(z, z') = \log\left\{ (z - z')(\bar{z} - \bar{z}') \right\} = \log \frac{(z - z')(R^2 - z\bar{z}')}{z}. \quad (10.33)$$

This function can be regarded as a complex potential function for a flow, which has a source at z' and a sink at z and for which the circle $|z| = R$ acts as an impermeable boundary.

As in the case of Green's function, Neumann's function $N(z, z')$ can be characterized by the following conditions:

(i) $N(z, z')$ is a regular analytic function of z on a region D except for a logarithmic singularity at $z = z'$ and at another fixed point $z = z^*$, i.e.,

$$N(z, z') = \log \frac{z - z'}{z - z^*} + n(z, z'), \quad (10.34)$$

where $n(z, z')$ is a regular analytic function on D.

(ii) $N(z, z')$ as a function of z has a boundary value which is continuous everywhere on the boundary Γ of D, and $\Im\{N(z, z')\} = 0$ on Γ. If these conditions are satisfied, then

$$u(z') = u(z^*) - \frac{1}{2\pi} \oint_\Gamma N(z, z') \frac{\partial u}{\partial u}\, ds. \tag{10.35}$$

If the region D is mapped conformally onto the unit disk by the function $w = f(z)$, then $N(z, z')$ can also be defined by

$$N(z, z') = \log \frac{[f(z) - f(z')]\overline{f(z')}}{f(z)}, \tag{10.36}$$

where z^* under this map goes into $f(z^*) = 0$.

Note that the Neumann problem is solvable only if the condition (10.29) is satisfied. For a multiply connected region the contour Γ in (10.29) must include the exterior and all interior paths. The reason why this condition does not hold for each individual path is that the function u may be multiple-valued, which does not let the integral of $\partial u/\partial n$ around Γ vanish.

10.4. Green's and Neumann's Functions

Analogous to Green's functions, the solution of the Neumann problem of potential theory is Neumann's function which also possesses an integral representation, and bears a close relationship with Green's functions. We will also determine Green's and Neumann's functions for an ellipse, and annulus, and a parallel strip.

The Laplacian ∇^2 is invariant to translation as well as rotation of the coordinate system. This result is proved in §10.3.1. Green's function for the circle has been extensively studied and only a few explicit expressions are known. One such expression, due to Lanczos [1961], is given below. Since, in view of the Riemann mapping theorem, any simply connected region D with a piecewise smooth boundary can be conformally mapped onto the unit disk, the pertinent Green's function is also a conformal mapping according to the mapping equation (Courant and Hilbert, [1968:377]), but only a few examples are known to be described by explicit formulas. In this section we use some ideas developed in Borre [2001] with certain modifications.

10.4.1. Laplacian. Let a potential function in the z-plane $\phi(z)$ be mapped conformally onto a potential function $\psi(w)$ in the w-plane such that ψ at $(u(x, y), v(x, y))$ equals ϕ at (x, y). The potential ϕ is harmonic, i.e., it satisfies Laplace's equation $\nabla^2 \phi = 0$. The question is whether ψ is also harmonic. We will prove the following result:

Theorem 10.2. *Laplace's equation* $\nabla^2 \phi = 0$ *remains invariant under conformal transformation.*

PROOF. (Based on Henrici [1974: Ch. 5]) If we set $\phi(x, y) = \psi(u, v)$, then

$$\frac{\partial \phi}{\partial x} = \frac{\partial \psi}{\partial u} \frac{\partial u}{\partial x} + \frac{\partial \psi}{\partial v} \frac{\partial v}{\partial x}. \tag{10.37}$$

In the z- and w-planes the complex gradients are

$$\nabla_c \phi = \frac{\partial \phi}{\partial x} + i \frac{\partial \phi}{\partial y}, \quad \nabla_c \psi = \frac{\partial \psi}{\partial u} + i \frac{\partial \psi}{\partial v}. \tag{10.38}$$

Substituting (10.37) into (10.38) and using the Cauchy-Riemann equations (to replace $\dfrac{\partial u}{\partial y}$ by $-\dfrac{\partial v}{\partial x}$ and $\dfrac{\partial v}{\partial y}$ by $\dfrac{\partial u}{\partial x}$), we get

$$
\begin{aligned}
\nabla_c \phi &= \left(\frac{\partial \psi}{\partial u} \frac{\partial u}{\partial x} + \frac{\partial \psi}{\partial v} \frac{\partial v}{\partial x} \right) + i \left(\frac{\partial \psi}{\partial u} \frac{\partial u}{\partial y} + \frac{\partial \psi}{\partial v} \frac{\partial v}{\partial y} \right) \\
&= \left(\frac{\partial \psi}{\partial u} \frac{\partial u}{\partial x} + \frac{\partial \psi}{\partial v} \frac{\partial v}{\partial x} \right) + i \left(-\frac{\partial \psi}{\partial u} \frac{\partial v}{\partial x} + \frac{\partial \psi}{\partial v} \frac{\partial u}{\partial x} \right) \\
&= \left(\frac{\partial \psi}{\partial u} + i \frac{\partial \psi}{\partial v} \right) \left(\frac{\partial u}{\partial x} - i \frac{\partial v}{\partial x} \right) = \nabla_c \psi \cdot \overline{f'(z)}.
\end{aligned} \tag{10.39}
$$

To obtain the second partial derives, we start with (10.37) and get

$$\frac{\partial^2 \phi}{\partial x^2} = \frac{\partial^2 \psi}{\partial u^2} \left(\frac{\partial u}{\partial x} \right)^2 + 2 \frac{\partial^2 \psi}{\partial u \partial v} \frac{\partial u}{\partial x} \frac{\partial v}{\partial x} + \frac{\partial^2 \psi}{\partial v^2} \left(\frac{\partial v}{\partial x} \right)^2 + \frac{\partial \psi}{\partial u} \nabla^2 u + \frac{\partial \psi}{\partial v} \nabla^2 v. \tag{10.40}$$

A similar expression is obtained for $\dfrac{\partial^2 \psi}{\partial y^2}$. Then adding them we obtain

$$
\begin{aligned}
\frac{\partial^2 \phi}{\partial x^2} + \frac{\partial^2 \phi}{\partial y^2} &= \frac{\partial^2 \psi}{\partial u^2} \left[\left(\frac{\partial u}{\partial x} \right)^2 + \left(\frac{\partial u}{\partial y} \right)^2 \right] + \frac{\partial^2 \psi}{\partial v^2} \left[\left(\frac{\partial v}{\partial x} \right)^2 + \left(\frac{\partial v}{\partial y} \right)^2 \right] + \\
&\quad + 2 \frac{\partial^2 \psi}{\partial u \partial v} \left\{ \frac{\partial u}{\partial x} \frac{\partial v}{\partial x} + \frac{\partial u}{\partial y} \frac{\partial v}{\partial y} \right\} + \nabla^2 u \frac{\partial \psi}{\partial u} + \nabla^2 v \frac{\partial \psi}{\partial v},
\end{aligned} \tag{10.41}
$$

On the right side of (10.41) the first and second terms in the square brackets are equal to $|f'(z)|^2$; the third term in the braces is zero because of the Cauchy-Riemann equations, and $\nabla^2 u = 0 = \nabla^2 v$. Thus, (10.41) reduces to

$$
\begin{aligned}
\nabla^2 \phi(x, y) &= \nabla^2 \psi(u, v) \, |f'(z)|^2, \\
\text{or} \quad \nabla^2 \psi(u, v) &= \nabla^2 \phi(x, y) \, |f'(z)|^{-2}.
\end{aligned} \tag{10.42}
$$

Hence, $\nabla^2 \psi = 0$ when $\nabla^2 \phi = 0$ so long as $|f'(z)| \neq 0$, a condition which is satisfied by analytic functions except at their singular points. Note that we can substitute $f'(z) = \dfrac{1}{f'(w)}$ in (10.42). ∎

10.4.2. Green's Function for a Circle. The Poisson integral representation (10.25) is one form of Green's function for a circle. Other representations are given below by some examples.

Example 10.1. Consider the following problem with the Neumann boundary condition:

$$\nabla^2 h(z) = f(z) \quad \text{in } D, \tag{10.43a}$$

$$\frac{\partial h(z)}{\partial n} = 0 \quad \text{on } \partial D. \tag{10.43b}$$

Using the Gauss theorem (§1.7) we get

$$\iint_D \nabla h \, dS = \int_{\partial D} \frac{\partial h}{\partial n} \, ds = 0,$$

which after comparing with (10.43a) yields the *compatibility condition*

$$\iint_D f(z) \, dS = 0. \tag{10.44}$$

Since the operator ∇^2 is singular, Green's function cannot be uniquely determined, but any Green's function yields

$$h(z') = \iint_D G(z, z') \, f(z) \, dS, \tag{10.45}$$

such that

$$\nabla^2 G(z, z') = \delta(z, z') - \phi(z), \tag{10.46a}$$

$$\frac{\partial G(z, z')}{\partial n_z} = 0, \tag{10.46b}$$

where $\phi(z)$ is a function that must satisfy the condition

$$\iint_D [\delta(z, z') - \phi(z)] \, dS = 0, \quad \text{or} \quad \iint_D \phi(z) \, dS = 1. \tag{10.47}$$

For the open unit disk U (see Fig. 10.3) we choose $\phi(z) = \text{const} = \dfrac{1}{\pi}$. Then the explicit solution of Eqs (10.46a) and (10.47) is given by (see Lanczos [1961: (8.4.66)])

$$
\begin{aligned}
G(z, z') &= \frac{1}{4\pi} \Big\{ \log |z - z'|^2 + \log \left[|z - z'|^2 + \left(|z|^2 - 1 \right) \left(|z'|^2 - 1 \right) \right] - \\
&\qquad\qquad - |z|^2 - |z'|^2 + \tfrac{3}{2} \Big\} \\
&= \frac{1}{4\pi} \Big\{ \ln \left(r^2 - 2 r r_0 \cos\theta + r_0^2 \right) + \ln \left(1 - 2 r r_0 \cos\theta + r^2 r_0^2 \right) - \\
&\qquad\qquad - r^2 - r_0^2 + \tfrac{3}{2} \Big\}. \ \blacksquare
\end{aligned}
\tag{10.48}
$$

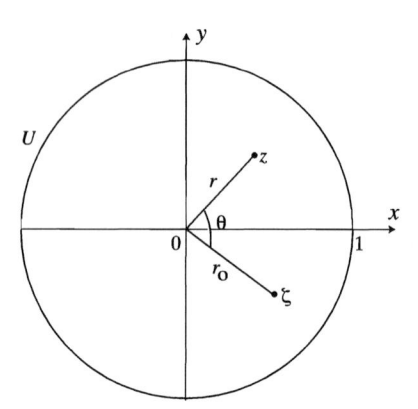

Fig. 10.3. Open Unit Disk U.

We will now consider the case of Green's function in $\mathbb{R}^2$ for a domain D which can be mapped conformally onto the unit circle. The following result holds.

Theorem 10.3. *Let* $w = f(z) = u + iv$, *where* $z = x + iy$, *represent the conformal mapping of a domain* D *in the* z-plane *onto the unit circle in the* w-plane, *where* $f(z)$ *is a simple analytic function of the complex variable* z. *Then Green's function for the domain* D *is given by*

$$G(x, y; x', y') = \frac{1}{2\pi} \Re \left\{ \log \frac{f(x' + iy') - f(x + iy)}{f(x' + iy')\, f(x + iy) - 1} \right\}. \tag{10.49}$$

PROOF. To show that $G = 0$ on ∂D, note that the boundary ∂D is mapped by f onto the boundary of the unit circle in the z-plane. Thus, $f(x + iy) = e^{i\theta}$, where the point (x, y) lies on ∂D. Set $z = x + iy$ and $z' = x' + iy'$. Then by (10.17b)

$$\text{On } \partial D: \quad G(x, y; x', y') = \frac{1}{2\pi} \Re \left\{ \log \frac{f(z') - e^{i\theta}}{\overline{f(z')}\, e^{i\theta} - 1} \right\}$$

$$= \frac{1}{2\pi} \Re \left\{ \log \frac{f(z') - e^{i\theta}}{\left(f(z') - e^{i\theta} \right) e^{i\theta}} \right\}$$

$$= \frac{1}{2\pi} \log(1) = 0.$$

Next, we will show that the above function $G(x, y; x', y')$ is a fundamental solution of Laplace's equation, i.e., we will show that it satisfies $\nabla^2 G = 0$ everywhere except at $(x, y) = (x', y')$, and that in the neighborhood of this point we have

$$2\pi G(x, y; x', y') = \log \left| (x, y) - (x', y') \right| + \text{a harmonic function.} \tag{10.50}$$

Since f is one-to-one, we have $f(z) - f(z') = 0$ only at $z = z'$. For any two pints $z = (x, y)$ and $z' = (x', y')$ in D, $z \neq z'$, we have $|f(z)| < 1$ and $|f(z')| < 1$. Thus, $f(z)\, \overline{f(z)} - 1 \neq 0$. The function

$$\log \left[\frac{f(z') - f(z)}{\overline{f(z')} \overline{f(z)} - 1} \right] \tag{10.51}$$

is analytical in z except for $z = z'$, and the real part of the function (10.51), being a harmonic function, satisfies Laplace's equation. In the neighborhood of the point z' we have the Taylor's series

$$f(z) - f(z') = \sum_{n=1}^{\infty} \frac{g^{(n)}(z')}{n!} (z - z')^n$$

$$= (z - z') \sum_{n=1}^{\infty} \frac{g^{(n)}(z')}{n!} (z - z')^{n-1} = (z - z') \, g(z),$$

where $g(z)$ is analytic. Since the mapping is conformal, the function $4666g(z) \neq 0$ in a neighborhood of z'. Thus,

$$G(z, z') = \frac{1}{2\pi} \left[\log |z - z'| + \Re\left\{ \log \left[1 - f(z)\overline{f(z')} \right] - \log \frac{f(z) - f(z')}{z - z'} \right\} \right]$$

$$= \frac{1}{4\pi} \log \left[(x - x')^2 + (y - y')^2 \right] + \gamma,$$

where γ is the harmonic function defined by (8.41). Hence, the required Green's function is defined by (10.49). ∎

10.4.3. Green's Function for an Ellipse. As we have seen, Green's function $G(z, z') \equiv G(x, y; x', y')$ and Neumann's function $N(z, z') \equiv N(x, y; x', y')$ are both harmonic functions in any region D, and possess the following properties:

1. Both $G(z, z')$ and $N(z, z')$ are harmonic for fixed $z' \in D$ and $z \in \bar{D}$ for all $z \neq z'$.

2. In the neighborhood of $z = z'$ the functions $G(z, z') + \ln r$ and $N(z, z') + \ln r$ are harmonic, where $r = \sqrt{(x - x')^2 + (y - y')^2}$.

3. At a fixed boundary point $z \in \partial D$, we have $G(z, z') = 0$, $\dfrac{\partial N(z, z')}{\partial n} = $ const for any $z' \in D$.

Note that the function $N(z, z')$ is normalized on the boundary ∂D by the condition

$$\int_{\partial D} N(z, z') \, ds = 0, \tag{10.52}$$

and any harmonic function $h(z)$ defined on $\bar{D}$ can be represented in the form

$$h(z') = \frac{1}{2\pi} \int_{\partial D} \frac{\partial G(z, z')}{\partial n_z} h(z) \, ds_z, \tag{10.53}$$

$$\text{or} \quad h(z) = \frac{1}{2\pi} \int_{\partial D} N(z, z') \frac{\partial h(z)}{\partial n} \, ds_z, \tag{10.54}$$

where ds_z is an element of the arclength along the boundary ∂D, and (10.54) holds only if the function $h(z)$ is normalized by $\displaystyle\int_{\partial D} h(z) \, ds = 0$.

Let D denote the interior of an ellipse with eccentricity e. Consider the conformal mapping $z = \cosh w$, where $z = x + i\,y$ and $w = u + i\,v$. Then

$$x = \cosh u \cos v, \quad y = \sinh u \sin v. \tag{10.55}$$

The lines $u = \text{const}$ are mapped onto a confocal family of ellipses:

$$\frac{x^2}{\cosh^2 u} + \frac{y^2}{\sinh^2 u} = 1, \tag{10.56}$$

with foci at $z = \pm 1$, semi-major axis $a = \cosh u$, semi-minor axis $b = \sinh u$, and eccentricity $e = \dfrac{\sqrt{a^2 - b^2}}{a} = \dfrac{1}{\cosh u}$. The lines $v = \text{const}$ are mapped onto a family of hyperbolas:

$$\frac{x^2}{\cos^2 v} - \frac{y^2}{\sin^2 v} = 1,$$

with foci at $z = \pm 1$. The element ds of the arclength is given by

$$ds^2 = dx^2 + dy^2 = (\sinh u \cos v\, du - \cosh u \sin v\, dv)^2 +$$
$$+ (\cosh u \sin v\, du + \sinh u \cos v\, dv)^2 \tag{10.57}$$
$$= \left(\cosh^2 u - \cos^2 v\right)\left(du^2 + dv^2\right).$$

Let us consider a particular ellipse defined by $u = u_0$ and a particular hyperbola by $v = v_0$. Then, by (10.57), it is obvious that the ellipse with $u = u_0$ and the hyperbola with $v = v_0$ are orthogonal to each other. Let dn and dt denote the arclength elements at the node (u_0, v) along the normal and the tangent to the ellipse, respectively. Then

$$dn = \sqrt{\cosh^2 u_0 - \cos^2 v}\; du, \quad dt = \sqrt{\cosh^2 u_0 - \cos^2 v}\; dv,$$

$$\frac{\partial}{\partial n} = \frac{1}{\sqrt{\cosh^2 u_0 - \cos^2 v}} \frac{\partial}{\partial u},$$

$$\nabla^2 = \frac{\partial^2}{\partial x^2} + \frac{\partial^2}{\partial y^2} = \frac{1}{\cosh^2 u - \cos^2 v}\left(\frac{\partial^2}{\partial u^2} + \frac{\partial^2}{\partial v^2}\right).$$

Thus, Green's function for this particular ellipse, defined with $u = u_0$ and $w_{1,2} = u_{1,2} + i\,v_{1,2}$, is given by

$$G(w_1; w_2) = u_0 - \ln 2 - \ln|z_1 - z_2| -$$

$$- \sum_{n=1}^{\infty} \frac{2}{n} e^{-n u_0} \left[\frac{\cosh n u_1 \cosh n u_2 \cos n v_1 \cos n v_2}{\cosh n u_0} + \right.$$

$$\left. + \frac{\sinh n u_1 \sinh n u_2 \sin n v_1 \sin n v_2}{\sinh n u_0} \right] \tag{10.58}$$

$$\equiv u_0 - \ln 2 - \ln|z_2 - z_1| - \sum_{n=1}^{\infty} \frac{2}{n} e^{-n u_0} \left[a_n + b_n\right].$$

We will now verify if this Green's function satisfies the three properties listed at the beginning of this section. The properties 1 and 2 are satisfied if we apply the Laplacian, defined above, on both terms under the summation in (10.58) and show that this sum is uniformly convergent. For the first term in the sum we get the estimate

$$\sum_{n=1}^{\infty} a_n \le \sum_{n=1}^{\infty} |a_n| \le \sum_{n=1}^{\infty} \frac{2}{n} e^{-nu_0} \frac{\cosh nu_1 \cosh nu_2}{\cosh nu_0} \equiv \sum_{n=1}^{\infty} \alpha_n.$$

For any $\varepsilon > 0$ we can find an n_0 such that $\dfrac{\alpha_{n+1}}{\alpha_n} < (1 + \varepsilon) e^{-(2u_0-u_1-u_2)}$ for all $n \ge n_0$, i.e., for every $u_1 < u_0$ the series $\sum \alpha_n$ is uniformly convergent for any $u_2 \le u_0$. A similar argument can be applied to the second term of series $\sum b_n$. To prove the property 3, we use the following formulas: $\cosh x = \dfrac{e^x + e^{-x}}{2}$; $\ln(1 -$

$x) = -\sum_{n=1}^{\infty} \dfrac{x^n}{n}$, $-1 \le 1 \le 1$; $\cosh(ix) = \cos x$, and $\sinh(ix) = i \sin x$. Since

$z_1 - z_2 = \cosh w_1 - \cosh w_2 = -\dfrac{1}{2} e^{w_2} (1 - e^{-w_1-w_2})(1 - e^{w_1-w_2})$, we have for $u_2 > u_1$:

$$\ln(\cosh w_2 - \cosh w_1) = -\ln 2 + w_2 - \sum_{n=1}^{\infty} \frac{1}{n} (e^{n(-w_1-w_2)} + e^{n(w_1-w_2)}).$$

Thus, for $u_2 > u_1$,

$$\ln|z_1 - z_2| = u_2 - \ln 2 - \sum_{n=1}^{\infty} \frac{2}{n} e^{-nu_2} (\cosh nu_1 \cos nv_1 \cos nv_2 +$$
$$+ \sinh nu_1 \sin nv_1 \sin nv_2).$$

In particular, for $u_2 = u_0$, we get $G(u_1, v_1; u_2, v_2) = 0$, which proves that the property 3 is satisfied.

Neumann's function for the ellipse is given by

$$N(w_1; w_2) = u_0 - \ln 2 - \ln|z_1 - z_2| +$$
$$+ \sum_{n=1}^{\infty} \frac{2}{n} e^{-nu_0} \left[\frac{\cosh nu_1 \cosh nu_2 \cos nv_1 \cos nv_2}{\sinh nu_0} + \right.$$
$$\left. + \frac{\sinh nu_1 \sinh nu_2 \sin nv_1 \sin nv_2}{\cosh nu_0} \right]. \tag{10.59}$$

Using the same arguments as used above in the case of Green's function we can establish that this Neumann's function satisfies the properties 1 through 3 (see Exercise 6.3).

10.4.4. Green's Function for an Infinite Strip. If we keep the semi-minor axis b of the ellipse in §6.3.3 fixed and take the limiting process as the semi-major axis

$a = \cosh u_0 \to \infty$, i.e., as $u_0 \to \infty$, then we obtain an infinite strip $-\infty < x < \infty$, $-b < y < b$. This limiting process can be visualized in Fig. 10.4. However, we will show that Green's function, and Neumann's function also, for this infinite strip cannot be derived from (10.58) and (10.59), respectively, under this limiting process.

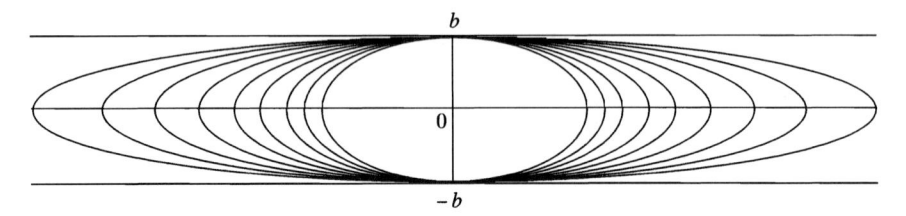

Fig. 10.4. Limiting Process of Confocal Ellipses to an Infinite Strip.

To carry out this limiting process successfully, we will first modify the conformal mapping $z = \cosh w$ by introducing a real scaling factor $\mu > 0$, which need not be equal to unity, such that $x + iy = \mu \cosh(u + iv)$. Then $a = \mu \cosh u_0$ and $b = \mu \sinh u_0$. To maintain the geometry of the infinite strip, we keep b fixed; then the limiting process will require that only $u_0 \to \infty$. Since $\lim\limits_{u_0 \to \infty} \dfrac{e^{-nu_0}}{\cosh nu_0} = 0$ and $\lim\limits_{u_0 \to \infty} \dfrac{e^{-nu_0}}{\sinh nu_0} = \lim\limits_{u_0 \to \infty} \dfrac{e^{-nu_0}}{b} = 0$, Green's function for the infinite strip is

$$
\begin{aligned}
G\left(w_1; w_2\right) = \lim_{u_0 \to \infty} &\left\{ u_0 - \ln \mu - \ln 2 - \ln |z_1 - z_2| - \right. \\
&- \sum_{n=1}^{\infty} \frac{2}{n} e^{-nu_0} \left[\frac{\cosh nu_1 \cosh nu_2 \cos nv_1 \cos nv_2}{\cosh nu_0} + \right. \\
&\qquad\qquad\qquad\qquad \left.\left. + \frac{\sinh nu_1 \sinh nu_2 \sin nv_1 \sin nv_2}{\sinh nu_0} \right] \right\} \\
= {}& -\ln \mu - \ln 2 + \lim_{u_0 \to \infty} u_0 - \ln |z_1 - z_2|,
\end{aligned}
$$

which does not exist because it is not finite. Similarly, Neumann's function, given by

$$
\begin{aligned}
N\left(w_1; w_2\right) = \lim_{u_0 \to \infty} &\left\{ u_0 - \ln \mu - \ln 2 - \ln |z - z'| - \right. \\
&- \sum_{n=1}^{\infty} \frac{2}{n} e^{-nu_0} \left[\frac{\cosh nu_1 \cosh nu_2 \cos nv_1 \cos nv_2}{\sinh nu_0} + \right. \\
&\qquad\qquad\qquad\qquad \left.\left. + \frac{\sinh nu_1 \sinh nu_2 \sin nv_1 \sin nv_2}{\cosh nu_0} \right] \right\} \\
= {}& -\ln \mu - \ln 2 + \lim_{u+0 \to \infty} u_0 - \ln |z - z'|,
\end{aligned}
$$

does not exists. ∎

In order to determine Green's function for the infinite strip, we will use the methods of conformal mapping and first consider the case of Green's function for the upper half-plane, as in the following example.

Example 10.2. Green's function for the upper half-plane $y \geq 0$ is given by

$$G(\mathbf{x}, \mathbf{x}') = \frac{1}{2\pi} \log \frac{1}{|\mathbf{x} - \mathbf{x}'|} - \frac{1}{2\pi} \log \left| \mathbf{x} - \frac{R^2}{|\mathbf{x}'|^2} \mathbf{x}' \right|. \tag{10.60}$$

Another interesting result for Green's function for the upper half-plane $\{\Im\{z\} \geq 0\}$ is given in Wayland [1970:333] as

$$G(z, z') = \log \left| \frac{z + \bar{z}'}{z - z'} \right|. \ \blacksquare \tag{10.61}$$

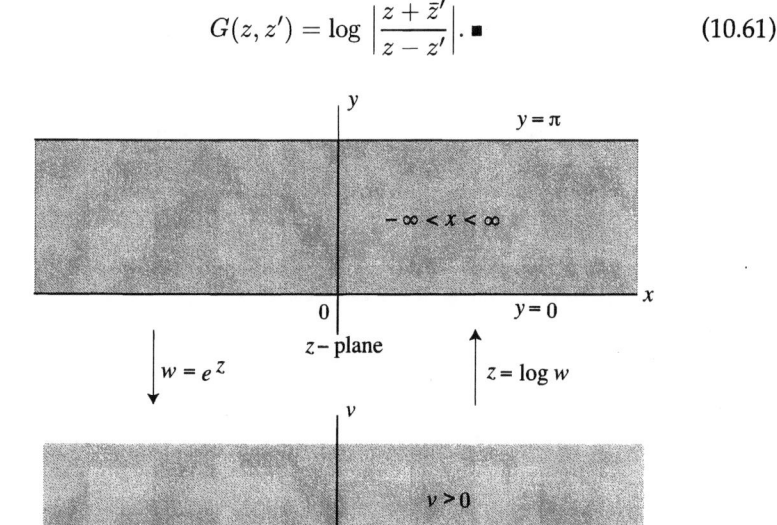

Fig. 10.5. Mapping of the Upper Half-Plane onto the Parallel Strip.

Example 10.3. To find Green's function for the infinite strip $0 < y < \pi$, note that the function $w = e^z$, $z = x + i y$, maps this strip conformally onto the upper half-plane $\Im\{w\} > 0$, $w = u + i v$ (see Fig. 10.5). Thus, Green's function for this strip is obtained from (10.60) with $r = 1$ as

$$G(x, y; , x', y') = \frac{1}{2} \log \frac{e^{2x} - 2 e^{x+x'} \cos(y + y') + e^{2x'}}{e^{2x} - 2 e^{x+x'} \cos(y - y') + e^{2x'}}. \tag{10.62}$$

If u is harmonic in this strip, then

$$u(x', y') = \frac{\sin y'}{2\pi} \int_{-\infty}^{\infty} \left[\frac{u(x, 0)}{\cosh(x - x') - \cos y'} + \frac{u(x, \pi)}{\cosh(x - x') + \cos y'} \right] dx.$$

is the solution of Laplace's equation $\nabla^2 u = 0$ in this strip. ∎

10.4.5. Green's Function for an Annulus. Consider an annulus D with the outer radius r_1 and the inner radius r_2, respectively. Then, using cylindrical polar coordinates with $z = r\,e^{i\theta}$ and $z' = \rho\,e^{i\phi}$, Green's function for the Dirichlet problem for $r < \rho$ is given by

$$
G\,(r, \theta; \rho, \phi) = \frac{(\ln r - \ln r_2)\,(\ln r_1 - \ln \rho)}{\ln r_1 - \ln r_2} +
$$

$$
= + \sum_{n=1}^{\infty} \frac{\left[\left(\frac{r_2}{r}\right)^n - \left(\frac{r}{r_1}\right)^n\right]\left[\left(\frac{\rho}{r_1}\right)^n - \left(\frac{r_1}{\rho}\right)^n\right]}{n\left[\left(\frac{r_1}{r}\right)^n - \left(\frac{r_2}{r}\right)^n\right]}\cos n(\theta - \phi) \quad (10.63)
$$

Example 10.4. (Green's function for a sphere of radius R.) Let $B(R,0)$ be a ball (sphere) of radius R and center at the origin in $\mathbb{R}^n$, and let $\mathbf{x}' \in B(R,0)$ be an arbitrary point. Then Green's function for this sphere is given by

$$
G(\mathbf{x}, \mathbf{x}') = F(r) - F\left(\frac{r_1}{R}\,|\mathbf{x}'|\right),
$$

where $S_n(1) = \dfrac{2\pi^{n/2}}{\Gamma(n/2)}$ is the surface area of $B(1,0)$ (unit sphere in $\mathbb{R}^n$), and

$$
F(r) = \begin{cases} \dfrac{1}{(n-2)\,S_n(1)}\,|\mathbf{x} - \mathbf{x}|^{2-n}, & \text{if } n > 2, \\[2ex] \dfrac{1}{2\pi}\log\dfrac{1}{|\mathbf{x} - \mathbf{x}'|}, & \text{if } n = 2; \end{cases}
$$

$$
r = |\mathbf{x} - \mathbf{x}'|, \quad |\mathbf{x}'|^2 = x_1^2 + \cdots + x_n^2, \quad r_1^2 = \left|\mathbf{x} - \frac{R^2}{|\mathbf{x}'|^2}\,\mathbf{x}'\right|.
$$

Note that r_1 is the distance of the point $\mathbf{x}$ from the reflected image of the point $\mathbf{x}'$ in the sphere. The function $G(\mathbf{x}, \mathbf{x}')$ satisfies all the requirements of being Green's function since (i) it is of the form $s(\mathbf{x}, \mathbf{x}') + \phi(\mathbf{x})$, where $s(\mathbf{x}, \mathbf{x}')$ is the so-called singularity function defined by (2.55), and $\phi(\mathbf{x})$ is regular in $B(R,0)$ and continuous in $B \cup \partial B$, and (ii) it vanishes on ∂B, since $r = \dfrac{r_1}{R}\,|\mathbf{x}'|$ on ∂B. ∎

Example 10.5. We will determine Green's function for the positive (right) half-plane $x_1 \geq 0, n > 2$, in $\mathbb{R}^n$. Let $\mathbf{x}'$ be an arbitrary point in the positive half-plane. Then Green's function is

$$
G(\mathbf{x}, \mathbf{x}') = \frac{1}{(n-2)S_n(1)}\,|\mathbf{x} - \mathbf{x}'|^{2-n} - p(\mathbf{x}), \quad n > 2,
$$

where

$$
p(\mathbf{x}) = \frac{1}{(n-2)S_n(1)}\left[(x_1 + x_1')^2 + \sum_{k=2}^{n}(x_k - x_k')^2\right]^{(2-n)/2}.
$$

The function $p(\mathbf{x})$ is obtained by taking the image of $\mathbf{x}'$ in the boundary plane $x_1 = 0$. The solution of the Laplace equation $\nabla^2 u = 0$ in this half-plane is given by

$$u(\mathbf{x}') = \int_{x_1=0} f(\mathbf{x})\left\{\frac{\partial}{\partial x_1}G(\mathbf{x},\mathbf{x}')\right\}\Big|_{x_1=0} dS.$$

Example 10.6. (Green's function for a cardioid) Consider the conformal mapping $z = f(w) = (w+1)^2$, with inverse $w = \sqrt{z} - 1$, where $-\pi < \arg\{z\} < \pi$. Then $f'(w) = 2(w+1) \neq 0$, but $f'(-1) = 0$. The unit disk $|w| < 1$ goes into the domain $D : |\sqrt{z} - 1|^2 < 1$, which in polar cylindrical coordinates is defined by $r + 1 - 2\sqrt{r}\cos\frac{\theta}{2} < 1$, or $r < 4\cos^2\frac{\theta}{2}$; thus, $D : r < 2(1 + \cos\theta)$, which is the interior of a cardioid with cusp at $z = 0$ that corresponds to $w = -1$, where $f'(0) = 0$ (Fig. 10.6). Green's function for this cardioid D is

$$G(z, z') = \frac{1}{2\pi}\log\frac{\left|1 - (\sqrt{z}-1)(\sqrt{\bar{z}'}-1)\right|}{\left|\sqrt{z} - \sqrt{z'}\right|},$$

or in polar cylindrical coordinates

$$G(r,\theta;r',\theta') = \frac{1}{4\pi}\log\frac{r + r' + rr' + 2\sqrt{rr'}\left[\cos\frac{\theta+\theta'}{2} - \sqrt{r}\cos\frac{\theta'}{2} - \sqrt{r'}\cos\frac{\theta}{2}\right]}{r + r' - 2\sqrt{rr'}\cos\frac{\theta-\theta'}{2}}.$$

$$(10.64)$$

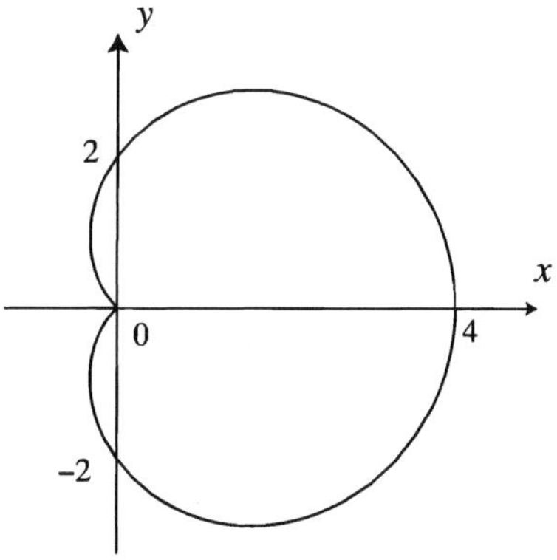

Fig. 10.6. Cardioid.

10.5. Computation of Green's Functions

Numerical approximation methods are important as they have extended the applications of conformal mapping in various areas of practical significance. These methods have a long history of development, starting with transforming complicated boundaries by successive mappings so that each approximation ultimately led to a near circle in the final complex plane. This smoothing or osculation process was developed by

Koebe [1915] who called the method 'Schmiegungsverfahren', but it suffered from
slow convergence for numerical computation, although it because a powerful technique for theoretical purposes. Many other methods were proposed during the late
half of the twentieth century, specially by Heinhold and Albrecht [1954], von Koppenfels and Stallmann [1959], Gaier [1964], and Henrici [1986]. Numerical methods
used today are based on either an approximation of the desired unknown mapping
function by power series expansions or variational techniques with orthogonalization,
or by methods of solving integral equations.

A domain D is said to be *convex*, if the line segment joining any two points in
the domain D is contained in D. A domain D is said to be *starlike* with respect to a
point $z_0 \in D$ if the line segment joining the point z_0 to any point $z \in D$ lies entirely
inside D. A convex domain D is starlike at every point in D. Also, if D is starlike
with respect a point $z_0 \in D$ and if $z \in D$, then $t z \in D$ for $0 < t < 1$.

10.5.1. Interpolation Method. We will develop a method, also known as
the method of simultaneous equations, developed by Kantorovich [1936] and Kantorovich and Krylov [1958], to numerically compute Green's function for a simply
connected domain $D \in \mathbb{R}^2$ that has the origin as an interior point. If the origin is
outside D, we carry out a suitable translation to bring the origin inside D. We will
use the notation $z = x + i y$ and $z' = x' + i y'$ instead of $\mathbf{x} = (x, y)$ and $\mathbf{x}' = (x', y')$,
respectively. The method is based on the following results from the theory of confor-
mal mapping: If Green's function $G(z; z')$ with a pole (singularity) at a point $z' \in D$
and an analytic function $f(z)$ each map the domain D conformally onto the unit disk,
then $G(z; z')$ and $f(z)$ are related by (10.17b), i.e.,

$$G(z; z') = \frac{1}{2\pi} \log |f(z)|, \quad z = x + iy,$$

and the mapping function $f(z)$ is given by (10.19), that is,

$$f(z) = (z - z') e^{-2\pi(g+ih)},$$

where g and h are harmonic functions in D. Since $\log |f(z)| = \log |z - z'| - 2\pi g(z)$,
the construction of Green's function $G(z; z')$ involves determining the harmonic
function $g(z) \equiv g(x, y)$ such that $G(z; z') = 0$ on the boundary ∂D. Thus, we solve
the Dirichlet problem for the domain D with the boundary condition $g(x, y)\big|_{\partial D} =$
$\frac{1}{2\pi} \log r, r = |z - z'|$. Hence, Green's function for the domain D is constructed by
the formula[2]

$$G(z; z') = \frac{1}{2\pi} \log r - g(x, y). \tag{10.65}$$

The interpolation method can be developed as follows: Assuming the series
representation for $g(z)$ as $g(z) = \sum_{k=0}^{\infty} c_k (z - z')^k$, $c_k = a_k + i b_k$, we approximate

---

[2] This formula is based on the assumption that a source of strength -1 exists at z'. If
this strength is taken as $+1$, formula (10.65) becomes $G(z, z') = \frac{1}{2\pi} \log \frac{1}{r} - g(x, y)$, which
is the same as in Kantorovich and Krylov [1958], and Kythe [1998].

it by the harmonic polynomial $p_n(r, \theta) = \Re\left\{\sum_{k=0}^{n} c_k(z - z')^k\right\}$. Thus, in the polar coordinates we have

$$g(x, y) \approx p_n(r, \theta) = a_0 + \sum_{k=1}^{n} r^k \left(a_k \cos k\theta - b_k \sin k\theta\right),$$

where $z - z' = re^{i\theta}$. Since this polynomial has $(2n+1)$ coefficients, we take $(2n+1)$ arbitrary points $z_1, \ldots, z_{2n+1}$ on the boundary ∂D and then choose the coefficients a_k and b_k such that at each of the points $z_j, j = 1, \ldots, 2n+1$, the polynomial $p_n(r, \theta)$ takes the same value as $g(z_j)$. Since $g(z)$ has the boundary value $g(x, y) = \dfrac{1}{2\pi} \log r$, the coefficients a_k, b_k are determined by solving the system of equations

$$a_0 + \sum_{k=1}^{n} r_1^k \left(a_k \cos k\theta_1 - b_k \sin k\theta_1\right) = \frac{1}{2\pi} \log r_1,$$

$$a_0 + \sum_{k=1}^{n} r_2^k \left(a_k \cos k\theta_2 - b_k \sin k\theta_2\right) = \frac{1}{2\pi} \log r_2,$$

$$\cdots \quad \cdots \quad \cdots \quad \cdots \quad \cdots \quad \cdots \quad \cdots \tag{10.66}$$

$$a_0 + \sum_{k=1}^{n} r_{2n+1}^k \left(a_k \cos k\theta_{2n+1} - b_k \sin k\theta_{2n+1}\right) = \frac{1}{2\pi} \log r_{2n+1},$$

where $z_j - z' = r_j e^{i\theta_j}, j = 1, \ldots, 2n + 1$. The determinant of the system (10.66) depends on the choice of the points z_j. Let us assume that these points lie on an equipotential line of a harmonic polynomial $Q_n(r, \theta)$ of degree at most n. This line has the equation

$$Q_n(r, \theta) = a_0 + \sum_{k=1}^{n} r^k \left(a_k \cos k\theta - b_k \sin k\theta\right) = 0,$$

where the coefficients a_k and b_k are all not zero at the same time. Then this line exists iff the homogeneous system of $(2n + 1)$ algebraic equations

$$a_0 + \sum_{k=1}^{n} r_j^k \left(a_k \cos k\theta_j - b_k \sin k\theta_j\right) = 0, \quad j = 1, \ldots, 2n + 1, \tag{10.67}$$

has a nonzero solution for $a_0, a_1, b_1, \ldots$. Hence, Green's function has the approximate representation

$$G(z; z') \approx \frac{1}{2\pi} \log r - a_0 - \sum_{k=1}^{n} r^k \left(a_k \cos k\theta - b_k \sin k\theta\right). \tag{10.68}$$

It is assumed here that the difference between $p_n(r, \theta)$ and $g(x, y)$ decreases as n increases and the above interpolation method becomes justified, provided that $\lim_{n \to \infty} p_n(r, \theta) = g(z)$. In view of the Riemann mapping theorem, the function $f(z)$ maps the domain D uniquely onto the unit disk iff $f(0) = 0$ and $f'(0) = 1$. The latter condition is used to compute the *distortion* in the mapping, and hence, the relative error in the computation of $G(z; z')$.

Example 10.7. We will approximate Green's function for the square $\{(x, y) : -1 \le x, y \le 1\}$ in the z-plane with the pole z', which is taken, without loss of generality, at the origin (see Fig. 10.7). In view of the symmetry about both coordinate axes, the values of $g(x, y)$ are arranged symmetric to the y-axis; thus, all $b_k = 0$ in (10.67). Also, since the values of $g(x, y)$ are symmetric to x-axis and bisectors of coordinate angles, the series (10.67) will contain cosine terms of angles $4n\theta$, $n = 0, 1, 2, \ldots$. Hence, the series expansion for $g(x, y)$ becomes

$$g(x, y) \approx a_0 + a_4 \, r^4 \, \cos 4\theta + a_8 \, r^8 \, \cos 8\theta + \cdots .$$

We take the points z_j on the boundary as follows: $z_1 = 1$, $z_2 = \dfrac{2}{\sqrt{3}} e^{i\pi/6}$, $z_3 = \sqrt{2} \, e^{i\pi/4}$. Then the coefficients a_0, a_4, and a_8 are obtained by solving the system (10.66), i.e.,

$$\begin{bmatrix} 1 & 1 & 1 \\ 1 & -\frac{8}{9} & -\frac{128}{81} \\ 1 & -4 & 16 \end{bmatrix} \begin{Bmatrix} a_0 \\ a_4 \\ a_8 \end{Bmatrix} = \frac{1}{2\pi} \begin{Bmatrix} 0 \\ \ln \frac{2}{\sqrt{3}} \\ \ln \sqrt{2} \end{Bmatrix} ,$$

which gives $a_0 = 0.120286 = 0.075578/2\pi$, $a_4 = -0.0117794 = -0.0740122/2\pi$, and $a_8 = -0.000249209 = -0.00156583/2\pi$. Note that additional points $z_4 = \dfrac{2}{\sqrt{3}} e^{i\pi/3}$ and $z_5 = e^{i\pi/2}$ add nothing to the solution.

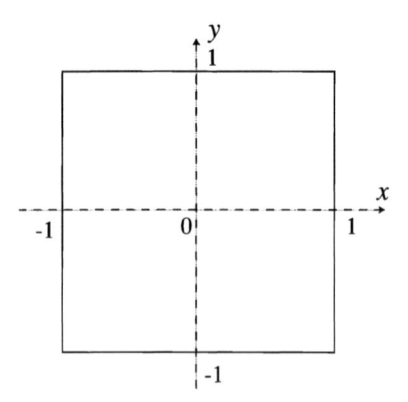

Fig. 10.7. The Unit Square.

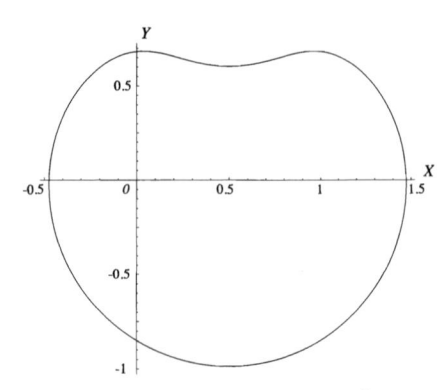

Fig. 10.8. Cardioid for $\alpha = 1$.

The approximate Green's function is given by

$$
\begin{aligned}
G(z;0) &= \frac{1}{2\pi} \log r - g(z) \\
&\approx \frac{1}{2\pi}\left[\log r - 0.075578 + 0.0740122\, r^4 \cos 4\theta + 0.00156583\, r^8 \cos 8\theta \right] \\
&= \frac{1}{2\pi} \Re \left\{ \log z - 0.075578 + 0.0740122\, z^4 + 0.00156583\, z^8 \right\}.
\end{aligned}
$$

Then, the function $f(z)$ that maps the square onto the unit disk is found from

$$
\log |f(z)| = \Re \left\{ \log z - 0.075578 + 0.0740122\, z^4 + 0.00156583\, z^8 \right\},
$$

up to a purely imaginary additive constant which we will ignore. Hence,

$$
f(z) = z\, e^{-0.075578 + 0.0740122\, z^4 + 0.00156583\, z^8}.
$$

Note that $f'(0) = e^{-2\pi a_0} = e^{-0.075578} = 0.927207$. The value is $f'(0) = \int_0^1 \frac{d\zeta}{\sqrt{1+\zeta^4}} \approx 0.927037$, which shows the error of about 0.017%. ∎

Example 10.8. To approximate Green's function for the ellipse

$$
x = \left(1+\lambda^2\right)\cos t, \quad y = \left(1-\lambda^2\right)\sin t, \quad 0 \le t < 2\pi,
$$

or, in complex notation, $z(t) = e^{it}\left(1 + \lambda^2 e^{-2it}\right), 0 \le t < 2\pi$, we consider the first quadrant because of the axial symmetry and take the points z_j for $t = 0, \pi/8, \pi/4, 3\pi/8$, and $\pi/2$. Thus,

$$
g(x,y) \approx a_0 + a_1 r_1 \cos 2t + a_2 r_2 \cos 4t + a_3 r_3 \cos 6t + a_4 r_4 \cos 8t.
$$

Then, for example, for $\lambda = 0.1$ we find that

$$
\begin{aligned}
g(z) = &-0.0000159163 + 0.00159187\, z^2 - 0.000023828\, z^4 \\
&+ 5.31236 \times 10^{-7}\, z^6 - 1.39292 \times 10^{-8}\, z^8,
\end{aligned}
$$

which gives

$$
\begin{aligned}
G(z;0) = \frac{1}{2\pi} \Re\Big\{ &\log z + 0.0001 - 0.010002\, z^2 + 0.00015\, z^4 \\
&- 3.3 \times 10^{-6}\, z^6 + 8.75 \times 10^{-8}\, z^8 \Big\}.
\end{aligned}
$$

Note that $f'(0) = e^{0.0001} \approx 1.0001$. For computational details for $\lambda = 0.1$ and also for $\lambda = 0.5$. Note that if $\lambda = 0$, then $g(x,y) = 0$, and Green's function $G(z;0)$ is the same as (8.1). ∎

Example 10.9. Consider the curve

$$F(x, y, \alpha) = \left[(x - 0.5)^2 + (y - \alpha)^2\right]\left[1 - y^2 - (x - 0.5)^2\right] = 0.1$$

for $\alpha = \infty, 1, 0.5, 0.3$, and 0.2746687749, which is plotted in Fig. 10.9. We will consider the curve for $\alpha = 1$ which is a cardioid shown in Fig. 10.8, and approximate Green's function for this region which is a nearly circular cardioid. Since the region is symmetric about the line $x = 0.5$, we shall choose seven boundary points z_j as follows: $z_1 = 0.5 - 0.98725\,i$, $z_2 = 0.5 + 0.60343\,i$, $z_3 = 0.8 - 0.94024\,i$, $z_4 = 0.8 + 0.65588\,i$, $z_5 = 1.2 - 0.69296\,i$, $z_6 = 1.2 + 0.59722\,i$, and $z_7 = 1.474003$. Green's function is given by

$$\begin{aligned} G(z; 0.5) \approx {}& \frac{1}{2\pi}\,\log r - 0.0431944 + 0.0404596\,r\,\cos\theta \\ & + 0.0107145\,r^2\,\cos 2\theta - 0.0114393\,r^3\,\cos 3\theta \\ & - 0.0569173\,r\,\sin\theta + 0.039137\,r^2\,\sin 2\theta - 0.00477567\,r^3\,\sin 3\theta, \end{aligned}$$

where $r = |z - 0.5|$. Since the origin is at the point $(-0.5, 0)$, the distortion, given by $f'(0.5) = e^{0.0431944} = 1.04415$. ∎

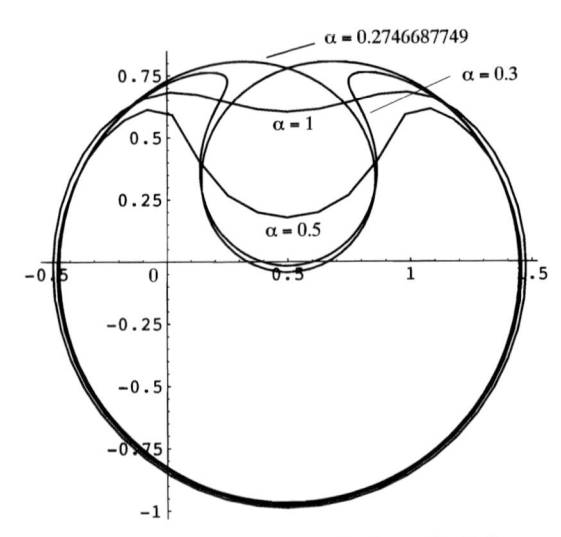

Fig. 10.9. Graph of the Curve in Example 10.9.

Example 10.10. Cassini's ovals, defined by

$$F(x, y, \alpha) = \left[(x + \alpha)^2 + y^2\right]\left[(x - \alpha)^2 + y^2\right] = 1,$$

are plotted for $\alpha = 0, 0.5, 0.9, 0.99$, and 1 in Fig. 10.10. For $\alpha = 0$ the curve becomes the unit circle. The region does does not remain simply connected for $\alpha = 1$. We will approximate Green's function for the Cassini's oval at $\alpha = 0.5$ which is bounded by a nearly circular ellipse. Because of the symmetry of the region about the coordinate axes, the function $g(z)$ has a series expansion of the form

$$g(z) = a_0 + a_2\, r^2 \cos 2\theta + a_4\, r^4 \cos 4\theta + \cdots .$$

Let the points z_j on the boundary be chosen as $z_1 = 1.11803$, $z_2 = 1.04942\, e^{i\pi/6}$, $z_3 = 0.98399\, e^{i\pi/4}$, $z_4 = 0.92265\, e^{i\pi/3}$, and $z_5 = 0.866\, e^{i\pi/2}$. Then compute $a_0 = -0.00513$, $a_2 = 0.02124$, $a_4 = -0.00282$, $a_6 = 0.0005$, $a_8 = -0.000096$. Thus, Green's function is given by

$$G(z;0) \approx \frac{1}{2\pi}\Re\Big\{\log z - 0.0322236 + 0.133465\, z^2 - 0.0177425\, z^4$$

$$+ 0.0031616\, z^6 - 0.000605\, z^8 \Big\},$$

with $f'(0) = e^{0.0322236} \approx 1.03276$, which has an error of about 3%. ∎

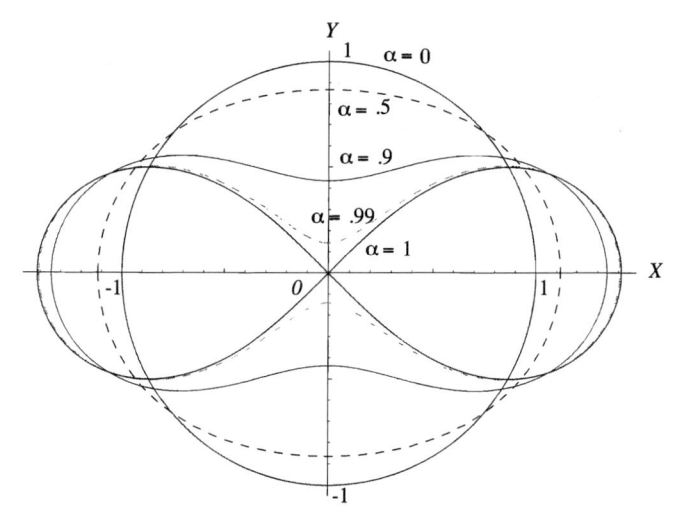

Fig. 10.10. Cassini's Ovals.

10.6. Exercises

10.1. Determine Green's function for the positive (right) half-plane $x_1 \geq 0$, in $\mathbb{R}^n$, $n > 2$.

SOLUTION. Let $\mathbf{x}'$ be an arbitrary point in the positive half-plane. Then Green's function is $G(\mathbf{x}, \mathbf{x}') = \dfrac{1}{(n-2)S_n(1)}|\mathbf{x} - \mathbf{x}'|^{2-n} - p(\mathbf{x})$, $\quad n > 2$, where

$$p(\mathbf{x}) = \frac{1}{(n-2)S_n(1)}\left[(x_1 + x_1')^2 + \sum_{k=2}^{n}(x_k - x_k')^2\right]^{(2-n)/2}.$$

The function $p(\mathbf{x})$ is obtained by taking the image of $\mathbf{x}'$ in the boundary plane $x_1 = 0$. The solution of Laplace's equation $\nabla^2 u = 0$ in this half-plane is given by $u(\mathbf{x}') = \int_{x_1=0} f(\mathbf{x}) \left\{ \frac{\partial}{\partial x_1} G(\mathbf{x}, \mathbf{x}') \right\} \bigg|_{x_1=0} dS.$

10.2. Show that Neumann's function (10.59) satisfies the properties listed in §10.3.

HINT. Properties 1 and 2 are satisfied by the same method as used for Green's function (10.58). Property 3 can be established by computing $\dfrac{\partial N}{\partial n_z}$ and showing that $\dfrac{\partial N}{\partial n_z}\bigg|_{u_2=u_0} = -\dfrac{1}{\sqrt{\cosh^2 u_0 - \cos^2 v_2}}$, which is independent of (u_1, v_1). Since all terms containing v_2 are involve the terms $\displaystyle\int_0^{2\pi} \left\{ \begin{matrix} \cos nv \\ \sin nv \end{matrix} \right\} dv$ which have zero value, the normalization condition (10.52) is satisfied.

10.3. Compute Green's function for an equilateral triangle of side 2 units, where the x-axis is parallel to the base of the triangle and $1/2$ units above it and the y-axis passes through the top vertex.

HINT. Use the symmetry and assume $g(x, y) \approx a_0 + a_1 r^2 \cos 2\theta + a_1 r^2 \cos 2\theta + a_2 r^4 \cos 4\theta + \cdots$. Consider the two sets of points z_j:

(i) Four points: $(0, -1/2)$, $(1, -1/2)$, $(1/2, (\sqrt{3} - 1)/2)$, $(0, \sqrt{3} - 1/2)$; and
(ii) Seven points: $(0, -1/2)$, $(1/2, -1/2)$, $(1, -1/2)$, $(3/4, (\sqrt{3} - 2)/4)$, $(1/2, (\sqrt{3} - 1)/2)$, $(1/4, (3\sqrt{3} - 2)/4)$, and $(0, \sqrt{3} - 1/2)$.

10.4. Show that Neumann's function for the annulus with the outer radius r_1 and the inner radius r_2, and using cylindrical polar coordinates with $z = r\, e^{i\theta}$ and $\zeta = \rho\, e^{i\phi}$, for $r < \rho$ is given by

$$N(r, \theta; \rho, \phi) = \frac{r_2 \ln r - r_1 \ln \rho}{r_1 + r_2} - \frac{r_1^2 \ln r_1 - r_2^2 \ln r_2}{(r_1 + r_2)^2} +$$
$$+ \sum_{n=1}^{\infty} \frac{\left[\left(\frac{r_2}{r}\right)^n + \left(\frac{r}{r_1}\right)^n \right] \left[\left(\frac{\rho}{r_2}\right)^n + \left(\frac{r_2}{\rho}\right)^n \right] \cos n(\theta - \phi)}{n \left[\left(\frac{r_1}{r_2}\right)^n - \left(\frac{r_2}{r_1}\right)^n \right]}.$$

HINT. See §10.4.4.

10.5. Compute Green's function for the regions bounded by the Cassini's ovals $\left[(x + 1)^2 + y^2\right] \left[(x - 1)^2 + y^2\right] = a^4$ (Fig. 10.11) for the following cases: $a = 1.1, 1.2,$ and 1.5.

10.6. Compute Green's function for the square with round corners $x^4 + y^4 = 1$ (Fig. 10.12).

10.7. Compute Green's function for the Limaçon $r = a - \cos\theta$, $a = 1.5, 2$, $0 \le \theta < 2\pi$ (Fig. 10.13).

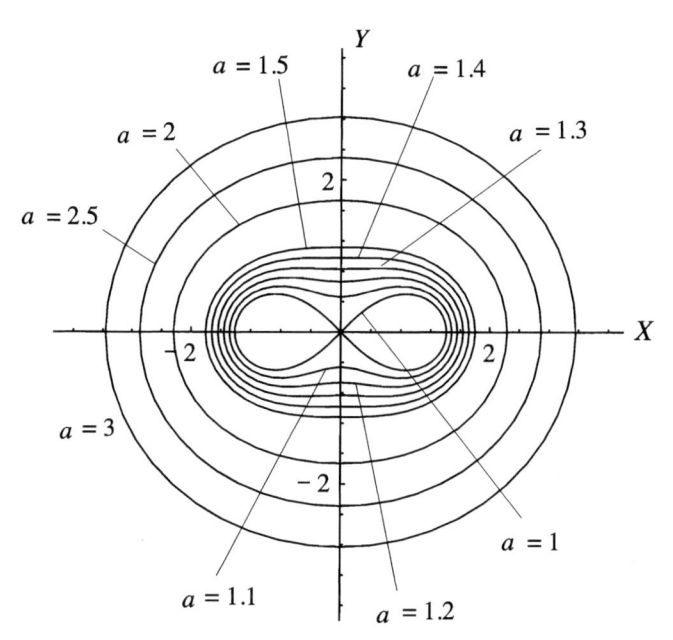

Fig. 10.11. Cassini's Ovals for $a = 1, 1.1, 1.2, 1.3, 1.4, 1.5, 2, 2.5, 3, 5$.

10.8. Compute Green's function for the upper half of the bean-shaped curve defined by $z = 2.25 \left[0.2 \cos s + 0.1 \cos 2s - 0.1 \right.$

$$+i \left(0.35 \sin s + 0.1 \sin 2s - 0.02 \sin 4s\right) \right], \quad -\pi \le s \le \pi,$$

and shown in Fig. 10.14.

10.9. Consider the region D bounded by two hyperbolic arcs

$$\Gamma_1 = \text{arc } \widehat{ABC} : \{z : z = x_0 - 2 \cosh s + i \left(y_0 - \sinh s\right), \; s_1 < s < s_2\}$$

$$\Gamma_2 = \text{arc } \widehat{CDA} : \{z : z = -x_0 + 2 \cosh s + i \left(-y_0 + \sinh s\right), \; s_1 < s < s_2\},$$

where $x_0 = \cosh s_1 + \cosh s_2$, $y_0 = \left(\sinh s_1 + \sinh s_2\right)/2$ are the coordinates of the center z_0 of the arc $\widehat{ABC}$ (see Fig. 10.15). Take $s_2 = 1$, and choose s_1 such that $4 \tanh s_1 \cdot \tanh s_2 + 1 = 0$. Compute Green's function for this region D.

10.10. Consider the quadrilateral in Fig. 10.16. Take the parametric equations of the sides as

$$\text{AC: } z = \left(2 + \frac{1}{\sqrt{3}}\right) t + z_A, \quad 0 \le t \le 2,$$

CE: $z = -\dfrac{4}{\sqrt{3}}\,(t-3)^2\,e^{2i\pi/3} + z_E, \quad 2 \le t \le 3,$

EG: $z = -(t-3)^2 + z_E, \quad 3 \le t \le 5,$

GA: $z = -2\,(t-6)\,i + z_A, \quad 5 \le t \le 6,$

and compute Green's function for this quadrilateral.

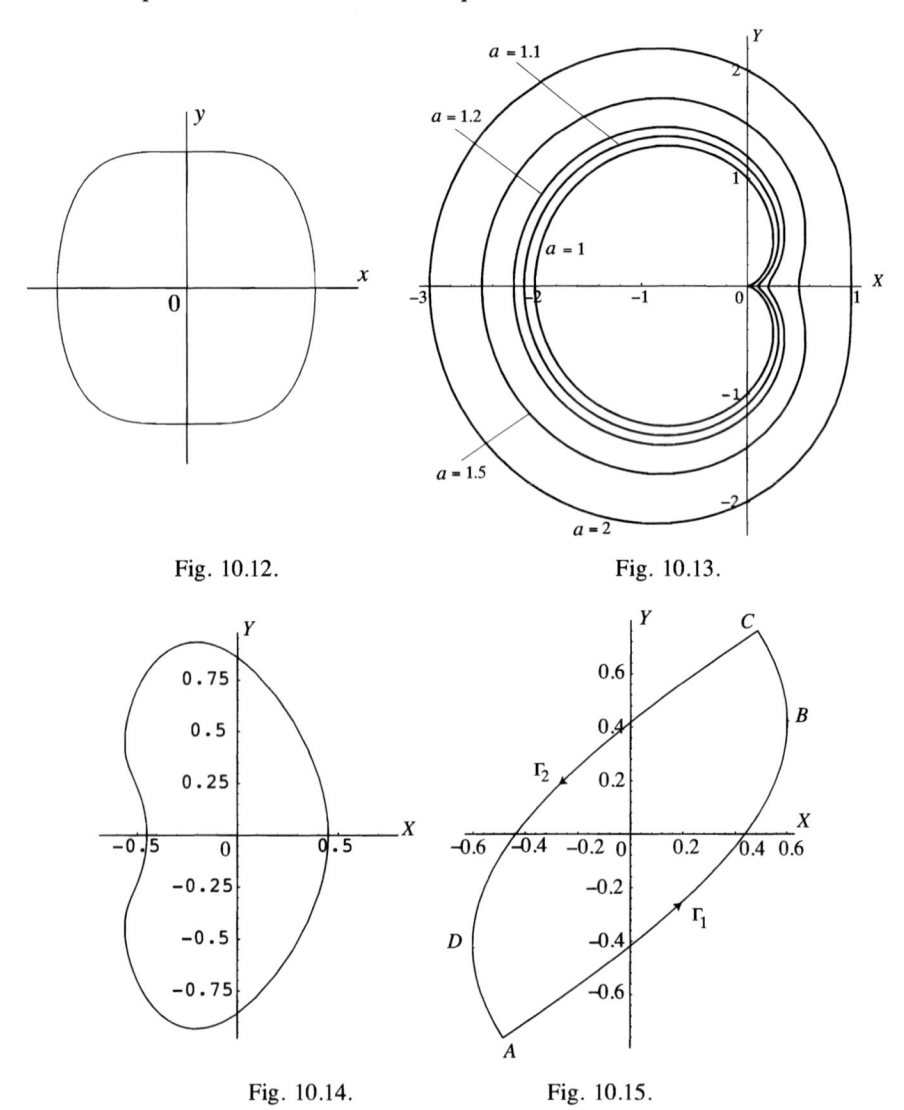

Fig. 10.12.   Fig. 10.13.

Fig. 10.14.   Fig. 10.15.

10.11. Consider the octagonal region in Fig. 10.17. Take the parametric equations of the sides as

$$LA: \ z = 2it^3 + z_1,$$
$$FG: \ z = -2it^3 + z_2, \quad -1 \le t \le 0;$$
$$AB: \ z = 2t^3 + z_1,$$
$$BD: \ z = t^3 z_D + (1-t)^3 z_B, \quad 0 \le t \le 1,$$
$$DF: \ z = t^3 z_F + (1-t)^3 z_D, \qquad GH: \ z = -2t^3 + z_2, \quad 0 \le t \le 1,$$
$$HK: \ z = t^3 z_K + (1-t)^3 z_H, \qquad KL: \ z = t^3 z_L + (1-t)^3 z_K, \quad 0 \le t \le 1.$$

Compute Green's function for this octagonal region.

HINT. This region is not convex; it is starlike with respect to any point on the line segment AG, but since it has fourfold symmetry about the origin, consider either half of the region.

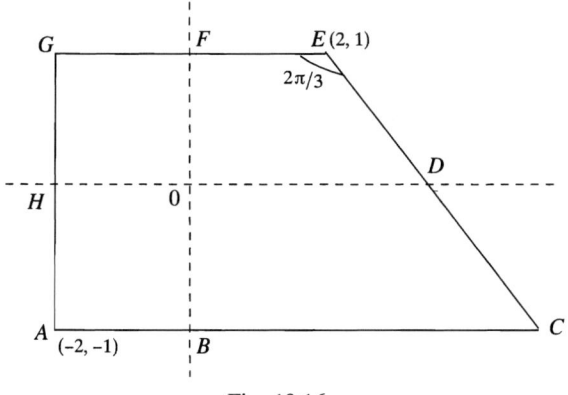

Fig. 10.16.

10.12. Consider the circular sector of radius 1 and angle $3\pi/2$ (Fig. 10.18). This region is not convex; it is starlike with respect to any point on the line segment AC, but it is symmetric about the origin along the line $P_1 P_2$, and each part is convex. Compute Green's function for each part.

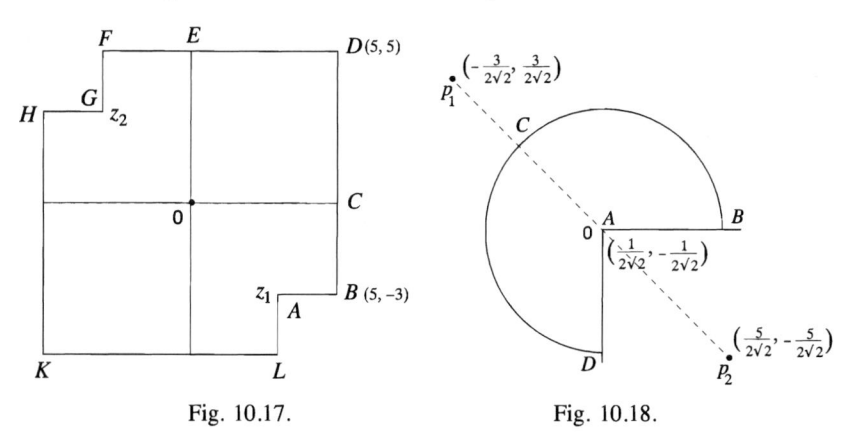

Fig. 10.17.   Fig. 10.18.

A

Adjoint Operators

A system of two first order linear homogeneous differential equations

$$y_1' = a_{11}(x)\, y_1 + a_{12}(x)\, y_2,$$
$$y_2' = a_{21}(x)\, y_1 + a_{22}(x)\, y_2,$$
(A.1)

where $a_{ij}(x)$ are real valued functions of x, can be written in matrix form as

$$\mathbf{y}' = A(x)\,\mathbf{y}, \quad \text{where } A(x) = \begin{bmatrix} a_{11} & a_{12} \\ a_{21} & a_{22} \end{bmatrix}.$$
(A.2)

A system of linear homogeneous differential equations is called *adjoint* to (A.1) if it has the form $\mathbf{y}' = B(x)\,\mathbf{y}$, where $B = -A^T$, and A^T denotes the transpose of A.

A second-order differential equation can always be reduced to a system of first order equations. For example, the general second order differential equation

$$y'' + a_1(x)\, y' + a_2(x)\, y = 0,$$
(A.3)

where a_i, $i = 1, 2$, are real valued functions of x, is reduced to a system of first order equations by setting $y = y_1$ and $y' = y_2$. Thus, (A.3) becomes

$$\begin{aligned} y_1' &= y_2, \\ y_2' &= -a_2 y - a_1 y, \end{aligned} \qquad \text{or } \mathbf{y}' = A(x)\,\mathbf{y},$$
(A.4)

where $A = \begin{bmatrix} 0 & 1 \\ -a_2 & -a_1 \end{bmatrix}$. Since $A^T = \begin{bmatrix} 0 & -a_2 \\ 1 & -a_1 \end{bmatrix}$, the adjoint system to (A.4) is $\mathbf{y}' = -A^T\,\mathbf{y}$, or

$$\begin{aligned} y_1' &= a_2(x)\, y_2, \\ y_2' &= -y_1 + a_1(x)\, y_2. \end{aligned}$$

After differentiating the second equation and substituting $a_2 y_2$ for y_1' from the first equation and writing y for y_2, this system reduces to

$$y'' - [a_1(x)\,y]' + a_2(x)\,y = 0,$$

which is the adjoint equation to (A.3). In the case when the coefficient of y'' in Eq (A.3) is $a_0(x) \neq 0$, i.e., if we have $a_0(x)\,y'' + a_1(x)\,y' + a_2(x)\,y = 0$, then its adjoint equation is

$$[a_0(x)\,y]'' - [a_1(x)\,y]' + a_2(x)\,y = 0.$$

DEFINITION. The operators

$$L \equiv a_0\,\frac{d^2}{dx^2} + a_1\,\frac{d}{dx} + a_2, \tag{A.5}$$

$$M \equiv \frac{d^2}{dx^2}\,(a_0\,\cdot) - \frac{d}{dx}\,(a_1\,\cdot) + a_2, \tag{A.6}$$

where the dot denotes the place holder for the function to which M is applied, are called *adjoint differential operators of second order*.

LAGRANGE'S IDENTITY. If L and M are adjoint differential operators of second order, there there exists a function $F(x, y, y', z, z')$ such that

$$z\,L[y] - y\,M[z] = \frac{d}{dx}\,F(x, y, y', z, z'). \tag{A.7}$$

PROOF. $zL[y] - yM[z] = z\,[a_0 y'' + a_1 y' + a_2] - y\,[(a_0 z)'' + y\,(a_1 z)' + a_2\,z]$
$= \frac{d}{dz}\,[a_0 y' z - y\,(a_0 z)' + y'\,(a_1 z)]$ ∎.

COROLLARY. If L and M are adjoint operators of second order, then

$$\int_{x_1}^{x_2} \left\{ zL[y] - yL[z] \right\}\,dx = \Big[a_0 y' z - y\,(a_0 z)' + y\,(a_1 z) \Big]_{x-1}^{x_2}. \tag{A.8}$$

A differential equation of second order can be written in the (Sturm-Liouville) form

$$L[y] \equiv \frac{d}{dx}\,[p(x)\,y']' + q(x)\,y = 0. \tag{A.9}$$

LEMMA 1. A second order differential equation $L[y] = 0$ of the form (A.9) is *self-adjoint* (i.e., adjoint to itself) iff the operator L has the form

$$L \equiv \frac{d}{dx}\,\left[p(x)\,\frac{d}{dx}\,(\cdot) \right] + q(x), \tag{A.10}$$

where $p(x)$ is differentiable.

PROOF. Since M is the adjoint operator to L, we have, in view of (A.6),

$$M \equiv \frac{d^2}{dx^2}\left[p(x)\,(\cdot)\right] - \frac{d}{dx}\left[p'(x)\,(\cdot)\right] + q(x) = p(x)\frac{d^2}{dx^2}(\cdot) + p'(x)\frac{d}{dx}(\cdot) + q(x),$$

i.e., $L = M$. Conversely, if $L = a_0\frac{d^2}{dx^2} + a_1\frac{d}{dx} + a_2$ and $M = \frac{d^2}{dx^2}\,(a_0\,\cdot) - \frac{d}{dx}\,(a_1\,\cdot) + a_2$ are equal, then $a_0' = a_1 = p'(x)$. $\blacksquare$

LEMMA 2. The Lagrange's identity for self-adjoint differential operators become

$$zL[y] - yL[z] = \frac{d}{dx}\left[p(x)\,(y'z - yz')\right]. \tag{A.11}$$

B

List of Fundamental Solutions

B.1. Linear Ordinary Differential Operator with Constant Coefficients that satisfies the equation

$$Lu^*(x) = \frac{d^n u^*}{dx^n} + a_1 \frac{d^{n-1}u^*}{dx^{n-1}} + \cdots + a_{n-1}\frac{du^*}{dx} + a_n u^* = \delta(x),$$

where $u^* = u^*(x,0)$, has fundamental solution:

$$u^* = H(x)\,w(x), \tag{B.1}$$

where $H(x)$ is the Heaviside function, and $w(x) \in C^n(\mathbb{R})$ is the solution of the homogeneous equation $Lw = 0$ with the initial conditions $w(0) = w'(0) = \cdots = w^{(n-2)}(0) = 0$, $w^{(n-1)}(0) = 1$.

B.2. Fundamental Solutions for the Operators $L_1 = \dfrac{d}{dx} + a$, and $L_2 = \dfrac{d^2}{dx^2} + a^2$ are, respectively, given by

$$u_1^*(x) = H(x)\,e^{-ax}, \tag{B.2}$$

$$u_2^*(x) = H(x)\,\frac{\sin ax}{a}. \tag{B.3}$$

B.3. Elliptic Operator. Fundamental solution for the Laplace operator or the Laplacian ∇^2 in $\mathbb{R}^n$, such that $-\nabla^2 u_n^*(\mathbf{x}) = \delta(\mathbf{x})$, $\mathbf{x} = \{x, \ldots, x_n\}$, is given by

$$u_n^*(\mathbf{x}) = \frac{1}{(n-2)S_n(1)}\,|\mathbf{x}|^{2-n}, \quad n \geq 3, \tag{B.4}$$

where $S_n(1) = \dfrac{2\pi^{n/2}}{\Gamma(n/2)}$ denotes surface area of the unit ball.

Special Cases: For $n = 2$: $u_2^*(\mathbf{x}, \mathbf{x}') = -\dfrac{1}{2\pi} \ln |\mathbf{x} - \mathbf{x}'|$;

        For $n \geq 3$: $u_3^*(\mathbf{x}, \mathbf{x}') = \dfrac{1}{4\pi |\mathbf{x} - \mathbf{x}'|}$.

        For $n = 1$: solve $-\dfrac{d^2 u_1^*}{dx^2} = \delta(x)$, which gives the fundamental

        solution: $u_1^*(x) = -\dfrac{1}{2}|x|$.

B.4. Helmholtz Operator.

Let $u_n^*(\mathbf{x})$ denote the fundamental solution for the Helmholtz operator $-(\nabla^2 + k^2)$ in $\mathbb{R}^n$, where k is complex, such that

$$- \left(\nabla^2 + k^2\right) u_n^*(\mathbf{x}) = \delta(\mathbf{x}). \tag{B.5}$$

For $n \geq 2$ and $k^2 \notin [0, \infty)$, the fundamental solution for (B.5) is given by

$$u_n^*(r) = \frac{i}{4} \left(\frac{k}{2\pi}\right)^{(n/2)-1} H_{(n/2)-1}^{(1)}(kr), \quad r = |\mathbf{x} - \mathbf{x}'|, \quad n \geq 2, \tag{B.6}$$

or, since $H_{(n/2)-1}^{(1)}(kr) = H_{(n/2)-1}^{(1)}(ihr) = \dfrac{2}{i\pi} K_{(n/2)-1}(hr)$, $h = -ik$, where $H_0^{(j)}$, $j = 1, 2$, are the Hankel functions, and $K_{(n/2)-1}$ are the modified Bessel functions, by

$$u_n^*(r) = \frac{1}{2\pi} \left(\frac{h}{2\pi r}\right)^{(n/2)-1} K_{(n/2)-1}(hr), \quad n \geq 2, \tag{B.7}$$

which holds whenever $-h^2 \notin [0, \infty)$, i.e., for all h with $\Re\{h\} > 0$.

For $n = 2$, the fundamental solution is

$$u_2^*(r) = \frac{i}{4} H_0^{(1)}(kr) = \frac{1}{2\pi} K_0(hr). \tag{B.8}$$

For $n = 3$, by using $H_{1/2}^{(1)}(z) = \dfrac{1}{i} \left(\dfrac{2}{\pi}\right)^{1/2} \dfrac{e^{iz}}{z^{1/2}}$, we get

$$u_3^*(r) = \frac{e^{ikr}}{4\pi r} = \frac{e^{-hr}}{4\pi r}. \tag{B.9}$$

For $n = 1$, the fundamental solution is found directly as

$$u_1^*(x) = \frac{1}{2k} \sin k|x|. \tag{B.10}$$

Note that if the Helmholtz equation is taken as $(\nabla^2 + k^2)u = 0$, where $k \neq 0$, real, denotes the wave number, then the fundamental solution in $\mathbb{R}$ is the same as (B.10), but in $\mathbb{R}^2$ and $\mathbb{R}^3$ they are, respectively,

$$u_2^*(r) = -\frac{i}{4} H_0^{(2)}(kr), \quad \text{and} \quad u_3^*(r) = \frac{e^{-ikr}}{4\pi r}. \tag{B.11}$$

Fundamental solution for the operator $\nabla^2 - k^2$ are given by

$$u_n^*(x, x') = \begin{cases} \dfrac{1}{2k} \sinh kr & \text{for } n = 1, \\[2mm] \dfrac{1}{2\pi} K_0(kr) & \text{for } n = 2, \\[2mm] \dfrac{1}{4\pi r} e^{ikr} & \text{for } n = 3, \end{cases} \tag{B.12}$$

where $r = |x - x'|$. Note that $\lim\limits_{x \to \pm\infty} u^*(x, x') = \infty$. These fundamental solutions are used in problems of neutron diffusion and reactor physics.

B.5. Fundamental Solution for the Cauchy-Riemann Operator $\dfrac{\partial}{\partial \bar{z}} = \dfrac{\partial}{\partial x} + i\dfrac{\partial}{\partial y}$, $z = x + iy$, which satisfies $\dfrac{\partial}{\partial \bar{z}} u^*(z) = \dfrac{\partial}{\partial \bar{z}} u^*(x, y) = \delta(x, y)$, is $u^*(z) = \dfrac{1}{\pi z}$.

B.6. Fundamental Solution for the Diffusion Operator $\dfrac{\partial}{\partial t} - a\nabla^2$, also called the heat conduction operator, which satisfies $\dfrac{\partial u^*}{\partial t} - a\nabla^2 u^* = \delta(\mathbf{x}, t)$, where $a > 0$ is a constant determined by specific heat and thermal conductivity of the medium, is

$$u^*(\mathbf{x}, t) = \frac{H(t)}{(4\pi a t)^{n/2}} e^{-|\mathbf{x}|^2/4at}. \tag{B.13}$$

Note that if the fundamental solution for the diffusion equation in $\mathbb{R}$ is taken at initial time t' and the instantaneous point heat source is located at x', then the fundamental solution is written as

$$u^*(\mathbf{x}, \mathbf{x}'; t, t') = \frac{1}{\sqrt{4\pi a(t - t')}} e^{-(\mathbf{x}-\mathbf{x}')^2/4a(t-t')}. \tag{B.14}$$

In this notation, the coordinates $(\mathbf{x}', t')$ express the role of the source point, whereas the coordinates $(\mathbf{x}, t)$ represent the field (or observation) point. In this sense, the value of u^* at the point $\mathbf{x}$ at time t is determined by the evolution of the distribution $\delta(\mathbf{x}')$ at time t'. Thus, the fundamental solution (B.14) can be regarded as the superposition of the evolutions of initial distributions of the δ-functions at points $\mathbf{x}'$ with amplitudes $f(\mathbf{x}')$.

B.7. Schrödinger Operator. The linear diffusion operator in $\mathbb{R}^n \times \mathbb{R}^+$ is sometimes written as

$$\frac{1}{a}\frac{\partial}{\partial t} - \nabla^2 = \frac{1}{a}\frac{\partial}{\partial t} - \sum_{j=1}^{n}\left(\frac{\partial}{\partial x_j}\right)^2.$$

But the fundamental solution is still given by (B.13). If we consider the non-homogeneous case, then the equation

$$\left(\frac{1}{a}\frac{\partial}{\partial t} - \nabla^2\right)u(\mathbf{x}, t) = f(\mathbf{x}, t) \tag{B.15}$$

is known as the Fourier heat equation. There are two interpretation of Eq (B.15). Firstly, if $a > 0$ is a real constant depending on specific heat and thermal conductivity of the medium, then $u(\mathbf{x}, t)$ determines the temperature distribution. The function $f(\mathbf{x}, t)$ on the right side describes local heat production minus absorption. Secondly, $u(\mathbf{x}, t)$ defines a particle density and a is the diffusion coefficient. If a is purely imaginary such that $a = \dfrac{i\hbar}{2m}$, where m is the mass of the quantum particle, and $\hbar = 1.054 \times 10^{-27}$ erg-sec is the Planck's constant, then Eq (B.15) defines the Schrödinger equation

$$i\hbar\frac{\partial u(\mathbf{x}, t)}{\partial t} + \frac{\hbar^2}{2m}\nabla^2 u(\mathbf{x}, t) = f(\mathbf{x}, t). \tag{B.16}$$

If the quantum particle of mass m is an external force field with potential $V(\mathbf{x})$, and $u(\mathbf{x}, t)$ denotes the wave function of the particle such that $|u(\mathbf{x}, t|^2 \Delta u$ is the probability of the particle being in the neighborhood $u(\mathbf{x})$ of the point $\mathbf{x}$ at time t, where Δu denotes the volume of $u(\mathbf{x}, t)$, then $f(\mathbf{x}, t) = V(\mathbf{x})u(\mathbf{x}, t)$, and Eq (B.16) becomes

$$i\hbar\frac{\partial u}{\partial t} + \frac{\hbar^2}{2m}\nabla^2 u = V u. \tag{B.17}$$

In the steady state, the energy E of the particle has a definite value, and the function u has the form

$$u(\mathbf{x}, t) = e^{-iEt/\hbar} u(\mathbf{x}). \tag{B.18}$$

Thus, in view of (B.17), the wave function $u(\mathbf{x})$ satisfies the steady state Schrödinger equation

$$\frac{\hbar^2}{2m}\nabla^2 u + E u = V u \tag{B.19}$$

If $V = 0$, the free particle Schrödinger equation reduces to homogeneous Helmholtz equation. As in the case of the Helmholtz equation, so for the fundamental solution we must require that Sommerfeld's radiation condition

$$V(\mathbf{x}) = O(|\mathbf{x}|), \quad \frac{\partial V(\mathbf{x})}{\partial |\mathbf{x}|} - i k V(\mathbf{x}) = o\left(|\mathbf{x}|^{-1}\right) \tag{B.20}$$

must be satisfied at infinity with $k = \sqrt{2mE}/\hbar$, $E \geq 0$. From (B.13) the fundamental solution for the Schrödinger operator $i\hbar\dfrac{\partial}{\partial t} + \dfrac{\hbar^2}{2m}\dfrac{\partial^2}{\partial x^2}$ in $\mathbb{R} \times \mathbb{R}^+$ is given by

$$u^*(x,t) = -H(t)\frac{1+i}{\hbar\sqrt{2}}\sqrt{\frac{m}{2\pi\hbar t}}\,e^{i\,mx^2/2\hbar t}. \tag{B.21}$$

B.8. Fundamental Solution for the Wave Operator $\Box_c = \dfrac{\partial^2}{\partial t^2} - c^2\nabla^2$ in $\mathbb{R}^n \times \mathbb{R}^+$, which satisfies the equation $\Box_c\,u_n^*(\mathbf{x},t) = \delta(\mathbf{x},t)$, where $\mathbf{x} \in \mathbb{R}^n$ and $t \in (0,t) \subset \mathbb{R}^+$, is given by

In $\mathbb{R}$: $u_1^*(x,t) = \dfrac{1}{2c}H(ct - |x|)$,

In $\mathbb{R}^2$: $u_2^*(\mathbf{x},t) = \dfrac{H(ct - |\mathbf{x}|)}{2\pi c\sqrt{c^2t^2 - |\mathbf{x}|^2}}$.

In $\mathbb{R}^3$: $u_3^*(\mathbf{x},t) = \dfrac{H(t)}{4\pi c^2 t}\delta_{S_{ct}}(\mathbf{x}) = \dfrac{H(t)}{2\pi c}\delta(c^2t^2 - |\mathbf{x}|^2)$.

B.9. Fundamental Solution for the Fokker-Plank Operator which is a modification of the diffusion operator in §B.6, with $a = 1$, i.e., for the Fokker-Planck equation $\dfrac{\partial u}{\partial t} = \dfrac{\partial}{\partial x}\left(\dfrac{\partial}{\partial x} + x\right)u$, is

$$u^*(x,x';t,t') = \frac{H(t-t')}{\sqrt{2\pi(1 - e^{-2(t-t')})}}e^{-(x-x'e^{-(t-t')})^2/2(1-e^{-2(t-t')})}. \tag{B.22}$$

This equation is used in statistical physics to describe the evolution of probability distribution functions.

B.10. Klein-Gordon Operator is defined by $\Box + m_0^2$, where m_0 is the mass of a free relativistic pseudo-scalar particle. Then the wave function $u(\mathbf{x}, ct)$ in $\mathbb{R}^3 \times \mathbb{R}^+$, where c is the speed of light, satisfies the equation $(\Box + m_0^2)\,u = 0$. For $m_0 = 1$, the fundamental solution for the Klein-Gordon operator is a solution of the equation

$$\frac{\partial^2 u^*}{\partial t^2} - \frac{\partial^2 u^*}{\partial x^2} + u^* = \delta(x - x')\delta(t - t'), \tag{B.23}$$

such that $u^*(x - x', 0) = 0 = u_t^*(x - x', 0)$. Fundamental solution for this operator is

$$u^*(x,x';t,t') = \frac{1}{2}J_0[\sqrt{(t-t')^2 - (x-x')^2}]\,H[t - t' - (x - x')]\times$$
$$\times\,H[t - t' + (x - x')]. \tag{B.24}$$

C

List of Spherical Harmonics

C.1. Legendre's Equation. Legendre's equation of order n is

$$\left(1 - x^2\right) \frac{d^2u}{dx^2} - 2x \frac{du}{dx} + n(n+1)\,u = 0. \tag{C.1}$$

If $n = 0, 1, 2, \ldots$, a solution of this equation is the Legendre polynomials $P_n(x)$ given by Rodrigue's formula

$$P_n(x) = \frac{1}{2^n\,n!} \frac{d^n}{dx^n} \left(x^2 - 1\right). \tag{C.2}$$

C.1.1. Special Legendre polynomials are:

$P_0(x) = 1,$ $\qquad\qquad P_4(x) = \frac{1}{8}\left(35x^4 - 30x^2 + 3\right),$

$P_1(x) = x,$ $\qquad\qquad P_5(x) = \frac{1}{8}\left(63x^5 - 70x^3 + 15x\right),$

$P_2(x) = \frac{1}{2}\left(3x^2 - 1\right),$ $\qquad P_6(x) = \frac{1}{16}\left(231x^6 - 315x^4 + 105x^2 - 5\right),$

$P_3(x) = \frac{1}{2}\left(5x^3 - 3x^2\right),$ $\qquad P_7(x) = \frac{1}{16}\left(429x^7 - 693x^5 + 315x^3 - 35x\right).$

C.1.2. If we set $x = \cos\theta$, then some of the above Legendre polynomials become

$P_0(\cos\theta) = 1,$

$P_1(\cos\theta) = \cos\theta,$

$P_2(\cos\theta) = \frac{1}{4}\left(1 + 3\cos 2\theta\right),$

$P_3(\cos\theta) = \frac{1}{8}\left(3\cos\theta + 5\cos 3\theta\right),$

$P_4(\cos\theta) = \frac{1}{64}\left(9 + 20\cos 2theta + 35\cos 4\theta\right),$

$P_5(\cos\theta) = \frac{1}{128}\left(30\cos\theta + 35\cos 3\theta + 63\cos 5\theta\right).$

C.1.3. The recurrence formulas for Legendre polynomials are:

$$(n+1)P_{n+1}(x) - (2n+1)xP_n(x) + nP_{n-1}(x) = 0,$$
$$P'_{n+1}(x) - xP'_n(x) = (n+1)P_n(x),$$
$$xP'_n(x) - P'_{n-1}(x) = nP_n(x),$$
$$P'_{n+1}(x) - P'_{n-1}(x) = (2n+1)P_n(x),$$
$$(x^2 - 1) L'_n(x) = nxP_n(x) - nP_{n-1}(x),$$

C.1.4. Orthogonality. Orthogonality property in the interval $-1 \leq x \leq 1$:

$$\int_{-1}^{1} P_m(x)\,P_n(x)\,dx = \begin{cases} 0, & m \neq n, \\ \dfrac{2}{2n+1}, & m = n. \end{cases}$$

C.1.5. Generating Function:

$$\left(1 - 2xt + t^2\right)^{-1/2} = \sum_{n=0}^{\infty} P_n(x)\,t^n, \quad |t| < 1 >$$

C.1.6. Some special results involving Legendre polynomials are:

$$P_n(1) = 1, \quad P_n(-1) = (-1)^n, \quad P_n(-x) = (-1)^n P_n(x),$$

$$P_n(0) = \begin{cases} 0, & n \text{ odd}, \\ (-1)^{n/2} \dfrac{1 \cdot 3 \cdot 5 \cdots (n-1)}{2 \cdot 4 \cdot 6 \cdots n}, & n \text{ even}, \end{cases}$$

$$P_n(x) = \frac{1}{\pi} \int_0^{\pi} \left(x + \sqrt{x^2 - 1}\,\cos\phi\right)^n d\phi,$$

$$\int P_n(x)\,dx = \frac{P_{n+1}(x) - P_{n-1}(x)}{2n+1},$$

$$|P_n(x)| \leq 1,$$

$$P_n(x) = \frac{1}{2^{n+1}\pi i} \oint_C \frac{(z^2 - 1)^n}{(z-x)^{n+1}}\,dz,$$

where C is a simple closed curve with x as an interior point.

There are Legendre functions of the second kind and order n, denoted by $Q_n(x)$, but we will not discuss them since they do not concern us.

C.2. Associated Legendre's Equation. We mentioned in §4.3.3 that Eq (4.15) is the associated Legendre equation. We will rewrite it in the form

$$\left(1 - x^2\right) u'' - 2xy' + \left\{ n(n+1) - \frac{\mu^2}{1-x^2} \right\} = 0. \tag{C.3}$$

Solutions of this equation are called the *associated Legendre functions*. We will restrict our discussion to the case where μ and n are nonnegative integers.

C.2.1. Associated Legendre Functions of the First Kind. These functions are defined by

$$P_m^n(x) = \left(1 - x^2\right)^{m/2} \frac{d^m}{dx^m} P_n(x) = \frac{\left(1 - x^2\right)^{m/2}}{2^n\, n!} \frac{d^{m+n}}{dx^{m+n}} \left(x^2 - 1\right), \quad (C.4)$$

where $P_n(x)$ are Legendre polynomials discussed above. We have

$$P_n^0(x) = P_n(x); \quad P_n^m(x) = 0 \quad \text{if } m > n.$$

C.2.2. Special Associated Legendre Functions of the First Kind:

$$P_1^1(x) = \left(1 - x^2\right)^{1/2}, \qquad\qquad P_3^1(x) = \tfrac{3}{2}\left(5x^2 - 1\right)\left(1 - x^2\right)^{1/2},$$

$$P_2^1(x) = 3x\left(1 - x^2\right)^{1/2}, \qquad\qquad P_3^2(x) = 15x\left(1 - x^2\right),$$

$$P_2^2(x) = 3\left(1 - x^2\right), \qquad\qquad P_3^3(x) = 15\left(1 - x^2\right)^{3/2}.$$

C.2.3. Recurrence Formulas:

$$(n + 1 - m)P_{n+1}^m(x) - (2n + 1)x P_n^m(x) + (n + m)P_{n-1}^m(x) = 0,$$

$$P_n^{m+2}(x) - \frac{2(m + 1)x}{\left(1 - x^2\right)^{1/2}} P_n^{m+1}(x) + (n - m)(n + m + 1)P_n^m(x) = 0.$$

C.2.4. Orthogonality: In the interval $-1 \le x \le 1$

$$\int_{-1}^{1} P_k^m(x) P_n^m(x) = \begin{cases} 0 & \text{if } k \ne n, \\ \dfrac{2}{2n + 1}\dfrac{(n + m)!}{(n - m)!} & \text{if } k = n. \end{cases}$$

There are associated Legendre functions of the second kind and order n, denoted by $Q_n^m(x)$, but we will not discuss them since they do not concern us.

C.2.5. Generating Function:

$$\frac{(2m)!(1 - x^2)^{m/2} t^m}{2^m\, m!\, (1 - 2xt + t^2)^{m+1/2}} = \sum_{n=m}^{\infty} P_n^m(x)\, t^n, \quad |t| < 1.$$

C.3. Relations with or without Condon-Shortley Phase Factor. Various kinds of relations concerning associated Legendre functions of the first kind with and without the Condon-Shortley phase factor are presented here.

C.3.1. Some Orthonormalized Laplace's Spherical Harmonics with the Condon-Shortley Phase Factor:

$$Y_0^0(\theta, \varphi) = \frac{1}{2}\sqrt{\frac{1}{\pi}},$$

$$Y_1^{-1}(\theta, \varphi) = \frac{1}{2}\sqrt{\frac{3}{2\pi}}\, e^{-i\varphi}\, \sin\theta,$$

$$Y_1^0(\theta, \varphi) = \frac{1}{2}\sqrt{\frac{3}{\pi}}\, \cos\theta,$$

$$Y_1^1(\theta, \varphi) = -\frac{1}{2}\sqrt{\frac{3}{2\pi}}\, e^{i\varphi}\, \sin\theta,$$

$$Y_2^{-2}(\theta, \varphi) = \frac{1}{4}\sqrt{\frac{15}{2\pi}}\, e^{-2i\varphi}\, \sin^2\theta,$$

$$Y_2^{-1}(\theta, \varphi) = \frac{1}{2}\sqrt{\frac{15}{2\pi}}\, e^{-i\varphi}\, \sin\theta\cos\theta,$$

$$Y_2^0(\theta, \varphi) = \frac{1}{4}\sqrt{\frac{5}{\pi}}\, \left(3\cos^2\theta - 1\right),$$

$$Y_2^1(\theta, \varphi) = -\frac{1}{2}\sqrt{\frac{15}{2\pi}}\, e^{i\varphi}\, \sin\theta\cos\theta,$$

$$Y_2^2(\theta, \varphi) = \frac{1}{4}\sqrt{\frac{15}{2\pi}}\, e^{2i\varphi}\, \sin^2\theta,$$

$$Y_3^0(\theta, \varphi) = \frac{1}{4}\sqrt{\frac{7}{\pi}}\, \left(5\cos^3\theta - 3\cos\theta\right).$$

C.3.2. Recurrence Relations of the Associated Legendre Functions P_n^m without Condon-Shortley Phase Factor:

$$(n+1-m)\, P_{n+1}^m(\cos\theta) - (2n+1)\cos\theta\, P_n^m(\cos\theta) + (n+m)P_{n-1}^m(\cos\theta) = 0;$$

$$P_n^{m+2}(\cos\theta) - \frac{2(m+1)\cos\theta}{\sin\theta}\, P_n^{m+1}(\cos\theta) + (n-m)(n+m+1)P_n^m(\cos\theta) = 0.$$

C.3.3. Orthogonality: In the interval $-1 \le x \le 1$

$$\int_{-1}^1 P_n^m(x)\, P_k^m(x)\, dx = 0 \text{ if } n \ne k;$$

$$\int_{-1}^1 [P_n^m(x)]^2\, dx = \frac{2}{2n+1}\frac{(n+m)!}{(n-m)!}, \text{ where } x = \cos\theta.$$

C.3.4. Without Condon-Shortley Phase Factor:

RECURRENCE RELATIONS OF ASSOCIATED LEGENDRE FUNCTIONS:

1. $P_n^0(\sin\theta) = 1$;

2. $P_n^m(\sin\theta) = (2n-1)\cos\theta\, P_{n-1}^{m-1}(\sin\theta)$ for all $n \ge 1$;

3. $P_n^{n-1}(\sin\theta) = (2n-1)\sin\theta\, P_{n-1}^{n-1}(\sin\theta)$ for all $n \ge 1$;

4. $P_n^m(\sin\theta) = \dfrac{1}{n-m}\sin\theta\, P_{n-1}^m(\sin\theta) - (n+m-1)P_{n-2}^m(\sin\theta)$ for all $m \ge 0$ and $m \le n-2$ and $n \ge 2$; where $|\sin\theta| \le 1$, $-\frac{\pi}{2} \le \theta \le \frac{\pi}{2}$, and θ is the elevation angle.

C.3.5. Calculated Associated Legendre Functions:

1. $P_0^0(\sin\theta) = 1$;

2. $P_1^0(\sin\theta) = \sin\theta$;

3. $P_1^1(\sin\theta) = \cos\theta$;

4. $P_2^0(\sin\theta) = \frac{1}{2}\left(3\sin^2\theta - 1\right)$;

5. $P_2^1(\sin\theta) = 3\cos\theta\sin\theta$;

6. $P_2^2(\sin\theta) = 3\cos^2\theta$;

7. $P_3^0(\sin\theta) = \frac{1}{2}\sin\theta\left(5\sin^2\theta - 3\right)$;

8. $P_3^1(\sin\theta) = \frac{3}{2}\cos\theta\left(5\sin^2\theta - 1\right)$;

7. $P_3^2(\sin\theta) = 15\cos^2\theta\sin\theta$;

8. $P_3^3(\sin\theta) = 15\sin^3\theta$.

C.4. Laguerre's Equation. This equation is: $xu'' + (1-x)u' + nu = 0$. If $n = 0, 1, 2, \ldots$, then a solution of this equation is the Laguerre polynomial $L_n(x)$ defined by Rodrigue's formula

$$L_n(x) = e^x \frac{d^n}{dx^n}\left(x^n e^{-x}\right). \tag{C.5}$$

C.4.1. Special Laguerre Polynomials:

$L_0(x) = 1$, $L_3(x) = -x^3 + 9x^2 - 18x + 6$,

$L_1(x) = -x + 1$, $L_5(x) = x^4 - 16x^3 + 72x^2 - 96x + 24$,

$L_2(x) = x^2 - 4x + 2$, $L_3(x) = -x^5 + 25x^4 - 200x^3 + 600x^2 - 600x + 120$,

$L_6(x) = x^6 - 36x^5 + 450x^4 - 2400x^3 + 5400x^2 - 4320x + 720$.

C.4.2. Recurrence Formulas:

$$L_{n+1}(x) - (2n + 1 - x)L_n(x) + n^2 L_{n-1}(x) = 0,$$
$$L_n'(x) - nL_{n-1}'(x) + nL_{n-1}(x) = 0,$$
$$xL_n'(x) = nL_n(x) - n^2 L_{n-1}(x) = 0.$$

C.4.3. Orthogonality: On the interval $[0, \infty)$

$$\int_0^\infty e^{-x} L_m(x)L_n(x)\, dx = \begin{cases} 0 & \text{if } m \neq n, \\ (n!)^2 & \text{if } m = n. \end{cases}$$

C.4.4. Special Results:

$$L_n(0) = n!, \qquad \int_0^x L_n(t)\, dt = L_n(x) - \frac{L_{n+1}(x)}{n+1},$$

$$L_n(x) = (-1)^n \left\{ x^n - \frac{n^2 x^{n-1}}{1!} + \frac{n^2(n-1)^2 x^2}{2!} - \cdots (-1)^n n! \right\},$$

$$\int_0^\infty x^k e^{-x} L_n(x)\, dx = \begin{cases} 0 & \text{if } k < n, \\ (-1)^n (n!)^2 & \text{if } k = n, \end{cases}$$

$$\sum_{k=0}^n \frac{L_k(x) L_k(y)}{(k!)^2} = \frac{L_n(x) L_{n+1}(y) - L_{n+1}(x) L_n(y)}{(n!)^2 (x - y)},$$

$$\sum_{k=0}^\infty \frac{t^k L_k(x)}{(k!)^2} = e^t J_0\left(2\sqrt{xt}\right),$$

$$L_n(x) = \int_0^\infty t^n e^{x-t} J_0\left(2\sqrt{xt}\right) dt.$$

C.5. Associated Laguerre's Equation: $xu'' + (m+1-x)u' + (n-m)u = 0$.

C.5.1. Associated Laguerre Polynomials: These are the solutions of the associated Laguerre's equation for integers m and n, defined by

$$L_n^m(x) = \frac{d^m}{dx^m} L_n(x),$$

where $L_n(x)$ are the Laguerre polynomials (§C.4), such that

$$L_n^0(x) = L_n(x); \qquad L_n^m(x) = 0 \quad \text{if } m > n.$$

C.5.1.1. Special Results:

$$L_1^1(x) = -1, \qquad L_3^3(x) = -6,$$
$$L_2^1(x) = 2x - 4, \qquad L_4^1(x) = 4x^3 - 48x^2 + 144x - 96,$$
$$L_2^2(x) = 2, \qquad L_4^2(x) = 12x^2 - 96x + 144,$$
$$L_3^1(x) = -3x^2 + 18x - 18, \qquad L_4^3(x) = 24x - 96,$$
$$L_3^2(x) = -6x + 18, \qquad L_4^4(x) = 24.$$
$$\int_0^\infty x^{m+1} e^{-x} \left[L_n^m(x)\right]^2 dx = \frac{(2n - m + 1)(n!)^2}{(n - m)!}.$$

C.5.2. Recurrence Formulas:

$$\frac{n - m + 1}{n + 1} L_{n+1}^m(x) + (x + m - 2n - 1)L_n^m(x) + n^2 L_{n-1}^m(x) = 0,$$

$$\frac{d}{dx}\{L_n^m(x)\} = L_n^{m+1}(x),$$

$$\frac{d}{dx}\{x^m e^{-x} L_n^m(x)\} = (m - n - 1)x^{m-1} e^{-x} L_n^{m-1}(x),$$

$$x\frac{d}{dx}\{L_n^m(x)\} = (x - m)L_n^m(x) + (m - n - 1)L_n^{m-1}(x),$$

$$L_n^m(x) = (-1)^n \frac{n!}{(n-m)!} \left[x^{n-m} - \frac{n(n-m)}{1!} x^{n-m-1} \right.$$

$$\left. + \frac{n(n-1)(n-m)(n-m-1)}{2!} x^{n-m-2} + \cdots \right].$$

C.5.3. Orthogonality: on the interval $[0, \infty)$

$$\int_0^\infty x^m e^{-x} L_n^m(x) L_k^m(x)\, dx = \begin{cases} 0 & \text{if } k \neq n, \\ \dfrac{(n!)^2}{(n-m)!} & \text{if } k = n. \end{cases}$$

D

Tables of Integral Transforms

D.1. Laplace Transform Pairs

	$f(t)$	$\mathcal{L}\{f(t)\} = F(s) = \bar{f}(s)$
1.	1	$\dfrac{1}{s}, \quad s > 0$
2.	e^{at}	$\dfrac{1}{s-a}, \quad s > a$
3.	$\sin at$	$\dfrac{a}{s^2 + a^2}, \quad s > 0$
4.	$\cos at$	$\dfrac{s}{s^2 + a^2}, \quad s > 0$
5.	$\sinh at$	$\dfrac{a}{s^2 - a^2}, \quad s > 0$
6.	$\cosh at$	$\dfrac{s}{s^2 - a^2}, \quad s > 0$
7.	$e^{at} \sin bt$	$\dfrac{b}{(s-a)^2 + b^2}, \quad s > a$
8.	$e^{at} \cos bt$	$\dfrac{s}{(s-a)^2 + b^2}, \quad s > a$
9.	$t^n \quad (n = 1, 2, \ldots)$	$\dfrac{n!}{s^{n+1}}, \quad s > 0$
10.	$t^n e^{at} \quad (n = 1, 2, \ldots)$	$\dfrac{n!}{(s-a)^{n+1}}, \quad s > a$
11.	$H(t - a)$	$\dfrac{e^{-as}}{s}, \quad s > 0$
12.	$H(t - a) f(t - a)$	$e^{-as} F(s) = e^{-as} \bar{f}(s)$
13.	$e^{at} f(t)$	$F(s - a) = \bar{f}(s - a)$
14.	$f(t) \star g(t)$	$F(s) G(s) = \bar{f}(s) \bar{g}(s)$

$f(t)$	$\mathcal{L}\{f(t)\} = F(s) = \bar{f}(s)$
15. $f^{(n)}(t)$	$s^n F(s) - s^{n-1} f(0) - \cdots - f^{n-1}(0)$
16. $f(at)$	$\dfrac{1}{a} F\left(\dfrac{s}{a}\right), \quad a > 0$
17. $\int_0^t f(t)\,dt$	$\dfrac{1}{s} F(s) = \dfrac{1}{s} \bar{f}(s)$
18. $\delta(t-a)$	e^{-as}
19. $t\,f(t)$	$-\dfrac{d}{ds} F(s)$
20. $\mathrm{erf}\left(\dfrac{a}{2\sqrt{t}}\right)$	$\dfrac{1 - e^{-a\sqrt{s}}}{s}$
21. $\mathrm{erfc}\left(\dfrac{a}{2\sqrt{t}}\right)$	$\dfrac{e^{-a\sqrt{s}}}{s}$
22. $f(t)$ with period $= T$ †	$\dfrac{\int_0^T e^{-st} f(t)\,dt}{1 - e^{-Ts}}$
23. $J_0(at)$	$\dfrac{1}{\sqrt{s^2 + a^2}}$
24. $I_0(at)$	$\dfrac{1}{\sqrt{s^2 - a^2}}$
25. $J_n(at)$	$\dfrac{1}{\sqrt{s^2 + a^2}}\left(\dfrac{a}{s + \sqrt{s^2 + a^2}}\right)^n, \; n > -1$
26. $I_n(at)$	$\dfrac{1}{\sqrt{s^2 - a^2}}\left(\dfrac{a}{s + \sqrt{s^2 - a^2}}\right)^n, \; n > -1$
27. $t\,J_1(at)$	$\dfrac{a}{(s^2 + a^2)^{3/2}}, \quad a > 0$
28. $t\,I_1(at)$	$\dfrac{a}{(s^2 - a^2)^{3/2}}, \quad a > 0$
29. † $t^n J_n(at)$	$\dfrac{(2n)!\,a^n}{2^n\,n!\,\left(\sqrt{s^2 + a^2}\right)^{2n+1}}, \; n > -1/2$
30. † $t^n I_n(at)$	$\dfrac{2^n}{\sqrt{\pi}}\dfrac{\Gamma(n + 1/2)\,a^n}{\left(\sqrt{s^2 - a^2}\right)^{2n+1}}, \; n > -1/2$
31. $t\,J_0(at)$	$\dfrac{s}{(s^2 + a^2)^{3/2}}, \quad a > 0$
32. $t\,I_0(at)$	$\dfrac{s}{(s^2 - a^2)^{3/2}}, \quad a > 0$

† $f(t)$ is continuous in $[0, T]$ and periodic with period T, $T > 0$.

$f(t)$	$\mathcal{L}\{f(t)\} = F(s) = \bar{f}(s)$
33. $\dfrac{(t-a)^{\mu-1}}{\Gamma(\mu)} H(t-a)$	$\dfrac{e^{-as}}{s^\mu}, \quad \mu > 0$
34. $\left(\dfrac{t}{a}\right)^{(\mu-1)/2} J_{\mu-1}\left(2\sqrt{at}\right)$	$\dfrac{e^{-a/s}}{s^\mu}, \quad \mu > 0$
35. $\left(\dfrac{t}{a}\right)^{(\mu-1)/2} I_{\mu-1}\left(2\sqrt{at}\right)$	$\dfrac{e^{a/s}}{s^\mu}, \quad \mu > 0$
36. $\dfrac{\cos 2\sqrt{at}}{\sqrt{\pi t}}$	$\dfrac{e^{-a/s}}{\sqrt{s}}, \quad a > 0$
37. $\dfrac{\sin 2\sqrt{at}}{\sqrt{\pi a}}$	$\dfrac{e^{-a/s}}{s^{3/2}}, \quad a > 0$
38. $\dfrac{e^{-a^2/4t}}{\sqrt{\pi t}}$	$\dfrac{e^{-a\sqrt{s}}}{\sqrt{s}}, \quad a > 0$
39. $\dfrac{ae^{-a^2/4t}}{2\sqrt{\pi t^3}}$	$e^{-a\sqrt{s}}, \quad a > 0$
40. $\dfrac{e^{-bt} - e^{-at}}{t}$	$\ln\left(\dfrac{s+a}{s+b}\right), \quad a,b > 0$
41. $\delta(t)$	1
42. $\delta(t-a)$	$e^{-as}, \quad a > 0$

† Note that $\Re\{s\} > \left|\Im\{a\}\right|$ (in 29), and $\Re\{s\} > \left|\Re\{a\}\right|$ (in 30).

D.2. Fourier Cosine Transform Pairs

	$f(x)$	$\mathcal{F}_c\{f(x)\} = \tilde{f}_c(\alpha)$
1.	$f(ax)$	$\dfrac{1}{a}\tilde{f}_c\left(\dfrac{\alpha}{a}\right)$
2.	e^{-ax}	$\sqrt{\dfrac{2}{\pi}}\dfrac{a}{\alpha^2 + a^2}, \quad a > 0$
3.	$x^{-1/2}$	$\dfrac{1}{\sqrt{\alpha}}$
4.	e^{-ax^2}	$\dfrac{1}{\sqrt{2a}}e^{-\alpha^2/4a}, \quad a > 0$
5.	$\dfrac{a}{x^2 + a^2}$	$\sqrt{\dfrac{\pi}{2}}e^{-a\alpha}, \quad a > 0$
6.	$x^2 f(x)$	$-\tilde{f}_c''(\alpha)$
7.	$\dfrac{\sin ax}{x}$	$\sqrt{\dfrac{\pi}{2}}H(a - \alpha)$
8.	$f''(x)$	$-\alpha^2 \tilde{f}_c(\alpha) - \sqrt{\dfrac{2}{\pi}}f'(0)$
9.	$\delta(x)$	$\sqrt{\dfrac{2}{\pi}}$
10.	$H(a - x)$	$\sqrt{\dfrac{2}{\pi}}\dfrac{\sin a\alpha}{\alpha}$
11.	$\dfrac{a + \sqrt{x^2 + a^2}}{a^2 + x^2}$	$\dfrac{e^{-a\alpha}}{\sqrt{\alpha}}$
12.	$\begin{cases} (a^2 - x^2)^{1/2}, & 0 < x < a \\ 0, & x > a \end{cases}$	$\sqrt{\dfrac{\pi}{2}}\dfrac{aJ_1(a\alpha)}{\alpha}$
13.	$\begin{cases} (x^2 - a^2)^{-1/2}, & a < x < \infty \\ 0, & 0 < x < a \end{cases}$	$-\sqrt{\dfrac{\pi}{2}}Y_0(a\alpha)$
14.	$\sin\left(a^2x^2\right)$	$\dfrac{\pi}{4a}\left(\cos(\alpha^2/4a^2) - \sin(\alpha^2/4a^2)\right)$
15.	$\cos\left(a^2x^2\right)$	$\dfrac{\pi}{4a}\left(\cos(\alpha^2/4a^2) + \sin(\alpha^2/4a^2)\right)$

D.3. Fourier Sine Transform Pairs

	$f(x)$	$\mathcal{F}_s\{f(x)\} = \tilde{f}_s(\alpha)$
1.	$f(ax)$	$\dfrac{1}{a}\tilde{f}_s\left(\dfrac{\alpha}{a}\right)$
2.	e^{-ax}	$\sqrt{\dfrac{2}{\pi}}\dfrac{\alpha}{a^2+\alpha^2}$
3.	$x^{-1/2}$	$\dfrac{1}{\sqrt{\alpha}}$
4.	x^{-1}	$\sqrt{\dfrac{\pi}{2}}$
5.	$\dfrac{x}{x^2+a^2}$	$\sqrt{\dfrac{\pi}{2}}\,e^{-a\alpha}$
6.	$\arctan(a/x)$	$\sqrt{\dfrac{\pi}{2}}\dfrac{1-e^{-a\alpha}}{\alpha}$
7.	$x^2 f(x)$	$-\tilde{f}_s''(\alpha)$
8.	$\operatorname{erfc}\dfrac{x}{2\sqrt{a}},\quad a>0$	$\sqrt{\dfrac{2}{\pi}}\dfrac{1-e^{-a\alpha^2}}{\alpha}$
9.	$f''(x)$	$-\alpha^2\,\tilde{f}_s(\alpha) + \sqrt{\dfrac{2}{\pi}}\,\alpha f(0)$
10.	$H(a-x)$	$\sqrt{\dfrac{2}{\pi}}\dfrac{1-\cos a\alpha}{\alpha}$
11.	$x\,e^{-a^2 x^2}$	$\dfrac{\pi\alpha}{4\sqrt{2}\,a^3}\,e^{-\alpha^2/4a^2}$
12. †	$\dfrac{1}{\sqrt{x}}\,e^{-a/x}$	$\dfrac{1}{\sqrt{\alpha}}\,e^{-\sqrt{2a\alpha}}(\cos\sqrt{2a\alpha}+\sin\sqrt{2a\alpha}),$
13.	$\dfrac{\sin bx}{x^2+a^2}$	$\begin{cases}\sqrt{\dfrac{\pi}{2}}\dfrac{1}{a}\,e^{-ab}\sinh(a\alpha),\quad 0<\alpha<b\\[2ex]\sqrt{\dfrac{\pi}{2}}\dfrac{1}{a}\,e^{-a\alpha}\sinh(ab),\quad b<\alpha<\infty\end{cases}$

† $|\arg a| < \pi/2$.

D.4. Complex Fourier Transform Pairs

	$f(x)$	$\mathcal{F}\{f(x)\} = \tilde{f}(\alpha)$
1.	$f^{(n)}(x)$	$(-i\alpha)^n \, \tilde{f}(\alpha)$
2.	$f(ax), \quad a > 0$	$\dfrac{1}{a} \tilde{f}\left(\dfrac{\alpha}{a}\right)$
3.	$f(x - a)$	$e^{i a \alpha} \, \tilde{f}(\alpha)$
4.	$\delta(x - a)$	$\dfrac{1}{\sqrt{2\pi}} e^{-i a\alpha}$
5.	$e^{-a\lvert x\rvert}$	$\sqrt{\dfrac{2}{\pi}} \dfrac{a}{a^2 + \alpha^2}, \quad a > 0$
6.	$e^{-a^2 x^2}$	$\dfrac{1}{a\sqrt{2}} e^{-\alpha^2/4a^2}$
7.	$\begin{cases} 1, & \lvert x\rvert < a \\ 0, & \lvert x\rvert > a \end{cases}$	$\sqrt{\dfrac{2}{\pi}} \dfrac{\sin a\alpha}{\alpha}$
8.	$\dfrac{1}{\lvert x\rvert}$	$\dfrac{1}{\lvert \alpha\rvert}$
9.	$f(x) \star g(x)$	$\sqrt{2\pi} \, \tilde{f}(\alpha) \, \tilde{g}(\alpha)$
10.	$H(x + a) - H(x - a)$	$\sqrt{\dfrac{2}{\pi}} \dfrac{\sin a\alpha}{\alpha}$
11.	$x e^{-a\lvert x\rvert}, \quad a > 0$	$\sqrt{\dfrac{2}{\pi}} \dfrac{2i a\alpha}{(\alpha^2 + a^2)^2}$
12.	$\dfrac{a}{x^2 + a^2}$	$\sqrt{\dfrac{\pi}{2}} e^{-a\lvert\alpha\rvert}$
13.	$\dfrac{ax}{(x^2 + a^2)^2}$	$\dfrac{i}{2} \sqrt{\dfrac{\pi}{2}} \, \alpha \, e^{-a\lvert\alpha\rvert}$
14.	$\cos ax$	$\sqrt{\dfrac{\pi}{2}} \left[\delta(\alpha + a) + \delta(\alpha - a)\right]$
15.	$\sin ax$	$i\sqrt{\dfrac{\pi}{2}} \left[\delta(\alpha + a) - \delta(\alpha - a)\right]$
16.	$\begin{cases} \cos ax, & \lvert x\rvert < \pi/2a \\ 0, & \lvert x\rvert > \pi/2a \end{cases}$	$\sqrt{\dfrac{2}{\pi}} \dfrac{a}{a^2 - \alpha^2} \cos(\pi\alpha/2a)$

17.
$$\begin{cases} \dfrac{1}{\sqrt{a^2 - x^2}}, & |x| < a \\ 0, & |x| > a \end{cases} \qquad \sqrt{\dfrac{\pi}{2}}\, J_0(a\alpha)$$

18.
$$\begin{cases} 1 - |x|, & |x| < 1 \\ 0, & |x| > 1 \end{cases} \qquad 2\sqrt{\dfrac{2}{\pi}}\left(\dfrac{\sin(\alpha/2)}{\alpha}\right)^2$$

D.5. Finite Sine Transform Pairs

The tables are for the interval $[0, \pi]$. If the interval is $[a, b]$, then it can be transformed into $[0, \pi]$ by

$$y = \frac{\pi\,(x - a)}{b - a}. \tag{D.1}$$

	$f(x)$	$\tilde{f}_s(n)$
1.	$\sin mx, \quad m = 1, 2, \ldots$	$\begin{cases} \pi/2, & n = m \\ 0, & n \neq m \end{cases}$
2.	$\displaystyle\sum_{n=1}^{\infty} a_n \sin nx$	$\pi\, a_n/2$
3.	$\pi - x$	π/n
4.	x	$\dfrac{\pi}{n}(-1)^{n+1}$
5.	1	$\dfrac{1}{n}\left[1 - (-1)^n\right]$
6.	$\begin{cases} -x, & x \leq a \\ \pi - x, & x > a \end{cases}$	$\dfrac{\pi}{n}\cos na, \quad 0 < a < \pi$
7.	$\begin{cases} x(\pi - a), & x \leq a \\ a(\pi - x), & x > a \end{cases}$	$\dfrac{\pi}{n^2}\sin na, \quad 0 < a < \pi$
8.	e^{ax}	$\dfrac{n}{a^2 + n^2}\left[1 - (-1)^n e^{a\pi}\right]$
9.	$\dfrac{\sinh a(\pi - x)}{\sinh a\pi}$	$\dfrac{n}{a^2 + n^2}$
10.	$f''(x)$	$-n^2 \tilde{f}_s(n) + n\left[f(0) - (-1)^n f(\pi)\right]$

D.6. Finite Cosine Transform Pairs

The tables are for the interval $[0, \pi]$. If the interval is $[a, b]$, then it can be transformed into $[0, \pi]$ by formula (B.36).

	$f(x)$	$\tilde{f}_c(n)$
1.	$\cos mx, \quad m = 1, 2, \ldots$	$\begin{cases} \pi/2, & n = m \\ 0, & n \neq m \end{cases}$
2.	$\dfrac{a_0}{2} + \displaystyle\sum_{n=1}^{\infty} a_n \cos nx$	$\pi\, a_n/2$
3.	$f(\pi - x)$	$(-1)^n\, \tilde{f}_c(n)$
4.	1	$\begin{cases} \pi, & n = 0 \\ 0, & n = 1, 2, \ldots \end{cases}$
5.	x	$\begin{cases} \pi^2/2, & n = 0 \\ \dfrac{1}{n^2}\left[(-1)^n - 1\right], & n = 1, 2, \ldots \end{cases}$
6.	x^2	$\begin{cases} \pi^3/3, & n = 0 \\ \dfrac{2\pi}{n^2}(-1)^n, & n = 1, 2, \ldots \end{cases}$
7.	$\begin{cases} 1, & 0 < x < a \\ -1, & a < x < \pi \end{cases}$	$\begin{cases} 2a - \pi, & n = 0 \\ \dfrac{2}{n}\sin na, & n = 1, 2, \ldots \end{cases}$
8.	$\dfrac{e^{ax}}{a}$	$\dfrac{(-1)^n\, e^{a\pi} - 1}{a^2 + n^2}$
9.	$\dfrac{\cosh\left(c(\pi - x)\right)}{\sinh(\pi c)}$	$\dfrac{c}{c^2 + n^2}$
10.	$f''(x)$	$-n^2 \tilde{f}_c(n) + (-1)^n\, f'(\pi) - f'(0)$

D.7. Zero-Order Hankel Transform Pairs

	$f(x)$	$\hat{f}(\sigma) \equiv \mathcal{H}_0(\sigma) = \int_0^\infty x\, J_0(\sigma x)\, f(x)\, dx$
1.	$\dfrac{\delta(x)}{x}$	1
2.	$H(a-x)$ †	$\dfrac{a}{\sigma} J_1(a\sigma)$
3.	$(a^2 - x^2)\, H(a-x)$	$\dfrac{4a}{\sigma^3} J_1(a\sigma) - \dfrac{2a^2}{\sigma^2} J_0(a\sigma)$
4.	e^{-ax}	$a\left(a^2 + \sigma^2\right)^{-3/2}$
5.	$\dfrac{1}{x} e^{-ax}$	$\left(a^2 + \sigma^2\right)^{-1/2}$
6.	$a\left(a^2 + \sigma^2\right)^{-3/2}$	$e^{-a\sigma}$
7.	$\dfrac{1}{x} \cos(ax)$	$\left(\sigma^2 - a^2\right)^{-1/2} H(\sigma - a)$
8.	$\dfrac{1}{x} \sin(ax)$	$\left(a^2 - \sigma^2\right)^{-1/2} H(a - \sigma)$
9.	$\dfrac{1}{x^2}(1 - \cos ax)$	$\cosh^{-1}\left(\dfrac{a}{\sigma}\right) H(a - \sigma)$
10.	$\dfrac{1}{x} J_1(ax)$	$\dfrac{1}{a} H(a - \sigma),\ a > 0$
11.	$Y_0(ax)$	$\dfrac{2}{\pi}\left(a^2 - \sigma^2\right)^{-1}$
12.	$K_0(ax)$	$\left(a^2 + \sigma^2\right)^{-1}$
13.	$(x^2 + a^2)^{-1/2} e^{-a\sqrt{x^2 + b^2}}$	$\left(a^2 + \sigma^2\right)^{-1/2} e^{-b\sqrt{a^2 + \sigma^2}}$
14.	$(x^2 + a^2)^{-1/2}$	$\dfrac{1}{\sigma} e^{-a\sigma}$
15.	$\dfrac{\sin x}{x^2}$	$\begin{cases} \dfrac{\pi}{2}, & \sigma < 1, \\[2mm] \arcsin\left(\dfrac{1}{\sigma}\right), & \sigma > 1. \end{cases}$

† H denotes the Heaviside function.

E

Fractional Derivatives

In the case of the Cauchy problem, the space-time fractional diffusion equation is derived from the standard diffusion equation by replacing the second-order space derivative with a Riemann-Liouville derivative of order β, $0 < \beta \leq 2$, and the first-order time derivative with Caputo derivative of order α, $0 < \alpha \leq 1$. These derivatives are defined as follows:

Riemann-Liouville derivative of order β, $0 < \beta \leq 2$, for a differentiable function f on the interval $[a, b]$:

$$D^\beta \left[f(x) \right] = \frac{1}{\Gamma(1 - \beta)} \frac{d}{dx} \int_a^x \frac{f(s)}{(x - s)^\beta} \, ds, \quad a < x < b, \ 0 < \beta \leq 2.$$

Caputo derivative of order α, $0 < \alpha \leq 1$, of a differentiable function f on the interval $[0, T]$:

$$D^\alpha \left[f(t) \right] = \begin{cases} \dfrac{1}{\Gamma(1 - \alpha)} \displaystyle\int_0^t \dfrac{f'(s)}{(t - s)^\alpha} \, ds, & 0 \leq t \leq T, \ 0 < \alpha < 1, \\ \dfrac{df(t)}{dt}, & 0 \leq t \leq T, \ \alpha = 1. \end{cases}$$

The semiderivative corresponds to $\alpha = 1/2$. The Green's function (or the fundamental solution) for the Cauchy problem can also be interpreted as a spatial probability density function evolving in time. In many cases it leads to ill-posed problems which are solved numerically by a suitable regularization method. Some formulas for fractional derivatives are given below.

$$D^\alpha \left[f(t) \right] = D^m \left[D^{-(m-\alpha)} f(t) \right], \quad m \geq \lceil \alpha \rceil \text{ is an integer;}$$

$$D^\alpha \left[t^\lambda \right] = \frac{\Gamma(\lambda + 1)}{\Gamma(\lambda - \alpha + 1)} t^{\lambda - \alpha}, \text{ for } \lambda > -1, \ \alpha > 0;$$

$$D^\alpha \left[c \right] = \frac{c t^{-\alpha}}{\Gamma(1 - \alpha)}, \quad \text{where } c \text{ is a constant;}$$

$$D^\alpha \left[e^{kx} \right] = k^\alpha e^{kx};$$

$$D^\alpha \left[\sin x \right] = \sin \left(x + \frac{\pi \alpha}{2} \right); \quad D^\alpha \left[\cos x \right] = \cos \left(x + \frac{\pi \alpha}{2} \right);$$

$$D^\alpha \left[\sin(kx)\right] = k^\alpha \sin\left(kx + \frac{\pi\alpha}{2}\right); \quad D^\alpha \left[\cos(kx)\right] = k^\alpha \cos\left(kx + \frac{\pi\alpha}{2}\right);$$

$$D^\alpha \left[e^{i\,x}\right] = i^\alpha\, e^{i\,\alpha} = e^{i\,(x+i\,\pi/2)\alpha)} = \sin\left(x + \frac{\pi\alpha}{2}\right) + i\,\cos\left(x + \frac{\pi\alpha}{2}\right);$$

$$D^\alpha \left[x^p\right] = \frac{p!}{(p-n)!}\, x^{p-n} = \frac{\Gamma(p+1)}{\Gamma(p-\alpha+1)}\, x^{p-\alpha};$$

$$D^\alpha \left[\sum_{n=0}^{\infty} a_n x^n\right] = \sum_{n=0}^{\infty} a_n \frac{\Gamma(n+1)}{\Gamma(n-\alpha+1)}\, x^{n-\alpha}, \text{ by differentiating}$$

term-by-term;

$$D^\alpha \left[e^x\right] = \sum_{n=0}^{\infty} \frac{x^{n-\alpha}}{\Gamma(n-\alpha+1)}, \text{ using } e^x = \sum_{n=0}^{\infty} \frac{x^n}{n!};$$

$$D^\alpha \left[\sum_{n=-\infty}^{\infty} c_n\, e^{i\,nx}\right] = \sum_{n=-\infty}^{\infty} c_n (i\,n)^\alpha\, e^{i\,nx}, \text{ by fractionally differentiating}$$

term-by-term;

$$D^{-n} \left[f(x)\right] = \frac{1}{(n-1)!} \int_0^x f(s)(x-s)^{n-1}\, ds;$$

$$D^\alpha \left[f(x)\right] = \frac{1}{\Gamma(-\alpha)} \int_0^x \frac{f(s)\, ds}{(x-s)^{\alpha+1}}.$$

For integration from b to x, we have

$$_b D_x^\alpha \left[f(x)\right] = \frac{1}{\Gamma(-\alpha)} \int_b^x \frac{f(s)\, ds}{(x-s)^{\alpha+1}};$$

$$_b D_x^\alpha \left[(x-c)^p\right] = \frac{\Gamma(p+1)}{\Gamma(p-\alpha+1)}\, (x-c)^{p-\alpha};$$

$$_b D_x^{-1} \left[e^{ax}\right] = \int_b^x e^{ax}\, dx = \frac{1}{a}\, e^{ax}, \text{ when } \frac{1}{b}\, e^{ab} = 0, \text{ or when } ab = -\infty;$$

so if a is positive, then $b = -\infty$. This kind of integral with a lower limit of $-\infty$ is known as the Weyl fractional derivative, and $_{-\infty} D_x^\alpha \left[e^{ax}\right] = a^\alpha\, e^{ax}$;

$$_b D_x^{-1} \left[x^p\right] = \int_b^x x^p\, dx = \frac{x^{p+1}}{p+1} - \frac{b^{p+1}}{p+1}; \text{ so if } b = 0, \text{ we get}$$

$$_0 D_x^\alpha \left[x^p\right] = \frac{\Gamma(p+1)}{\Gamma(p-\alpha+1)}\, x^{p-\alpha};$$

$$_b D_x^\alpha \left[(x-c)^p\right] = \frac{\Gamma(p+1)}{\Gamma(p-\alpha+1)}\, (x-c)^{p-\alpha}; \text{ and}$$

$$_b D_x^\alpha \left[f(x)\right] = \frac{1}{\Gamma(-\alpha)} \int_b^x \frac{f(s)\, ds}{(x-s)^{\alpha+1}}.$$

We write $_0 D_x^\alpha$ or $_0 D_t^\alpha$ simply as D^α. Note that $D^{(-\alpha)} D^{(-\beta)} = D^{-(\alpha+\beta)}$ always, but $D^\alpha D^\beta = D^{\alpha+\beta}$ sometimes. The functional derivative of an elementary function is usually a higher transcendental function. For more details on fractional derivatives, see Oldham and Spanier [1974], Miller and Ross [1993], and Debnath [2005].

.

F

Systems of Ordinary Differential Equations

F.1. Systems of First-Order Equations. We will discuss the one-sided Green's function for a system of n linear first-order ordinary differential equations subject to n initial conditions. Consider a system of first-order homogeneous ordinary differential equations

$$u_i' = \sum_{j=1}^{n} a_{ij}(x)u_j, \quad i = 1, \ldots, n, \tag{F.1}$$

where $a_{ij}(x)$ are defined and continuous on an interval $I : [a, b]$. The following result holds:

Theorem F.1. *There exist n solutions*

$$
\begin{array}{ccc}
u_{11}(x) & \cdots & u_{n1}(x) \\
u_{12}(x) & \cdots & u_{n2}(x) \\
\cdots & \cdots & \cdots \\
u_{1n}(x) & \cdots & u_{nn}(x)
\end{array}
$$

of the system (F.1) such that

$$W(x) = \begin{vmatrix} u_{11}(x) & \cdots & u_{n1}(x) \\ u_{12}(x) & \cdots & u_{n2}(x) \\ \cdots & \cdots & \cdots \\ u_{1n}(x) & \cdots & u_{nn}(x) \end{vmatrix} \neq 0 \ \ on \ a \leq x \leq b, \tag{F.2}$$

and $W(x)$ satisfies the differential equation $W' - \left[\sum_{j=1}^{n} a_{ij}(x)\right]W = 0.$

A proof can be found in Murray and Miller [1954:118].

Now, consider the nonhomogeneous system of equations

$$u_i' = \sum_{j=1}^{n} a_{ij}(x)u_j + b_i, \quad i = 1, \ldots, n. \tag{F.3}$$

An extension of the method of variation of parameters will enable us to solve this system and also lead us to the one-sided Green's function matrix. Suppose $a_{ij}(x), b_i(x) \in C(I : [a, b])$, and to solve Eq (F.3) subject to the initial conditions $u_i(x_0) = u_{i,0}$ for

$x_0 \in I$, we further assume that we have functions $\phi_i(x)$ such that $\phi_i(x_0) = 0$ and

$$\phi_i' = \sum_{j=1}^{n} a_{ij}(x)u_j + b_i, \quad i = 1, \ldots, n. \tag{F.4}$$

Thus, if the set $\{u_{jk}(x)\}$ is the primitive of solutions of the homogeneous equations corresponding to Eqs (F.3) with the initial conditions $u_{jk}(x_0) = \delta_{jk}$, then

$$\psi_i = \sum_{k=1}^{n} u_{k,0} u_{ik}(x) + \phi_i, \quad i = 1, \ldots, n,$$

where $u_{k,0} = u_k(x_0)$, is a solution of Eq (F.3) with the prescribed initial conditions. Hence, if we have a solution $\phi_i(x)$ of the nonhomogeneous system (called the *principal integral*) with the initial conditions $\phi_i(x_0) = 0, i = 1, \ldots, n$, then the solution of Eq (F.3) is $\phi_i(x)$ plus a linear combination of the primitives. This solution is unique by Picard's existence and uniqueness theorem.[1]

Theorem F.2. *The solution $\phi_i(x)$ of Eq (F.3), where $a_{ij}(x), b_i(x) \in C(I : [a, b])$, with the initial conditions $\phi_i(x_0) = 0$, $x_0 \in I$, can be expressed in terms of quadratures involving the primitives of the corresponding homogeneous equations.*

PROOF. Let $u_{ij}(x)$ be the primitives of the homogeneous equations (F.1). Let $h_i(x)$ denote the n continuously differentiable functions of x to be determined later. Consider the functions

$$\phi_i(x) = \sum_{m=1}^{n} h_m(x)u_{im}(x), \quad i = 1, \ldots, n, \tag{F.5}$$

and substitute in Eq (F.3):

$$\sum_{m=1}^{n} h_m' u_{im} + \sum_{m=1}^{n} h_m u_{im}' = \sum_{j=1}^{n} a_{ij} \sum_{m=1}^{n} h_m u_{jm} + b_i,$$

or

$$\sum_{m=1}^{n} h_m \left(u_{im}' - \sum_{j=1}^{n} a_{ij} u_{jm} \right) + \sum_{m=1}^{n} h_m' u_{im} = b_i,$$

Since u_{im} are solutions of Eq (F.1), the terms within the parentheses are zero. Now, we determine h_m such that

$$\sum_{m=1}^{n} h_m' u_{im} = b_i. \tag{F.6}$$

---

[1] The theorem states: If $f(u, x)$ is a real-valued function of two real variables u, x, which are defined and continuous on an open domain D of the plane $\mathbb{R}^2$ and satisfy a Lipschitz condition in D, then for every point $(u_0, x_0) \in D$ we can find a $\delta > 0$ and a function $u(x)$ which has a continuous first derivative in the neighborhood $|x - x_0| \leq \delta$ such that $\dfrac{du}{dx} = f(u(x), x)$ and $u(x_0) = u_0$ in this neighborhood, and $u(x)$ is unique.

Since u_{ij} are the primitives and their Wronskian is defined by (F.2), we can regard Eq (F.6) as a system of linear algebraic equations on the unknowns h'_m with determinant $W(x) \neq 0$ on I. Hence, by Cramer's rule

$$
h'_m(x) = \frac{1}{W(x)}
\begin{vmatrix}
u_{11}(x) & u_{21}(x) & \cdots & u_{n1}(x) \\
\cdots & \cdots & \cdots & \cdots \\
u_{1,m-1}(x) & u_{2,m-1}(x) & \cdots & u_{n,m-1}(x) \\
b_1 & b_2 & \cdots & bn \\
u_{1,m+1}(x) & u_{2,m+1}(x) & \cdots & u_{n,m+1}(x) \\
\cdots & \cdots & \cdots & \cdots \\
u_{1n}(x) & u_{2n}(x) & \cdots & u_{nn}(x)
\end{vmatrix}
$$

$$
= \frac{1}{W(x)} \sum_{j=1}^{n} b_i(x) Y_{kj}(x),
$$

where Y_{kj} is the cofactor[2] of u_{jk} in $W(x)$. Thus,

$$
h_k(x) = \sum_{j=1}^{n} \int_{x_0}^{x} \frac{1}{W(x)} b_j(s) Y_{kj}(s)\, ds + C_k,
$$

where C_k are constants of integration. From Eq (F.5)

$$
\phi_i(x) = \sum_{m=1}^{n} h_i(x) \sum_{j=1}^{n} \int_{x_0}^{x} \frac{1}{W(x)} b_j(s) Y_{kj}(s)\, ds + \sum_{m=1}^{n} C_m u_{im}(x)
$$

$$
= \sum_{j=1}^{n} \int_{x_0}^{x} \left[\sum_{m=1}^{n} \frac{u_{jm}(x) Y_{mj}(s)}{W(s)} \right] b_j(s)\, ds + \sum_{m=1}^{n} C_m u_{im}(x).
$$

The constants C_m can be determined by the initial conditions; thus, $\phi_i(x_0) = 0$, $i = 1, \ldots, n$, we get $0 = \phi_i(x_0) = 0 + \sum_{m=1}^{n} C_m u_{im}(x_0), i = 1, \ldots, n$, which is a system of n linear algebraic equations on C_m with determinant $W(x_0) \neq 0$. Hence $C_m = 0$ for $m = 1, \ldots, n$, and the solution becomes

$$
\phi_i(x) = \sum_{j=1}^{n} \int_{x_0}^{x} \left[\sum_{m=1}^{n} \frac{u_{jm}(x) Y_{mj}(s)}{W(s)} \right] b_j(s)\, ds, \quad i = 1, \ldots, n. \tag{F.7}
$$

The functions $g_{ij}(x, s) = \frac{1}{W(s)} \sum_{m=1}^{n} u_{im}(x) Y_{mj}(s)$, are called *one-sided Green's functions*. Note that Eq (F.7) can be written in matrix form as

$$
\phi(x) = \int_{x_0}^{x} \mathbf{g}(x, s) + \mathbf{b}(s), \tag{F.8}
$$

---

[2] Let $A = [a_{ij}]$ be an $n \times n$ matrix. If the ith row and jth column of A are deleted, the remaining $(n-1)$ rows and $(n-1)$ columns form another matrix M_{ij}, and $\det[M]$ is called the *minor* of a_{ij}. The cofactor of $[a_{ij}]$ is defined as $c_{ij} = (-1)^{i+j} \det[M]$, such that $\sum_{i=1}^{n} a_{ij} c_{ik} = 0$ for $j \neq k$, and $\sum_{i=1}^{n} a_{ij} c_{kj} = 0$ for $i \neq k$. These relations can also be written as $\sum_{i=1}^{n} a_{ij} c_{ik} = \det[A]\, \delta_{jk}$ and $\sum_{i=1}^{n} a_{ij} c_{ik} = \det[A]\, \delta_{ik}$.

where

$$\boldsymbol{\phi}(x) = \left\{ \begin{array}{c} \phi_1(x) \\ \phi_2(x) \\ \vdots \\ \phi_n(x) \end{array} \right\}, \quad \mathbf{b}(x) = \left\{ \begin{array}{c} b_1(x) \\ b_2(x) \\ \vdots \\ b_n(x) \end{array} \right\},$$

$$\mathbf{g}(x, s) = \begin{bmatrix} G_{11}(x, s) & \cdots & G_{1n}(x, s) \\ G_{21}(x, s) & \cdots & G_{2n}(x, s) \\ \cdots & \cdots & \cdots \\ G_{n1}(x, s) & \cdots & G_{nn}(x, s) \end{bmatrix}.$$

The matrix $\mathbf{g}$ is the one-sided Green's function, which can also be written as

$$g_{ij}(x, s) = \frac{(-1)^{j-1}}{W(s)} \begin{vmatrix} u_{11}(x) & u_{21}(x) & \cdots & u_{n1}(x) \\ u_{11}(s) & u_{12}(s) & \cdots & u_{1n}(s) \\ \cdots & \cdots & \cdots & \cdots \\ u_{j-1,1}(s) & u_{j-1,2}(s) & \cdots & u_{j-1,n}(s) \\ u_{j+1,1}(s) & u_{j+1,2}(s) & \cdots & u_{j+1,n}(s) \\ \cdots & \cdots & \cdots & \cdots \\ u_{n1}(s) & u_{n2}(s) & \cdots & u_{nn}(s) \end{vmatrix} . \blacksquare \qquad \text{(F.9)}$$

Formula (F.9) is similar to (F.8), although each is useful in its respective situation.

Example F.1. Consider the system $u_1' = u_2$, $u_2' = 4u_1 - 3u_2$, which has the general solution: $u_1 = c_1 e^{-4x} + c_2 e^x$, $u_2 = -4c_1 e^{-4x} + c_2 e^x$. The Wronskian is

$$W(x) = \begin{vmatrix} u_{11} & u_{21} \\ u_{12} & u_{22} \end{vmatrix} = \begin{vmatrix} e^{-4x} & -4e^{-4x} \\ e^x & e^x \end{vmatrix} = 5e^{-3x} \neq 0.$$

Then, since $a_0(x) = 1$, by (F.9) the one-sided Green's function is given by

$$g_{22}(x.s) = \frac{-1}{5\,e^{-3s}} \begin{vmatrix} e^{-4x} & e^x \\ e^{-4s} & s^s \end{vmatrix} = \frac{1}{5} \left[e^x\, e^{-s} - e^{4x}\, e^{4s} \right].$$

Tho check this result, note that the given system of equations reduces to the second order equation $u'' + 3u' - 4u = 0$, where by setting $u = u_1$, $u' = u_2$ we get the given system. Since for this second-order equation $a_0(x) = 1$ and $W(t) = 5e^{-3x}$, the one-sided Green's function $g(x, s)$ obtained by using (3.8) is the same as the above $g_{22}(x, s)$. $\blacksquare$

Note that Green's functions defined by (F.9) convert a system of first-order linear ordinary differential equations together with the initial conditions into Volterra integral equations of the first kind. The classical Green's functions which are developed below for the Sturm-Liouville systems with boundary conditions lead to Fredholm integral equations of the first kind. However, these developments into integral equations are outside the scope of this book; the interested readers should consult literature on integral equations, some of which is available in the Bibliography.

Bibliography

Abramowitz, M. and I. A. Stegun (Eds.). 1965. *Handbook of Mathematical Functions*. New York: Dover.

Akhiezer, N. I. 1990. *Elements of the Theory of Elliptic Functions*. AMS Translation of Mathematical Monographs, Vol. 79. Providence, RI: American Mathematical Society.

Arfken, G. 1985. *Mathematical Methods in Physics*, 3rd edition. Orlando, FL: Academic Press.

Bateman, H. 1944. *Partial Differential Equations*. New York: Dover.

————. 1959. *Partial Differential Equations of Mathematical Physics*. Cambridge: Cambridge University Press.

Borre, K. 2001. *Plane Networks and Their Applications*. Boston: Birkhäuser.

Boyce, W. E. and R. C. DiPrima. 1992. *Elementary Differential Equations*, 5th ed. New York: Wiley.

Brebbia, C. A. and J. Dominguez. 1992 *Boundary Elements*. Southampton: Computational Mechanics Publications, and New York: McGraw-Hill.

Brauer, F. and J. A. Nohel. 1989. *Qualitative Theory of Ordinary Differential Equations: An Introduction*. New York: Dover.

Brychkov, Y. A. 2008. *Handbook of Special Functions: Derivatives, Integrals, Series and Other Formulas*. Boca Raton, FL: CRC Press.

Carrier, G., M. Krook and C. E. Pearson. 1966. *Functions of a Complex Variable: Theory and Technique*. New York: McGraw-Hill.

Churchill, R. V. 1972. *Operational Methods*, 3rd edition. New York: McGraw-Hill.

Coddington, Earl A. 1989. *An Introduction to Ordinary Differential Equations*. New York: Dover.

————. and N. Levinson. 1955. *Theory Ordinary Differential Equations*. New York: McGraw-Hill.

Costabel, M. 2004. Time-dependent problems with the boundary integral equation method. In *Encyclopedia of Computational Mechanics*, eds. E. Stein, R. Borst and T. J. R. Hughes. New York: Wiley.

Courant, R. and D. Hilbert. 1968. *Methods of Mathematical Physics*, Vol. 1, 2. New York: Interscience.

Davies, B. 1978. *Integral Transforms and Their Applications*. New York: Springer-Verlag.

Debnath, L. 2005. *Nonlinear Partial Differential Equations for Scientists and Engineers*, 2nd edition. Boston: Birkhäuser.

Dirac, P. A. M. 1926-27. The physical interpretation of the quantum dynamics. *Proc. Royal Society, A, London* 113: 621–41.

―――― 1947. *The Principles of Quantum Mechanics*. Oxford: Clarendon Press.

Duffy, D. G. 1994. *Transform Methods for Solving Partial Differential Equations*. Boca Raton, FL: CRC Press.

Duru, H. and H. Kleinert. 1979. Solution of the path integral for the H-atom. *Physics Letters B* 84: 185–88.

Epstein, B. 1962. *Partial Differential Equations*. New York: McGraw-Hill.

Erdélyi, A., W. Magnus, F. Oberhettinger and F. G. Tricomi. 1954. *Tables of Integral Transforms*, Vol. 1. New York: McGraw-Hill.

Farlow, S. J. 1982. *Partial Differential Equations for Scientists and Engineers*. New York: Wiley.

Fox, C. 1961. The G and H-functions as symmetrical Fourier kernels. *Trans. Amer. Soc.* 98: 395–429.

Friedlander, F. G. 1982. *An Introduction to the Theory of Distributions*. Cambridge: Cambridge Univ. Press.

Gaier, D. 1964. *Konstruktive Methoden der konformen Abbildung*. Berlin: Springer-Verlag.

Gladshteyn, I. S. and I. M. Ryzhik. 2007. *Tables of Integrals, Series and Products*, eds. Alan Jeffrey and Daniel Zwillinger, 7th edition. New York: Academic Press.

Goldsmith, Paul F. 1998. *Quasioptical Systems*. Piscataway, NJ: IEEE Press.

Griffiths, David, J. 1995. *Introduction to Quantum Mechanics*. Upper Saddle River, NJ: Prentice Hall.

Groetsch, C. W. 1984. *The Theory of Tikhonov Regularization for Fredholm Integral Equations of the First Kind*. Boston: Pitman Advanced Publishing Program.

Hecht, K. T. 2000. *Quantum Mechanics*. New York: Springer-Verlag.

Heinhold, J. and R. Albrecht. 1954. Zur Praxis der konformen abbildung. *Rendiconti Cirulo Math. Palermo* 3: 130–48.

Henrici, P. 1974-1986. *Applied and Computational Complex Analysis*, Vols. 1–3. New York: John Wiley.

Kanwal, R. 1983. *Generalized Functions: Theory and Technique*. New York: Academic Press.

Kantorovich, L. V. and V. I. Krylov. 1936. *Methods for the Approximate Solution of Partial Differential Equations* (Russian). Leningrad-Moscow: Gostekhizdat.

―――― and V. I. Krylov. 1958. *Approximate Methods for Higher Analysis*. New York: Interscience.

Kemppainen, J. and K. Ruotsalainen. 2008. Boundary integral solution of the time-dependent fractional diffusion equation. In *Integral Methods in Science and Engineering*, ed. C. Constanda and S. Potapenko, 141-48. Boston: Birhäuser.

Koebe, P. 1915. Abhandlungen zur Theorie der konformen Abbildung, I. *Journal reine und angew. Math.* 145: 177–223.

Korn, G. A. 1961. *Mathematical Handbook for Scientists and Engineers: Definitions, Theorems, and Formulas.* New York: McGraw-Hill.

Kulshrestha, P. K. 1973. Generalized convexity in conformal mappings. *J. Math. Analysis and Appls.* 43:441-49.

————. 1974. Coefficient problem for alpha-convex univelaent functions. *Archive Ratnl. Mech. Anal.* 54:204-11.

————. 1976. Coefficient problem for a class of Mocanu-Bazilevic functions. *Annales Polon. Math.* 31: 291-299.

———— and P. Puri. 1969. An exact solution of hydromagnetic rotating flow. In *Developments in Mechanics* 5: 265-71. Proc. 11th Midwestern Conf. Ames: Iowa State Univ. Press.

———— and P. Puri. 1983. Wave structure in oscillatory Couette flow of a dusty gas. *Acta Mechanica* 46: 127–35.

Kythe, P. K. 1995. *An Introduction to Boundary Element Methods.* Boca Raton, FL: CRC Press.

————. 1996. *Fundamental Solutions for Differential Operators and Applications.* Boston: Birkhäuser.

————, P. Puri and M. R. Schäferkotter. 1997. *Partial Differential Equations and Mathematica* Boca Raton, FL: CRC Press, ISBN 1-58488-314-6.

————. 1998. *Computational Conformal Mapping.* Boston: Birkhäuser.

———— and P. Puri. 2002. *Computational Methods for Linear Integral Equations.* Boston: Birkhäuser.

————, P. Puri and M. R. Schäferkotter. 2003. *Partial Differential Equations and Boundary Value Problems with Mathematica,* 2nd edition. Boca Raton, FL: Chapman & Hall/CRC.

———— and M. R. Schäferkotter. 2005. *Handbook of Computational Methods for Integration.* Boca Raton, FL: Chapman & Hall/CRC.

Lanczos, C. 1961. *Linear Differential Operators.* New York: Van Nostrand.

MacRobert, T. M. 1967. *Spherical Harmonics: An Elementary Treatise on Harmonic Functions, with Applications.* New York: Pergamon Press.

Miller, K. S. and B. Ross. 1993. *An Introduction to the Fractional Calculus and Fractional Differential Equations.* New York: John Wiley.

Murray, F. J. and K. S. Miller. 1954. *Existence Theorems for Ordinary Differential Equations.* New York: New York University Press.

Oberhettinger, F. 1990. *Tables of Fourier Transforms and Fourier Transforms of Distributions.* Berlin: Springer-Verlag.

Oldham, K. B. and J. Spanier. 1974. *The Fractional Calculus.* New York: Academic Press.

Pauli, W. 1926. Über das Wasserstoffspektrum von Standpunkt der neuen Quantenmechanik. *Zeitschrift für Physik* 36: 336–363.

Pierpont, J. 1959. *Functions of a Complex Variable.* New York: Dover.

Porter, D. and D. S. G. Sterling. 1993. *Integral Equations.*Cambridge: Cambridge

University Press.

Pontriagin, L. S. 1962. *Ordinary Differential Equations*. Reading, MA: Addison-Wesley.

Prudnikov, A. P., Y. A. Brychkov and O. Marichev. 1990. *Integrals and Series, Vol. 3, Special Functions*. Amsterdam: Overseas Publishers Association.

Puri, P. and P. K. Kythe. 1988. Wave structure in unsteady flows past a flat plate in a rotating medium. In *Proc. SECTAM XIV, Developments in Theor. Appl. Mech.* 14: 207–213.

———— and P. K. Kythe. 1988. Some inverse Laplace transforms of exponential form. *ZAMP* 39: 150–156; 954.

Roach, G. F. 1982. *Green's Functions*, 2nd edition. Cambridge: Cambridge University Press.

Ross, S. L. 1964. *Differential Equations*. Waltham: Bleisdell.

Ruehr, Otto G. 2002. Analytical-Numerical treatment of the one-phase Stephan problem with constant applied heat flux. In *Integral Methods in Science and Engineering*, ed. P. Schiavone, C. Constanda and A. Mioduchowski, 215–220. Boston: Birkhäuser.

Sagan, Hans. 1989. *Boundary and Eigenvalue Problems in Mathematical Physics*. New York: Dover.

Sneddon, I. N. 1957. *Partial Differential Equations*. New York: McGraw-Hill.

————. 1978. *Fourier Transforms and Their Applications*. Berlin: Springer-Verlag.

Stakgold, I. 1968. *Boundary Value Problems of Mathematical Physics*, Vol. II. New York: Macmillan.

————. 1979. *Green's Functions and Boundary Value Problems*. New York: Wiley.

Unsöld, A. 1927. Beiträge zur Quantummechanik der Atome. *Annalen der Physik* 387: 355–393.

von Koppenfels, W. and F. Stallmann. 1959. *Praxis der konformen Abbildung*. Berlin: Springer-Verlag.

Vladimirov, V. S. 1984. *Equations of Mathematical Physics* (English Translation). Moscow: Mir Publishers.

Wayland, H. 1970. *Complex Variables Applied in Science and Engineering*. New York: Van Nostrand Reinhold.

Weinberger, H. F. 1965. *A First Course in Partial Differential Equations*. New York: John Wiley.

Whittaker, E. T. and G. N. Watson. 1962. *A Course of Modern Analysis*. 4th edition. Cambridge: Cambridge University Press.

Index